FUTURE COMMUNICATION TECHNOLOGY AND ENGINEERING

PROCEEDINGS OF THE 2014 INTERNATIONAL CONFERENCE ON FUTURE COMMUNICATION TECHNOLOGY AND ENGINEERING (FCTE2014), SHENZHEN, CHINA, 16–17 NOVEMBER 2014

Future Communication Technology and Engineering

Editor

Kennis Chan

Advanced Science and Industry Research Center, Hong Kong

CRC Press
Taylor & Francis Group
Boca Raton London New York

CRC Press is an imprint of the
Taylor & Francis Group, an **informa** business
A BALKEMA BOOK

Published by:
CRC Press/Balkema
Schipholweg 107C, 2316 XC Leiden, The Netherlands

CRC Press/Balkema is an imprint of the Taylor & Francis Group, an informa business

ISBN: 978-1-138-02777-0 (hbk)
ISBN: 978-1-315-69045-2 (ebk)

Visit the Taylor & Francis Web site at
http://www.taylorandfrancis.com

and the CRC Press Web site at
http//www.crcpress.com

Typeset by MPS Limited, Chennai, India

Future Communication Technology and Engineering – Chan (Ed.)
© 2015 Taylor & Francis Group, London, ISBN: 978-1-138-02777-0

Table of contents

Future Communication Technology and Engineering – Chan (Ed.)
© 2015 Taylor & Francis Group, London, ISBN: 978-1-138-02777-0

Preface

On November 16 to 17, 2014 in Shenzhen, one of the most developed and beautiful cities in China, the 2014 International Conference on Future Communication Technology and Engineering (FCTE2014) was successful held. It is a great pleasure for me to see so many people come together for the advance of and progress of the technology in a specific field.

From hand-written mails to telegraphs, fixed-line telephones and today's mobile phones which could basically provide a connection to any one at any place in the world, we humans have made tremendous progress in communications technology.

Instead of coming to an end of development, communication technologies are actually at its fast growing stage; there are more possibilities for future communication technologies provided by the achievements we have made rather than limits. For this reason, we have decided to organize this international conference on communication technologies. It is intended that the innovative ideas and inspiring studies presented at FCTE2014 will benefit other researches in the future by other people. With the publication of this book, I am sure that this purpose will be within our reach.

On November 16–17 at the convention, some innovative and inspiring ideas were presented; certain controversial topics were discussed like what the most efficient/convenient methods for information communication are and what the most possible prospect for future communication technology is. It is truly a difficulty to make any definite conclusions from those presentations and discussions, but the desire and passion for improvement and development shown by our participants are surely remarkable and respectable.

In this book, 70 inspiring papers are included, which were chosen from hundreds of submissions contributed by people from various countries and regions after careful reading and discussing by our diligent reviewers. These carefully selected papers studied almost every possible aspect of communication technology; the depth of the researches' is also thorough and complete. The discussed aspects include communication systems, automation and control engineering, electrical engineering, AI algorithms, signal processing, data mining, knowledge-based systems and so on.

Finally, the successful opening of FCTE2014 and publishing of this book is the collective work of many people's. As a member of the organizing committee, I would like specially thank these people with all my heart, including all our participants and contributors, the reviewing committee, the CRC Press and other people who have shown their kindness to FCTE2014. I earnestly look forward to another chance to work with these people again in the future.

Kennis Chan

Future Communication Technology and Engineering – Chan (Ed.)
© 2015 Taylor & Francis Group, London, ISBN: 978-1-138-02777-0

Organizing Committee

Honor Chair

M. Pal, Center for Advanced Materials Processing, India
J. Yeh, Tallinn University of Technology, Estonia

General Chair

Q. Rajesh, National Institute of Technology, India

Program Chair

K. Chan, Advanced Science and Industry Research Center, Hong Kong

International Scientific Committee

K.S. Rajesh, Defence University College, India
M.M. Kim, Chonbuk National University, Korean
X. Ma, University of Science and Technology of China, China
X. Lee, Hong Kong Polytechnic University, Hong Kong
J. Xu, Northeast Dianli University, China
Z.Y. Jiang, University of Wollongong, Australia
V.K. Jain, Indian Institute of Technology, India
Q. Zhang, Shenzhen University, China
M.S. Chen, Da-Yeh University, Taiwan
M.V. Raghavendra, Adama Science & Technology University, Ethiopia
J. Ye, Hunan University of Technology, China
Q. Yang, University of Science and Technology Beijing, China
Z.Y. Jiang, University of Wollongong, Australia
V.K. Jain, Indian Institute of Technology, India
O.P. Rishi, University of Kota, India

Communication systems and network technologies

Future Communication Technology and Engineering – Chan (Ed.)
© 2015 Taylor & Francis Group, London, ISBN: 978-1-138-02777-0

Key techniques and implementations of low-cost Radio Frequency (RF) communication

K.J. Hou
School of Computer & Communication Engineering, University of Science & Technology Beijing, Beijing, China

F.Y. Hou
School of Information Technology, Deakin University, Melbourne, Australia

ABSTRACT: Radio Frequency (RF) communication is widely used in various fields such as remote control, data exchanges, and sensor network. It greatly expands the application fields of computing and brings more facilities to various applications, especially in the consumer electronics filed. As one kind of widely used technology, the low-cost radio frequency communication system plays an important role. This paper describes the scheme of low-cost radio frequency communication, including its hardware and software architecture, analyses the important design issues, such as circuit, protocol, code, security, power consumption, and provides a reference implementation built from typical radio frequency communication chips.

1 INTRODUCTION

With the development of computer and communication technologies, Radio Frequency (RF) communication provides a facilitated way to exchange data (Beasley & Miller 2007, Beasley 2008). Thus, it is widely used in many fields such as remote control, sensor networks, supermarkets, logistics management, fieldwork, as well as various embedded and consumer electronics devices. Generally, the radio frequency ranges from around 3 kHz to 300 GHz, which corresponds to the frequency of radio waves. The advantage of RF communication is that it establishes a wireless connection for data exchange.

Although RF provides great convenience, it requires careful design and proper implementation to ensure the data exchange is available, reliable, and secure (Agrawal & Zeng 2011, Rappaport 2011). As these techniques/products are often oriented to consumer electronics and various widely deployed embedded devices, low-cost becomes an important factor when such systems are designed and realized. Usually, a low-cost communication is built from the single chip based solution. A single chip based solution means that the RF sub-system has a few hardware components, which are typically composed of one main transceiver chip and several other circuit elements. There are many such single chip based solutions. For example, TR3000 (2014) applies amplitude modulation and XE1202 (2014) uses frequency modulation, as well as more complex spread spectrum communication. Although different kinds of solutions require different working schemes, there exist common issues that must be well solved. This paper analyses some important and common issues for low-cost RF communication, and then describes how to design and implement an actual RF communication system. Through comprehending the hardware and software related key technologies, it provides valuable facilities.

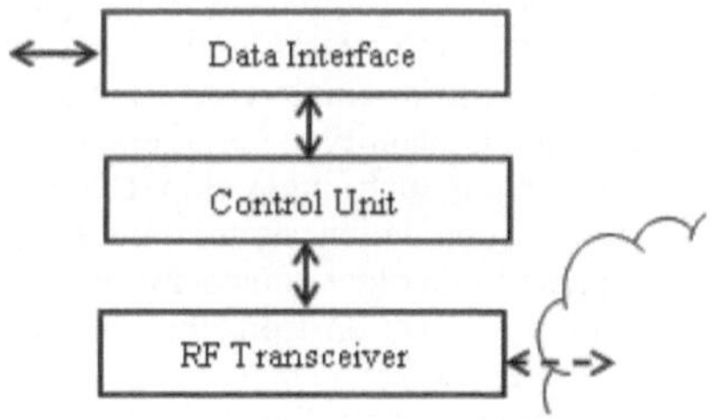

Figure 1. Typical architecture of RF communication system.

The rest of this paper is organized as follows. Section 2 provides an analysis of related key techniques. Section 3 describes the specific design and realization of an RF communication system. Section 4 concludes this paper.

2 KEY TECHNIQUES ANALYSE

2.1 *System architecture*

The typical RF communication system has the architecture of Figure 1.

In Figure 1, Data Interface manages the data transferred between the RF systems and is the main data processing system. For instance, the main data processing system is a computer, and there are several selective solutions to implement such interfacing. Different solutions bring different cost, complexity and data rate. The traditional RS-232 provides the most simple and the cheapest interfacing solution to both high-end modern computers and low-end embedded devices. However, if the required data rate is much high, the USB interface may be necessary. Other interface manners are also possible, for example the plug-in PCI card. However, this is more complex.

Control Unit does the necessary processing work for data preparation, packet assemble, protocol management and other control work. Usually, it is made with an embedded microprocessor with restricted memory capacity. The traditional MCS-51 series single-chip controller is more widely used than other microcontrollers. However, if a big computing ability or a fast computing speed is required, another advanced microcontroller such as an ARM based 32-bit microcontroller should be used to undertake the required tasks.

An RF Transceiver is the core component of the RF data communication system. It converts the base band data signals into wave band signals and transmits the signals through an antenna. It also undertakes the task of wave band signals receiving, converting the signals into base band signals to be recognized by back-end data processing parts. Usually, a low-cost RF Transceiver is built from single-chip RF transceiver, together with a small quantity of peripheral circuit elements. There are many such single-chip RF transceiver solutions. Range, modulation mode, transfer speed, a working frequency, and other considerations will play a part in making a selection from the options available. For example, the Frequency Modulation (FM) can provide a more reliable link than an Amplitude Modulation (AM) for mobile applications, because the device may experience a lot of sudden drops in amplitude. Frequency modulation is not as sensitive as amplitude modulation to such influences.

2.2 *Protocol*

Protocol is indispensable for any communication system. Usually, data to be transferred are assembled into separate packets. Protocol is used by both the transmitter and the receiver to establish and maintain the wireless data link. Hand shaking, data length, data speed, data code, error control, flow control, as well as any other details to understand and control the data packets, are all the content of communication protocol.

There are several common points that should be carefully considered for protocol design: the packet length, data speed, data code, error verification, and collision detection. Considering the RF working environment, the data packet should not be very long. Besides, as for how the data speed can be reached, the working environment should be taken into account. Generally, a short data length and a slow data speed are helpful to establish a stable RF data link. If a high probability of data error is inevitable, error correcting code should be appended to ensure the validity of data transmission. Sometimes, special data code may be required due to the characters of the under-layer circuit. For example, a Direct Current (DC) balance coding may be required for some RF transceiver chips. Otherwise, the RF circuit cannot work properly. Although DC balance coding can be fulfilled easily through Manchester's code, it will decrease the data speed by half. Therefore, other ways like some customized coding solutions should be applied to overcome such problem (Hou & Dai 2002). Collision detection must be considered for protocol, because most of the low-cost RF system cannot transmit signals together with others during the same period and by using the same frequency. Usually, a simple carrier sensing and random delay is an available way to resolve the problem of collision.

2.3 *Security*

RF communication makes data signal exposed to wide space, which implies that it is easier to be intercepted or snooped by others. If sensitive information is contained in the data packet, corresponding measures should be applied to prevent adversaries from understanding the meaning of the data. Methods like Chaffing and winnowing can be used to realize such purpose (Rivest 1998). However, the real safe way is to encrypt the data through cryptograph operations (Boncella 2002, Denis 2006). Data integrity should also be protected for secure purpose besides data privacy. That is to say, data confidentiality and integrity must be assured through cryptograph method when security is required for RF communication system.

Based on cryptography operations, encryption can convert the transferred plaintext data into its cipher text form to maintain confidentiality, while Message Authentication Code (MAC) can be attached to check data integrity. Although there are many mature and reliable cryptography algorithms, they all have the property of intensive computing. This property requires the processing unit for cryptography execution must have powerful computing ability, which is un-available for low-end microcontrollers like the MCS-51 series single chip microcomputers. If the RF system is attached to a powerful computer (e.g., a laptop computer), data encryption and related integrity code calculation can be executed in the main processor of the computer. If data has to be processed in tiny microcontroller, complicated cryptography algorithms with high encryption intensity cannot be executed smoothly. In such situations, systems may enhance the encryption intensity through other remedial ways. For example, Data Encryption Standard (DES) can run in MCS-51 series single chip microcomputers. So designers can set multiple secret keys and run DES for multiple times. Although such measures cannot reach complete cryptography security, they can actually create greater difficulties for attackers.

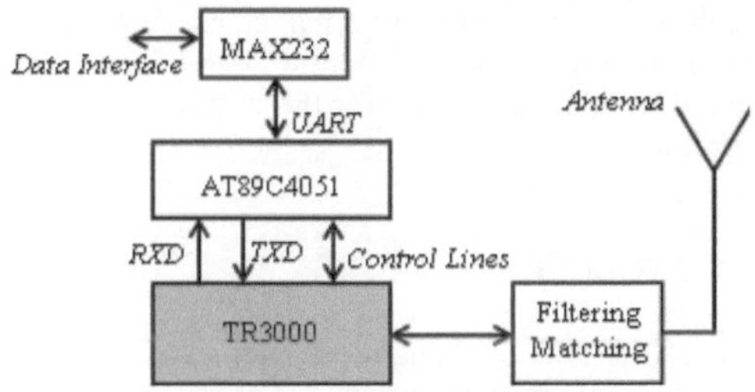

Figure 2. An instance of RF communication system.

2.4 *Low power consumption*

More and more consumer electronics and embedded devices are mobile devices, and their attached RF units are also mobile. One of the most important factors for a mobile device is its power consumption, which is a decisive factor in the feasibility of the devices. To decrease the power consumption, measures can be applied on circuit level or system level. Circuit level means to choose those circuit components with power save abilities. For example, a chip with SLEEP (or IDLE, POWER DOWN) working modes can significantly save power when it enters into idle status. System level measures mean that hardware and software measures should be comprehensively applied. For example, a signal amplification circuit can be turned off by a software instruction to minimize the static current, when there is no data signal to be transmitted.

2.5 *Antennae*

An antenna is often a neglected factor. However, it is very important for the quality of RF communication. The design and fabrication of antennae is a very professional task, which is beyond the skill of common computer and communication researchers and engineers. Different antennae have different gains and directionalities. For example, a whip antenna has good gain and is omnidirectional, but it has a certain length which makes it difficult to be contained into small devices. Some special integrated antennae are small, omnidirectional, and good at signal gains. However, they may increase the cost of the final product.

3 IMPLEMENTATIONS

3.1 *Hardware realizations*

An RF communication system based on TR3000 [5] single chip transceiver and AT89C2051 microcontroller is implemented as in Figure 2.

In Figure 2, TR3000 is the single chip RF transceiver, which provides both On-Off Keyed (OOK) and amplitude-shift keyed (ASK) modulations. The OOK modulation is better for power conservation and link reliability, while ASK allows more fast data rates. Several circuit components are installed around TR3000 chip. Among these peripheral components, Automatic Gain Control (AGC) and threshold sensitivity adjustment will affect the quality of the RF demodulation. When long distance communication is required, threshold sensitivity can be adjusted to much higher SNR to get bigger receiving gain. But this will incur more data errors, which should be removed by other applications such as Error Correction Code (ECC).

An RF signal is radiated and received through an antenna. No matter what kind of antenna is chosen, filter, matching and amplification should be deployed between the antenna and the transceiver chip. A Surface Acoustic Wave (SAW) filter provides receiver out-of-band rejection and suppresses output harmonics for transmitter. Impedance matching must be properly set for antenna. If necessary, such as long distance is required (Chow et al. 2013), radio amplifier or high frequency amplifier will be installed to improve the emissive radio wave power.

If the RF communication system should be attached to other computers, such as being attached to a laptop computer, the traditional RS-232 interface can link them together. However, due to the problem of voltage matching, the UART signals must convert their voltage before connecting with the laptop computer. Otherwise, damages may occur. Such converting can be finished by special single chip dual driver/receivers that include capacitive voltage generators, such as the MAX232.

3.2 *Software realizations*

Most of the related control work is executed through programs, such as packet buffering, flow control, and error detection, which run by the microcontroller of AT89C4051. It is an 8-bit microcontroller with 4 kbytes flash memory. If necessary, the off-chip memory should be installed to extend its memory space. Control work is various. Here we only describe packet formatting, error detection, DC balance coding and data encrypting.

For packet formatting, the total packet length should not be too long. A preamble bit stream should be applied, but its content needs to be specified according to specific requirements. For the TR3000 transceiver, the preamble bit stream takes the effect of slicing symmetry. Besides the effect of slicing symmetry, maybe the starting of a new packet need to be indicated by the preamble bit stream. To realize such a purpose, a barker code with sharp single peak characteristic of self-correlation function is much helpful.

In most wireless applications, error detection, even error correction must be applied to improve the data exchange reliability. To detect errors, the simple even-odd check is not very appropriate, because the probability of error may be much higher in wireless situations. To enhance the ability of error detection, a Cyclic Redundancy Check (CRC) based checking result should be calculated and attached to the packet. Fortunately, CRC calculation is not strong enough

Table 1. Dynamically DC balanced coding.

Original half-byte of 4 bits	6 bits coding result
0 0 0 0	0 1 0 1 0 1
0 0 0 1	1 1 0 0 0 1
0 0 1 0	1 1 0 0 1 0
0 0 1 1	1 0 0 0 1 1
…………	
1 1 1 1	0 1 1 1 0 0

for low-end microcontrollers. But CRC based checking cannot reach the level of cryptograph security. It should only be used for the purpose to detect the data link error, not for the purpose of resistance against malicious data hackers.

Due to the characteristics of the inner circuit of TR3000 chip, the bit stream to be transmitted should maintain DC balance. Although Manchester's coding is perfect at DC balance, it will reduce the data speed by half. As the DC balance requirement is not very strict for the chip (that is, it need not to be 1-0-1-0-by turns strictly; it just need to be equal in the number of bit "1" and the number of bit "0", during one bits slicing), a dynamically DC balanced coding can be applied as in Table 1.

This coding scheme replaces each one 8-bit byte with a 12-bit result. Although it is not strictly overturn each next bit, the total number of bit of '1' is equal to the total number of bit of '0', if they are counting during the slicing of sequential 12 bits. In such way, the requirement of DC balance is fulfilled, while the data speed is also maintained.

With the low-end microcontroller as the main processor for data processing, encryption is difficult to be implemented, especially for those advanced cryptographic algorithms like the AES Advanced Encryption Standard (AES) or RSA public key encryption. Without a high performance processor or co-processor, only relatively simple encryption algorithms can be implemented. Here, a DES based encryption can be programmed into AT89C4051 using assemble language of MCS-51 series instructions. DES uses 56 bits key sizes and operates on 64 bits block size. Although it is not very secure due to the short length of the secret key, it can still provide confidentiality protection for many situations. The entire DES realization involves many procedures, such as block expansion and key mixing. Among these procedures, the construction and the storage of the s-box of DES are crucial to the realization optimizing. To obtain fine performance, program can use more memory space to facilitate the storing, looking and fetching of the related transformations. To enhance encryption security, more rounds with different secret keys can be applied to make a remedy for the weakness of a DES algorithm.

4 CONCLUSION

As wireless data exchange gets more and more popular for consumer electronics devices and various embedded devices, building a low-cost RF communication has great application value. A low cost RF communication system has its special characteristics, which restrict their designs and realizations. It also faces special problems, which need different solutions from wired data linking. Key techniques involving transceiver chip, controller, protocol management, coding, security, and others, are analysed. Corresponding solutions or measures are described. Also, a specific instance of a low-cost RF communication system is implemented. With deeper applications and further demands emerging, further research work should be dedicated to improving the quality of communication links and enhancing the security of data protection.

REFERENCES

E.Y. Chow, M.M. Morris & P.P. Irazoqui. 2013. Implantable RF Medical Devices: The Benefits of High-Speed Communication and Much Greater Communication Distances in Biomedical Applications. Microwave Magazine, IEEE 14 (4): 64–73.

D.P. Agrawal & Q.-A. Zeng. 2011. Introduction to Wireless and Mobile Systems (3rd Edition). Stamford: Cengage Learning

F.Y. Hou & K. Dai. 2002. The Design and Realization of Computer RF-communication System. Application Research of Computers 19(12).

Jeffrey S. Beasley. 2008. Computer Networking (2nd Edition). Upper Saddle River: Prentice Hall.

Jeffrey S. Beasley & Gary M. Miller. 2007. Modern Electronic Communication (9th ed.). Upper Saddle River: Prentice Hall.

Robert J. Boncella. 2002. Wireless Security: An Overview. Communications of the Association for Information Systems 9 (Article 15): 269–282.

R.L. Rivest. 1998. Chaffing and winnowing: Confidentiality without encryption. CryptoBytes 4(1): 12–17.

Theodore S. Rappaport. 2011. Wireless Communications: Principles and Practice (2nd Edition). Upper Saddle River: Prentice Hall.

Tom St Denis. 2006. Cryptography for Developers. Rockland: Syngress.

TR3000 hybrid transceiver. 2014. http://www.rfm.com.

XE1202 FSK transceiver. 2014. http://www.semtech.com.

Future Communication Technology and Engineering – Chan (Ed.)
© 2015 Taylor & Francis Group, London, ISBN: 978-1-138-02777-0

A link evaluator: A channel quality estimator for Voice over Internet Protocol (VoIP) communications

M. Buzzanca, M. Malgeri & G. Schembra
Dipartimento di Ingegneria Elettrica, Elettronica e Informatica University of Catania, Catania, Italy

ABSTRACT: VoIP telephony has grown more and more widespread over the years, but it's still a relatively young technology which has not yet reached the maturity of traditional telephony, owing to several unresolved challenges. One major issue stems from the underlying UDP/IP protocol stack, which lacks an adequate QoS policy. This paper proposes an algorithm, named Link Evaluator, which evaluates channel quality, and analyses statistical parameters that are important to VoIP applications.

1 INTRODUCTION

Internet telephony poses several challenges which are still unsolved. Bandwidth available to end-users is getting wider, but, unfortunately, bandwidth alone does not guarantee a successful call. This is due to a number of factors, but it mostly involves how the underlying network works. While traditional telephony operates on a circuit switched network, VoIP relies on a packet switched network, which is not suitable for periodic streams of data. The UDP/IP stack and its best-effort nature contrasts the timeliness required by VoIP applications, making VoIP calls unfeasible under certain network conditions. Since there is no guarantee of a successful stream, it is important to tell when the state of the network is suitable for VoIP.

There are already several technologies and algorithms that estimate network conditions on the fly (discussed in Section 3), but there are not many available solutions that evaluate link quality before the stream begins. By doing so, it can be useful to fine tune the VoIP stream in order to deliver a better experience to the end-user according to the detected network conditions.

To answer the need for a solution that is able to provide useful statistical data to determine when a VoIP call can be successfully transmitted with tolerable degradation, this paper proposes an algorithm, called Link Evaluator, aiming at estimating network conditions.

This paper is structured as follows. Section 2 briefly introduces the problem of quality estimation. Section 3 describes the available technologies to estimate network conditions and voice quality. Section 4 describes the newly designed Link Evaluator algorithm. The paper concludes with a summary of Link Evaluator's limits and suggestions for further work.

Table 1. Comparison of common audio codecs used in telephony.

Codec name	Sample rate	Bitrate	Algorithmic delay
G.711	8 kHz	64 kbit/s	125 μs
AMR-WB	16 kHz	6.60 to 23.85 kbit/s	25 ms
G.729	8 kHz	8 kbit/s	15 ms
Opus (LP)	8 kHz	4 to 20 kbit/s	10 to 60 ms

2 AUDIO QUALITY DEGRADATION

Technologies for audio quality evaluation such as MOS (Mean Opinion Score), which became ITU-T recommendations in 1996 (ITU-T, 1996b), have been used for decades, despite being highly subjective.

In the VoIP scenario, we can find two main causes for degradation in audio quality: a poor choice of an audio codec and unsuitable network conditions.

An extremely compressing audio codec may result into lack of voice clarity, which can be seen as the impairment in the understanding of listened speech. Such audio quality degradation occurs before the audio data is sent through the network, because of the limitations of the audio acquisition hardware, and compression owing to the chosen audio codec.

Table 1 features a comparison of commonly used audio codecs in traditional with VoIP telephony: G.711 (ITU-T, 1988), AMR-WB (ITU-T, 2003b), G.729 (ITU-T, 1996a) and Opus (IETF, 2012).

Network traversal is a problem owing to the nature of the underlying network and protocol stack (UDP/IP).

Since the network is not ideal, events like packet losses, duplication, and non-sequential delivery can occur. In traditional telephony networks, all packets are delivered sequentially and in a timely manner, since they all travel in a specific dedicated circuit, and many of these issues are not present. Data consistency in packet switched networks can be still guaranteed using application level protocols, but due to the time-critical nature of the payload, if it is not delivered in a timely manner, several side effects might degrade the audio stream quality.

Recommendations concerning network conditions have been formulated in the past. A one-way end-to-end delay larger than 300 ms, according to ITU-T recommendation G.114 (ITU-T, 2003a), is considered not acceptable for most user applications. Interarrival jitter compensated by jitter buffers must not be higher than 100 ms (Cisco Systems, 2001).

Assuming a pre-set hardware set cannot be changed (which is often the case with embedded devices with VoIP capability, such as VoIP phones or smartphones), choosing the appropriate codec for the available network is therefore essential.

It is also very important how the VoIP client connects to the access point providing internet access. Ideally, a cable provides the most reliable bridge between the client and the access point, but it is not always possible to use wires, most notably in the mobile VoIP domain. At best, the mobile device is connected to an access point via 802.11, but in more realistic scenarios a mobile phone obtains internet access via 3G/pre-4G networks. Older generation networks like HDSPA/HUSPA, were proven to be unsuitable for VoIP (Arjona, 2009), but with the advent of LTE, the scenario is different. An approach consists in using the traditional VoIP technologies on top of LTE networks, with the same issues and constraints, but there is a new technology named Voice over LTE (VoLTE) which defines a real-time protocol built from scratch, tailored for mobile pre-4G networks.

3 RELATED WORK

There are at least two existing technologies which can be used to evaluate the VoIP QoS. The first one is RTPC, which is part of the RTP (IETF, 2003a) protocol defined in the same RFC. Its purpose is to provide statistical information relevant to QoS while the real-time data is being transmitted.

Information transmitted in the protocol payload includes means for estimating one-way delay, inter-arrival jitter, and packet loss, valuable parameters for any kind of QoS estimation.

There also exists an extension of RTPC, named RTPC XR (IETF, 2003b), which further extends the range of estimated network parameters by adding packet loss ratio, call quality evaluation in terms of an MOS score, round trip time, and SNR information. RTPC evaluates network conditions while the call is ongoing, and it is not possible to evaluate network conditions beforehand. Currently, no RTP extension provides such service to the user.

There are also attempts to objectively evaluate the conversational or listening quality.

A relatively new algorithm regarding this approach is Perceptual Objective Listening Quality Assessment (POLQA (ITU-T, 2011). It was standardized by ITU-T in 2011 as the successor of the more obsolete Perceptual Evaluation of Speech Quality (PESQ ITU-T, 2001), resolving a few issues with higher-bandwidth encoded audio and providing higher accuracy with modern mobile networks evaluation.

POLQA is a full-reference algorithm, which means it works by comparing the degraded audio signal to the original. This means it can only be used in a planned test environment, and it cannot be integrated into measurement tools that operate in real life situations.

4 LINK EVALUATOR

In certain scenarios it is important to know whether a VoIP call results in acceptable quality before such call is started. For example, knowing the network conditions before the call is initiated allows the setting up of more efficiently certain audio codecs. It can also avoid the frustration of attempting a VoIP call with bad network conditions.

As already stated, available solutions either are unsuitable for real-life applications, or operate while the call is ongoing. In order to satisfy the need to evaluate network conditions before call initiation, an algorithm named Link Evaluator was developed.

Link Evaluator was not designed to evaluate audio quality, as it does not inspect the packets' payload. It is tailored to work in conjunction with the well-known G.729 audio codec: it may not be as accurate as other codecs without proper adjustments, but such adjustments are relatively trivial once the structure of the audio frame is known.

The evaluation consists of two sequential steps: initially, statistical data is acquired using Quality Check Protocol (QCP), which is part of the link evaluator specification, and then an evaluation algorithm analyses the data and determines link quality.

4.1 *Quality Check Protocol*

Quality Check Protocol (QCP) is the name given to the network protocol responsible for the acquisition of all the QoS parameters involved into the evaluation phase. Its design revolves around the idea that a QCP packet must also be an RTP packet, implying that all the QCP data is encapsulated inside an RTP packet as payload.

This makes the QCP packet appear like it was an RTP packet, which is essential since statistical data is collected by analysing dummy traffic sent in QCP packets.

In order to make the estimations reasonably accurate, this traffic must resemble as much as possible the application traffic. Resemblance is also important if the packets must traverse certain nodes which

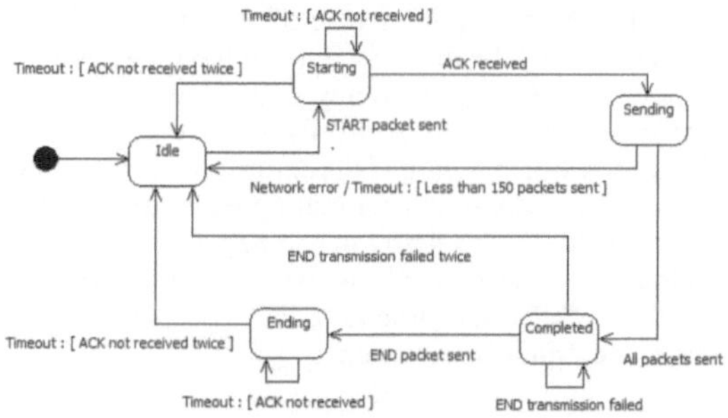

Figure 1. Link evaluator state machine diagram.

offer different QoS profiles to different packet types, because the probe packets could be treated differently if they do not appear as RTP packets.

The protocol can be used in P2P applications, although during its runtime there is always a 'client' node (which initiates the evaluation request) and a 'server' node (which replies to the evaluation requests).

4.1.1 *Protocol description*

The way the protocol works is described by the state machine diagram shown in Figure 1.

Idle state

The default state. Whenever the client wants to initiate network evaluation, the protocol changes from this state to the 'starting' state, and sends a START packet to the server. If any network error or two consecutive timeouts occur, the protocol will revert back to this state.

Starting state

In this state the client waits for an ACK packet. If the ACK packet is not received after one second, another START packet is sent, and the client waits again for an ACK. If the transmission of the ACK packet fails a second time, the client resets into the 'idle' state. Otherwise, if the ACK packet is received within 1 second, the client changes its state to sending'.

Sending state

In this state, DATA packets are exchanged between the client and the server. However, the server does not start sending DATA packets until the client successfully delivers the first one. The client sends exactly 150 DATA packets (the equivalent of 6 seconds of G.729 speech) at an interval of 40 ms in order to simulate real RTP packet exchange between the two nodes. After dispatching 150 packets, the client enters the completed' state if no critical error occurs (otherwise it would return to 'idle' state). The number of sent packets is an acceptable trade-off between algorithm accuracy and costs (in terms of battery, time and data plan) and corresponds to about 6 seconds of conversation. The appropriate algorithm duration was determined after empirical tests that were conducted on a real network.

It is worth noting that once the client enters the 'completed' state, all subsequently received DATA packets are discarded because the state expects an ACK packet. This can cause an approximation error in the calculation of the packet loss ratio if the network is so slow that is unable to send all his DATA packets before the client changes state.

In this state, all QoS parameters should be gathered and processed, although the protocol itself is independent of how these parameters should be calculated.

Completed state

This is a transitory state where the client sends an END packet to the server and puts itself into the 'ending' state. If the transmission of the END packet fails twice, the client goes back into the 'idle' state, however the computed parameters data is kept, and is available for the evaluation algorithm.

Ending state

In this state the client waits for an ACK packet. If the ACK packet is not received after one second, another END packet is sent, and the client waits again for an ACK. If the transmission of the ACK packet fails a second time, the client resets into the 'idle' state. Otherwise, if the ACK packet is received within one second, the client changes its state to 'idle', and also updates the evaluation data with the information about sent packets placed by the server in the packet payload.

4.1.2 *Packet format*

As QCP packets are basically RTP packets with QCP protocol payload, they are all structured so that they have the same size of an RTP packet carrying 40 ms of G.729 payload (40 bytes of IP/UDP/RTP headers + 40 bytes of payload). Of this payload, 12 bytes are occupied by QCP header data and the remaining 28 bytes are normally unused.

While most of the protocol packet types are essentially service packets used to control the protocol state, the DATA packets contain the parameters, taken into account during the evaluation, and are therefore worth describing in detail. Figure 2 displays the structure of the DATA packet excluding IP/UDP/RTP headers, which is composed by several fields:

Version Number (4 bit): protocol version.
Reserved (8 bit): reserved for future usage.
Packet Type (4 bit): packet type.
Cumulative Packets Lost (16 bit): total packets lost during the data exchange phase.
Jitter (16 bit): interarrival jitter of the previous packet in milliseconds.
Delay since last received packet (16 bit): time elapsed since the last packet was received in milliseconds.
Last received packet sequence number (16 bit): sequence number of the last received packet.
Unused (28 bytes): unused payload bytes.

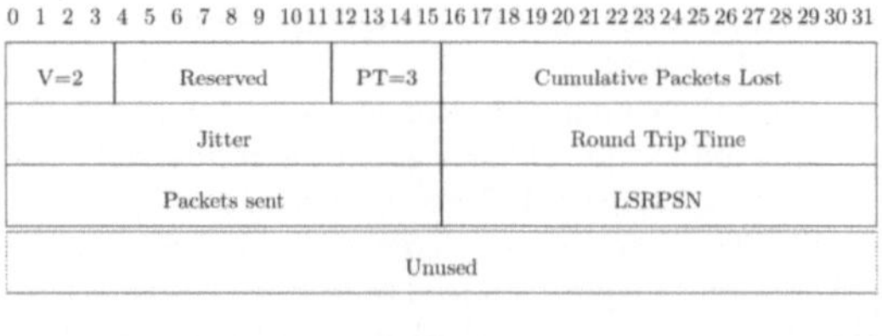

Figure 2. Structure of the DATA packet.

4.2 *Evaluation algorithm*

The evaluation algorithm challenges the collected QoS parameters against the following constraints:

1. Average interarrival jitter less than 100 ms
2. Maximum interarrival jitter less than 250 ms
3. Interarrival jitter must surpass the average jitter threshold (100 ms) at least once.
4. Average round trip time less than 300 ms
5. Packet loss ratio is less than 5%.

If any of these conditions is not satisfied, VoIP calls are deemed not feasible.

Note that there are two sets of QoS parameters that are tested by the evaluator: the local data set (locally computed parameters) and the server data set (received from the server in DATA packets). This is to test link reliability in both ways.

Constraints two and three are ways to control the amount and length of gaps that can occur during a conversation. Having a jitter higher than 100 ms means that jitter buffer is not large enough to compensate, causing a gap in the audio playback due to a buffer underrun; having more than one gap in six seconds of conversation results in an unacceptable call quality. Furthermore, if maximum jitter is too high, the caused gap is extended beyond acceptable levels, and, for this reason, the constraint 3 exists.

The average round trip time constraint is to ensure delay effects do not degrade call quality too much. If the average RTT is too large, the increased latency causes acoustic delay, which negatively affects call quality.

The last constraint concerning packet loss ratio is also used to control the amount of gaps, since high packet loss can be a cause of buffer underruns.

5 CONCLUSIONS

Link Evaluator is a system that is able to detect network conditions for a VoIP use case. However, it is currently limited to the usage of a particular codec, and it does not take into account bandwidth estimation techniques, which play an important role in scenarios where bandwidth is dynamically allocated.

Future work may extend the algorithm so that allows the usage of bandwidth estimation techniques to further improve the accuracy of the network conditions' evaluation. It also needs to be more flexible with regards to codec selection in order to be used within a larger spectrum of scenarios.

ACKNOWLEDGEMENTS

This work was developed under the project S.R.S. – Progetto di Formazione Integrato SINERGREEN (Smart Intelligent Energy Green), RES-NOVAE, SEM" supported by MIUR (Minister of Education, University and Research).

REFERENCES

Arjona, A. 2009. *A study of mobile VoIP performance in wireless broadband networks.*

Cisco Systems 2001. *Quality of Service for Voice over IP.* http://www.cisco.com/en/US/docs/ios/solutions_docs/qos_solutions/QoSVoIP/QoSVoIP.pdf

IETF 2003a. *RFC 3550: RTP: A Transport Protocol for Real-Time Applications.*

IETF 2003b. *RFC 3611: RTP Control Protocol Extended Reports (RTCP XR).*

IETF 2012. *RFC 6716: Definition of the Opus Audio Codec.*

ITU-T 1988. *ITU-T Recommendation G.711: Pulse code modulation (PCM) of voice frequencies.*

ITU-T 1996a. *ITU-T Recommendation G.729: Coding of speech at 8 kbit/s using conjugate-structure algebraic-code-excited linear prediction (CS-ACELP).*

ITU-T 1996b. *ITU-T Recommendation P.800: Methods for objective and subjective assessment of quality.*

ITU-T 2001. *ITU-T Recommendation P.862: Perceptual Evaluation of Speech Quality (PESQ): An Objective Method for End-To-End Speech Quality Assessment of Narrow-Band Telephone Networks and Speech Codecs.*

ITU-T 2003a. *ITU-T Recommendation G.114: One-way transmission time.*

ITU-T 2003b. *ITU-T Recommendation G.722.2: Wideband coding of speech at around 16 kbit/s using Adaptive Multi-Rate Wideband (AMR-WB).*

ITU-T 2011. *ITU-T Recommendation P.863: Perceptual Objective Listening Quality Assessment.*

Future Communication Technology and Engineering – Chan (Ed.)
© 2015 Taylor & Francis Group, London, ISBN: 978-1-138-02777-0

A study of the possibility of superluminal communications

Z.H. Zhang, H.A. Zhang & Z.Y. Zhong
Family Research Group in Superluminal Phenomena, Xitucheng Lu, Haidian District, Beijing, China

ABSTRACT: Based on the requirements of interstellar communication and high bit rate transmission of information, we discussed the possibility of the superluminal communications in future, the restraints of special relativity and SB theory and the time advance in negative dispersion media. Meanwhile, we have pointed out that the pulse reshape in high negative dispersion media can satisfy the requirement of bit error rate.

Keywords: superluminal communications, restriction of SR, velocity of energy transport

1 INTRODUCTION

With the development of the internet and astronautics, information needs to be exchanged both in amount and distance increasingly rapidly day by day. Therefore we still need to seek a new technique to enhance information transmission capacity and velocity. As we know in digital communications the information transmission is by virtue of coded pulses and obviously the transmission velocity of information is just that of the pulses and is determined by a refractive index and a dispersion of medium. At present, for wireless, the velocity is about c; for fibre is equal to c/n_1, where n_1 is index of fibre core. Because the short pulse easily suffers a serious distortion [1], in addition the transmission velocity of pulses is restricted. For getting a high transmission bit rate we have to employ a Dense Wavelength Division Multiplex (DWDM) or multi-core fibre technique. Right now, the channel numbers, as reported, are as high as several hundreds. As all channel transmit through one fibre at the same time, this complicates the techniques of the non-linearity and dispersion compensation of the fibre as well as the multiplex and de-multiplexing. It also makes the communication system very complex and expensive and even restricts the property of communication systems. Although the multi core fibre can help us reduce the channel numbers in the fibre, but it complicates the technique of the fibre manufacture. We can imagine that the communications just as the traffic, and the capacity of a communication system will be in direct ratio to the velocity of the signal transmission. For a system, the faster the information current, the higher the transmission bit rate and the less channel numbers are needed. Recently, a lot of superluminal phenomena of light pulses have been observed [2,3], even the light pulses have a negative velocity (about –c/310) [3], which means that the transport velocity of light pulses can be faster than infinite and the time advance happens. If these light pulses can be employed to transmit information, one channel is enough to transport several T bits signal, the system will be simplified, and the problems appearing in multi-channels can also be avoided and the velocity of information transmission can also be raised. Here we shall discuss the possibility of superluminal transmission of information.

2 SUPERLUMINAL PROPAGATION OF THE LIGHT PULSES AND SUPERLUMINAL OPTICAL COMMUNICATIONS

A light pulse contains a series of frequency components. Supposing the waveform of the pulse as U(r, t), we have [4]

$$U(r,t)=\frac{1}{\sqrt{2\pi}}\int_{-\infty}^{\infty} A(\omega)e^{i(\omega t-kr)}d\omega \tag{1}$$

$$A(\omega)=\frac{1}{\sqrt{2\pi}}\int_{-\infty}^{\infty} U(0,t)e^{-j\omega t}dt \tag{2}$$

here $k=n\omega/c$, $n=n_0+(dn/d\omega)\Delta\omega+(d^2n/d\omega^2)\Delta\omega^2$.

Instead of *n* and *k* in equ. (1), as a first approximation we can get [5],

$$\begin{aligned} U'(r,t) &= 2\frac{\sin(t-\frac{r}{c}(n_0+\frac{dn}{d\omega}\omega_0))}{t-\frac{r}{c}(n_0+\frac{dn}{d\omega}\omega_0)} A(\omega_o)\exp[i\omega_o(t-\frac{r}{c}n_o)] \\ &= C(r,t)\exp[i\omega_o(t-\frac{r}{c}n_0]. \end{aligned} \tag{3}$$

The waveform of reshaped pulse can be represented by the envelope function $C(r, t)$ and the pulse peak is at

$$t-\frac{r}{c}(n_0+\frac{dn}{d\omega}\omega_0)=0 \tag{4}$$

And moves with a group velocity

$$v_g = \frac{c}{n_o + \frac{dn}{d\omega}\omega_0} \tag{5}$$

This group velocity v_g connects with phase velocity v_p through the formula

$$n_g = n + \omega\frac{dn}{d\omega} \quad \text{or} \quad v_g = v_p - \lambda\frac{dv_p}{d\lambda} \tag{6}$$

Although the dispersion value $dn/d\omega$ of the medium is very small, but the angle frequency ω is much larger, and we can have $v_g < c$; or $v_g > c$; even $v_g < 0$ and the time advance happens for different ω [4]. Such as n_g can be much larger to make the group velocity reduce to 8 m/s, so that the superluminal propagation of light pulse and the photons trap is due to the interaction of light and matter. They are reshaped by the dispersion of the medium and are a natural phenomenon. At any time the reshaped light pulse is an assembly of electric fields or photons and detectable. The 'time advance' can be used to make a 'time compensator'.

We have proved the correctness of equ. (4) by virtue of computer stimulation. Supposing a pulse Y is consist of 21 cosine waves with a slight frequency difference $d\omega$.

$$Y = \sum_{i=-10}^{i=10} y(i) = \sum_{i=-10}^{i=10} \mathbf{Cos}[\,\omega(i)(\mathbf{t} - \mathbf{L} \times (\mathbf{no} + \mathbf{i} \times \frac{\mathbf{dn}}{\mathbf{d\omega}} \times \mathbf{d\omega})/\mathbf{c})]$$

$$\omega(i) = \omega_0 + i \times d\omega \qquad i = -10{:}10 \tag{7}$$

In the calculation we take the parameters as: $\omega_0 = 1$, $d\omega = 1 \times 10^{-2}$, $no = 1$, $c = 1$. The calculated results are shown in Figure 1. Such as at $L = 6$, because the velocity v_g changes with the dispersion value D ($dn/d\omega$), the arriving time t is different too. Put the values of n_0, D and ω_o in Equation (7), estimated t coincides with the result of Equation (4) shown in Figure 1. When the group velocity is negative, the time advance happens, and arriving time t at $L = 6$ is negative. We can see that when dispersion $D = -10$, calculated the arriving time $t = -54$ time unit; but when $D = -20$, $t = -114$ time unit. These results agree with those shown in Figure 1. The results of the computer simulation also show that the dispersion can cause broadening and splitting of pulse in the propagation as well as in the photon trap. The calculated result shows that the group velocity is different from the phase velocity due to the interaction between light and the medium, or the dispersion reshape. Therefore the superluminal propagation of light pulse or photon trap can be by virtue of either Electromagnetic Induced Transparency (EIT) or the Electromagnetic Induced Absorption (EIA). The propagation of light pulses without distortion has been discussed in another paper [6].

3 SOME THEORETIC PROBLEMS FACED

Although we have discussed the superluminal propagation of light pulses in a medium above, we still have

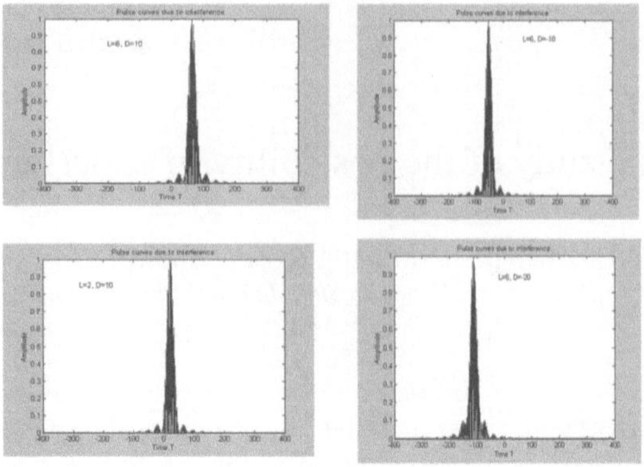

Figure 1. Propagation of a pulse in Media, L – propagation distance, D – dispersion of media, T – arriving time at distance L.

two problems that needed answering before using these pulses to transfer information.

3.1 *Are the superluminal light pulses virtual?*

According to Sommerfeld and Brillouin's (SB) theory [7] the velocity v_e of energy transport cannot be larger than c, so that some people think that the superluminal pulse is virtual and contains no energy and it cannot be used to transfer information. As in experiments these pulses were detected, it means that they are real. We also have found that the SB theory is not correct, and the curve v_e/c in Brillouin's figure perhaps corresponds to the contour of the absorption line. A confusion of the transmission efficiency and transmission velocity of energy was made in SB theory. The velocity of energy transport v_e was defined as P/ω [7], where P is average density of power flow, and ω is total density of transmission energy. We think that the density of the power flow in a medium can be represented by Poynting vector **S**, and for monochromatic light is [4]

$$S = RE\,(E \times H) = \frac{E^2}{\eta}\vec{k} \tag{8}$$

For light pulse **U**(r, t),

$$S = \frac{U^2}{\eta'}\vec{k} \tag{8'}$$

where, $\vec{\mathbf{k}}$ is the unit vector in the propagation direction.

$$\eta = \sqrt{\frac{\mu}{\varepsilon}}$$

is wave impedance, η' is the effective value of wave impedance, and U(r,t) given in equ. (1). Obviously, the velocity of energy transport v_e is determined by **E** or **U**. The problem is what is the total transmission of energy density? We think the inner energy of the transmission medium does not take part in the energy transmission and it only produces a loss (or gain) of the transmission energy. So that for monochromatic light $\omega = \varepsilon E^2$, we can get $v_e = v_p$; and for light pulse $v_e \approx v_g$. In such case, the centre of gravity of the pulse may not be consistent with the peak owing to the broadening of the pulse may be asymmetric. The conclusion that v_e must be less than c is wrong. As we know both of v_p and v_g can be larger than c certainly. In such case, we think that the group velocity v_g is more precise than the velocity v_e to describe the velocity of energy transmission and still represents the velocity of a signal in negative dispersion region for digital communications. The pulse is an assembly of photons or electric fields in time-space, and pulse reshape under the action of dispersion of medium (or interference) can make the transmission velocity go up or slow down. These pulses are detectable and show that they tote the energy and can be used to transfer information.

3.2 *Is the propagation of the information faster than c?*

According to the theory of special relativity, Einstein said: no body (including a signal) can move faster than c [8]. For developing superluminal communications we must answer this question: is Einstein's conclusion correct? Therefore, we studied the special relativity carefully and found that Einstein's conclusion was incorrect too, and pointed out that the superluminal motion was a natural phenomenon. Here introducing our study result briefly.

First, we think there are two kinds of epistemology for knowing the external world,

1. Objective epistemology: external body exists objectively and independent of our observation. Classical mechanics and quantum mechanics belong to this category, they study the real law of body motion (in macro or micro scope) and recognize the superluminal motion;
2. Subjective epistemology. We cognize the existence of an external body through observation and relativity and astronomy belong to such a theory of knowledge. They only deal with the image of the body and the study of the visual laws of body motion. We also can perceive relativity as a visual mechanics.

Since light's speed is limited, there are differences between real and visual law. So that, the conclusion of special relativity may not be suitable for the real situation.

Second, we also found there is a serious mistake in special relativity that is based on two principles: 1. The relativity principle: (include special and general relativity principle); 2. The principle of light velocity invariance: at any time the light velocity in vacuum is c. Albert Einstein thought the light velocity was independent of both the light source and the observer since it is the solution of the Maxwell equation. He thought (Einstein's paradox) that if you follow a light beam, go forward and looked at other light beam the result should as a observer which is still relative to the earth, since light velocity is independent of the observer. Therefore, he asserted that all physical laws should be at covariance under the Lorentz transformation. But what is the real situation? The experimental result shows that this assert made by Einstein is wrong. The interference of two light beams shows that the other beam is rest relative to you following a light beam. If other beam travels with c, yet, the phase difference between two beams will change rapidly and the result is that no interference fringe appears after combining two beams. Of course that is not true. The interference experiment shows that the light velocity depends on the state of the observer certainly even if in vacuum. The time – space covariance should be $s^2 - (ct)^2 = s'^2 - (c't')^2$. Generally $c \neq c'$, this equation has no solution; therefore the Lorentz transformation does not exist. But when $v \ll c$, we can consider $c \approx c'$, only in this situation, the Lorentz transformation can be used approximately. So that relativity is only valid for $v \ll c$ approximately. We also suggested [9] a new time-space transformation instead of the Lorentz transformation. That is

$$x' = x - (x_0 + vt), y' = y, z' = z, t' = (1 - v/c)t - x_0/c;$$

$$x = x' - (x_0' + vt'), y = y', z = z', t = (1 - v/c')t' - x_0'/c'.$$

Here x_0 is an initial distance between two systems, v is relative velocity, and defined $v > 0$ if distance x increases and $v < 0$ if x decreases. We have discussed this in paper [9]. Real visual mechanics also recognizes that superluminal motion is a natural phenomenon [10]. Therefore information can propagate faster than c certainly in theory.

4 PROSPECT OF SUPERLUMINAL OPTICAL COMMUNICATIONS

4.1 *Advantage of the superluminal communications*

At present, the information that needs to be exchanged increases rapidly. Because of the limitation of the capability of the Opto-electric techniques, the bite rate transmitted by each channel is about 10–20 Gb/s. In order to fit the requirement of the need to increase information exchange, a Dense Wavelength Division Multiplex (DWDM) and multi-core fibre technique have been adopted, all of them increasing the complexity of the system. If the velocity of a signal is faster than c, if suppose its velocity can reach as high as 100c, obviously one channel can has a several Tb/s information capacity. A single channel will simplify the technique of dispersion and non-linearity compensation, as well as the exchange and multiplex techniques. It will also lessen the need for the fabrication of fibre and the cost of the system will decrease. In the $v_g < 0$ situation, since the time advance of output pulse happens, we can manufacture a 'time compensator' to shorten the transmission time of information. From the construction of the information highway, only after getting faster signal transmission speed to lealize the

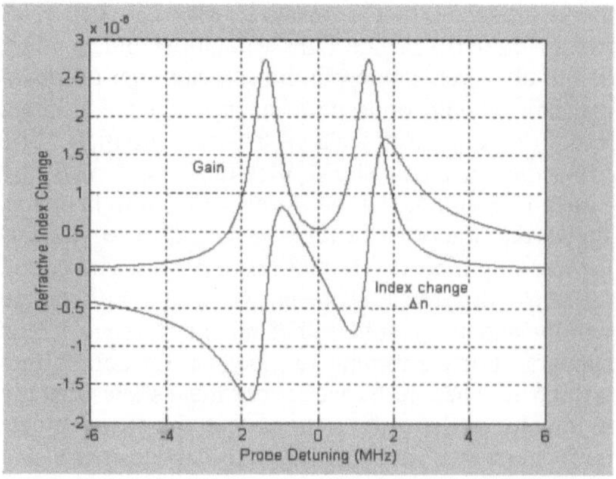

Figure 2. Simulated Curves of refractive index n and gain coefficient in WKD Experiment [1].

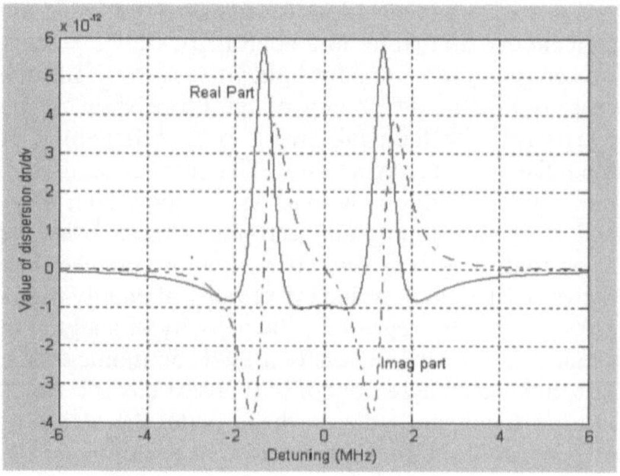

Figure 3. Curve of deviation of index n. — Real; --- Imag.

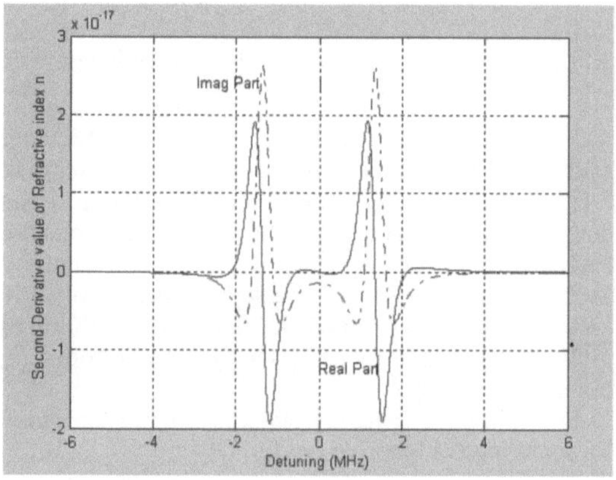

Figure 4. Curve of second derivation of index n. — Real; --- Imag.

superluminal communication can we say that we have constructed a real information highway exactly. So that to develop the superluminal communication is very attracting. We are facing the new task of how to transmit a series of coded light pulses in a medium over long distances.

4.2 *Realization of the superluminal communications*

Since the information is transferred by the coded pulses that propagate in a medium with the group velocity, if we want $v_g > c$, the dispersion of the medium must be negative, it means that we need the pulses to propagate in anomalous region. In theory, we can choose a suitable $dn/d\omega$ to make the $0 < n_g < 1$, and $v_g = c/n_g > c$, before the experiment [3]. Most scientists obtained the larger dispersion by means of the gain or absorption of a single spectral line. Because of the frequency superposition of signals and spectral lines, the distortion of the signal is serious; right now scientists still study how to get a larger negative dispersion. The measure of gaining aided superluminal propagation of light pulses suggested by Chiao et al [11] as in experiment [3] is one of the valuable suggestions. Using two pumping Raman lasers with very near frequencies ($\nu_{1,2} = \nu_0 \pm \Delta\nu$) induces two gain peaks, When the signal with a centre frequency ν_0 propagates in this medium, the signal also can get a gain to compensate the loss of the medium When the intensity of two pumping Raman laser beam is equal, the polarization of medium for that signal beam is [3,7]

$$\chi(\nu) = \frac{M}{\nu - \nu_1 + i\gamma} + \frac{M}{\nu - \nu_2 + i\gamma}. \tag{6}$$

where M is a parameter connected with the medium character and the power of pumping beams. Based on the relation between n and χ, we can get n, $dn/d\omega$ and $d^2n/d\omega^2$, varies with frequency ω as in Figures 2–4.

The pulse propagation equation ignoring absorption and non-linear effect,

$$\frac{\partial A}{\partial z} = -\beta_1 \frac{\partial A}{\partial t} - \frac{i}{2}\frac{\partial^2 A}{\partial t^2} + \cdots$$

here

$$\beta_1 = \frac{1}{v_g} = \frac{n_g}{c} = \frac{1}{c}\left[n + \omega\frac{dn}{d\omega}\right],$$

$$\beta_2 = \frac{d\beta_1}{d\omega} = \frac{1}{c}\left[2\frac{dn}{d\omega} + \omega\frac{d^2n}{d\omega^2}\right] = \frac{\lambda^3}{2\pi c^2}\frac{d^2n}{d\lambda^2},$$

In experiment [3] group index measured $n_g = 310$ ($\pm$ 5), observed that the group velocity is negative, and is faster than infinite and appears the time advance. By the property selection of the parameters, we can get $1 < n_g > 0$, and $v_g > c$. We also demonstrated that by calculation the superluminal light pulses can propagate long distance without distortion [6].

4.3 *The practical problems faced by superluminal communications*

Although the superluminal propagation of light pulses has been observed, and the theoretic research has been done, but in practice it maybe meet some new problems. So it still has a long way to go before practical application. At first, since the medium has larger negative dispersion, the distortion of light pulses is larger and limits the long distance transmission of light pulses. Fortunately as we see from Fig. 3, 4 $dn/d\omega$ and $d^2n/d\omega^2$ are not too large and are relatively flat in the middle region, so that this distortion may be not be serious. On the other hand, this distortion can be corrected by some approaches of the waveform control technique. Secondly, since dispersion $dn/d\omega$ depends on parameter M, therefore v_g also depends on the intensity of pumping laser beam and the density of medium atoms devoted to χ, so that the time jitter is important, since the clock information also follows the signal transmission. In such case the problem of time jitter will be another research items. At last, in superluminal communications, the requirement of frequency stabilization of light sources and the narrow width of spectral lines increase; to raise the laser quality is also important. On the other hand, how to keep superluminal pulse propagates long distance without distortion is still the primary task. Superluminal communication systems in future can only be employed in trunk or special lines. But the techniques of time compensation can be applied in remote control and remote measurement in space navigation.

Our research [4] also shows that the velocity of information transmission is phase velocity v_p of the load wave in analogy communications and is group velocity v_g in digital communications.

5 SUMMARY

In this paper we discussed the possibility of the superluminal communications. Our conclusion is that the superluminal communications are certainly possible and the theoretic problem has been solved. We think it is time to start experimental research immediately. In the future the research may still meet some problems needed to solve. But compared with the theoretic barrier, these difficulties cannot figure out what we need, and we would like to cooperate with all people interesting in superluminal communication around the world and push the research work of superluminal communications forward.

REFERENCES

[1] G.P. Agrawal. Nonlinear Fiber Optics, 2nd. ed. (Academic Press, 1995).

[2] Huang Zhi-xun. Recent Advances in Faster Than Light Research. National Defence Industry Press 2002). Huang Zhi-xun; Theory and Experiments of Faster than Light Research; (Academic Press 2005).

[3] Wang L.J., Kuzmich A., Dogariu A. Gain-assisted superluminal light propagation, Nature, 2000, 406, 277–279

[4] Zhang Zihua, Zhang Huaan. Superluminal Phenomena and its influence on Physics and Informatics. 6th symposium of Photonics of China, Oct. 2008, Chongqing.

[5] Zhang zi hua, Zhang Hua an, Cause of superluminal transmission of light pulse and Photon Capture (AOM 2010 – OSA-IEEE Topical Conference, Advances in Optoelectronics & Micro/nano-Optics, 3–6 Dec, 2010, Guang zhou, China) Zhang zi hua, Reshape and Supeluminal propagation of light pulses in media with Negative dispersion, Chinese Scientists 2011, No. 1, pp. 1–7.

[6] Zhang zi hua, the distortion-less propagation of the Superluminal light pulses, "Shen Shi Zhi Guang", the Literature of reform and innovation of China. (Press of Science and Literature of China 2009, Beijing).

[7] Brillouin. L. Wave propagation and group velocity, (Academic Press, New York 1960).

[8] A. Einstein, Relativity, (Methuen & Co Ltd Methuen & Co Ltd, 1916)

[9] Zi-hua Zhang, Hua-an Zhang, Zhi-ying Zhong, Correcting the mistake of the Principle of light invariance and its effect. ICOELC. June 2014, Taiyuan, China. Aplied Mechanics and Materials Vols 651-653 p.2287 (TTP Ltd. Switzerland).

[10] Cyclopedia of China (Electrical and Electronic technology, 2002) (Press of Cyclopaedia of China, 2002 Beijing).

[11] R. Y. Chiao, Phys. Rev. A 48, R34 (1993).

Future Communication Technology and Engineering – Chan (Ed.)
© 2015 Taylor & Francis Group, London, ISBN: 978-1-138-02777-0

Transmission performance optimization of multi-wavelength repeaterless undersea optical fibre cable

J.T. Yu & W.Y. Liu
Luoyang University of Foreign Languages, Luoyang, Henan, China

X. Li
Unit 62041, Changchun, Jilin, China

H.Y. He
Luoyang University of Foreign Languages, Luoyang, Henan, China

ABSTRACT: Multi-wavelength repeaterless undersea optical fibre cable transmission is a credible communication method with big capability to satisfy the requirement of communication across the sea. This paper discusses the influences from chromatic dispersion, polarization mode dispersion and nonlinearity. According to the simulation results on the platform of OptSim, we selected the wavelength of 1546.12 nm, 1547.72 nm, 1549.32 nm, 1550.92 nm and CSRZ code to optimize the wavelength and coding simultaneously. As a result, the eye diagram of each channel is clear and their performances have little difference between them. The system performance has been improved well and the transmission distance is further extended.

1 INTRODUCTION

Multi-wavelength repeaterless undersea optical fibre cable transmission becomes necessary in view of limitation of chromatic dispersion and Polarization Mode Dispersion (PMD) and long-distance transmission without repeater. Chromatic dispersion, PMD, nonlinearity and noise are factors that limit the repeaterless transmission distance. A Raman amplifier is introduced in the environment of long distance transmission. In this paper, the influence of dispersion and nonlinear effects is studied.

200 GHz (1.6 nm) is selected as channel spacing. The wavelengths are 1550.12 nm, 1551.72 nm, 1553.33 nm, 1554.94 nm. Enhanced Forward Error Correction (EFEC) technology is adopted, and its redundancy is 23%.

BER and Q factors are used to evaluate system performance. Also the eye diagram and system margin are used as assistant measures. Generally, the system margin is estimated by Q factor in dB [1]. System margin is required in system design to be greater than 3 dB considering that a variety of factors in the actual environment cannot be reflected in simulation system.

2 EFFECTS AND OPTIMIZATION OF CHROMATIC DISPERSION

2.1 *Effects and simulation of chromatic dispersion*

Chromatic dispersion can accumulate and will increase linearly with the increase of transmission distance. The smaller cumulative dispersion at the receiving end will maintain pulse shape well and reduce inter-symbol interference. Large local Group Velocity Dispersion (GVD) can inhibit the cross-phase modulation (XPM) and Four Wave Mixing (FWM) effectively [2]. A signal cannot be transmitted normally without dispersion.

When using a NRZ code, the system with chromatic dispersion can achieve more than 450 km (468 km) repeaterless transmission. The BER is 2.01×10^{-5}, system margin is 4.07 dB. But the worst channel in the system is all error without chromatic dispersion and the eye diagram is closed completely. The reason is that chromatic dispersion will cause pulse broadening and crosstalk between the codes. But FWM affects the system transmission performance so seriously that the system does not work properly.

2.2 *Simulation of chromatic dispersion compensation technology*

In the most distant transmissions, chromatic dispersion of G.652 fibre at 1550 nm is very large after overcoming nonlinearity and the effects of the transmission distance and performance. Dispersion Compensating Fibre (DCF) post-compensation is chosen as the output power of power amplifier at sender is large. Dispersion compensation ratio η is defined as follows:

$$\eta = \left| \frac{D_{DCF} L_{DCF}}{D_{SMF} L_{SMF}} \right| \tag{1}$$

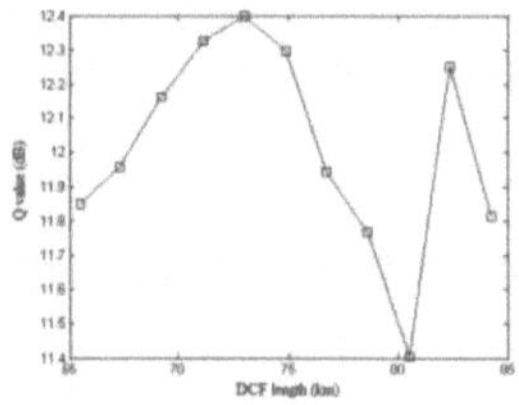

Figure 1. Change curve of Q factors of channel 3 with the change of DCF length.

D_{DCF}, D_{SMF} are respectively dispersion coefficient of DCF and Single-Mode Fibre (SMF). L_{DCF}, L_{SMF} are respectively length of DCF and SMF. $\eta = 1$ is fully-compensation. $\eta < 1$ is owed-compensation. $\eta > 1$ is over-compensation. The owed-compensation is chosen in the system. Dispersion Compensation Module (DCM) is used to compensate on the receiving shore. DCF and loss simulator are used to simulate the DCM. η changes from 0.7 to 0.9 by changing the length of DCF. The channel 1 and 4 can achieve error-free transmission. Figure 1 shows the change curve of Q factors of channel 3 (the worst channel) with the change of DCF length.

As is shown in Figure 1, Q factor of channel 3 is the largest when the DCF length is 73 km (η is 78%). System margin of channel 3 improves 0.2 dB. There may be two reasons: one is that wavelength selection makes the channel 3 have more nonlinear product, original channel power reduce, and noise increase and performance become poor. The other is that the amplifier gain is not flat which makes the Optical Signal Noise Ratio (OSNR) of channel 3 lower. As a result, we change the wavelength to 1549.32 nm, 1550.92 nm, 1552.52 nm, 1554.13 nm without any dispersion compensation and the system margin of each of the channels are shown as follows: channel 1 is 13.34 dB, channel 2 is 13.18 dB, channel 3 is 1.12 dB, and channel 4 is 10.96 dB. Channel 1, 2 and 4 are free-error. Channel 3's performance is the worst and differs from the other channels more. It is necessary to optimize the choice of channel wavelength.

3 EFFECTS AND OPTIMIZATION OF PMD

In undersea optical fibre cable transmission system, the PMD will become one of the major limiting factors in long-distance transmission because of the uncertainty of applications environment. It broadens the pulse, limits the transmission rate, and affects the system's transmission performance [3].

The PMD will cause pulse broadening and distortion and interference between codes in a WDM system. By analysing the experimental eye diagram, the eye opening of the worst channel has increased a little. The BER reduces to 2.95×10^{-6} and the system margin is 4.93 dB. System margin has improved 0.86 dB relatively to the system considering PMD (system margin with PMD is 4.07 dB). Overall, the impact of PMD on the system is not great and can be neglected. However, since the simulation cannot simulate complex undersea environments, various environmental factors will influence PMD randomly, and the PMD produced by external factors may not be higher than 38 ps [4].

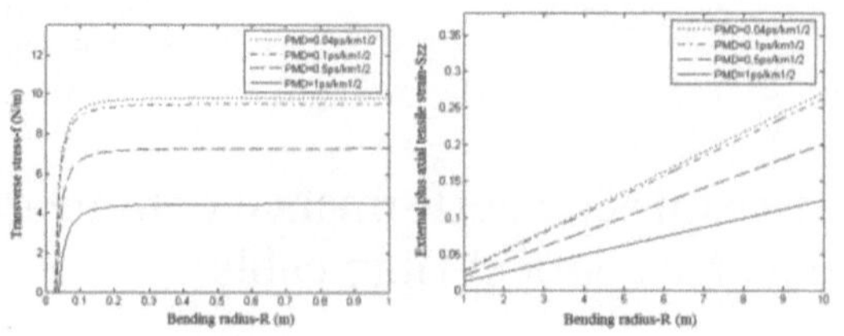

(a) (b)

Figure 2. Attainable range of transverse stress and bending radius (a) and attainable range of external plus axial tensile strain and bending radius (b).

To further illustrate the influence of the stress and bending on PMD, Figure 2(a) shows the attainable range of transverse stress and bending radius and Figure 2(b) shows the attainable external range plus axial tensile strain and bending radius.

As is shown in Figure 2(a), the transverse stress that the system can endure decreases gradually with the increase of the PMD coefficient. When the bending radius is less than 0.3 m, the transverse stress that the system can endure increases sharply with the increase of the bending radius. When the bending radius is more than 0.3 m, the transverse stress has no change basically with the increase of the bending radius. At this time, birefringence induced by transverse stress plays a decisive role. It can be concluded that when the PMD coefficient is $0.04\,ps/km^{1/2}$, the transverse stress and bending radius should be taken under the curve range to ensure the system is not affected by PMD, while eliminating the PMD compensation. For example, transverse stress is required not to be greater than 8.9 N/m when bending radius is 0.3 m.

As is shown in Figure 2(b), the tensile strain that the system can endure decreases gradually with the increase of the PMD coefficient. The tensile strain increases linearly with the increase of bending radius. When the PMD coefficient is $0.04\,ps/km^{1/2}$, the value of the bending radius is larger than that in Figure 4. When the bend radius changes from 1 m to 10 m, the range of the tensile strain is 0 to 0.3 basically. As long as taking the value under the curve range, we can ensure the system is not affected by the PMD while eliminating any PMD compensation. For example, when the PMD coefficient is $0.04\,ps/km^{1/2}$ and the bending radius is 0.3 m, the tensile strain is required to be not greater than 0.05.

4 EFFECTS AND OPTIMIZATION OF NONLINEARITY

4.1 *Effects and simulation of nonlinearity*

Fibre nonlinearity can be divided into two categories [5]. One is a scattering effect including Stimulated

Raman Scattering (SRS) and Stimulated Brillouin Scattering (SBS). The other is the Kerr effect including Self-Phase Modulation (SPM), Cross-Phase Modulation (XPM) and Four-Wave Mixing (FWM).

Through analysing the worst channel simulation, the BER is 2.01×10^{-5} and the system margin is 4.07 dB when considering SPM, XPM and FWM. The BER is 1.05×10^{-10} and the system margin is 7.87 dB without FWM. The BER is $9.11 \times 10^{-17}$and the system margin is 10.12 dB without SPM, XPM or FWM. It can be seen that the system margin can improve by 3.8 dB without FWM, and the BER is lower and the eye diagram is clearer. The system then adds 2.25 dB redundancy, system performance improves, eye opening is bigger, jitter amplitude decreases and the eye becomes clearer. Therefore, the Kerr effect has a great impact on the system, making the system margin reduce 6.05 dB. It is necessary to reduce the impact of the Kerr effect on the system performance.

When considering SBS and SRS, the BER is 2.01×10^{-5}. And the BER reduces to 3.51×10^{-6} without SBS. We can see that SBS has little impact on the system. Furthermore, seen from the transmission power loss, the signal power in 75 km section of the fibre loses 0.022 dBm due to SBS and on the 280 km section the loss of signal power is up to 0.769 dBm. Moreover, the higher the input power, the longer the transmission fibre, the greater will be the line loss caused by SBS. Therefore, choosing the right input power is very important to reduce the power loss caused by SBS. When the Raman Effect in a Raman amplifier is not changed, SRS and SBS are closed at the same time in order to study the effects of SRS on system. Then we obtain the BER at 7.88×10^{-6} and the system only has 0.4 dB redundancy. Eye definition has no improvement basically. Therefore, the impact of SRS on the system is negligible.

4.2 *Optimization of the main point of system*

The input power of fibre will directly affect the nonlinearity in the case of fixed fibre parameters. The amplifier gain is closely related to the pump intensity and erbium-doped fibre (EDF) length. There is an optimum EDF length corresponding to the maximum gain. Figure 3(a) shows the system margin changes of the worst channel (channel 3) by using different codes (NRZ, RZ, CSRZ) and FEC when nonlinearity, dispersion. and noise are considered.

From the analysis of Figure 3(a), when the system performance is optimal, the optimal EDF length in power amplifier module and corresponding fibre input power, using different codes, are different.

Similarly, the output power of a remote-booster gain module is also enough to cause a large nonlinearity. Set the EDF length of power amplifier module at the optimal value, change the EDF length of the remote-booster gain module and the system margin of the worst channel is shown in Figure 3(b).

From the analysis of Figure 3(b), when we use different codes, the optimal EDF length in the remote-booster module and the corresponding fibre input power are different. When we use the CSRZ, system, the performance improves well and the system margin increases. Therefore, when we optimize the system, we can select the CSRZ code to extend the transmission distance.

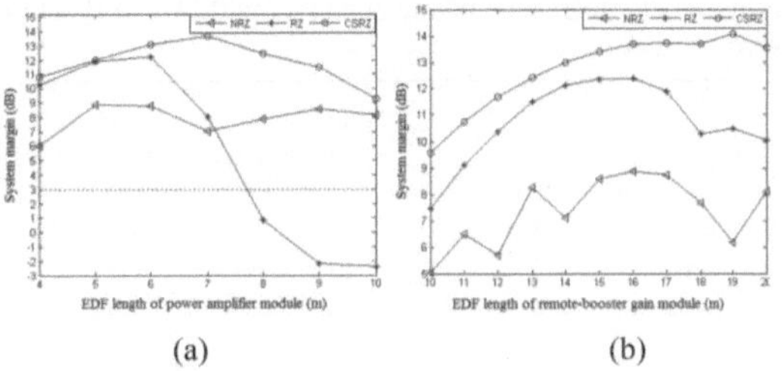

Figure 3. Relation between system margin and EDF length in power amplifier module (a) and relation between system margin and EDF length in remote-booster gain module (b) (with FEC).

4.3 *The optimization of adjacent channel spacing and channel wavelength*

4.3.1 The optimization of adjacent channel spacing
The channel spacing, the light power density, the power into fibre, dispersion, etc. are the major limiting factors affecting the WDM system. And these factors have impacts on nonlinearity. The unequal channel spacing will reduce the impact of FWM greatly, thereby increasing the maximum optical power of each channel.

Referring to the ITU-T G.692 recommendation, we select the NRZ code, and 4 waves with the wavelength of 1556.7 nm, 1558 nm, 1559.7 nm, 1560.7 nm as a simulator. The system margin of each channel is 11.67 dB, 12.14 dB, 11.62 dB, 11.95 dB. Though some of the channels are poorer than that of 200 GHz channel spacing, the performances of four channels are basically similar to each other. Moreover, nonlinearity is controlled well and the system margin of the worst channel (channel 3) improves 7.5 dB, making the system utmost transmission distance can be further extended.

From the above analysis, in order to better suppress the influence of nonlinearity on the system and overcome the unbalancing between channel performances, the system uses unequal channel spacing for transmission. Thus the system performance will be improved greatly.

4.3.2 *Optimization of channel wavelength*
Simulation results from the previous analysis shows that, when we select the wavelength as 1550.12 nm, 1551.72 nm, 1553.32 nm, 1554.92 nm (A) and 1549.32 nm, 1550.92 nm, 1552.52 nm, 1554.13 nm (B), and don't consider dispersion compensation, the performance of each channel is shown in Table 1.

Table 1. Performance of each channel with A and B set of wavelengths.

Performance		BER	Q value (dB)	System margin (dB)
A	ch 1	8.88×10^{-34}	21.62	13.43
	ch 2	1.35×10^{-24}	20.15	11.95
	ch 3	2.01×10^{-5}	12.27	4.07
	ch 4	1.14×10^{-29}	21.02	12.83
B	ch 1	4.34×10^{-33}	21.53	13.34
	ch 2	5.46×10^{-32}	21.37	13.18
	ch 3	1.73×10^{-3}	9.32	1.12
	ch 4	5.63×10^{-20}	19.16	10.96

As is shown in Table 1, with respect to the A set of wavelengths, channel 3 performance of B set of wavelengths is still the worst. So we change the wavelengths to 1546.12 nm, 1547.72 nm, 1549.32 nm, 1550.92 nm and do not consider the dispersion compensation. The BER of channel 1 is 1.24×10^{-29} and system margin is 12.82 dB. The BER of channel 2 is 1.24×10^{-35} and system margin is 13.68 dB. The BER of channel 3 is 1.08×10^{-35} and system margin is 13.69 dB. The BER of channel 4 is 4.33×10^{-33} and system margin is 13.34 dB. The performance of each channel is almost the same. There is no error in the condition of existing transmission distance and the number of multiplexing wavelengths. It is possible to further extend the transmission distance.

As for the analysis of changing the channel wavelength, it is better to avoid the wavelength of 1552.52 nm–1553.32 nm when select the wavelength.

Synthesizing all the above analysis results, we optimize the wavelength and coding simultaneously. That is selecting the wavelength of 1546.12 nm, 1547.72 nm, 1549.32 nm, 1550.92 nm and CSRZ code. The system margin of channel 1 is 18.75 dB. The channel 2 is 17.43 dB. The channel 3 is 16.54 dB, and the channel 4 is 16.20 dB. The eye diagram of each channel is clear. The performance has some improvement over that using unequal channel spacing. The system margin increases. Transmission distance is further extended.

When using CSRZ and wavelength optimization, the system performance improves more than the one using unequal channel spacing. But the two kinds of optimization are both good. From the perspective of wavelength utilization, we suggest selecting wavelength optimization when there are less wavelengths and using unequal channel spacing when the number of wavelength is more.

5 CONCLUSIONS

This paper studies the effects to multi-wavelength repeaterless undersea optical fibre cable transmission performance, caused by chromatic dispersion, PMD, and nonlinearity. It proposes a series of optimization methods to improve the system's performance and extend the transmission distance from the aspects of dispersion compensation, fibre input power, channel spacing, wavelength, and coding technology. (1) The use of the DCF post-compensation and owed-compensation compensates the influence of chromatic dispersion. (2) We get the range of transverse stress, tensile strain and a bending radius that the system can endure to avoid PMD compensation. (3) It is better to avoid the wavelengths of 1552.52 nm–1553.32 nm when we select a wavelength. (4) When we select the wavelengths of 1546.12 nm, 1547.72 nm, 1549.32 nm, 1550.92 nm and the CSRZ code, the eye diagram of each channel is clear and their performances have little difference. The system performance has been improved well and the transmission distance is further extended.

What's more, we can obtain the research methods about extending transmission distance in a WDM system. Due to limited conditions, there are some issues to be further resolved: for example, with the 10 Gbit/s and even the 40 Gbit/s system, the effects of PMD and PMD compensation will be researched continuously.

REFERENCES

[1] OptSim 4.5 Models Reference Volume II Block Mode. RSoft Design Group, Inc. [EB/OL]. http://www.resoft design.com

[2] Sultana, Nasrin & Islam, M.S. 2012. Analysis of XPM effect with SPM and GVD in WDM fiber optic transmission system. 2012 International Conference on Computer and Communication Engineering, ICCCE 2012: 263–266

[3] Lignie, M.C. & Nagel, H.G.J. 1994. Large polarization mode dispersion in fiber optic cables. Lightwave Technol., 1994,12(8): 1325–1329

[4] ITU-T G.650

[5] Kaur, Gurmeet & Singh, M.L & Patterh, M.S. 2010. Effect of fiber nonlinearities in a WDM transmission system. Optik, 2010(6): 889–896

Future Communication Technology and Engineering – Chan (Ed.)
© 2015 Taylor & Francis Group, London, ISBN: 978-1-138-02777-0

The research of a joint channel coding based on LDPC and physical layer network coding

X.M. Lu, F. Yang, J.L. Wang, L. Su & B. Pan
Northwestern Polytechnical University, Xi'an, Shanxi Province, China

ABSTRACT: This paper combined LDPC channel coding technology with physical layer network coding, and, based on the principle of superposition by the code word, put forward a new joint channel coding technology. The technology treated the network coding signal in the relay node as a redundant code and combined it with the original coded signal to form a joint coding signal. Finally, it is simulated for bit error rate on the Matlab which proves that technology can effectively reduce the BER and improve transmission performance.

1 INTRODUCTION

Network coding, which is put forward in recent years, is a new theory. Its basic principle is that relay node deals with information of a transmission link before sending it, and the destination node can receive the information finally [1], [2]. The theory proves that network coding can realize the maximum transmission of a network. References come up with Physical Network Coding (PNC) and prove that PNC has a higher performance gain as compared with a direct transmission [3], [4].

As the channel is not independent in the actual network and references [1] indicate that the separation theorem of network coding and channel coding may be false in the theory of random linear codes [5], so network coding and channel coding often require a joint design in the wireless network [6].

Research of LDPC technology indicates that LDPC performance of 1/2 code rate in the BPSK modulation mode is only 0.0045 dB worse than the Shannon limit. The common algorithm of LDPC is a BP decoding algorithm based on iterative arithmetic which is widely used in communication technology.

This paper comes up with a joint channel coding technology based on LDPC and the physical layer network coding which is applied in a wireless communication network and the technology can reduce the bit error rate effectively so as to improve the transmission performance of system.

2 JOINT CHANNEL CODING SYSTEM MODEL

It assumes that an ideal wireless communication network system includes two user source nodes: one relay node, and one destination node. In the network, the user source node sends information (a, b) to the relay node, and then the relay node uses network coding technology to deal with the information and sends it (a, b, $a \oplus b$) to the destination node. According to the Shannon noisy channel coding theorem, the decoding error probability tends to zero with the coding length increasing ($L \rightarrow \infty$). Thus, compared with the traditional network coding, decoding error probability of the joint channel coding is the lower. The network model is displayed in Figure 1.

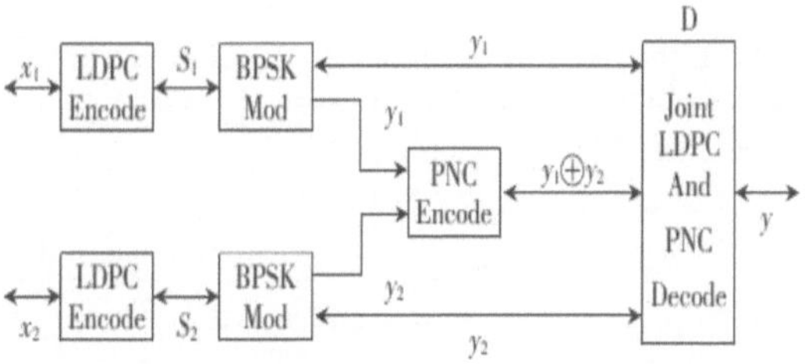

Figure 1. Joint channel coding system model based on LDPC and PNC.

S1 sends the information flow (k_{S1}) to the relay node and the node receives at the same time, and then S2 also sends the information flow (k_{S2}) to them. After receiving the information flow ((k_{S1}, k_{S2}), the relay node XORs the two packets and broadcasts the XOR-ed version to D. D can obtain its packet by XOR-ing again with its own packet. The system uses LDPC and BPSK modulation.

It assumes that S1 and S2 use the same coding, the generated matrix of S1 and S2 is G1 and G2 respectively and their check matrix are H1 and H2.

2.1 *The coding decoding scheme of the joint channel*

It uses LDPC and BPSK modulation to deal with the signals at S1 and S2.

$y_k = l_k + n_k$ when $l_k = G_k + x_k$, we can obtain

$$y_k = G_k x_k + n_k, k = 1,2,..n_k \tag{1}$$

is the additive white Gaussian noise of channel.

The relay nodes XORs the two packets and we can obtain $y_{xor} = y_1 \oplus y_2$.

The receiving node D can obtain its packet by XOR-ing again with its own packet (y_1, y_2).

2.2 *The check matrix H of joint channel coding system*

The node D receives the sequence ($y = [y_1, y_2, y_{xor}]$) and if decoding it, it must have the check matrix H_{xor} ($H_{xor}^T = 0$).

When $[y_1, y_2, y_{xor}]H_k^T = 0$, we can obtain

$$\begin{aligned}(y_1 + y_2 + y_{xor})H_k^T &= (y_1 + y_2 + y_1 \oplus y_2)H_k^T \\ &= (G_1x_1 + G_2x_2 + G_1x_1 \oplus G_2x_2)H_k^T \\ &= G(x_1 + x_2 + x_1 \oplus x_2)H_k^T\end{aligned} \tag{2}$$

With the features of model 2 and XOR operation, we can obtain $x_1 + x_2 + x_1 \oplus x_2 = 0$. In this paper, the check matrix is

$$H_{xor} = \begin{bmatrix} H & & \\ & H & \\ I & I & I \end{bmatrix} \tag{3}$$

3 ANALYSIS OF BIT ERROR RATE

3.1 *Error rate of pair-wise codewords*

Instantaneous SNR of each channel is r_k respectively. The error rate of pairing codeword is shown below.

$$P(c \to e \mid \gamma) = Q(\sqrt{2\sum_{n \ni \eta} \gamma(n)}) \tag{4}$$

η is error code.

$$Q(x) = p(z \ge x) = \int_x^{\infty} \frac{1}{\sqrt{2\pi}} e^{-z^2/2} dz \tag{5}$$

In the fading channel, the error rate of pairing codeword is PEP $(d|\gamma_1, \gamma_2)$, d is Hamming distance.

$$\begin{aligned}PEP(d) &= \int_0^{\infty}\int_0^{\infty} PEP(d \mid \gamma_1, \gamma_2) P_{\gamma_1}(\gamma_1) P(\gamma_2) d\gamma_1 d\gamma_2 \\ &= \frac{1}{\pi}\int_0^{\infty}\int_0^{\frac{\pi}{2}} \exp(-\frac{d_{\gamma_1}}{\sin^2(\theta)}) d\theta * P_{\gamma_1} d\gamma_1 * \int_0^{\infty}\int_0^{\frac{\pi}{2}} \exp(-\frac{d_{\gamma_2}}{\sin^2(\theta)}) d\theta P_{\gamma_2} d\gamma_2 \\ &= \frac{1}{\pi}\int_0^{\frac{\pi}{2}} (1 + \frac{\overline{\gamma_1}}{\sin^2(\theta)})^{-d} (\frac{\overline{\gamma_2}}{\sin^2(\theta)})^{-d}\end{aligned} \tag{6}$$

We can obtain

$$PEP(d) \le \frac{1}{2}(\frac{1}{1+\gamma_1})^{-d}(\frac{1}{1+\gamma_2})^{-d} \tag{7}$$

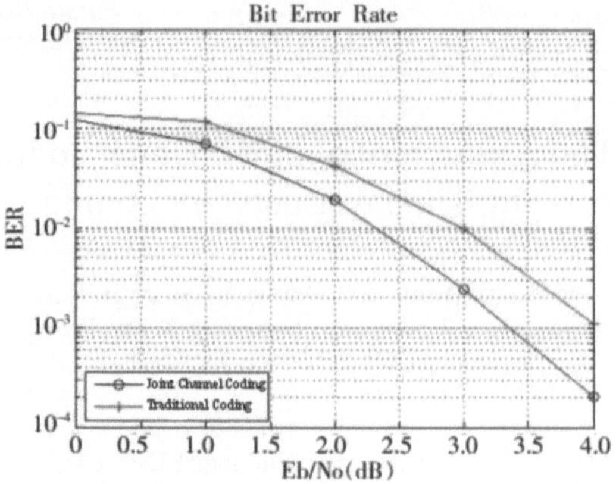

Figure 2a. Bit error rate.

3.2 *Bit error rate*

Firstly, we analyse the interruptive probability of transmission. In the Figure 1, source nodes send information to the destination node respectively through two links (L_1, L_2). Assuming that channel capacity of the two links is $C(\gamma_1)$ and $C(\gamma_2)$, the interruptive probability is

$$P_0 = \{(C(\gamma_1) + C(\gamma_2)) < r\}. \tag{8}$$

r is the maximum transmission rate of information.

In Figure 1, the modulation system of the network system is BPSK and the channel is AWAN, so bit error rate is:

$$\begin{aligned}P_b(\gamma_1,\gamma_2) &\le \frac{1}{k_c}\sum_{d=d_{free}}^{\infty} C_d PEP(d \mid \gamma_1,\gamma_2) \\ P_b(\gamma_1,\gamma_2) &\le \int_{\gamma_1}\int_{\gamma_2} \min[1, \frac{1}{2}\sum_{d=d_{free}}^{\infty} C_d PEP(d \mid \gamma_1,\gamma_2)] P(\gamma_1) P(\gamma_2) d\gamma_1 d\gamma_2\end{aligned} \tag{9}$$

4 PERFORMANCE SIMULATION

Using computer simulation software (Matlab), we can obtain the performance curve with the traditional network coding and the joint channel coding. In Figure 2a, compared with the traditional network coding, the bit error rate of the joint channel coding has been significantly reduced. And as the SNR increases, the bit error rate reduction becomes greater. From Figure 2b, the frame error rate of the joint channel coding is also greatly different. When the bit error rate is 10^{-3}, the performance gain of the joint channel coding is greater than 5 dB.

The experimental results demonstrate that the joint channel coding can reduce the bit error rate and frame error rate effectively.

From the above results and analysis, we can see that, compared to the network coding, the transmission capability of a system based on LDPC and PNC technology has been greatly improved.

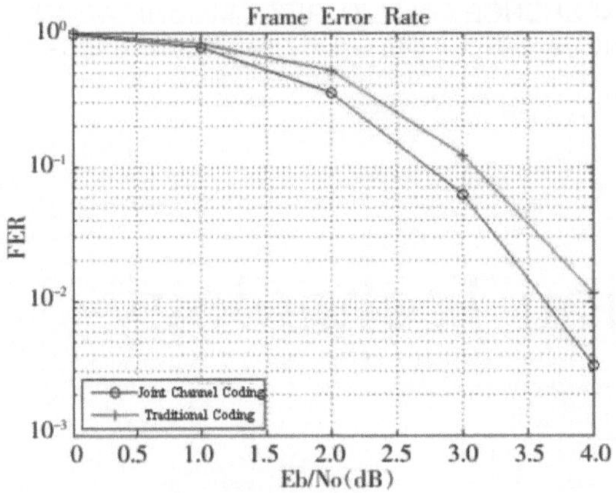

Figure 2b. Frame error rate.

ACKNOWLEDGMENT

Combining network coding theory with network communication technology, a new joint channel coding based on LDPC and PNC is put forward in this paper. And through mathematical analysis and simulation results, it shows that this coding technology can effectively reduce the bit error rate and frame error rate effectively, which can greatly improve the transmission capability of a system.

From practical application, when we use network coding strategies to increase the network throughout, at the same time the coding strategies can reduce the bit error rate and frame error rate effectively. However, with network technology development, its promotion throughout the network is becoming more and more important and we can use the new joint channel coding based on LDPC and PNC to improve the network throughout.

This paper is supported by Northwestern Polytechnical University Foundational Research Fund 2011 (JC201143).

REFERENCES

[1] Li S.Y.R., Sun Q.T. & Z.Y. Shao. *Linear network coding: theory and algorithms* [J]. Proceedings of the IEEE. 2011, 99(3):372–387.

[2] Hao, K. *Research of Network Coding Key Technology and Applications* [D]. Tianjin University, 2011

[3] Yu H.H. & Zhang J. *Physical-layer network coding in wireless multi access relay channel* [C]. *Applied Information and Communication*. Springer Berlin Heidelberg, 2011, 227(1):312–320.

[4] Yeung R.W., Lis Y.R. & Cai N., et al. *Network coding Theory (1st ed)* [M]. Now Publishers 2005.

[5] Song L., Yeung R.W. & Cai N. *A separation theorem for single-source network coding* [J]. *IEEE Transactions Information Theory*, 2006, 52(5):1861–1871.

[6] Zhou L.C. & Xiu C.D. *A joint design of physical layer network coding and channel coding for wireless networks* [J]. *Computer Science and Electronics Engineering (ICCSEE), 2012 International Conference*, 2012(3):312–316.

PROCEEDINGS OF THE 2014 INTERNATIONAL CONFERENCE ON FUTURE COMMUNICATION TECHNOLOGY AND ENGINEERING (FCTE2014), SHENZHEN, CHINA, 16–17 NOVEMBER 2014

Future Communication Technology and Engineering

Editor

Kennis Chan
Advanced Science and Industry Research Center, Hong Kong

CRC Press
Taylor & Francis Group
Boca Raton London New York

CRC Press is an imprint of the
Taylor & Francis Group, an **informa** business
A BALKEMA BOOK

Future Communication Technology and Engineering – Chan (Ed.)
© 2015 Taylor & Francis Group, London, ISBN: 978-1-138-02777-0

Hardware network protocol stack design for FPGA-based FBG data transmission

Q. Wang
China National Materials Group Corporation, China

Z.Q. Xu, Q. Liu, Q.J. Huang & Z.Y. Li
School of Information Engineering, Key Laboratory of Fiber Optic Sensing Technology and Information Processing, Ministry of Education, Wuhan University of Technology, Wuhan, China

ABSTRACT: In this paper, with respect to low data transmission rate for the existing FBG demodulation system, a data transmission method based on FPGA hardware network protocol stack is proposed. This paper designs a state machine-based LAN91C111 hardware driver, crops the standard TCP/IP protocol, and takes into account the characteristics of system data transmission. Timing simulation software is applied to functional simulation in the design, so as to ensure the reliability of the design. In hardware tests, a wireshark packet capture tool is employed to obtain real-time network communication data, and to analyse the data in detail, so as to ensure the accuracy of the design. Test results show that a high-speed and stable network transmission of sensing data can be assured by the design when the electric circuit remains unchanged.

1 INTRODUCTION

Since FBG sensors are competitive in many superior characteristics (such as small size, immunity to electromagnetic interference, etc. (Macro Esposito 2013)), they are widely used in structural health monitoring, heavy-duty machinery, equipment condition monitoring, and such fields (L. Wang 2011; Jeannot Frieden 2010). In order to achieve high-speed parallel demodulation of multi-channel FBG sensors, existing FBG demodulation systems have utilized FPGA-based data acquisition and transmission circuits, running LwIP protocol via NiosII soft core to achieve network communication. Test results show that the design proposal can ensure 4 kHz stable demodulation of 4-channel FBG sensor (Li Zheng-ying 2013).

However, with the development of technology, the FBG sensor system is becoming large in capacity (Liu M.Y. 2012; Zhou Zu-de 2013), and a large-capacity FBG sensor may produce a huge amount of data during high-speed real-time synchronous demodulation, causing the data transmission rate to become a bottleneck problem in the system. For instance, when a demodulation channel is extended to eighteen, the system runs at a demodulation rate of 4 kHz. That is to say, it is connected to 576 sensors at full load (32 FBG sensors per channel) and each frame data is as long as 1,192 bytes. Data transmission rate will reach 38.14 Mbps, far beyond the maximum transmission rate specified in the above design proposal. To solve this problem, in this paper, by making full use of FPGA flexible and reprogrammable features on the basis of unchanged circuit structure, a FPGA-based pure hardware UDP protocol is thus designed to improve the network transmission rate of a demodulation system. Researches in this topic can solve the data high-speed network transmission problems of a FBG demodulation system, and further make sure that a demodulation system can achieve the stable and high-speed demodulation of high-capacity FBG sensors. As a result, it has great research value and huge engineering significance.

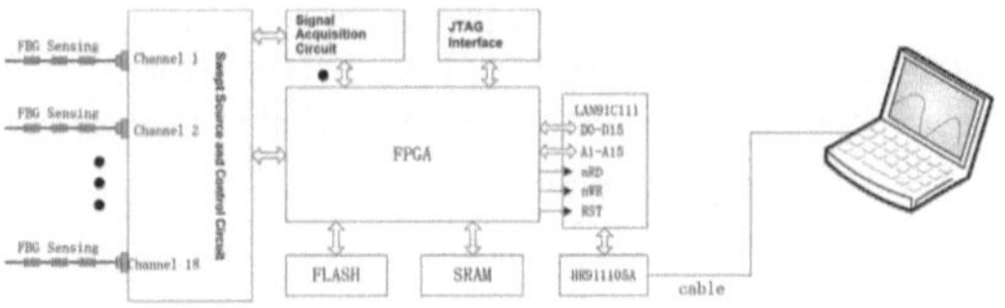

Figure 1. Schematic diagram of system hardware structure.

2 SYSTEM HARDWARE DESIGN

2.1 *System hardware structure*

Figure 1 shows the hardware structural schematic diagram of a FBG demodulation system. A FBG demodulation system relies on FPGA as the core to achieve large-capacity and high-speed FBG demodulation. FBG sensing is accessed to the system through eighteen channels, and the FPGA controls sweep the source and signal acquisition circuit to collect and calculate FBG wavelength values, and then transmits

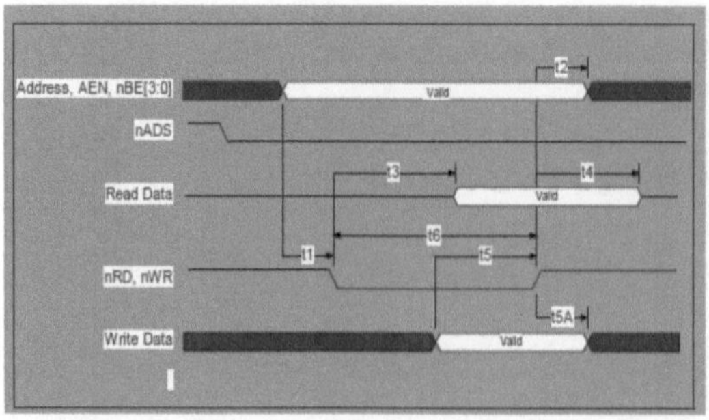

Figure 2. LAN91C111 interface timing diagram.

```
`define WrLANReg(RegAddr,RegData)  \
begin case(wrstate) \
    4'd0:begin HardAddr<=RegAddr;HardWrData<=RegData;wrstate<=4'd1;end \
    4'd1:begin  HardWr<=1'b0;wrstate<=4'd2;end \
    4'd2:begin  HardWr<=1'b1;wrstate<=4'd0;state<=state+1'b1;\
    end default:; \
    endcase \
end

`define RdLANReg(RegAddr) \
begin    \
    case(wrstate) \
    4'd0:begin  HardAddr<=RegAddr;wrstate<=4'd1;end \
    4'd1:begin  HardRd<=1'b0;;wrstate<=4'd2;end \
    4'd2:begin  wrstate<=4'd3;end \
    4'd3:begin  HardRdData<=DATA;wrstate<=4'd4;end \
    4'd4:begin  HardRd<=1'b1;wrstate<=4'd0;state<=state+1'b1;end \
    default:; \
    endcase \
end
```

Figure 3. Verilog-based LAN91C111 interface driver code.

them to PC for further processing through the network interface chip LAN91C111. HR911105A refers to a network isolation transformer, provided with network filters and indicators.

2.2 *LAN91C111 introduction and realization of hardware driver*

LAN91C111 is a non-PCI 100 M/10 M adaptive network communication chip integrated with MAC and PHY (LAN91C111 Ref). The system requires initial configuration in design, and then sends and receives network data. LAN91C111 operations are realized by means of reading or setting its MAC and PHY registers, and the read and write timings for its interface pins are shown in Figure 2.

In order to facilitate the use of Verilog to achieve timing operations for the network interface chip pin, a macro definition with parameters is adopted in design, as shown in Figure 3. The read/write register is a small state machine to be embedded into the main network data reception-transmission state machine by the macro definition, as shown in Figure 6.

3 PROTOCOL CROPPING AND REALIZATION

3.1 *Network communication protocol*

In order to simplify development, the standard TCP/IP protocol suite has been cropped taking into account the actual demands of the project.

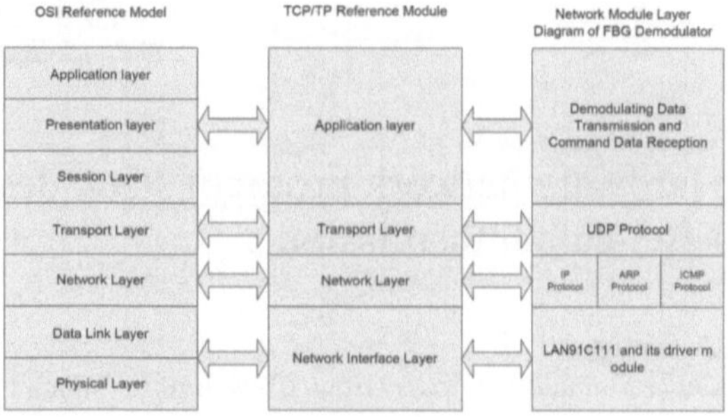

Figure 4. A network protocol layer diagram.

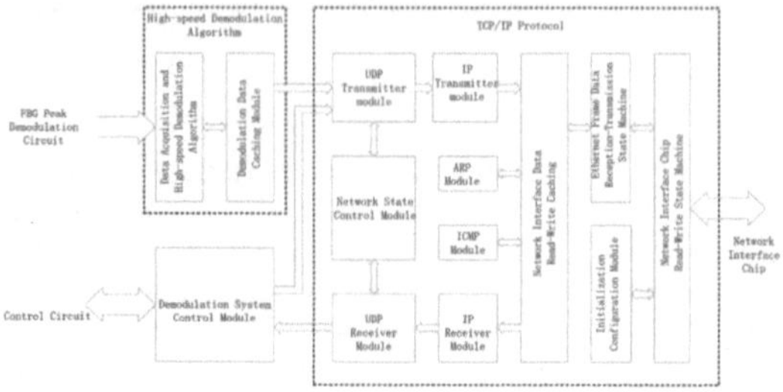

Figure 5. Design structure chart of TCP/IP protocol in FPGA.

Figure 4 is the network protocol layer diagram. As the figure shows, on the left is the Open Systems Interconnection Reference model (OSI). It is the reference architecture for the world-wide network communication standard, proposed by the International Organization for Standardization. In practice, TCP/IP protocol suite has been widely used, as shown in figure below (Richard S.W. 2000). On this topic, the network communication is designed to point-to-point single connection in the LAN, with a PC serving as the client and FBG demodulation system as the server. The server is required to frequently send a large amount of FBG sensor data to the client, but the client only sends command control word to the server. Therefore, the standard TCP/IP has been greatly simplified in this paper, in order to markedly reduce the complexity of the protocol. As shown in the figure below, in the network module layer diagram of the FBG demodulation system, the network interface layer contains the functions of initialization and data reception-transmission for LAN91C111. On the network layer, only ARP/ICMP/IP protocols are retained. The transport layer employs the relatively simple UDP protocol; and the application layer is the data transmission and command word receiver module for the demodulation system.

3.2 *Design proposal*

Figure 5 is the design structure chart of TCP/IP protocol in FPGA. LAN91C111 read-write timing control is achieved by network interface chip read-write state machine; LAN91C111's initial configuration and data

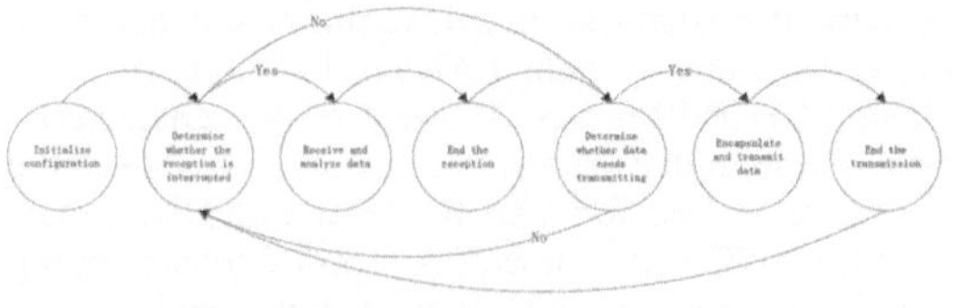

Figure 6. Data reception-transmission state machine.

collection functions are achieved by initialization configuration module and Ethernet frame data collection module through read-write state machine.

Figure 6 is a data reception-transmission state machine. The system will enter data reception-transmission state upon the completion of LAN91C111 initialization configuration. Since it is complicated for it to achieve interrupted responses for LAN91C111 receiving data by hardware description, a query-reception mode shown in the figure is adopted in the design to receive and transmit data alternately.

Response to an ARP request from the client is achieved by an ARP module, as shown in the figure. When an ARP request is received, the ARP module will be booted by network control module for data analysis. If the destination IP address is in conformity with the server IP address, an ARP response frame will be transmitted to the client. When the ICMP is received, an ICMP module will be booted by network control module for data analysis. If the ICMP message is a loopback request and the destination IP and MAC address are both in conformity with that of the server, a response frame for ICMP loopback request will be transmitted.

If the IP frame data of the client is received by data reception-transmission state machine and the protocol type of IP frame header is UDP, a UDP receiver module will be booted by network control module for data analysis. If the destination port number is in conformity with the receiver port number, the frame header of UDP protocol will be filtered out and the frame data will be transmitted to a demodulation system control module, which will conduct command parsing for the data transmitted by the client and then control the relevant hardware of a FBG demodulator.

If the data storage of demodulation data caching module exceeds 1,400 bytes in high-speed demodulation algorithm, UDP transmitter module will be booted by a network control module. Since the network frame data length is 46–1,500 bytes, the transmitting threshold has been set as 1,400 bytes in the design, which can make the network frame include data with maximum length and improve the data transmission rate. A UDP transmitter module will encapsulate FBG sensing data in a UDP protocol frame, set the destination port, calculate, verify, and fill in the UDP frame and then boot the IP transmitter module, which will encapsulate the UDP data received in IP protocol frame, calculate, verify, and fill in an IP frame, write in transmitting cache and wait for transmission by the data reception-transmission state machine.

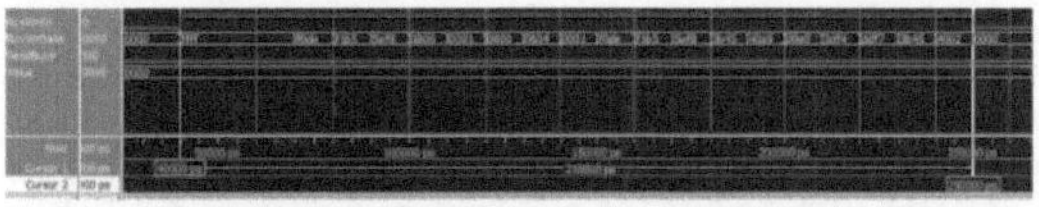

Figure 7. ARP request simulation received.

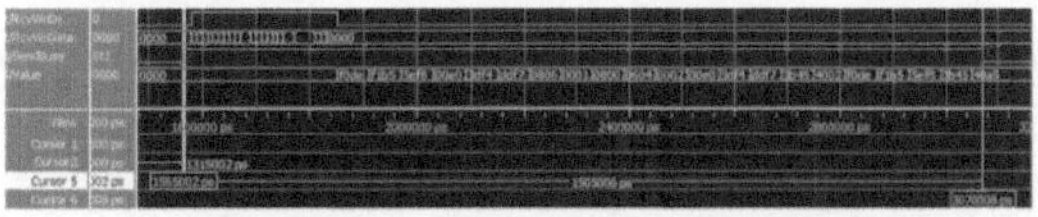

Figure 8. ARP response simulation transmitted.

4 TEST RESULT

4.1 *Software simulation*

Altera-Modelsim software is adopted for function simulation in the design and the parameters in the test-bench document have been set as the actual values to ensure the simulation results conform to the actual result. MAC address and IP address of LAN91C111 are set as 00 E0 3D F4 DD F7 and 59.69.64.2 respectively (hexadecimal: 3B 45 40 02) with laptop as the client in the test, whose MAC address and IP address are F0 DE F1 B5 5E F8 and 59.69.64.169 (hexadecimal: 3B 45 40 A9).

Figure 7 shows an ARP request simulation received. Signal list is on the left of simulation timing diagram, namely: enable signal written in reception caching RcvWrEn (active high), RcvWrData (16 bit mode) written in reception caching by LAN91C111, signal send busy (high level means being sent), sending data value (16 bit mode) written by LAN91C111. As shown in Figure 7, during RcvWrEn high level (from 40 ns shown in Cursor 1 to 250 ns shown in Cursor 2), 42 bytes are written in with the destination address as the broadcast address (0×FFFFFF) and the protocol type as ARP request. The simulation result shows that the function simulation of data reception state machine and reception caching is normal, which is in conformity with the expected design goal.

Figure 8 is an ARP response simulation transmitted, which is the response signal for ARP request of Figure 7. As shown in Figure 8 below, during send busy high level (from 1,565 ns shown in Cursor 5 to 3,070 ns shown in Cursor 6), 42 bytes are written in LAN91C111 by the system with the destination address as the client and the protocol type as ARP response. The simulation result shows that the function simulation of data transmission state machine, the transmission caching, the ARP module, and the network control module are normal, which is in conformity with the expected design goal.

For function simulation shown in Figure 7 and Figure 8, most of modules are verified in the design, and the simulations of ICMP and UDP transceiver modules are similar to the simulation process above.

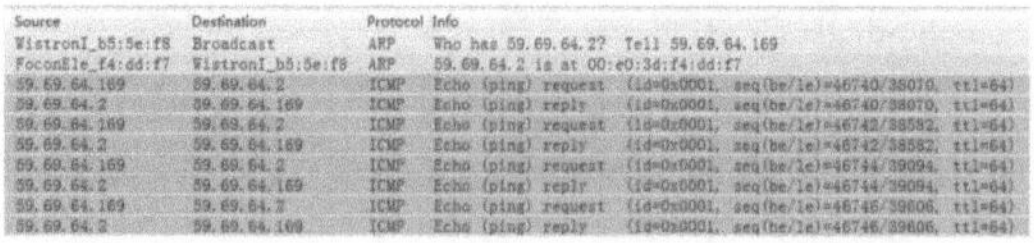

Figure 9. Packet capture data of ARP and ICMP protocol.

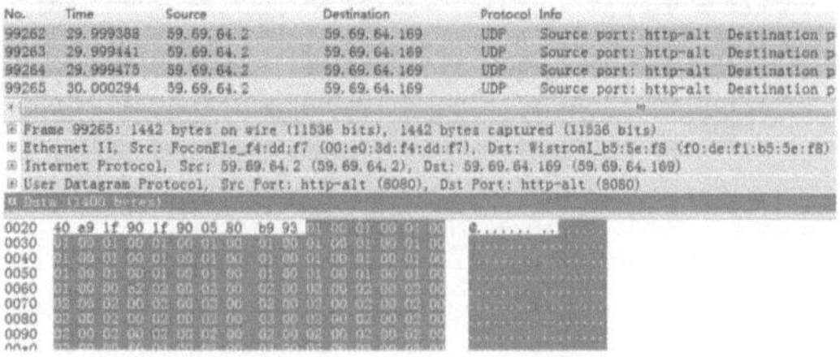

Figure 10. Packet capture data of UDP protocol.

4.2 *Hardware test*

Connect FBG demodulation system to laptop with cable; download the generated .s of document compiled by QuartusII to FPGA; packet capture network data with wireshark.

The packet capture data of the ARP protocol and ICMP protocol, which is shown in Figure 9, can be obtained upon the input of command ping 59.69.64.2 in the cmd command prompt window in the laptop. The first line shown is ARP request data sent by the computer to the demodulator; the second line is ARP response data sent back by the demodulator; the following eight lines are successively ICMP loopback data of the computer and the response data of the demodulator.

Figure 10 shows the packet capture data of a UDP protocol. A UDP transmitter module will be booted to send data to the computer when the FBG sensing caching data of the demodulator reaches 1,400 bytes. The length of the network frame shown in the figure is 1,442 bytes with the data length of 1,400 bytes. The FBG sensing value (with channel number as simulation numerical value to substitute sensing value) and the communication code value of a demodulation system are shown in the data area. Data with 99,265 frames is sent within 30.000294 s with the data length of 1,442 bytes for each frame and with the network transmission rate calculated of 38.17 Mbps, which is in conformity with the maximum transmission rate required for the system (38.14 Mbps), calculated previously (since frame header data, except for sensor data, is also in the network frame, the measured transmission rate value is bigger than the calculation value).

5 CONCLUSION

This paper introduces a FPGA-based network data transmission method of a FBG demodulation system. It designs a state machine-based hardware driver for network chip LAN91C111 and crops the standard TCP/IP protocol, based on the characteristics of system data transmission, and it completes the hardware designs for ARP, ICMP, IP and UDP protocols for a TCP/IP protocol. Altera-Modelsim timing simulation software is applied to functional simulation in the design, so as to ensure the reliability of the design. In hardware tests, a wireshark packet capture tool is employed to obtain real-time network communication data, and to analyse the data in detail, so as to ensure the accuracy of the design. Test results show that a stable network transmission of FBG sensing data can be assured by the design.

ACKNOWLEDGEMENTS

This paper is supported by the National High Technology Research and Development Program (863 plan) of China (No. 2012AA040106).

REFERENCES

Jeannot Frieden, Cugnoni Jo, Botsis John, et al. High-speed internal strain measurements in composite structures under dynamic load using embedded FBG sensors [J]. Composite Structures, 2010, 92(8): 1905–1912.

LAN91C111 Reference Manual [Z]. http://www.microchip.com/wwwproducts/Devices.aspx?product

Liu M.Y., Zhou Z.D., Tan Y.G., et al. Fiber Bragg grating sensors-based mechanical structure damage identification. [J]. Mechanika, 2012, 1:108–112.

Li Zheng-ying, Xu Zhi-qiang, Tang Zhi-hao, et al. Research of high-speed FBG demodulation system for distributed dynamic monitoring of mechanical equipment [J]. Advances in Mechanical Engineering, 2013, 2013(107073): 1–8.

Macro Esposito, Salvatore Buontempo, Angelo Petriccione, et al. Fiber Bragg Grating sensors to measure the coefficient of thermal expansion of polymers at cryogenic temperatures [J]. Sensors and Actuators A: Physical, 2013, 189:195–203.

Richard S.W. TCP/IP Illustrated Volume 1: Protocol [M]. Translated by Fan Jianhua. Beijing: China Machine Press, 2000.

Wang L., Li D., and Ou J. Fiber Bragg grating temperature sensors system on a twin-deck continuous rigid frame bridge for long term monitoring [J]. Advanced Materials Research, vol. 148–149, pp. 1611–1618, 2011.

Zhou Zu-de, Tan Yue-gang. Fiber Bragg Grating distributed dynamic monitoring and damage identification of mechanical system [M]. Beijing: Science Press, 2013, 305–343.

Future Communication Technology and Engineering – Chan (Ed.)
© 2015 Taylor & Francis Group, London, ISBN: 978-1-138-02777-0

The development of multihomed node model interworking with WLAN and WiMAX networks

Y. Wang & F. Ling
Department of Electronic and Information Engineering, Lishui University, Zhejiang Province, China

Y.M. Cheng
Department of Electric and Information Engineering, Beihua University, Jilin Province, China

ABSTRACT: This paper addresses the mobility simulation model of the multihomed node for supporting a MIPv6 function in an interworking of Worldwide Interoperability for Microwave Access (WiMAX) and IEEE 802.11 WLAN. The multihomed node model with two air interfaces is developed and based on WiMAX and WLAN workstation node model in simulation software. The main point of this paper is to support the MIPv6 function and provide a new network selection policy for multihomed nodes to make a handover decision between the WiMAX and the WLAN network. Finally, the simulation scenario and results are shown for testing and developing multihomed node models and MIPv6 functions.

1 INTRODUCTION

With the exponential growth of wireless communications, it is well accepted that next generation wireless networks will be heterogeneous, providing users of multi-interface devices with the ability to roam between different access networks. It is worth mentioning that the increasing popularity of Wireless Local Area Network (WLAN) enabled devices to have trigged interest in the development of systems that will integrate WLAN and cellular wide area data network technologies such as WiMAX or Universal Mobile Telecommunications System (UMTS) with support for vertical handoffs between the different access networks [1]. Because of the demand for efficient mechanism, mobile users will be able to experience ubiquitous data services and very high data rates across heterogeneous networks by using WLANs as a complementary technology for next generation cellular data networks, while providing the end user continuous access [2].

WiMAX Forum [3] believes that Mobile WiMAX (IEEE 802.16e–2005) services will complement existing and future broadband technologies such as WiFi, because both WLAN and WiMAX aim to provide ubiquitous low cost broadband wireless internet access. WLAN offers high data rates of 54 Mbps within a 100 m range. It is mostly used within buildings, whereas WiMAX offers high bandwidth 70 Mbps wireless back haul in a 5 km range, generally covering a large outdoor environment.

However, it should be noted that, at the time of writing, the main problem for WiMAX and WiFi is that the IPv6 is available only on WiFi and virtual interfaces, but not on the cellular interface (e.g., WiMAX interface) [4].

With Karl Andersson and ANM Zaheduzzaman [5], the authors provided network selection and handover timing with multihomed MIPv6 by suing Round-Trip Times (RTT) and RTT jitter values, but without OPNET full TCP/IP stacks. Deguang Le and Xiaoming Fu [6] evaluated MIPv6 performance only in WLAN network with wlan_wkstn_adv and wlan_server_adv model. However, this paper creates a multihomed node model based on wimax_ss_wkstn_adv and wlan_wkstn_adv node model in OPNET [7]. Unlike other paper's introduction, the network selection method utilizes the mathematical expression based on the Received Signal Strength Indicator (RSSI). We just modify the OPNET original C code to achieve two final goals, one to support a MIPv6 function, the other to provide the network selection policy. In our network selection policy, we put the RA (Router Advertisement) message together with interface type obtained from a received RA into the queue list in an IP layer once the RA is received. In comparison to the interface type, the IP layer determines the default network interface. It always prefers WLAN toWiMAX during the overlapping area between WiMAX and WLAN network.

2 OVERVIEW OF NETWORK TECHNOLOGIES

2.1 *IEEE 802.16e mobile WiMAX*

The IEEE 802.16e standard [3] is a technology proposed to offer wireless access to network stations in a metropolitan area environment. These networks are designed to operate at high data rates and to deal with several applications, resulting in different types of traffic profiles and demands. The motivation is to one day provide at last a mile of broadband wireless access to the general population; it is seeking to implement a single standard for fixed broadband wireless access and mobility with high scalability and a low cost of deployment.

2.2 *IEEE 802.11 wireless LAN*

IEEE 802.11 Wireless LAN is a flexible data communication system implemented as an extension to, or as an alternative for, a wired LAN within a building or campus using wireless radio technology. In WLAN it combines data connectivity with user mobility, through simplified configuration, and enables moveable LANs, which allows users to take advantage of the flexibility, convenience, and portability that WLAN can provide.

3 SYSTEM MODEL AND NETWORK SELECTION POLICY

3.1 *The system model*

The multihomed node model has been developed for MIPv4/MIPv6 mobility simulation in WiMAX and WLAN network. The developed node model has the structure shown in Figure 1, in which there are two WiMAX and WLAN interfaces. In Figure 1, in the protocol stack of the multihomed model, the network interface selection algorithm is implemented in an IP layer to determine one of the interfaces through which a higher layer transmits packets while the handover happens.

For testing the developed multihomed node model and MIPv6 function, the multihomed node runs in the interworking of WiMAX and WLAN network, as shown in Figure 2.

3.2 *The network selection policy*

In our multihomed node model, it is equipped with two access interfaces. Whenever handover happens between WiMAX and WLAN, the IP layer should decide from which interface the higher data packet should be transmitted. In order to distinguish the multiple interfaces, we bind the received RA (Router Advertisement) message together with the received interface number into the ordered list. By comparing the received RA with the interface number, the network access interfaces can be efficiently managed in ordered list, which is illustrated in Figure 3. Hereafter, the flow chart of the network selection policy between WiMAX and WLAN is described.

In Figure 3, there are four cases from which to choose WLAN and WiMAX access networks. While the multihomed node receives rx_RA:

- After receiving rx_RA, check whether the list is empty or not.

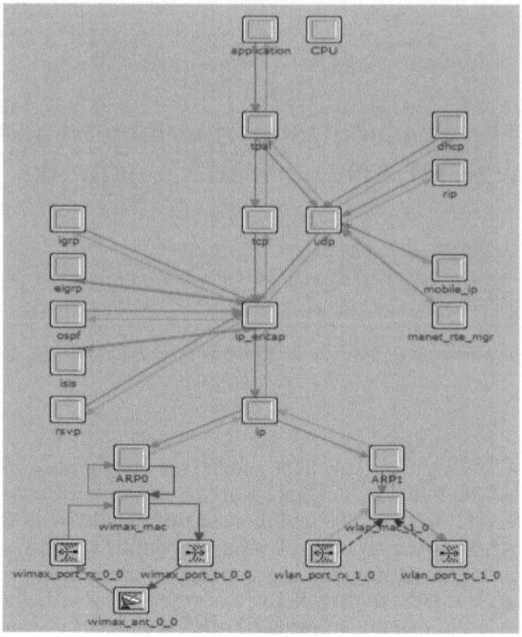

Figure 1. Overview of the developed Multihomed node model.

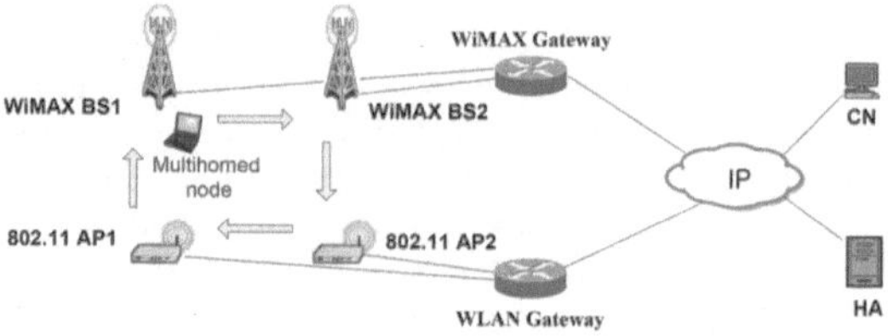

Figure 2. Multihomed node tested network architecture.

- If it is empty, insert the rx_RA in the first position of the list, and then start the timer.
- Or else, check the list that already has the same rx_RA or not. If yes, just restart the timer of the corresponding rx_RA_entry in the list. If no, it is the different rx_RA we got. By comparing the rx_interface number of the new rx_RA with the rx_interface number of each entry in the list, we find out the correct position and put this new rx_RA into this position in the ordered list.

It is worth highlighting that whether the list already has the same rx_RA or not in Figure 3, which can be decided in IP layer by multihomed node. For instance, the mobile node in MIPv6 network obtains its care-of-address by receiving a router advertisement message. From the receipt of the router advertisement message, the mobile node determines that it has connected with a foreign link because the router advertisements contain new network prefixes. Hence, according to this received network prefix, the IP layer will determine whether this received RA is the same as the previous received RA or not.

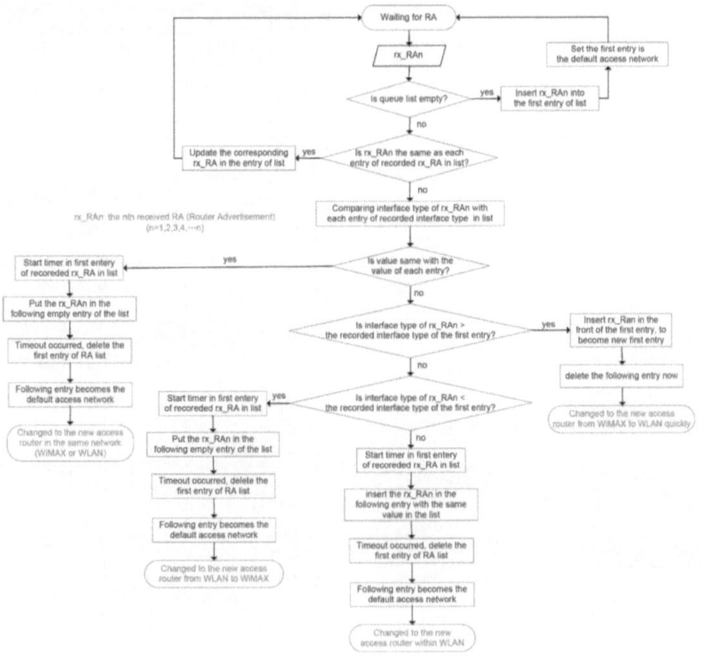

Figure 3. Illustration of network selection policy.

Here we should say, the multihomed node can be attached to multiple networks simultaneously (WiMAX/WLAN) while staying in the overlapping area between WiMAX and WLAN networks. In general network design, the multihomed node will, by default, be chosen from an ordered list of available interfaces. However, in our designed network selection policy, it is worth highlighting that the first entry is always the default access network in the ordered list. Note that, the default network connection policy will always prefer WLAN over WiMAX in our design, which can be seen in the following Figure 4.

In order to better understanding this network selection policy, an example of how to manage the access interface in ordered list is shown in Figure 4. Note that in Figure 4, we assume that multihomed node (MN) is moving from point A to point B throughout the entire WiMAX and WLAN network. The right side of Figure 4 is the updated rx_RA list.

- Put the first BS1_RA in the ordered list.
- When getting the new RA from AP1, we compare AP1_RA with BS1_RA, then put AP1_RA in the first position. Then we change to WLAN network. Even though BS1_RA timeout has not occurred, we prefer the WLAN over WiMAX, so it should be changed.
- When the MN moves out of WLAN, the AP1_entry would be deleted because of AP1_RA timeout and the default access network becomes WiMAX network again.
- When the MN gets into the overlapped area of BS1 and BS2, BS2_RA is attached after BS1_RA in the list.

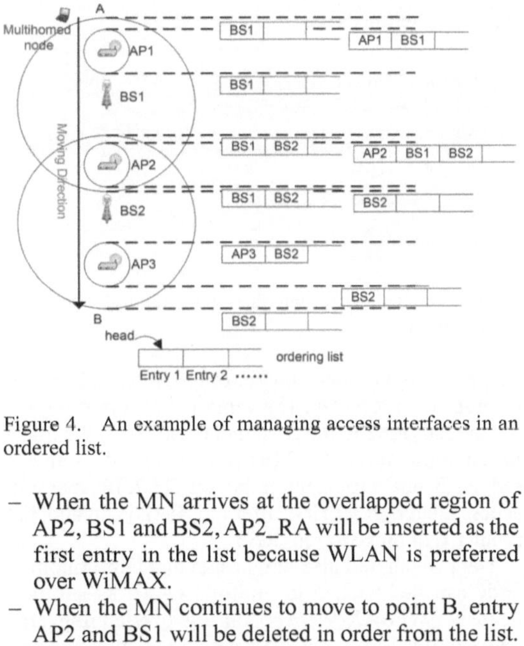

Figure 4. An example of managing access interfaces in an ordered list.

- When the MN arrives at the overlapped region of AP2, BS1 and BS2, AP2_RA will be inserted as the first entry in the list because WLAN is preferred over WiMAX.
- When the MN continues to move to point B, entry AP2 and BS1 will be deleted in order from the list.
- Finally, when the MN moves out of BS2, the rx_RA for BS2 would be deleted.

An example of network selection policy is shown in Figure 5. Once the IP layer received RA, it would record a receiving interface type together with RA, and then inserted it into the sorted list. After continuing to receive RA, this could be the same as or different from the first recorded RA, and the IP layer will compare these two RAs. As Figure 5 explains, for

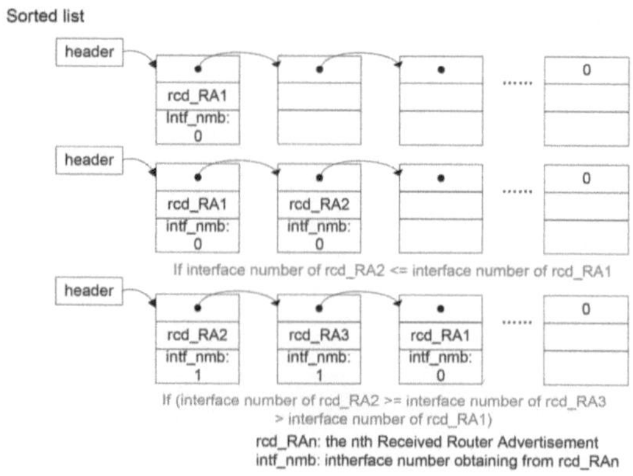

Figure 5. An example of new network selection policy.

```
// inserted by KJG 2012.02.02
    op_ima_obj_attr_get_str (iprmd_ptr->node_id, "name", 128, KJG_node_name_str);
    if (strncmp(KJG_node_name_str, "multihomed", 10) == 0)
        {
      // User Policy to choose the prefered interface : Prefer WLAN(IF1) than WiMAX(IF0).
      if (router_list_entry_ptr1->KJG_intf_recvd_index > router_list_entry_ptr2->KJG_intf_recvd_index)
            {// Entry1 has the larger KJG_intf_recvd_index value. Prefer it.
            FRET (1);
            }
      else if (router_list_entry_ptr1->KJG_intf_recvd_index < router_list_entry_ptr2->KJG_intf_recvd_index)
            {// Entry2 has the larger KJG_intf_recvd_index value. Prefer it.
            FRET (-1);
            }
            // Both entries have the same KJG_intf_recvd_index value.
            // The list will be ordered by OPNET original rules.
    }
// inserted by KJG 2012.02.02
```

Figure 6. Network selection policy code.

```
// inserted by KJG 2012.02.02
    op_ima_obj_attr_get_str (iprmd_ptr->node_id, "name", 128, KJG_node_name_str);
    if (strncmp(KJG_node_name_str, "multihomed", 10) == 0)
      {
       // multihomed node must know the RA_received_interface when it changes the default router.
       // so, retrive the received itnerface index of RA message and then save this information with
       // router_address in the router list.

       intf_ici_ptr = op_pk_ici_get (ip_pkptr);
       op_ici_attr_get (intf_ici_ptr, "rte_info_fields", &intf_ici_fdstruct_ptr);
       router_list_entry_ptr->KJG_intf_recvd_index = intf_ici_fdstruct_ptr ->intf_recvd_index;
      }
// inserted by KJG 2012.02.02
```

Figure 7. Obtained Rx interface index through which the RA was received.

instance, the interface number of rcd_RA2 is 1, which is larger than rcd_RA1 (its value is 0). Therefore, we put the rcd_RA2 in the first queue list. Next, the interface number of rcd_RA3 (its value is 1) is the same as rcd_RA2, but we put it after the rcd_RA2. IP layer will access the rcd_RA2 first, after rcd_RA2 is invalid, the pointer will move to rcd_RA3.

Here we should also emphasize that a multihomed node can be attached to multiple networks simultaneously (WiMAX/WLAN) during overlapping area between WiMAX and WLAN network. In this case, the multihomed node will, by default, be chosen from an ordered list of available interfaces. In our designed network selection policy, the default network connection policy will prefer WLAN over WiMAX.

Part of the network selection policy code is also shown in Figure 6. From Figure 6, we can see that the user policy to choose the preferred interface is WLAN first, and then it chooses the WiMAX interface according to the interface index value. The code

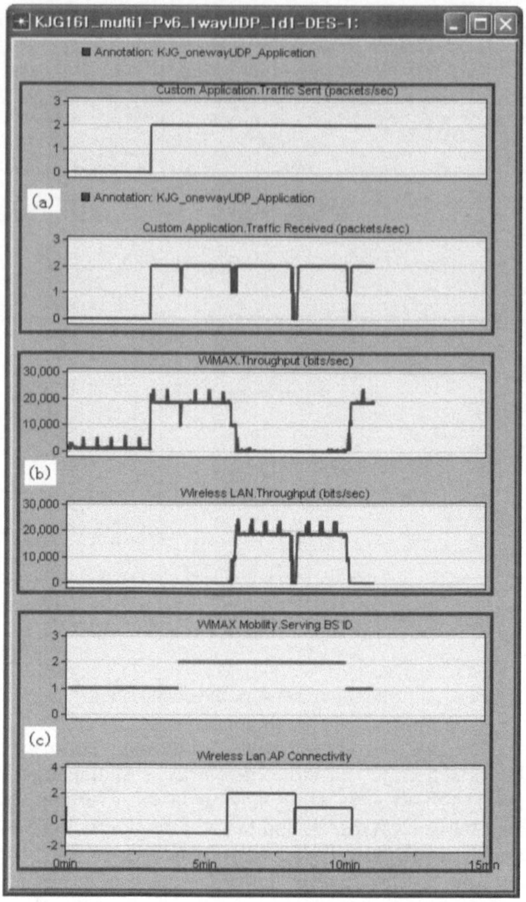

Figure 8. Handover in WiMAX and WLAN in MIPv6 network.

for obtaining the received interface index, which set in the ARP layer, is shown in Figure 7. We modified the ARP layer code to add intf_recvd_index field information in an ICI packet format. When the IP layer gets the packet from the ARP layer, it will parse this packet to get the intf_recvd_index value in this ICI field.

4 SIMULATION RESULTS AND DISCUSSION

In order to verify that our developed multihomed node can support both MIPv4 and MIPv6 mobility functions, we create two BSs (Base Station) in WiMAX and two APs (Access Point) in WLAN as shown in Figure 2. The multihomed node starts moving from WiMAX network, and then it goes through the WLAN network.

In the MIPv6 simulation results, Figure 8a depicts that the UDP traffics are almost received compared with UDP traffic sent, but it still experiences very few packet losses during the handover period in the WiMAX and WLAN networks. In Figure 8b, the transmission of data packets through the WLAN interface starting after the handover from WiMAX to WLAN has happened. Both WiMAX throughput and WLAN

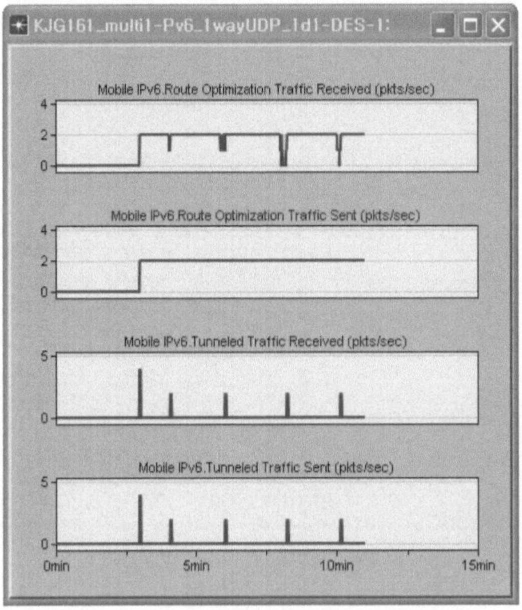

Figure 9. Traffic sent/received in MIPv6 network.

throughput in MIPv6 show better performance than the throughput in the MIPv4 network, because WLAN is preferably selected in our network selection policy. In Figure 8c, the numbers of BS ID and AP present the points of the attachment, which are changed as the BS1, BS2, AP2 and AP1.

Figure 9 depicts that the handover performs better in the WiMAX network than that in the WLAN network, during handover period. Furthermore, MIP tunnelled traffic only sent and received when multihomed node switches the network. Because of MIPv6 route optimization procedure, the MIP tunnel is not continuous. From this Figure we can see that network can be smoothly switched between WiMAX and WLAN because WLAN is preferably selected in our network selection policy.

5 CONCLUSIONS

In this paper, we try to address the implementation of multihomed node models in interworking with WiMAX and WLAN. By modifying the OPNET original code, the multihomed node can support MIPv6 as well as MIPv4 mobility function. It is worth noting that the network selection policy is performed well by comparing the interface type obtained from the router advertisement. It can better control which interface should be accessed firstly during the handover period between WiMAX and WLAN network. Both the simulation results of MIPv4 and MIPv6 show that the developed simulation model for the multihomed node works correctly while moving in WiMAX and WLAN networks.

In further research, the load sharing method between multi-interfaces will be considered in order to utilize both wireless networks.

REFERENCES

[1] Mandalaparty Bhaskara Rama Murthy & Fredson Alfred Phiri, "Performance Analysis of Downward Handoff Latency in a WLAN/GPRS Interworking System", Journal of Computer Science 1 (1): 24–27, 2005, ISSN 1549–3636, Malaysia, 2005.

[2] Kumudu S. Munasinghe & Abbas Jamalipour, "A 3GPP-IMS based approach for converging Next Generation Mobile Data Networks", 1-4244-0353-7/07 IEEE, 2007.

[3] IEEE 802.16e-2005, IEEE standard for local and Metropolitan Area Network- Part 16: Air Interface for Fixed Broadband Wireless Access Systems, Amendment 2: Physical and Medium Access Control Layers for Combined Fixed and Mobile Operation in Licensed Bands, February, 2005.

[4] M. Wasserman & P. Seite, "Current Practices for Multiple-Interface Hosts", Internet Engineering Task Force, RFC 6419, ISSN 2070-1721, pages 11, November 2011.

[5] Karl Andersson, ANM Zaheduzzaman Sarker, Christer Ahlund, "Multihomed Mobile IPv6: OPNET Simulation of Network Selection and Handover Timing in Heterogeneous Network Environments", HybriNet@Skellefteå project Sweden, 2008.

[6] Taeyeon Park & Arek Dadej, "OPNET simulation Modeling and Analysis of Enhanced Mobile IP", 0-7803-7700-1/03, page 1017- page 1024, 2003.

[7] OPNET Simulator, http://www.opnet.com, 2014.

Future Communication Technology and Engineering – Chan (Ed.)
© 2015 Taylor & Francis Group, London, ISBN: 978-1-138-02777-0

Research and implementation of MSC in TD-LTE system

F.S. Bai & R.M. Zhang
College of Electronic and Information Engineering, Inner Mongolia University, Hohhot, China

ABSTRACT: This paper begins with a brief introduction on Time Division-Long Term Evolution (TD-LTE) system network architecture, TD-LTE system handover, measurement model structure and functions of Mobility Settings Change (MSC) sub-layer in dynamic synchronous parameters measurement. And then it specifically focuses on implementation principles and design method of MSC sub-layer for satisfying the technical requirements of protocol standards.

1 INTRODUCTION

Long Term Evolution (LTE) is seen as the critical step of 3G to 4G evolution, and it is the long term evolution project of Universal Mobile Telecommunications System (UMTS) by the 3rd Generation Partnership Project (3GPP). LTE has been designed to give peak data rates in the downlink communication from a Base Station (BS) to a subscriber station or User Equipment (UE) of >100 Mbps whilst in the uplink communication from the UE to the BS > 50 Mbps. LTE creatively uses techniques of Orthogonal Frequency Division Multiplexing (OFDM) and Multiple Input Multiple Output (MIMO) as the sole criterion of its evolution and multiple-access technique instead of CDMA used for a long time by 3GPP. The evolution procedure is as following: GSM $\rightarrow$ GPRS/EDGE $\rightarrow$ WCDMA/HSDPA/HSPA+ $\rightarrow$ LTE.

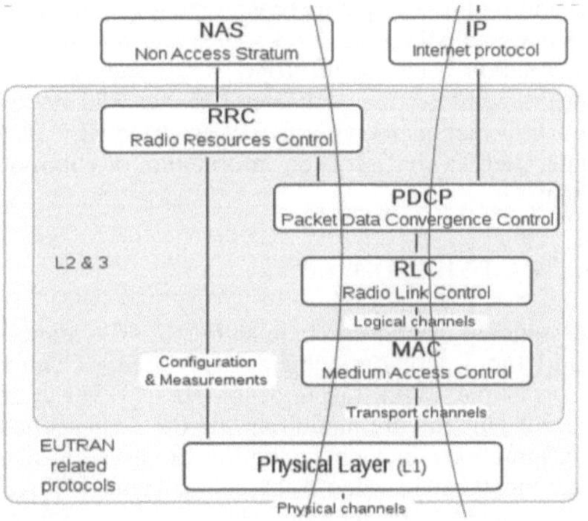

Figure 1. Network structure and functional partitioning of LTE protocol.

2 LTE PROTOCOL ARCHITECTURE

LTE protocol model consists of Access stratum (AS) and Non Access Stratum (NAS). AS includes uplink and downlink scheduling of Medium Access Control (MAC), Hybrid Automatic Repeat Request (HARQ), header compression and security of radio interface of Packet Data Convergence Protocol (PDCP), lossless data transmission of Radio Link Control (RLC), establishment of radio bearer and system broadcast message of Radio Resource Control (RRC). NAS is an important component of mobile communication networks and is mainly used to achieve the goals of user mobility and automatic management for mobility.

The network structure and functional partitioning of LTE Protocol are shown in Figure 1. As is shown in Figure 1, the area of left line is control plane data and the right is user plane data.

The main function of Layer 1 (Physical Layer) is to provide data transmission services for the upper.

Multiple access mainly performs in the cyclic prefix of OFDM in downlink and cyclic prefix of SC-FDMA in uplink.

Layer 2 is divided into MAC, RLC and PDCP sub-layers. MAC configured and managed by RRC supports the following functions: mapping of logical channels and transport channels, which can make some MAC Service Data Units (SDUs) of the same or different logical channels be multiplexed into the transmission block and transmitted to the transmission channel of physical layer; separating some MAC SDUs belonging to the same or different logical channels from the transmission block in the transmission channel of physical layer; scheduling report information by using HARQ for error correction; achieving priority between multi-UE through dynamic scheduling;

the priority of logic channel and transmission format selection.

RLC entity configured and managed by RRC gets RLC SDU through the upper RLC. After RLC entity processing, the RLC PDU is transmitted to the RLC entity in the UE on the other end by lower layer and vice versa. For transmitting data, RLC entity primarily uses the following modes: Transparent Mode (TM), Unacknowledged Mode (UM) and Acknowledge Mode (AM).

PDCP entity is located in its sub-layer and an UE can define multiple PDCP entities. PDCP supports the function of Robust Header Compression (ROHC) protocol used for the compression and decompression header of IP data stream for transmitting and receiving entities.

The system structure of RRC located in Layer 3 of LTE includes an UE state and state transition inside Radio Access Technology (RAT). When RRC connection is established, the UE is in RRC_CONNECTED state, otherwise, the UE is in RRC_IDLE state. Main functions of RRC layer are broadcasting system information, including the NAS's general information; information used for the RRC IDLE state of UE, such as cell selection parameter; channel information and information used for UE RRC_CONNECTED state, such as configuration information of common channels.

3 TD-LTE HANDOVER

Message of handover command is RRC Connection Reconfiguration including IE Mobility Control Information. Main functions of RRC Connection Reconfiguration are modifying the RRC connection configurations, including Mobility Control Information, measConfig, radio resource managment (Physical Channel Configuration, Main Configuration of MAC layer, RBs), NAS message and authentication configuration.

When the UE is in the RRC_CONNECTED condition and the current radio link quality is poor in access network, the LTE begins handover procedures first to ensure the continuity of services. The handover process is that under the control of mobile station auxiliary network, UE performs the measuring and reports the summary to the eNB and then eNB decides to perform handover operations or not by handover algorithm according to the report of the relevant parameters measurement.X2 interface is the key link between adjacent eNBs, which is designed as an efficient and smooth handover and a high quality of link guarantee. Its advantage is to reduce the user data and related handover signaling transfer delay between the eNBs for the purpose of improving the network performance. The most essential relationship between handover and measuring is that eNB selects the strongest signal area as the target area in handover, and the strongest signal information is provided by the largest RSRP/RSRQ value through the measuring. The eNB handover process is shown as Figure 2.

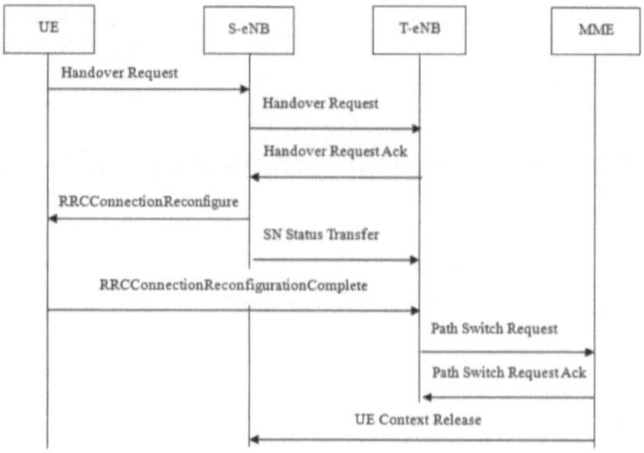

Figure 2. Process of eNB internal handover.

4 TD-LTE MEASUREMENT STRUCTURE MODEL

As is shown in figure 3, A is a measuring result of the internal physical layer, and UE will periodically measure the RSRQ and RSRP value of the around target area and its service area whether in the state of RRC_connect or RRC_idle. When the UE assesses and measures reference signals, the measuring values which cannot be used directly are all unstable and momentary value due to channel mutations or measuring error, so these instabilities should be eliminated in Layer 1 and Layer 3 filtering. Layer 1 filtering is the first filtration for the measurements provided by point A. B is the measurement report from L1 to L3. Layer 3 filtering is the second filtration for the measurements provided by point B, and the previous two filtering parameters are all provided by RRC signaling. C is the measuring result through L3 filtration, and its report of filter period and the rate are the same with B. The measuring report is used as an evaluation criteria of one or more criterions, and evaluation process can provide one or more different reference measuring flow for point C. UE Should do a reporting standards of evaluation at least when it receives new measuring results from C or C', and the reporting standards will be provided by the RRC signaling. Finally, the measuring report will be outputted through D.

In LTE measuring information can be divided into eNB configuration for UE and maintaining information for itself. Firstly measuring process is that eNB distributes measurement configuration to UE, then according to configuration information, the UE triggers measurement report under the condition of trigger. Measurement configuration process is a link of eNB configuration for UE and the UE maintaining measurement information itself. The eNB defines measurement identifier, measurement object, report configuration, measured interval configuration in the measurement configuration to achieve integrity and clarifications of information maintained by UE. In the process of measurement configuration, UE will get RSRP(s) value from its own service area and RSRP(s) value of its adjacent area as a measurement report.

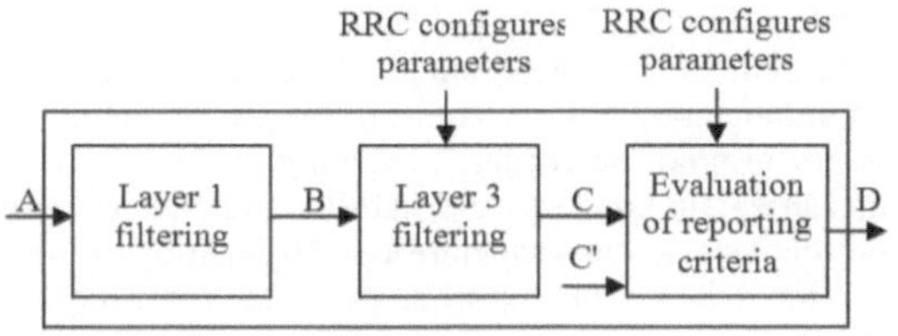

Figure 3. LTE measuring model.

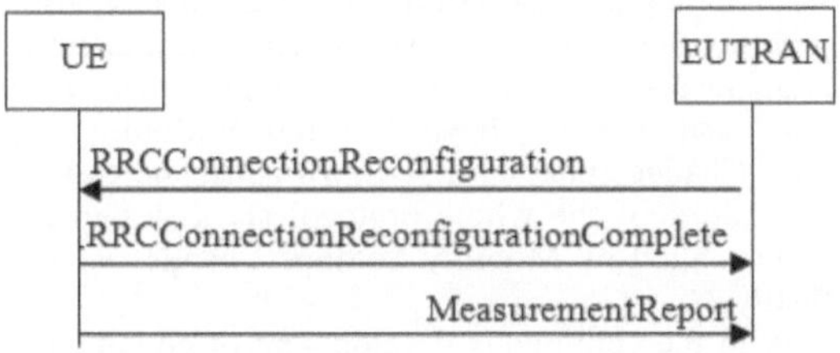

Figure 4. Measurement model.

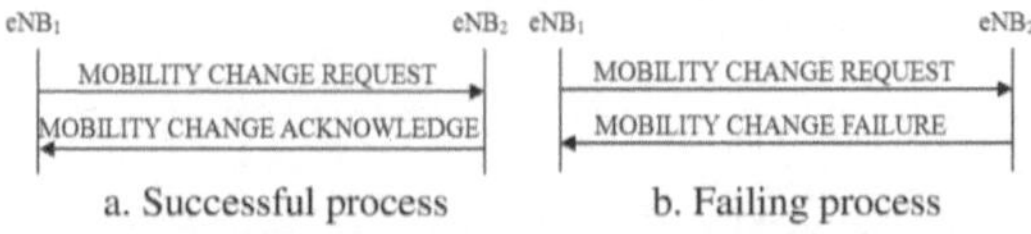

Figure 5. Mobility parameters management process.

It needs Mobility Settings Change function to synchronize if the two values changes. The measurement process model is shown as Figure 4.

5 OBILITY SETTINGS MANAGEMENT PROCESS

5.1 *Implementation principle*

This function is aimed at coordinating the fitness of movement parameters of eNB and the equivalent one, i.e., to switch a trigger settings. The source area that activates loading balance assesses whether it needs to change the mobility configuration of target area. If necessary, the process will be activated. Source area will inform target area new settings and provide reasons for change. And the change mentioned is in the form of the difference in value with the current value. Handover trigger is a particular offset value of cell corresponding to the switch trigger threshold. The process is shown in Figure 5.

Physical significance for Mobility Settings Change is that UE can get the RSRP(s) value of its own serving cell and the RSRP(t) value of its neighboring cells for being a measurement report.

RSRP value is a variables for UE and in the measurement decisions the base stations can switch according to the difference value $\Delta RSRP = RSRP(t) - RSRP(s)$. Each eNB has its own trigger value as a standard to choose when it needs to switch. When

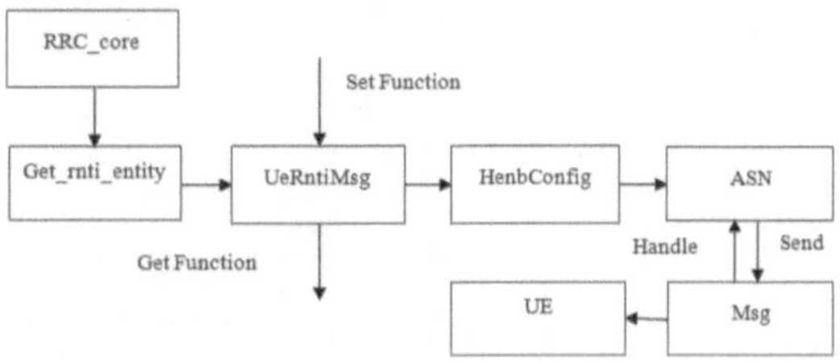

Figure 6. Software simulation flow.

$\Delta RSRP >= trigger$, it allows to toggle. Different trigger values of different eNBs constitute the switch selection level and load balance will synchronize each eNBs trigger value dynamically. Function for Mobility Settings Change is to send each eNBs trigger value to other stations in order to make a switch, in other words, the dynamically changed RSRP value (s) of its own serving cell and RSRP(t) of its neighbor cells will be synchronized to eNB.

Switching adaptive allocation function is aimed at switching dynamically the response of target area. Initialized source area assesses whether it needs to change its mobility configuration or the configuretion of target area. And if needed, source cell will launch mobile parameters configuration process on initial target cell. Source cell will inform target cell the new mobile parameters and the reason for changing. These parameters mentioned are in the performance of the difference between original value and the variations.

5.2 *Design and implementation*

The design and implementation process is shown in Figure 6. Data structure in the system contains a Henbcore which contains RRC_core, then it takes out this data structure and reserves it into their own buffer by using Get_rnti_entity function to get an empty data structure about current UeRntiMsg; Fill the Henb_Config data structure by means of RRM parameters configuration process. The filled data list is taken as the parameter input of Create function and configured in UeRntiMsg, which is equivalent to activate the current UE information. Then the UeRntiMsg can use set and get functions to access and also can use the ASN Codec to interact with the opposite end.

On receiving the message of Mobility Change Request, X2 interface information are called for getting adjacent eNBs information and these data are used on whether matching to eNB2. Running parameters are accepted through X2 entity of eNB1 and the process will quit if parameters are out of range or other reasons, otherwise parameters will be saved.

By receiving Mobility Change Acknowledge message, the ASN module will be called to decode messages for getting adjacent eNB cells information and can be saved to the local. If it receives Mobility Change Failure message, it will make different processing according to Cause content. It should wait for a period of time after re-launch the X2 interface process if the message contains TimeToWaitIE. Receiving module

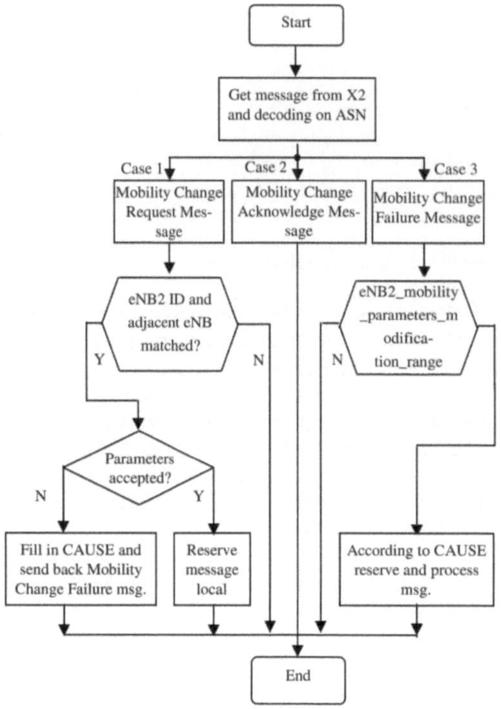

Figure 7. Signaling receiving process.

should also be able to handle some of abnormal situation. The signaling receiving flow is as shown in Figure 7.

6 CONCLUSIONS

In the present evolution process of communication system, equipment manufacturers and operators all begin research and development of related technologies on a large-scale. With the issuance of 4G licenses in China on Dec. 2013, there will be a large number of procurement and formation about equipment for LTE communication system. And a large-scale commercial is just around the corner. Switching in the design of measurement concerns the stability and efficiency of protocol stack, and real-time synchronization of measurement data is the important pursuit of robustness in the system.

And on this basis, the analysis of standard protocols for Mobility Settings Change layer and the research for signaling process determine the characteristics of Mobility Settings Change subsystem, and interaction principle and design thought are also presented . Each functional module of the architecture of Mobility Settings Change subsystem is drawn under the existing framework of the whole protocol stack design and implemented by Mobility Settings Change function finally.

With the continuous development of communication systems and constantly improving of LTE system technology, its commercial value and research value can be more extensive.

REFERENCES

3GPP TS 36.300: *Evolved Universal Terrestrial Radio Access (E-UTRA) and Evolved Universal Terrestrial Radio Access Network (E-UTRAN)*, overall description. 2008

3GPP TS 36.331: *Evolved Universal Terrestrial Radio Access (E-UTRA), Radio Resource Control (RRC) protocol specification.* The 3rd Generation Partnership Project, 2011

3GPP TS 36.421: *Evolved Universal Terrestrial Radio Access (E-UTRA), X2 layer 1.* The 3rd Generation Partnership Project, 2011

3GPP TS 36.422: *Evolved Universal Terrestrial Radio Access (E-UTRA), X2 signaling transport.* The 3rd Generation Partnership Project, 2011

3GPP TS 36.401: *Evolved Universal Terrestrial Radio Access Network (E-UTRAN), architecture description.* The 3rd Generation Partnership Project, 2011

Future Communication Technology and Engineering – Chan (Ed.)
 ISBN: 978-1-138-02777-0

The design and realization of a handheld terminal for industrial control purposes, based on RFID and embedded Linux

T. Zhang, Y.H. Xu & Z.Q. Ye
Key Laboratory of Intelligent Computing & Signal Processing Ministry of Education, Anhui University Hefei, P.R. China

ABSTRACT: Embedded systems and computational intelligence have been greatly developed in recent years. It is necessary to consider the design of a handheld terminal for industrial control purposes because there is still some work done by hand in the industrial process, and it is inconvenient and inefficient. In this paper, a design and achievement of a handheld terminal used in the industrial inspection process is proposed in order to make industrial control more efficient and safer. The handheld terminal hardware consists of ARM11 S3C6410 and RFID module together with a wireless transmission module. As for the software program, we use a QT platform in Fedora to design software programs including serial communication, RFID, and wireless transmission. We also design a useful Chinese input method. Finally, the test shows that the handheld terminal can work well in an industrial process and does have great application prospects.

1 INTRODUCTION

In the last two decades, with the growing development of embedded systems and the internet, there are more and more handheld terminals widely used in the commercial, medical, and civil fields, etc. [1]. Compared with the traditional PC terminal, a handheld terminal has some great advantages such as its small size, portability, easy operation and so on. Especially, a handheld terminal, integrated with RFID, can solve many problems easily and quickly. However, in the industrial process, there is still some work done by hand and it is inconvenient and inefficient, so it is necessary to design a handheld terminal which can be easily used to simply and efficiently solve industrial problems such as obtaining data, and inspecting industrial devices. On the other hand, RFID is a kind of information acquisition and recognition technology with high efficiency. Because it can identify and receive data without connection, read, and multiple objectives at the same time, it is widely used in industrial inspection, transportation management, and other aspects. So maybe it will be a great idea to integrate RFID with a handheld terminal. In this paper, a design of RFID handheld terminal based on Linux, which can be greatly used in industrial process, is proposed.

In this design, the terminal has advantages such as portability and easy operation and is expected to work well in industrial situations which require deployment in the remote and dangerous areas, which are difficult to reach. The work process of the handheld terminal is as follows: firstly, the RFID module reads the corresponding property value of the RFID tag. In response, a dialog box will appear on the LCD screen which is provided for inputting data for the inspector; finally, the wireless transmission module will communicate with the server, which can receive, save and process the data. It is thus easy to realize the function of inspecting and monitoring an industrial process remotely as is shown in Figure 1.

This paper is organized as follows: our system structure will be introduced in detail in section 2. In this part, we will also recommend hardware designs respectively. And the software design will be described in section 3. In the end, the test results of the system and the conclusion will be given in section 4.

2 HARDWARE DESIGN FOR THE TERMINAL

2.1 *The embedded processor*

As a start, we should establish the hardware platform of the system, and then we can design the software programs to realize the function on the platform.

As shown above, the hardware platform includes an embedded processor and a peripheral circuit, and the embedded processor is the core of the intelligent terminal. This topic chooses S3C6410 as the core processor, which is based on ARM11 architecture, and has advantages such as its small size, rich interfaces, and low power consumption.

2.2 *The peripheral circuit*

In order to realize the industrial inspection, considering the design of peripheral circuit is essential, and it includes a RFID, a wireless transceiver module, LCD

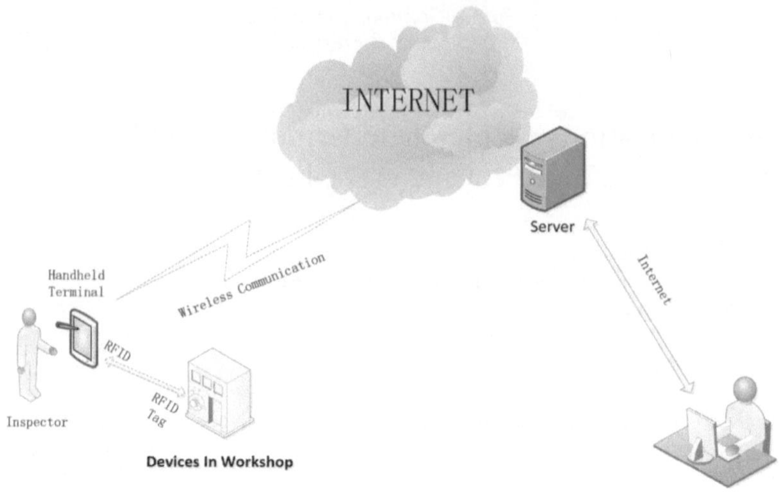

Figure 1. Industrial inspection system.

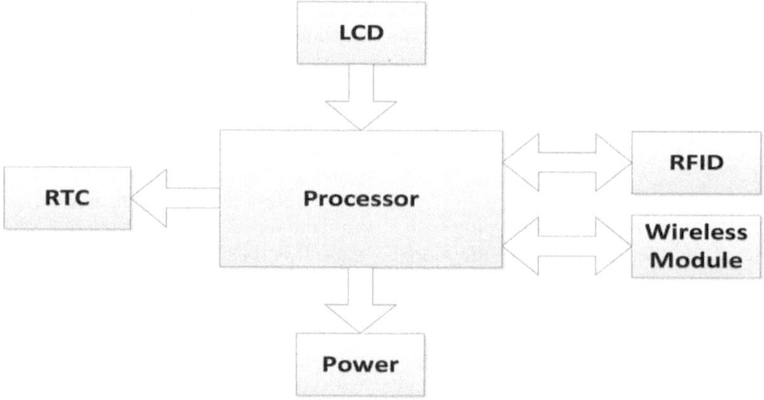

Figure 2. Hardware platform architecture.

screen, power circuit and Real-Time Clock (RTC) as shown in Figure 2.

Considering the problem of stability, interface types, and modules, etc. we select the UHF RFID module, which has the following characteristics: low voltage operation, modular package, supporting power saving, sleep, normal three working modes [2].

As for the wireless module, we have to consider its complete transmission and the rate, stability, portability and so on, and this paper selects the wireless transmission module which does not need to make any changes to the data to achieve transparent data transmission, and the module has the advantages of a small size, transmission over long distance, high stability, low power consumption, and fast transmission, etc.

Considering that we need to input the digital information, so it is necessary to design a 4*4 keyboard placed under S3C6410 plate. The small keyboard is used to input numbers between 0–9 besides a SCAN key when it is necessary to obtain data from an industrial process. The terminal package is shown in Figure 3.

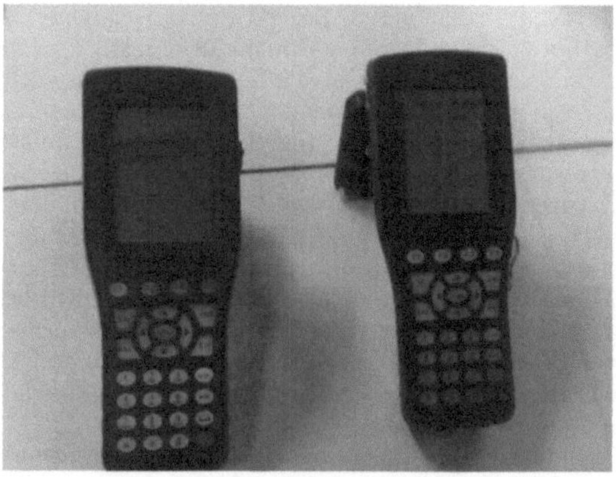

Figure 3. Handhold equipment after package.

3 SOFTWARE DESIGN FOR THE TERMINAL

The terminal software programming is based on Qt Creator in Linux, including application program, a database program, serial communication and a Chinese input method. It is shown in Figure 4.

3.1 *Application program*

The idea of the program design is that the RFID module begins with reading the ID of tag placed in device, and then the handheld terminal receives ID from the serial port, sending it to the processor to recognize which device it is, and then the inspector will record and save the real-time data from the device of the corresponding ID. The inspector will check all devices one by one

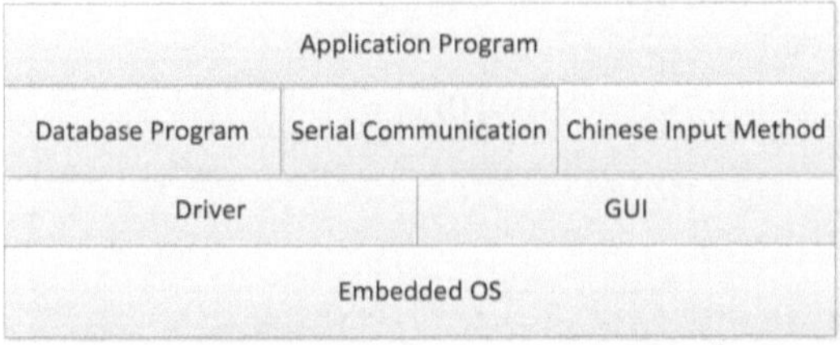

Figure 4. Software architecture.

according to the steps, and the inspection flow chart is shown in Figure 5.

After recording the data, we also need to upload the data to a PC through a wireless communication module. As a direct and pellucid transceiver, it is important to define a kind of data frame format in order to achieve the accuracy of the data transmission, as shown Figure 6 [3]. The data frame begins with 68 and ends with 16. The data type indicates that the frame is the data to upload or confirming data from PC. Beginning with '68 20…' is the recorded data uploading to PC, and '68 21…' is the data from PC to confirm received data. If the terminal does not receive confirming information from thence, it will continue to send the data frame every minute. The data byte is the device's parameters recorded by the inspector. The date and time is quoted from S3C6410 by a RTC circuit in order to inspect in real-time. The data frame's length is 31 bits.

A friendly UI is also a needed part of the software. First is the start interface for the inspector to input a number, then click on the reading card button to get the tag's ID. And then it automatically jumps to the

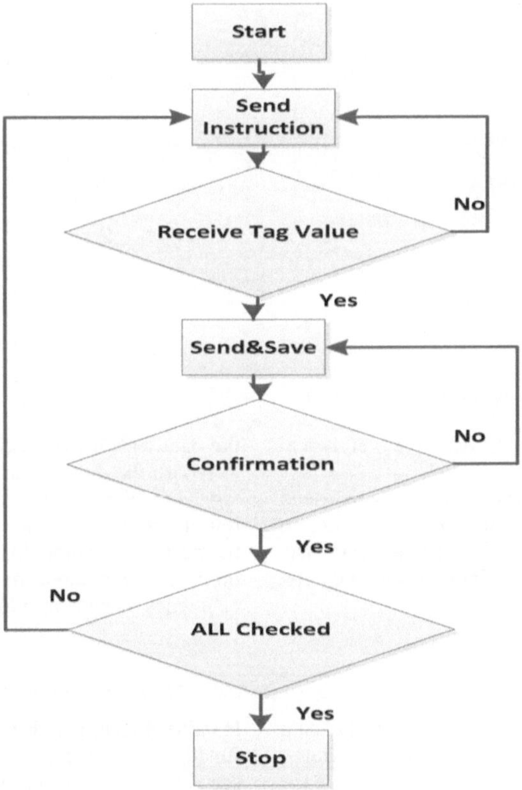

Figure 5. Inspection software flow chart.

Head	Type	Terminal No.	Work No.	Device No.	Data Bit	Date	Tail
68	0x20 0x00	0xXX	XX	XX	15 位	2013.6.6	16

Figure 6. Data frame.

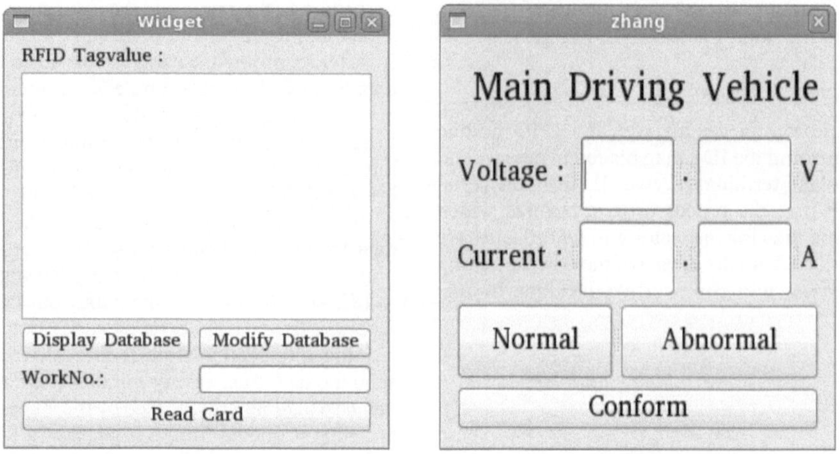

Figure 7. Start and input interfaces.

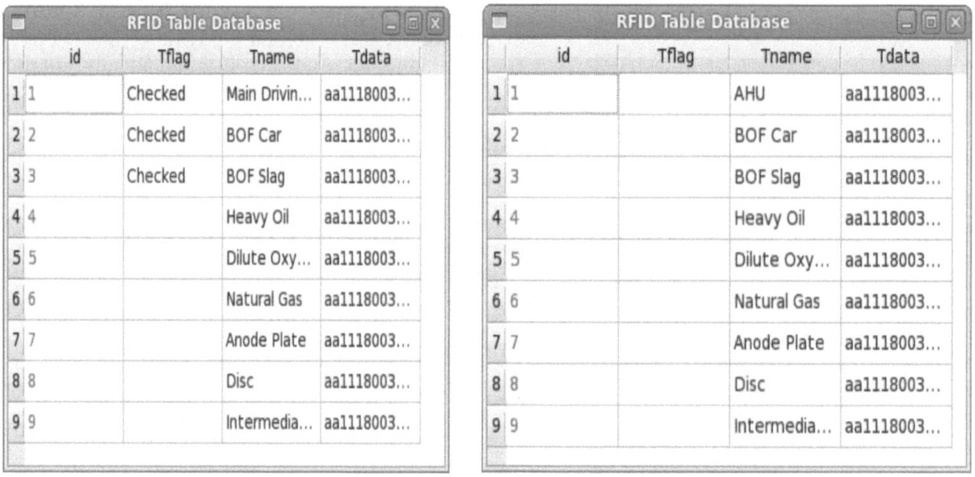

	id	Tflag	Tname	Tdata
1	1	Checked	Main Drivin...	aa1118003...
2	2	Checked	BOF Car	aa1118003...
3	3	Checked	BOF Slag	aa1118003...
4	4		Heavy Oil	aa1118003...
5	5		Dilute Oxy...	aa1118003...
6	6		Natural Gas	aa1118003...
7	7		Anode Plate	aa1118003...
8	8		Disc	aa1118003...
9	9		Intermedia...	aa1118003...

	id	Tflag	Tname	Tdata
1	1		AHU	aa1118003...
2	2		BOF Car	aa1118003...
3	3		BOF Slag	aa1118003...
4	4		Heavy Oil	aa1118003...
5	5		Dilute Oxy...	aa1118003...
6	6		Natural Gas	aa1118003...
7	7		Anode Plate	aa1118003...
8	8		Disc	aa1118003...
9	9		Intermedia...	aa1118003...

Figure 8. Display and modify interfaces.

parameter inputting interface. An example is shown in Figure 7 [4].

When having completed the data inputting, we need to check if all the devices have been checked or not. So we add two buttons to display and modify the database. For example, we can modify 'Main Driving Vehicle' to 'AHU'.

3.2 *Database program*

In this paper, handheld a terminal needs to process data information from the tag label and the host computer, so in the end it's necessary to consider how to store the information in S3C6410. First of all, we need to invoke the function to open the database. Because we have to do a variety of corresponding operations including building, adding, deleting, and updating the database, all the operations are embedded into a click button function. When the system starts up, it is necessary to click the button to read the label data, comparing the data with the one saved in the database, and then to record the device parameters of the corresponding tag label. Finally, the data must be uploaded to the remote server. The embedded database is an essential part of the handheld terminal, and its performance directly affects the terminal's reliability, stability and efficiency as shown in Figure 8.

3.3 *Serial communication based on Linux*

Handheld terminals have to realize the reading card function and wireless communication function by a serial port. Therefore, the serial communication program is very important in inspection software. In this paper, it reads and writes the serial port by polling and is event driven. In the program, we name the serial port 2 as my Com pointing to the processor on the floor, using the polling to query, and it checks the flag every five minutes, to see whether it needs to continue to send corresponding information. On the other hand,

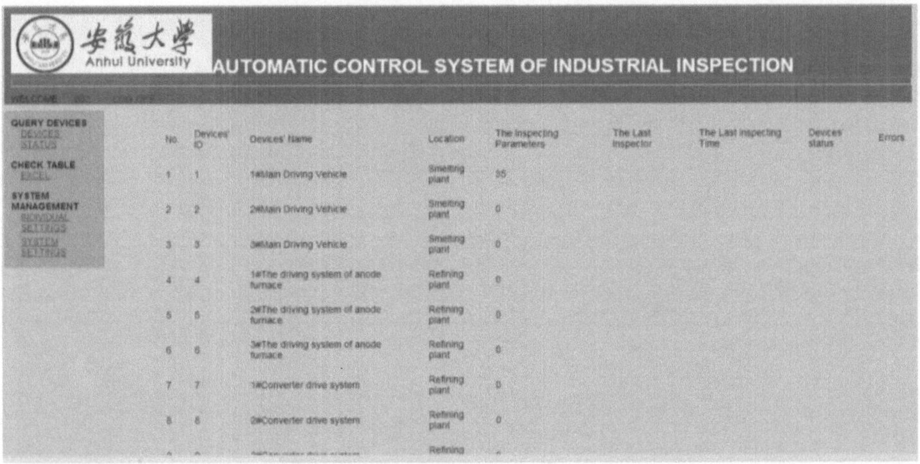

Figure 9. Web inspecting remotely.

serial port 1 reads and writes data using Linux standard I/O functions by an event driven mode. In this way, Serial port 1 does not use any polling, because there is a graphical interface in our program and the polling might cause an interface obstruction [5].

3.4 *Chinese input method*

Considering it necessary to input and modify data in some cases in Chinese, we need to transplant a Chinese input method. The whole process of transplantation is mainly divided into two parts: transplanting the database and compiling it into Qt program. First of all, we should copy the file of input method and unzip it. And then the input method will be compiled into a Qt program and added to the QT object, followed by creating and setting the functions to install the Chinese input method.

4 FUNCTION TEST & SUMMARY

After finishing all the work, we need to functionally test the handheld terminal. The whole industrial inspection system has been shown in Figure 1. The inspector checks the devices in the workshop, inputs the corresponding data to the handheld terminal, and then uploads it to the INTERNET. On the other hand, the manager can browse the web server to check and inspect it remotely, as shown in Figure 9.

This paper studies a handheld intelligent terminal which can be deployed in an industrial inspection process, and it can receive and transmit data completely, so we can inspect any industrial process efficiently. In this design, we have completed the needed functions, including reading the RFID tag value, adding a Chinese input method, uploading data, database storage, and serial port communication successfully. In a word, we have achieved the expected results, and the handheld intelligent terminal can work well. But there is still work to be done to improve the stability and power consumption of the terminal.

ACKNOWLEDGMENT

The work is supported by National 863 Project (2013AA041101) and the 211 Project of Anhui University.

REFERENCES

[1] Ning Yong-hai, Shen Sen, Ruan Wen-hui, Si Wei, "Application of Wireless Data Transfer in Intelligent Monitoring and Controlling System," Communications Technology, Vol. 43, No. 01, 2010, pp. 102–104.

[2] S. Dayekh, S. Affes, N. Kandil, and C. Nerguizian, "Cooperative localization in mines using fingerprinting and neural networks," in Wireless Communications and Networking Conference (WCNC), 2010 IEEE, 2010, pp. 1–6.

[3] J. Ralston, C. Hargrave, and D. Hainsworth, "Localization of mobile underground mining equipment using wireless ethernet," in Industry Applications Conference, 2005. Fourtieth IAS Annual Meeting. Conference Record of the 2005, vol. 1, 2005, pp. 225–230.

[4] Q.Q. Zhang, "The KT101 underground wireless communication system in the Nanshan mine," Energy technology and management, 2011 (3): 151–153.

[5] Salvatore, C., Bocchino, S., Petracca, M., Pelliccia, R., Ghibaudi, M., Pagano, P., "WSN and RFID integrated solution for advanced safety systems in industrial plants," in Proc. of Conference on Software, Telecommunications and Computer Networks, 2012, pp. 1–5.

Future Communication Technology and Engineering – Chan (Ed.)
© 2015 Taylor & Francis Group, London, ISBN: 978-1-138-02777-0

A novel algorithm of multi-target localization in sensor networks

Y.H. Zhang, B.N. Pei, X.Y. Zhang & N.N. Cao
Department of Information Engineering of Dalian, Dalian, Liaoning, China

ABSTRACT: In this paper, we propose a novel algorithm of multi-target localization in sensor networks, which can solve the problem of complex data association, expensive computational cost and deviation in the process of solving large number of equations. With the sparse characteristics of distributed sensors and targets, this algorithm brings together the idea of the back projection (BP) algorithm and the sparse signals reconstruction. Firstly, we perform the operation of projection with the echo information to the three-dimensional measurement space respectively. Then we transform the localization problem into the peak value extracting in the measurement space by the idea of sparse signals reconstruction. Finally, we analyse the performance of this method by a series of experiments with coherent accumulation and amplitude accumulation. The simulation experiments show that this method is suitable for multi-target localization with the coherent accumulation.

1 INTRODUCTION

A wireless sensor network is a network composed of a large number of inexpensive sensor nodes, which can do the real-time monitoring and data collection of the objects in the detection area, and transmit the processed data to the monitor. Furthermore, it is of broad prospect to many application fields such as military, civil, and commercial. Among them, the target localization is a supporting technology, and is the basis of a variety of applications [1, 2].

Due to the good localization accuracy and the low requirements for equipment, TDOA (Time Difference of Arrival) technology [3] is widely used in multiple base station target localization. It defines nonlinear hyperbola equations, based on the time difference of arrival and the position of the target is the solution of these equations. However, its solving process is very complex and needs a huge amount of computation. In recent years, domestic and foreign scholars have put forward some approaches by introducing an intermediate variable, such as the spherical interpolation method [4], the two-step weighted least squares method [5] and the constrained weighted least squares method [6]. But these methods bring a new problem of a large approximation error. On the other hand, when there are multiple targets or an amount of false targets, which result from the existence of noise in the detection area, it is necessary to determine whether the received echo data of different sensors come from the same target, namely data association [7]. Data association is also a difficult and a core issue in multi-targets localization. The common recent approaches to data association are probabilistic data association and joint probabilistic data association [8]. Their shortcomings are that the computation is complex and large.

In order to solve the above problems, this paper proposes a novel method of multi-target localization. The back projection algorithm can be used to solve easily the data association problem, then the sparse signals reconstruction strategy can be used to do the multi-target localization and false target elimination. Finally, the results of the simulation demonstrate the feasibility of this algorithm in multi-target localization.

2 SYSTEM MODEL

Multi-target positioning, based on distributed sensors uses radar signals as emitting sources and receives echoes from the detection area with sensor receiving equipment. Take a single target for example, the scenes of multi-target localization in wireless sensor networks are determined as shown in Figure 1. $G(x, y, z)$ is the location coordinate of the target. $T(x_0, y_0, z_0)$ represents the location coordinate of the radar transmitter. $R_n(a_n, b_n, c_n)$ is the location coordinate of the nth sensor receiver which is known. The model illustrates the distance of the transmitter to the target as

$$d_t = \sqrt{(x_0 - x)^2 + (y_0 - y)^2 + (z_0 - z)^2} \quad (1)$$

The distance of the target to the nth sensor receiver is obtained as

$$d_n = \sqrt{(x - a_n)^2 + (y - b_n)^2 + (c - c_n)^2} \quad (2)$$

So a time delay equation is established by (1), (2), as

$$d_n + d_t = c\tau_n \quad (3)$$

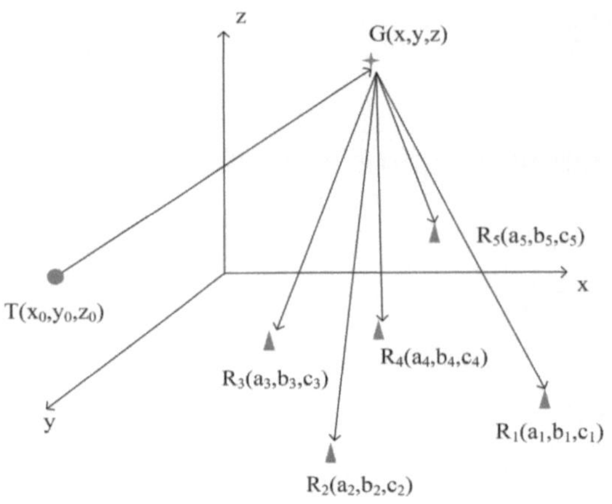

Figure 1. The system model.

where τ_n is the collected time delay information by the *nth* sensor receiver. When there are multiple targets, there are a number of such equations.

3 THE PROJECTION OF ECHO INFORMATION

The idea of a BP algorithm is that there is coherent superposition of the echo data along the curves of time delay respectively [9]. This paper applies this idea to multi-targets positioning, through projecting the echo information of the detection area to the three-dimensional measurement space, to avoid a complex data association process.

First, the detection area is divided into grid cells of equal size, and the size of the grid cell is associated with the radar resolution generally. The centre position (x_i, y_i, x_i) of each cell is set as the representative point, then the time delay of the representative points from the transmitter to each sensor receiver are calculated respectively by equation (3). If the calculated results belong to a time delay collection from the related sensor receiver, we accumulate the echo data to the three-dimensional measurement space.

The entire detection area is traversed and the time delay of the each representative point to sensor receiver calculated. When the time delay of the representative point of i (x_i, y_i, x_i) to the nth sensor receiver belongs to the time delay collection of the nth receiver, it is assumed that the radar launches a linear frequency modulation signal as

$$S(t) = Aa(t)\exp(j2\pi f_c t) \tag{4}$$

where $a(t) = \frac{1}{\sqrt{T}} rect\left(\frac{t}{T}\right) e^{j\pi\mu t^2}$, frequency modulation slope is $\mu = B/T$. T is the width of the transmitted pulse. f_c is a carrier frequency, and A is the signal amplitude.

The echo of the nth receiver from the ith representative point is as

$$\begin{aligned} S_{n,i}(t) &= \sigma_i S(t-\tau_{n,i}) \\ &= A\sigma_i a(t-\tau_{n,i})\exp(j2\pi f_c(t-\tau_{n,i})) \\ &= A\sigma_i a(t-\tau_{n,i})\exp(j2\pi f_c t)\exp(-j2\pi \frac{d_t + d_{n,i}}{\lambda}) \end{aligned} \tag{5}$$

where σ_i is a scattering coefficient of the ith representative point. $\tau_{n,i} = (d_t + d_{n,i})/c$ denotes the time delay of the transmitting signal through target i to the receiver n. $d_{n,i}$ is distance of the ith representative point to the nth receiver.

After pulse compression and removing the carrier frequency of the echo signal, it is expressed as

$$y_{n,i}(t) = A\sigma_i c(t-\tau_{n,i})\exp(-j2\pi \frac{d_t + d_{n,i}}{\lambda}) \tag{6}$$

where $c(t) = \int a(v)a^*(t-v)dv$.

Then, compensate the phase and perform the operation of coherent accumulation of target i, and the coherently integrated echo information of target i can be expressed as

$$\begin{aligned} P(x_i, y_i, z_i) &= \sum_n y_{n,i}(t)\exp(j2\pi \frac{d_t + d_{n,i}}{\lambda}) \\ &= \sum_n A\sigma_i c(t-\tau_{n,i}) \end{aligned} \tag{7}$$

The amplitude accumulation can be expressed as

$$\begin{aligned} P(x_i, y_i, z_i) &= \sum_n |y_{n,i}(t)| \\ &= \sum_n \left| A\sigma_i c(t-\tau_{n,i})\exp(-j2\pi \frac{d_t + d_{n,i}}{\lambda}) \right| \end{aligned} \tag{8}$$

where $P(x_i, y_i, z_i)$ denotes the projection value of the target signal in measurement space (x_i, y_i, z_i).

4 THE POSITIONING OF MULTIPLE TARGETS

Because the side-lobe level of a strong target may mask the weak ones nearby, in the process of echo information accumulation, false targets exist. With the sparse characteritics of distributed sensors and targets, the strategy of sparse signals restruction can be used to execute multi-target positioning and false target eliminating. Here, multiple targets are reconstructed by finding the peak point of the measurement space, as was constructed in section 3.

Based on the above idea, the steps of multi-target positioning are as follows:

Step 1. Divide the detection area into grid cells according to the radar spatial resolution, and set a representative for every cells;

Step 2. Calculate the time delay of every representative from the transmitter to each sensor and find

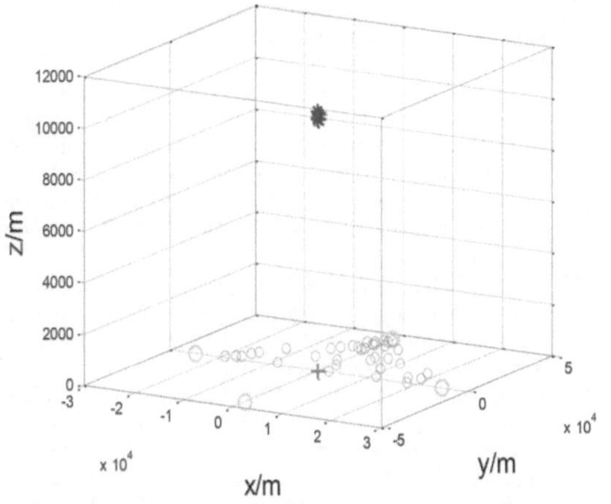

Figure 2. The system distribution.

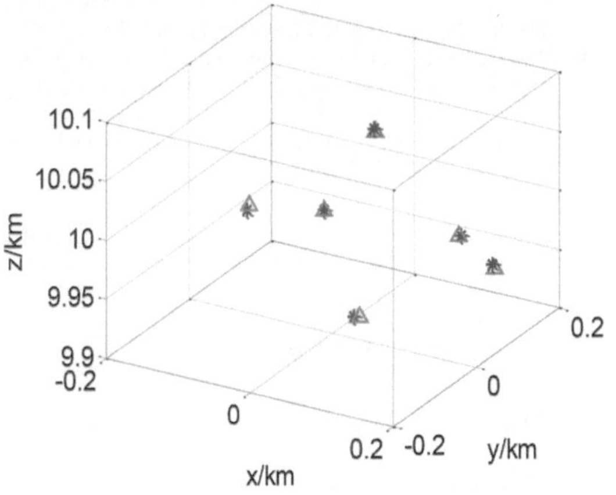

Figure 3. Localization results based on coherent accumulation.

the echo information for the representative and compensate the phase as in (7) or get the amplitude as in (8). Then project the information into the corresponding three-dimensional measurement space until all representatives and sensors are processed;

Step 3. Find out the peak value in the measurement space, and mark the position of this peak value as the target's position;

Step 4. Calculate the echo information of the position to all sensors and eliminate the information from the measurement space;

Step 5. Repeat steps 3 and 4 until all the targets are found. When the number of targets is unknown, regard the noise energy as the termination condition.

5 SIMULATION RESULTS

Three-dimensional simulation of the detection area is 200 m × 200 m × 200 m and there are six targets existing in this area randomly by uniform distribution. The location of the radar transmitter is [0, 0, 0] km, there are forty sensor receivers, and the locations of four are

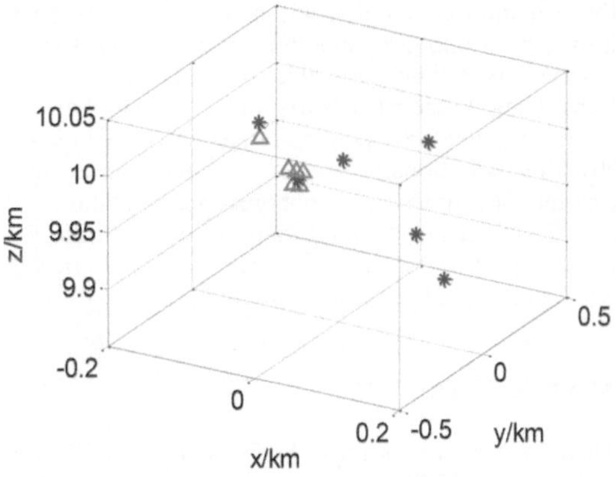

Figure 4. Localization results based on amplitude accumulation.

Table 1. MSE for multi-target localization.

MSE (m)	Coherent accumulation	Amplitude accumulation
Target 1	0.6037	1.4791
Target 2	0.3548	23.5012
Target 3	1.4102	145.7830
Target 4	1.4092	212.1214
Target 5	2.1622	199.6241
Target 6	4.5438	154.2732

[−25, 0, 0] km, [25, 0, 0] km, [0, −25, 0] km, [0, 25, 0] km. Others exist among them randomly by uniform distribution. In this simulation, the LFM waveform is adopted, and the carrier frequency $f_c = 2.4$ GHz, $B = 200$ MHz, time-width $T = 4\,\mu$s. Gaussian white noise with variance of 0.04 is added. And the system distribution, as is shown in Figure 2, where the "$*$" denotes the locations of the targets, and the "$\circ$" denotes the locations of the sensor receivers. "+" denotes the location of a radar transmitter. In this situation, the localization results based on coherent accumulation and amplitude accumulation are shown is Figure 3 and Figure 4. "*" is the actual targets location, and Δ are the estimated targets' locations in the algorithm of this paper. It shows that although errors caused by the interference and noise of the system, this algorithm still can locate multiple targets based basically on coherent accumulation. However, it is not suitable for amplitude accumulation. Under the same condition, through a hundred experiments, the Mean Square Errors (MSE) are shown in table 1.

Although the computation cost of traversal is expensive, when there are multiple targets in the detection area, this method can find the data association simultaneously, and its advantages are obvious.

6 CONCLUSION

This paper has investigated a novel method of multiple target localization in sensor networks. It divides the detection area into grids and projects the echo information into corresponding measurement spaces without first establishing all kinds of non-linear equations. Then it reconstructs the multi-target and cleans the false target with the idea of sparse signals reconstruction. Additionally, in a Gaussian simulation environment, a comparison of coherent accumulation and amplitude accumulation is made to demonstrate the effectiveness of this method. The proposed method can work well for coherent accumulation.

ACKNOWLEDGMENT

Thanks for the supporting of NSFC with No. 61271379 and No. 61301258.

REFERENCES

[1] Huang Yue, Wu Chengdong, Zhang Yunzhou, Cheng Long & Sun Yao 2013. Multi-objective localization method based on K-means clustering in binary sensor networks, Control and Decision, Vol. 28, No. 10.

[2] Patwari N., Ash J.N. & Kyperountas S, et al. 2005. Locating the nodes: cooperative localization in wireless sensor networks [J] IEEE Signal Processing Magazine, 22(4): 54–69.

[3] Su Feng, Wang Changhai & Xu Zheng, 2013. TDOA Localization Algorithms Based on the Least Squares, Radar Science and Technology, Vol. 11, No. 6.

[4] Smith J.O. & Abel J.S. 1987. Closed-Form Least-Squares Source Location Estimation from Range-Difference Measurement [J]. IEEE Trans on Acoustics, Speech and Signal Processing, 35(12):1661–1669

[5] Ho K.C., 2012. Bias Reduction for an Explicit Solution of Source Localization Using TDOA [J]. IEEE Trans on Signal Processing, 60(5):2101–2114

[6] Cheung K.W., So H.C. & Ma M.K., et al. 2006. A constrained least squares approach to mobile positioning algorithms and optimality [J]. EURASIP Journal on Applied Signal Processing: 1–23

[7] Wang Guohong, Bai Jing & Yu Shenggao, 2007. A New Emitter Position Location Algorithm, Journal of naval aeronautical engineering institute, 22(3): 321–325

[8] Si Xicai, Chen Yukun & Li Zhigang, 2007. Research on data association algorithms, Journal of Harbin Engineering University, Vol. 28, No. 7

[9] Wang Huaijun, Huang Chunlin, Lu Min & Su Yi, 2010. Back projection imaging algorithm for MIMO radar, System Engineering and Electronics, Vol. 32, No. 87

[10] Bowu Zhang, Xiuzhen Cheng, Nan Zhang, Yong Cui, Yingshu Li & Qilian Liang, 2011. Sparse target counting and localization in sensor networks based on compressive sensing. INFOCOM, Proceedings IEEE, DOI: 10.1109/INFCOM. 5935041

Future Communication Technology and Engineering – Chan (Ed.)
© 2015 Taylor & Francis Group, London, ISBN: 978-1-138-02777-0

A novel differential position code spread spectrum system with high data transmission capability

B.M. Li & L.L. Guo
College of Information and Communication Engineering, Harbin Engineering University, Harbin, China
Science and Technology on Information Transmission and Dissemination in Communication Networks Laboratory, Shijiazhuang, China

J. Wang & S.H. Qi
College of Information and Communication Engineering, Harbin Engineering University, Harbin, China

Z.G. Sun
College of Information and Communication Engineering, Harbin Engineering University, Harbin, China
Science and Technology on Information Transmission and Dissemination in Communication Networks Laboratory, Shijiazhuang, China

ABSTRACT: A novel scheme for Differential Position Code Parallel Combinatory Spread Spectrum (DPC/PC/SS) communication system is proposed in this paper, and performances are analysed. DPC/PC/SS system can be designed to have higher data transmission capability, even several times compared to ordinary PC/SS system on Gaussian channel. In the proposed system, modulation of differential position code is applied to each simultaneously transmitting PN sequences. The symbol error rate, average bit error rate and multiple access performance are analysed to evaluate the performance of the proposed systems. Furthermore, DPC/PC/SS system has excellent secret property than PC/SS and M-ary/SS communication systems.

1 INTRODUCTION

In future wireless communication systems for personal multimedia systems, for example 4G in wireless LAN, high-speed data transmission capability is required in the wild frequency bandwidth [l]–[3]. In the case of using a conventional direct sequence spread spectrum (DS/SS) technique, an excessively wide frequency bandwidth is occupied for high-speed data transmission. The authors have investigated a PC/SS communication system [4]–[7] as another DS-SS technique for high-speed data transmission. In this paper, we propose a novel Differential Position Code (DPC/SS) communication system for providing much higher data transmission capability. In the DPC/PC/SS system, modulation of the position code is applied to the transmitting PN sequences. To clarify the fundamental properties, we focus on a discussion on symbol error rate (SER), average bit error rate (BER) and multiple access performance performances in an additive white gaussian noise (AWGN) channel.

2 DIFFERENTIAL POSITION CODE PARALLEL COMBINATORY SPREAD SPECTRUM COMMUNICATION SYSTEM

Figure 1 shows the transmitter and receiver model of the DPC/PC/SS system. For simplicity, we show the baseband model. In the transmitter, a set of M orthogonal PN sequences with chip duration Tc are assigned for a specified user. In the PC/PC/SS system, above all we take a particularly PN sequence (e.g. 1st sequence is $a_{Bm}(t)$) as the benchmark sequence between the transmitter and receiver, and then take a combination of r sequences out of (M – 1) sequences (except $a_{Bm}(t)$ sequence).

In the novel DPC/PC/SS system, we do not take a particular PN sequence (fix 1st sequence is not the benchmark sequence) as the benchmark sequence between the transmitter and receiver, but we select the minimum sequence number as the benchmark sequence each time there is a data transmission. The proposed system of the M PN sequence set is given by

$$B = \{PN_1(0), \ldots, PN_M(0)\} = \begin{bmatrix} PN_0^1, PN_1^1, \ldots, PN_{N-1}^1 \\ PN_0^2, PN_1^2, \ldots, PN_{N-1}^2 \\ \ldots \\ \ldots \\ PN_0^M, PN_1^M, \ldots, PN_{N-1}^M \end{bmatrix} \quad (1)$$

The proposed system transmits r sequences. The PN sequence set is given by

$$B' = \{PN_1(0), \ldots, PN_j(p_j), \ldots, PN_{r_p}(p_{r_p})\} = \begin{bmatrix} PN_{p_1}^0, PN_{p_1+1}^0, \ldots, PN_{p_1-1}^0 \\ PN_{p_2}^2, PN_{p_2+1}^2, \ldots, PN_{p_2-1}^2 \\ \ldots \\ \ldots \\ PN_{p_{r_p}}^{r_s}, PN_{p_{r_p}+1}^{r_s}, \ldots, PN_{p_{r_p}-1}^{r_s} \end{bmatrix} \quad (2)$$

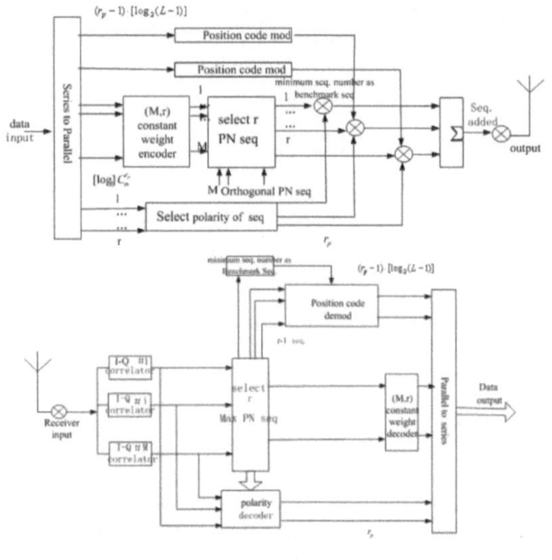

Figure 1. Transmitter and receiver model of DPC/PC/SS system.

N stands for the length of assigned sequence. $P_{Tc}(t)$ represents the chip waveform. $PN_{p_1}^0$ (the minimum sequence number) stands for the benchmark sequence in this transmission.

A transmitter input data stream with bits duration Td are converted to K streams with bits duration T (=KTd=NTc) in parallel. The data of K branches are divided into three parts: the first part of the input data specifies the combination of transmitting PN sequences; the second specifies the polarity of each PN sequences to be transmitted; the third specifies the starting position of every transmitting PN sequences.

At first, data in the same time-slot are encoded into a Constant Weight Code (CWC) of length M and weight R (=r) by a suitable encoding method. This is referred to as the (M, R) constant weight code. An element of a CWC corresponds to the on-off sign of assigned r PN sequences. Since an (M, R) CWC has r nonzero elements, a set of r transmitting PN sequences can be determined by choosing PN sequences that correspond to the nonzero elements and the benchmark sequence. It is expressed by

$$v=[v_1(t),v_2(t),v_3(t),\cdots,v_R(t)],\quad v_u(t)=\sum_{v=1}^{N-1}a_{u,v}P_{Tc}(t-vT_c) \tag{3}$$

$v_1(t)$ stands for the benchmark sequence in the transmitter. The first part carries the data of $[\log_2 C_M^r]$ bits. [x] stands for a maximum integer which is smaller than or equal to x.

The second part of the input data specifies the polarity of r PN sequences that is represented by

$$S=[s_1(t),s_2(t),s_3(t),\cdots,s_r(t)] \quad ; \quad s_i\in\{+1,-1\} \tag{4}$$

The second part carries the data of r bits.

The third part specifies the starting position of r − 1 transmitting PN sequences (except the benchmark sequence). According to the agreement between the transmitter and receiver, we define a certain array of M sequences, which is relative to the benchmark sequence as the $PC(v_i(t);j)$-th starting position. An element in Eq. 2 expresses the starting position of a transmitting PN sequence which carries the information of $[\log_2(N-1)]$ bits. R position code and the benchmark sequence are summed to form a transmitting signal and carry the data of $(r-1)\bullet[\log_2(N-1)]$ bits. $PC(v_i(t);j)$-th starting position is expressed by

$$PC(v_i(t);j)=\begin{bmatrix} a_{Bm,1}(t) & a_{Bm,2}(t) & \cdots & a_{Bm,k}(t) & a_{Bm,k+1}(t) & \cdots & a_{Bm,N}(t)\\ \cdots & \cdots & \cdots & \cdots & \cdots & \cdots & \cdots\\ a_{i,j+1}(t) & a_{i,j+2}(t) & \cdots & a_{i,N}(t) & a_{i,1}(t) & \cdots & a_{i,j}(t)\\ \cdots & \cdots & \cdots & \cdots & \cdots & \cdots & \cdots\\ a_{M-1,1}(t) & a_{M-1,2}(t) & \cdots & a_{M-1,k}(t) & a_{M-1,k+1}(t) & \cdots & a_{M-1,N}(t) \end{bmatrix} \quad ; \quad j=1,2,\ldots,N-1 \tag{5}$$

At the receiver, the received signal passes through M matched filters (or correlators). Each matched filter detects the assigned PN sequence that is the same with the counterpart in a transmitter.

In the receiver besides the benchmark sequence, larger R elements of M output of matched filters, in terms of the absolute value, are decoded to '1', and the others are decoded to '0'. From r maximal sequence, we determine minimum sequence number as the benchmark sequence each time according to the agreement. Non coherent detection is assumed, and this procedure forms an (M, R) CWC to estimate r PN sequences in transmission. The first part data is obtained by decoding the CWC estimate. The second part data of the polarity of the transmitting sequences is obtained from the matched filter outputs demodulator. The third is obtained from the relatively positioned relation between larger R elements and benchmark sequences. Finally, the receiver output – with bit duration Td-is reproduced through parallel to serial conversion. The perfect synchronization of PN sequences between a transmitter and a receiver is assumed.

The number of data bits per PN sequence period with r transmitting sequences out of M sequences are expressed by:

$$K_x=r_p+[\log_2 C_m^{r_p}]+(r_p-1)\cdot[\log_2(L-1)] \tag{6}$$

If N = 1023, M = 16 and r = 4, the number of information of DPC/PC/SS is 43(bits), however the number of the ordinary PC/PC/SS which also transmits 4 sequences carries only 39 bits in the same condition. Thus the DPC/PC/SS system can be designed to have a higher data transmission capability, even several times higher compared with an ordinary PC/SS system.

3 ERROR RATE ANALYSIS

In this section, error rate performances of DPC/PC/SS systems are analysed in an AWGN channel. The benchmark sequence is changed for each transmission.

First, the correct decision probability of transmitting PN sequences is investigated. Every transmitting PN sequence is assumed to have an equal power. According to the computation in Reference [8], the correct decision probability for identifying an active PN sequence is expressed by

$$P_{cc1} = \sum_{n=1}^{M-r-1} (-1)^n \binom{M-r-1}{n} \frac{1}{n+1} \exp\left(\frac{-n\gamma_0}{n+1}\right) \tag{7}$$

γ_0 stands for the signal-to-noise ratio (SNR) per transmitting a PN sequence and is related to the SNR per information bit γ_b as follows:

$$\gamma_0 = (K/r)\gamma_b \tag{8}$$

The second information that we need to know is the conditional probability of detecting the polarity of active PN sequences. It is obtained by the correct decision probability equal to a BPSK signal and its expression is in the form [8]

$$P_{cc2} = Q(\sqrt{2\gamma_0}) \tag{9}$$

The third information is the conditional probability of detecting the position code of an active PN sequences. It is obtained by the correct decision probability equal to an N orthogonal sequence detection and its expression is in the form [8]

$$P_{cc3} = \frac{1}{\sqrt{2\pi}} \int_{-h}^{\infty} e^{-x^2/2} \bullet \left[1 - erfc\left(\frac{x+r_0}{\sqrt{2}}\right)\right]^{(N-1)} dx \tag{10}$$

R transmitted PN codes are estimated correctly after i transmitted PN codes have been estimated with an error in the previous symbol. The correct decision probability of r data for polarity, corresponding to wrong-estimated transmitted PN codes, is expressed by $(1/2)^i$. The correct decision probability of r data for position codes, corresponding to the wrong-estimated transmitted PN codes, is expressed by $(1/[\log_2(N-1)])^i$.

Considering that the correct decision probability of r data for polarity and position codes, corresponding to wrong-estimated transmitted PN codes, the correct decision probability of the symbol in this case is expressed by

$$P_c = \sum_{i=1}^{r} C_r^i P_{cc1}^{2r-i} (1-P_{cc1})^i \left(\frac{1}{2}\right)^i (1-P_{cc2})^{r-i} \left(\frac{1}{[\log_2(N-1)]}\right)^i (1-P_{cc3})^{r-i} \tag{11}$$

Then total SER is expressed by

$$p_e = 1 - \sum_{i=0}^{r} C_r^i P_{cc1}^{2r-i} (1-P_{cc1})^i \left(\frac{1}{2}\right)^i (1-P_{cc2})^{r-i} \left(\frac{1}{[\log_2(N-1)]}\right)^i (1-P_{cc3})^{r-i} \tag{12}$$

Concerning the average Bit Error Rate (BER), it is necessary to take into consideration the BER for the three parts.

First, let us consider the BER for the data in the first part. As previously discussed, the BER for the first part is expressed by

$$P_{eb1} = \frac{2^{[\log_2 C_M^r]-1}}{2^{[\log_2 C_M^r]} - 1} (1 - P_{cc1}^{\ r}) \tag{13}$$

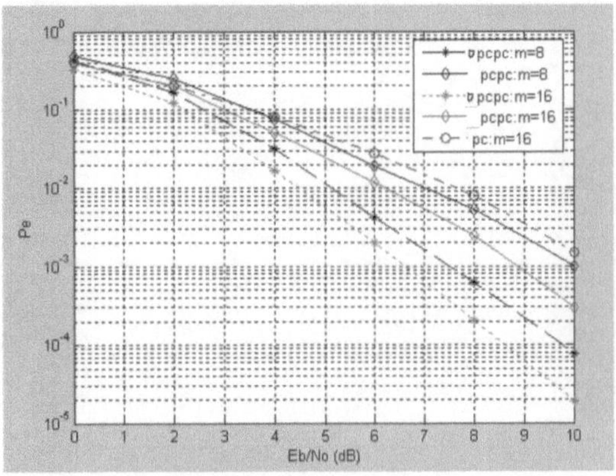

Figure 2. BER comparison between DPC/PC/SS and PC/SS (M = 16 and M = 16, N = 1023, r = 3).

The bit error probability of polarity data corresponding to a well-estimated active PN code is obtained byP_{cc2}. The average bit error probability of the polarity data is computed by

$$P_{eb2} = \frac{1}{r} \sum_{i=1}^{r} \{i(1-P_{cc1})^i P_{cc1}^{\ r-i} + (r-i) P_{cc2} \{1 - (1-P_{cc1})^i P_{cc1}^{\ r-i}\}\} \tag{14}$$

The average bit error probability of the position code data is computed by

$$P_{eb3} = \frac{1}{r} \sum_{i=1}^{r} \{i(1-P_{cc1})^i P_{cc1}^{\ r-i} + (r-i) P_{cc3} \{1 - (1-P_{cc1})^i P_{cc1}^{\ r-i}\}\} \tag{15}$$

Finally, average BER is obtained as follows:

$$P_{eb} = \frac{[\log_2 C_M^r]}{K} P_{eb1} + \frac{r}{K} P_{eb2} + \frac{(r-1) \bullet [\log_2(N-1)]}{K} P_{eb3} \tag{16}$$

4 NUMERICAL RESULTS

We consider a particular situation in which the number of assigned PN sequences is M, the information bit rate per PN period is K, and the number of the sequences length are given to be constant. The BER against SNR per information bit is shown in Figure 2.

We set M = 16, N = 1023 and r = 3, and select K = 31 (bits) information transmitted, respectively. In this case, the information bit rate per PN period K increases as r does. If a higher BER, greater than 10^{-4}, is acceptable, a choice of r = 3 DPC/PC/SS gives superior performance to the case of the three sequences PC/PC/SS. However if the required BER is less than 10^{-4}, the BER in the case of r = 3 DPC/PC/SS becomes worse than that in the case of the three sequences PC/PC/SS. However, in the case of r = 3 PC/PC/SS, higher SNR per bit is needed to give the same performance. At fixed M, N and K, the r of transmitting PN sequences of DPC/PC/SS can be decreased. Then, the SNR per transmitting sequence

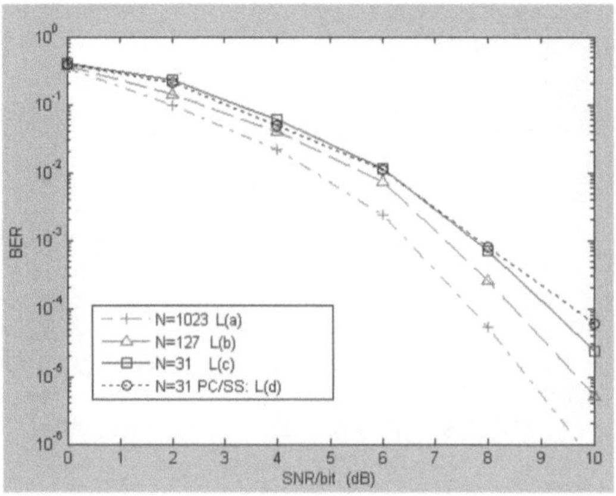

Figure 3. BER comparison between different length sequences of DPC/PC/SS and PC/SS (M = 16, r = 3 transmitted seq.).

increases, and the estimation error probability can decrease.

Figure 3 shows the comparison of the BER on DPC/PC/SS systems with different length sequences. For comparison, the BER of PC/SS systems is also shown in the same figure. In this figure, we consider the systems of M = 16 and r = 3 transmitted sequences.

(a) M = 16, r = 3, N = 1023, DPC/PC/SS system;
(b) M = 16, r = 3, N = 127, DPC/PC/SS system;
(c) M = 16, r = 3, N = 31,DPC/PC/SS system;
(d) M = 16, r = 3, N = 31, PC/SS system.

If the length of PN sequence is 31, the transmitted information rates in (c) and (d) are almost equal if higher BER greater than almost 10^{-3} is acceptable. The case of N = 1023 DPC/PC/SS gives us superior BER performance compared with the case of N = 127 and N = 31 DPC/PC/SS. In this case, information bit rate per PN period K increases as N does. If higher BER greater than 10^{-3} is acceptable, a choice of N = 31 DPC/PC/SS gives superior performance to the case of PC/SS. If N exceeds 127, the BER of DPC/PC/SS gets better as r increases. In the case of N = 1023 DPC/PC/SS, the SNR per bit required to yield the 10^{-4} BER is almost 1 dB lower than that in the case of N = 127. However, in the case of shorter length sequence of DPC/PC/SS, higher SNR per bit is needed to give the same performance. At fixed M, r and K, the N of transmitting PN sequences of DPC/PC/SS can be decreased.

The reason of that is as follows. According to Eqation 6, the DPC/PC/SS system in the case of N = 1023 is capable of overstepping 10 bits by using three PN sequences during one PN period. In the same transmitting sequences condition, the gain of the position code caused by a longer sequence in the DPC/PC/SS system is several times more beneficial than that in the PC/SS system.

5 MULTIPLE ACCESS PERFORMANCE

There are e users in multiple access systems to DPC/PC/SS systems, where every user makes use of M Gold sequences of N periods and transmits r sequences. When M ≫ 1, M ≪ N, it cannot proceed because e users make use of a set of Gold sequence simultaneously. The cross-correlation of these sequences can be presented by Reference [8] as follows

$$|R_{pq}(m)| \le \frac{2^{\frac{n+1}{2}}+1}{2^n-1} \tag{17}$$

Average signal-to-noise ratio of the DS-SS system has been analysed by Pursley [7]. Using similar calculations, assuming with a baseband signal, signal-to-noise ratio per information bit can be obtained approximately

$$\frac{E_b}{N_0'} \approx \left(\frac{4(K_s-1)}{3N}+\frac{N_0}{E_b}\right)^{-1} \tag{18}$$

In which, E_b is energy per information bit, N_0 and N_0' represent respectively noise spectral density without or with interference from other users. K_s is the number of users, and N is the length of the PN sequence (equal to spreading factor).

According to Eq. 17, the analysis can be easily extended to the DPC/PC/SS system. Since an orthogonal sequence set is used for each user in the DS-SS, there is no interference from the other channels of the same user. From the other users, all transmitted $r^*(e^{-1})$ sequences interfere with desired signal. Furthermore we take the equivalent gain of position code compared to DPC/PC/SS system into considerations (gain of polarity code is ignored). Taking into these considerations, assuming random sequences are used, average signal-to-noise ratio for correlation output can be obtained as

$$\frac{E_b}{N_0'} \approx \left(\frac{4(r)(e-1)}{3N}+\frac{N_0}{E_b}-[CG]_{PC}\right)^{-1} \tag{19}$$

$$e = \left(\frac{N_0'}{E_b}+[CG]_{PC}-\frac{N_0}{E_b}\right)\frac{3N}{4(r)}+1 \tag{20}$$

$[CG]_{PC}$ stands for the equivalent gain of a position code compared with a PC/PC/SS system.

6 CONCLUSION

The BER and multiple access performance of DPC/PC/SS are investigated. By using alterable benchmark sequences and position code modulations, the information bit rate can be increased during one PN period. The BER can be improved by applying longer lengths of sequence schemes to the DPC/PC/SS system compared with the conventional PC/PC/SS systems in the cases where the number of assigned PN sequences and information bit rates are constant. Furthermore, the DPC/PC/SS system has an excellent security property.

ACKNOWLEDGEMENTS

This work is supported by Science and Technology on Information Transmission and Dissemination in Communication Networks Laboratory (ITD-U13006/KX132600013), the Fundamental Research Funds for the Central Universities (HEUCF140818), and the National Natural Science Fund Project (61101141).

REFERENCES

[1] R. Kohno, R. Meidan, & L. B. Milstein, "Spread-spectrum access methods for wireless communications," IEEE Commun. Mag., pp. 58–67, Jan. 1995.

[2] R. van Nee, G. Awater, M. Morikura, H. Takanashi, M. Webster, & K. W. Halford, "New high-rate wireless LAN standards," IEEE Commun. Mag., pp. 82–88, Dec. 1999.

[3] F. Adachi, K. Ohno, A. Higashi, and Y. Okumura, "Coherent multicode DS-CDMA mobile radio access," lEICE trans. Commun., vol. E79-B, No. 9, pp. 1316–1325, 1996.

[4] J. Zhu, S. Sasaki & G. Marubayashi, "Proposal of parallel combinatory spread spectrum communication system," Trans. of IEICE., J74-B-II, No. 5, pp. 207–214, May 1991.

[5] K. Takizawa & S. Sasaki, "Iterative demodulation and decoding far parallel combinatorial SS systems," IEICE Trans. Fundamentals, vol. E84-A, no. 12, pp. 2991–2999, Dec. 2001.

[6] S. Sasaki, H. Kikuchi, H. Watanabe & J. Zhu, "Performance of parallel combinatoly spread spectrum communication systems using multiphase modulation," Proc. of SlTA '94, Hiroshima, pp. 421–424, Dec. 1994.

[7] J.G. Proakis, "Digital Communications. 4nd Ed.", McGraw-Hill, 1998.

[8] R.C. Dixon, "Spread Spectrum Systems (2ed)", New York, 1984.

Future Communication Technology and Engineering – Chan (Ed.)
© 2015 Taylor & Francis Group, London, ISBN: 978-1-138-02777-0

Network coding video transmission research and a design based on USRP

X.M. Lu, Y.J. Zhang, C.F. Li, H.D. Zhang & J. Yang
Northwestern Polytechnical University, Xi'an, Shaanxi Province, China

ABSTRACT: Theories of Physical-layer Network Coding and its application are hot topics in current research. However, the real situation of an experimental platform construction is still very rare. This paper realizes the function of the physical layer network coding video transmission by using the USRP + PC software radio platform in a wireless two-way relay network scenario. According to the simulation results, compared with traditional methods, the application of network coding of video transmission can effectively improve the quality of video transmission and its efficiency.

1 INTRODUCTION

Network coding is an information exchange technique which consists of routing and coding. The core idea is to apply a linear or non-linear process to the received information received from each channel at each node. Then the information is transmitted to the downstream node, while the middle node acts as an encoder or signal processor. Large numbers of theoretical studies have shown that network coding has become an effective way to improve network throughput, it being robust, load-balancing, and secure, etc.

Although the network coding can obviously increase the data throughput per unit time and effectively reduce the network load, the traditional hardware network infrastructure doesn't support this algorithm. The research has been mainly limited to theory so far. Some key techniques have not been effectively verified in practical systems [1]. Therefore, the implementation of a simulation platform for network coding has become more and more decisive.

In recent years, with the improvement of the general hardware processing performance, the software radio solution and prototype platform are constantly emerging. The Universal Software Radio Peripheral (USRP) proposed by the GNU Radio is the first scheme implementing theory into practice. The scheme is based on FPGA and DSP programmable hardware, which improves real-time performance and reconfiguration ability greatly. As a result, this paper studies the realization of a wireless two-way relay network coding algorithm on the USRP platform. And the performance of the transmission under this network environment is verified.

2 NETWORK TRANSMISSION MODAL BASED ON USRP

This system mainly contracts the wireless two-way relay transmission system based on USRP. The TDMA based network coding is achieved in this system. The transmission model of system is shown in figure 1.

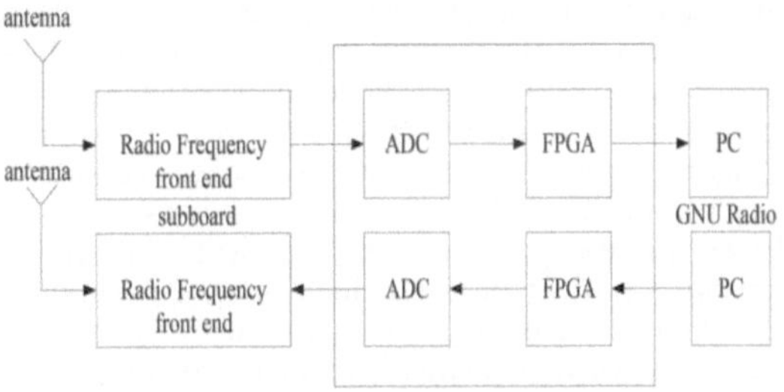

Figure 1. Network transmission model based on USRP.

2.1 *GNU Radio*

The GNU Radio was developed by Eric Blossom. The transmission and reception of the wireless electromagnetic waves are defined by computer software. Thus, the wireless communication system [2-3] is constructed. Therefore, the problem of digital modulation in high performance radio equipment has become the problem of software. The GNU Radio programming is based on the python scripting language and C++. C++ is used to write a variety of signal processing module while python is used to write the script graph block, which connects each block to a complete signal processing.

2.2 *USRP*

A USRP is composed of a motherboard and up to four subboards. The main functions of the motherboard are the intermediate frequency sampling and the transformation between intermediate frequency signal and baseband signal [4]. The subboard is responsible for receiving and transmitting the radio frequency signals as well as the transformation of the intermediate frequency signals. In essence, USRP acts as a digital baseband and intermediate frequency to a radio communications system. In this system, it converts analog signals into digital intermediate frequency signals, while the rest of the signal processing is completed in the GNU Radio.

2.2.1 *The motherboard*

The main function of the USRP motherboard is to convert analog signals in the intermediate frequency band into digital signals by ADC (Analog-to-Digital Converter) and send the received digital intermediate frequency data to the processing module in computer software.

2.2.2 *The subboard*

The subboard is used to load the radio frequency, to receive as an interface or tuner and act as an RF transmitter. Each board slot can visit two of the four high speed AD/DA converters (DAC (Digital-to-analog Convertor) output is used to transmit, the ADC input is used to receive), which makes each sampling subboard have two independent parts of radio frequency when using (not orthogonal), and two antenna (a total of 4 in the whole system). When multiple quadrature sampling is used, each sheet supports a single part. The whole system has a total of two. In general, each subboard has two SMA (SubMiniature version A) connectors, which are used to connect the input or the output signal. In order to maximize flexibility when planning the subboard frequency, the USRP motherboard does not provide anti-aliasing or a reconstruction filter.

3 NETWORK CODING DATA TRANSMISSION BASED ON USRP

3.1 *Data receive and transmit*

In the system platform proposed in this paper, the physical layer consists of three parts: a transmitter, a receiver, and a carrier sensor. The physical layer can achieve the transformation from information bits to baseband waveform and judge whether the current channel is free or not through the energy test. The MAC (Media Access Control) layer completes the data packing and adds the CRC (Cyclic Redundancy Check) check function. The physical layer adds packet information to data packets of the MAC layer, then completes any data processing such as modulation [5]. After that, the data is sent to the USRP. The physical layer transceiver block diagram is shown in Figure 2.

In Mod-Pkt, Send-Pkt is processed first. The data in the MAC layer is packed and the access code is inserted. Then, the CRC check is performed as well as adding white noise. The data is sent to the USRP for further transmission after modulation. The data stays in a form of message queue before modulation, and the connection to the MAC layer is achieved by message queue. However, the data turn out to be in a

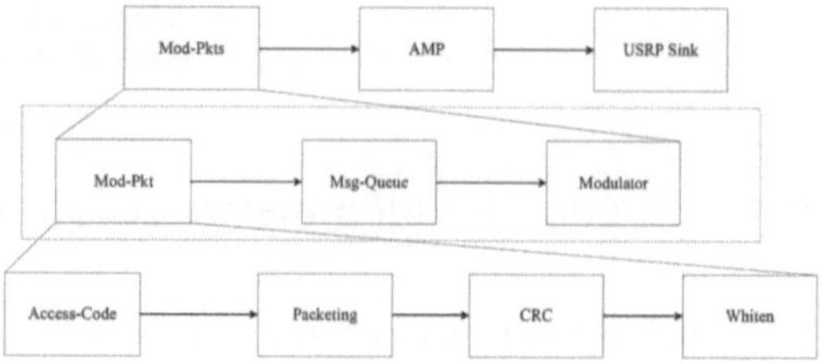

Figure 2. Data transmission process.

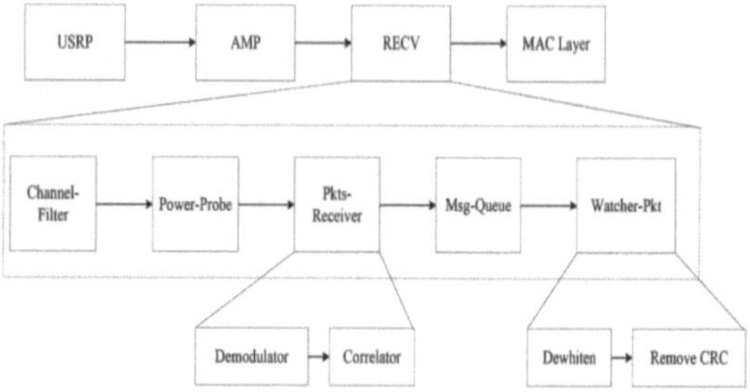

Figure 3. Data reception process.

form of flow graph after modulation which connects asynchronous and variable-length MAC data packet to the physical layer in synchronization with the system clock. The process is shown in figure3.

The signal received from the USRP is transmitted through a channel filter and a power detector. When the power of the received signal is greater than a certain threshold, the received signals are judged to be valid. Then the signal is sent to the pkts-receiver to complete corresponding demodulation and error detection. Data outputs are in the form of a message queue. The multithreaded Watcher-Pkt is responsible for sending the messages back to the MAC layer or the application layer, as explained above. It is relatively simple to perform MAC data packing.

The CRC checks on data packets is performed first. The length of the check digit is 4 bytes. Data, CRC check bits and tail bits (0x55) are whitened in order to obtain the capability of stochastic uniform distribution. Finally, a 4 bytes of package header is added. The package header contains two kinds of information: 4 bits of whiten parameters and 12 bits of packet length. Package header is transmitted repeatedly to increase the reliability. A length-fixed Preamble (16 bit) is added before the length-unfixed Burst. Time synchronization and frequency synchronization are relying on the Preamble. With the 64 bits of the access code, the receiver can resolve the data frame. Thus, an integrated MAC data packing is finished. The process is shown in Figure 4.

3.2 *Network coding data transmission based on TDMA*

As is shown in Figure 5, it takes three time slots to complete a transmission by a network coding technique.

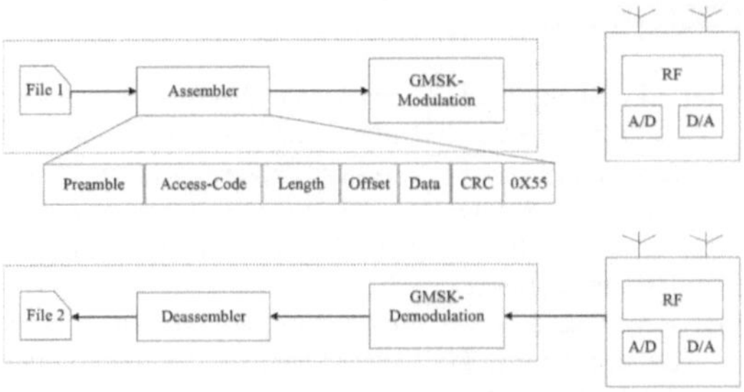

Figure 4. Frame structure and data processing.

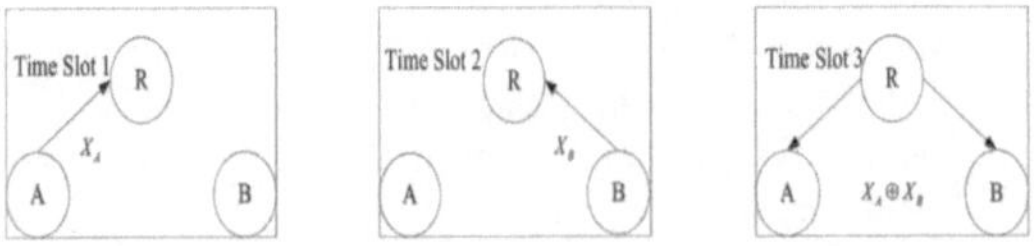

Figure 5. NC transmission scheme based on TDMA.

In time slot 1, node A will send information to the relay node R. In the time slot 2, node B sends its own information to relay node R. In time slot 3, relay node performs XOR operation with the received and stored signal in the previous two time slots. Then, the operated signal is transmitted according to the encoding way of node A and B.

(1) Node A sends data to relay node R in time slot 1, node B sends data to relay node R in time slot 2, relay node R sends data to node of A and B in time slot 3, and that cycle repeats itself.
(2) Relay node R performs demodulation and decoding to data received from node A and node B [6]. In addition, XOR operation is taken. If it does not match the length of the data package, the short data package will take a zero padding operation firstly before any XOR (exclusive or) operation.
(3) Owing to the characteristic of the USRP itself, it will cost some time in the settings of the transmission of the flow graph. In fact, the time that it takes is greater than the length of the time slot. Thus, a certain protective bandwidth is set up.

4 SIMULATION AND RESULT ANALYSIS

During transmission, the way to achieve time synchronization between three nodes is making the three machines synchronize to the internet time. The accuracy is poor. Therefore, the protective bandwidth for each time slot is very wide in the experiment and half of each time slot is set as the effective transmission time. The simulation conditions are set as follows: the transmission rate is 1 Mbps and the transmission data between node A and B is 4 M bytes. When NC (Network Code) code processing is not used, a cycle takes up four time slots and the effective slot equals one time slot. The time for the protective bandwidth in each time slot should be subtracted. Therefore, the transmission rate comes to 128 Kbps, while the transmission time is 256 s. When NC code processing is used, a cycle takes up three time slots and the effective slot equals one time slot. The transmission time is reduced to 192 s. The comparison between network coding video transmission performances is shown in Table 1.

In the two-way relay transmission system, we transmit a video based on TDMA (Time Division Multiple Access). We compare the transmission performance with/without the network coding, namely node A and node B, send a video to each other by relay node, and play the video in real-time to make an intuitive comparison. The result is shown in Figures 6 and 7.

Table 1. Performance contrast of NC video transmission.

	Not use NC	Use NC
The file size	4	4
Transmission rate	128	171
Time	256	192

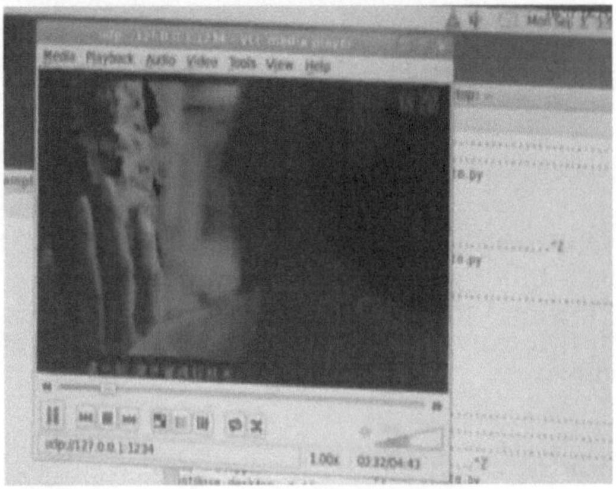

Figure 6. Video transmission without network coding algorithm.

Figure 7. Video transmission with network coding algorithm.

In the traditional transmission mode, it will be less coherent in real-time video transmission. However, after network coding is added, the video quality is relatively better.

The results indicate that, compared to traditional transmission scheme without networking coding, the throughput improvement in the new scheme is 25%.

By transmission and playing the same video, the network coding scheme turns out to have the capability of smooth playing and better quality. Thus, networking coding has a big advantage and, in improving the system, the throughput is verified.

5 CONCLUSION

This paper proposes a new video transmission scheme on a USRP platform with a network coding algorithm [8]. The simulation result shows that the video transmission quality and efficiency can be improved effectively.

ACKNOWLEDGEMENTS

This paper is supported by Northwestern Polytechnical University Foundational Research Fund 2011(JC201143).

REFERENCES

[1] H.L. Duan. 2013. *Network coding based on software radio research and implementation of simulation platform* [D]. Xi'an: Northwestern Polytechnical University.

[2] Y.B. Jiang & S.Q. You. 2007. *Principle and engineering application of software radio*. Beijing: Mechanical Industry publishes.

[3] H.W. Cao & W.B. Wang. GNU Radio: Open software radio platform [J]. *Telecommunications Information* 2007. (4): 31–34.

[4] J.W. Huang & X.F. Zhong & Jing Wang. A GSM 900M road tester design with USRP and GNU Radio [J]. *Communications Technology*, 2011, 44(4): 158–160.

[5] Y.Z. Ding. 2013. *Network coding based on software radio implementation and performance analysis* [D]. Beijing: Beijing University of Posts and Telecommunications.

[6] Y.Z. Ding & G.Y. Li. Multiple-Access Relay Channel with Direct Network Coding [C]. IEEE, *2012 IEEE 14th International Conference on Communication Technology* 2012: 1191–1195

[7] M.F. Zhao. 2011. *Research collaboration based on network coding in wireless communication system* [D]. Chengdu: University of Electronic Science and Technology of China.

[8] H.Y. Chen & S.B. Zhu & C.Q. Li. Research of network coding construction algorithm [J]. *Modern Electronics Technique* 2011(19): 11–14.

Future Communication Technology and Engineering – Chan (Ed.)
© 2015 Taylor & Francis Group, London, ISBN: 978-1-138-02777-0

The complex network Fokker-Plank equation

B.H. Guo, S.Y. Zhou & W.J. Yan
Electronic Information Engineering Institute of Anshun University, Guizhou, China
Avionics and Information Network Engineering Center, Guizhou, China

S.L. Wu
Library of Anshun University, Guizhou, China

ABSTRACT: The Complex Network (CN) dissemination dynamics model can be described by the Fokker-Plank Equation (FPE). With the non-equilibrium fluctuation occurring in complex network parameters, the Complex Network Fokker-Plank's Equation (CNFPE) solution will be significantly different from deterministic equations. It indicates that the fluctuations of the network parameter will produce an important effect on communication.

1 INTRODUCTION

Complex Networks (CNs) that include a large number of individuals and the interactions between them are highly generalize to real complex systems. Other, CN is a chart which used to describe the phenomenons and relationships of the real system, in which the nodes (individuals) idealized by the phenomenons and the sides by the interactions.

Complex networks that include a large number of individuals and the interactions between them are highly generalized. and they are abstract to real complex systems; it is a chart which is used to describe the phenomenon and relationship, in which some phenomena are abstract to individuals (nodes) and the interaction between individuals is as sides.

D.J. Watts and S.H. Strogatz of Cornell University discovered in 1999 that they could construct a new network between regular networks and random networks if they changed the way of network connection by adding long range correlation to very small probabilities. It has both large clustering coefficients and small average path lengths, while they revealed the small world features of CNs [1, 2], inaugurating a new era in CN areas. Then, it has become an extremely important challenge to scientifically understanding the quantitative and qualitative characteristics of CNs in the network, even being referred to as "the new science of network" [3].

Many real networks have small-world characteristics (for example, it was found that any two chemicals within the cell can almost be connected through 3 chemical compositions [2]; the average path length of WWW is 19, the average path length between two routers on the Internet is about 10 [3] etc.), so it has stimulated a great deal of interest in study on the nature and transmission dynamics of Small-world Networks (SWN). M.E.J. Newman, S.H. Strogatz and D.J. Watts [2], & C.F.Moukarzel [5] use a linear differential equations of SWN model, study an event on a small-world network propagation, and respond. However, there is usually a time-delay and nonlinear interactions in the dissemination and response, as X.S. Yang presented in a SWN model, with delay and nonlinearly, in his research [6]. And then, he studied this model's chaos and control in one-dimensional nonlinear difference equation in his literature [6]. In his literature [7], a time–delay linear fractal dynamics of SWN model was studied. In his literature [10] he described the local stability and hopf bifurcation of this model.

For CN is made up of a large number of individuals (nodes), there are always internal spontaneous fluctuations and external random disturbances in real networks. These fluctuations inevitably have an impact on a network structure and the interaction parameters, so that the communication and responses are influenced by these random disturbances and fluctuations, especially when the network is in a certain critical state, and the impact is very huge [8]. In paper [8], we use the method of non-equilibrium statistical and stochastic theory to research the model which was presented in paper [7], FPE can be used to describe the effects on communication and response caused by parameter fluctuations. The influence of only a few parameters was simplified in paper [8], but the impact on the fluctuations of each basic parameter could not be discovered. In this paper, on the basis of what was mentioned above, we deeply study the fluctuation of network nodes and edges, get the steady state solutions of the FPE if the random fluctuations occur in N-W length scale [3] and nonlinear interaction coefficient

ζ, and λ. The results obtained by this method are in agreement with the literature [7].

2 THE FOKKER-PLANK EQUATION OF COMPLEX NETWORK

2.1 *NW length scale*

In [4] ζ is the length scale:

$$\zeta = \frac{1}{(pdk)^{1/d}} \tag{2.1}$$

where p is the randomized reconnect probability, and k is the average degree of SWN, and d is the dimensions of the small-world network, λ is the coefficient of nonlinear interaction on network.

Without taking into account the parameter fluctuation and the time-delay in [7], a one-dimensional non-linear model can be expressed by the deterministic equation of transmission dynamics of CN [8]:

$$\frac{dx(t)}{dt} = \zeta + x(t) - \lambda\zeta x^2(t) \tag{2.2}$$

where $x(t)$ is the measure of the overall effect of the CN dynamic procedure, ζ is a Newman–Watts length scale, and t is the transmission time.

If fluctuations appear in nodes and edges of complex networks, it is inevitable that it results in ζ, λ fluctuations. In terms of reference [8, 9] we can get the FPE.

2.2 *Fokker-Plank equation with λ fluctuations*

At time t, assuming the parameter λ in (2.2) was influenced by the fluctuations, λ can be expressed as:

$$\lambda(t) = \bar{\lambda} + \xi(t) \tag{2.3}$$

where $\bar{\lambda}$ is the average, $\xi(t)$ is the fluctuation occurs near $\bar{\lambda}$ which caused by λ perturbation. Then the equation (2.2) is: $\xi(t)$is a perturbation it is nearly caused by the fluctuation (2.2) is:

$$\frac{dx}{dt} = \zeta + x - \bar{\lambda}\zeta x^2 - \xi(t)\zeta x^2 \tag{2.4}$$

In terms of [8], if $\xi(t)$ satisfies the Gaussian distribution, the distribution function $P\ (x,\ t)$ in (2.4) corresponding to the FPE can be written as:

$$\frac{\partial P(x,t)}{\partial t} = -\frac{\partial}{\partial x}[(\zeta + x - \bar{\lambda}\zeta x^2)P(x,t)] + \frac{1}{2}\sigma^2\frac{\partial^2}{\partial x^2}[(\zeta x^2)^2 P(x,t)] \tag{2.5}$$

where σ^2 is the fluctuation factor.

2.3 *Fokker-Plank equation with ζ fluctuations*

Likewise, if the ζ was influenced by fluctuation, it can be obtained:

$$\zeta(t) = \bar{\zeta} + \xi(t) \tag{2.6}$$

If $\xi(t)$ satisfies the Gaussian distribution, the FPE of the distribution function $P\ (x,\ t)$ is:

$$\frac{\partial P(x,t)}{\partial t} = -\frac{\partial}{\partial x}[(x + \bar{\zeta}(1 - \lambda x^2))P(x,t)] + \frac{1}{2}\sigma^2\frac{\partial^2}{\partial x^2}[(1 - \lambda x^2)^2 P(x,t)] \tag{2.7}$$

3 THE STATIONARY SOLUTION OF FOKKER-PLANK EQUATION

3.1 *Deterministic equations*

(2.2) is the deterministic equation of a CN propagation dynamic. It can be obtained in steady state:

$$\zeta + x - \lambda\zeta x^2 = 0 \tag{3.1}$$

For (3.1), two stable solutions are:

$$x = \frac{1 \pm \sqrt{1 + 4\lambda\zeta^2}}{2\lambda\zeta} \tag{3.2}$$

3.2 *The λ fluctuation*

The steady state distribution function $P_s(x)$ in the formula (2.5) should be:

$$\frac{\partial}{\partial x}[(\zeta + x - \bar{\lambda}\zeta x^2)P_s(x,t)] = \frac{1}{2}\sigma^2\frac{\partial^2}{\partial x^2}[(\zeta x^2)^2 P_s(x,t)] \tag{3.3}$$

In terms of (3.3) and its normalization condition, the steady state distribution function is:

$$\frac{1}{2}\sigma^2\frac{\partial}{\partial x}[(\zeta x^2)^2 P_s(x,t)] = (\zeta + x - \bar{\lambda}\zeta x^2)P_s(x,t) \tag{3.4}$$

At the most probable steady state distribution:

$$\frac{\partial P_s(x)}{\partial x} = 0 \tag{3.5}$$

According to (3.4), (3.5) the most probable steady state condition is:

$$\frac{1}{2}\sigma^2\frac{\partial}{\partial x}[(\zeta x^2)^2] = \zeta + x - \bar{\lambda}\zeta x^2 \tag{3.6}$$

The most probable steady state equations can be obtained from (3.6) if the fluctuation is in (2.3):

$$2\sigma^2\zeta^2 x^3 + \bar{\lambda}\zeta x^2 - x - \zeta = 0 \tag{3.7}$$

If the system has no nodes fluctuations, comparing with (3.7) and (3.1) two results are consistent.

Further, transform (3.7) as the following form:

$$x = y - \frac{\bar{\lambda}}{6\sigma^2\zeta} \tag{3.8}$$

So (3.7) can be:

$$\delta_1 y - y^3 + (\delta_1 - \delta_2) = 0 \tag{3.9}$$

where:

$$\delta_1 = \frac{\bar{\lambda} - 6\sigma^2}{12\sigma^4\zeta^2} \tag{3.10}$$

$$\delta_1 - \delta_2 = \frac{\bar{\lambda}^3}{108\sigma^6\zeta^3} + \frac{\bar{\lambda}}{12\sigma^4\zeta^3} - \frac{1}{2\sigma^2\zeta} \tag{3.11}$$

Considering (2.1), (3.11) can be zero by way of adjusting the parameters of the network, (3.9) change into a classic bistable structure equation:

$$\delta_1 y - y^3 = 0 \tag{3.12}$$

If fluctuation $\sigma^2 \neq 0$, (3.12) shows that there is an essential change in the stability of the system, which is an unstable bifurcation $y = 0$ and two stable bifurcations:

$$y = \pm\sqrt{\delta_1} \qquad (\delta_1 > 0) \tag{3.13}$$

That is:

$$x^{\pm} = \frac{-\bar{\lambda} \pm \sqrt{3\bar{\lambda} - 18\sigma^2}}{6\sigma^2\zeta} \tag{3.14}$$

while:

$$\zeta = \sqrt{\frac{\bar{\lambda}^3 + 9\bar{\lambda}\sigma^2}{54\sigma^4}} \qquad (\zeta > 0) \tag{3.15}$$

or:

$$k = \sqrt{\frac{54}{\bar{\lambda}^3 + 9\bar{\lambda}\sigma^2}\frac{\sigma^2}{p}} \tag{3.16}$$

If (3.11) is zero, (2.1) can get (3.15), (3.16).

It can be seen from Figure 3.1, that, if the fluctuation of network nodes only affects the interaction coefficients, it will make the system be in one of the two stable solutions driven by the fluctuation.

Figure 3.2 shows that NW length scale ζ changes with σ^2 if the fluctuations occur in λ ($\zeta = 1$, $\lambda = 1$).

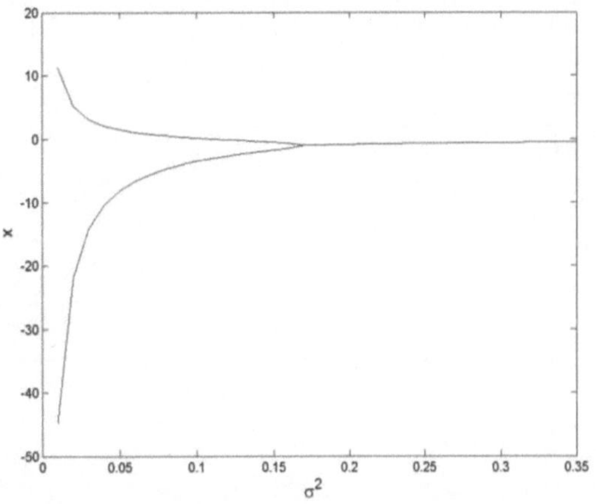

Figure 3.1. x changes with σ^2 if the fluctuations occur in λ (where $\zeta = 1$, $\lambda = 1$).

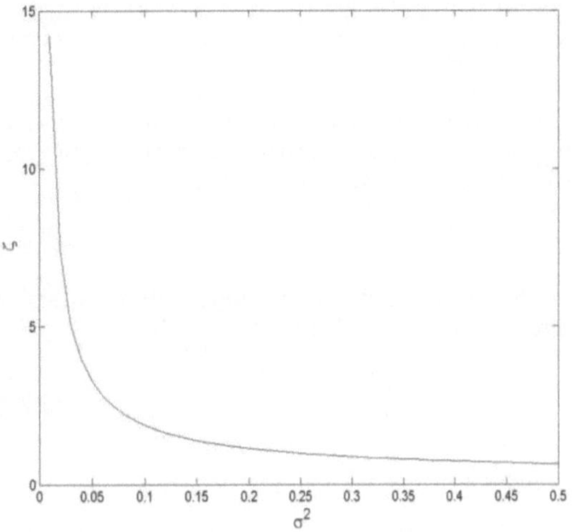

Figure 3.2. NW length scale ζ changes with σ^2 if the fluctuations occur in λ (where $\zeta = 1$, $\lambda = 1$).

3.3 *The ζ fluctuation*

Likewise, the most probable steady state equations can be obtained by the above formula (2.8):

$$2\sigma^2\lambda^2 x^3 + \bar{\zeta}\lambda x^2 - (2\sigma^2\lambda + 1)x + \bar{\zeta} = 0 \tag{3.17}$$

Transforming (3.17) as the following form:

$$x = y - \frac{\bar{\zeta}}{6\sigma^2\lambda} \tag{3.18}$$

We can get similar result with (3.7):

$$\delta_1' y - y^3 + (\delta_1' - \delta_2') = 0 \tag{3.19}$$

where:

$$\delta_1' = \frac{\bar{\zeta}^2 + 6\sigma^2 + 12\sigma^4\lambda}{12\sigma^4\lambda^2} \tag{3.20}$$

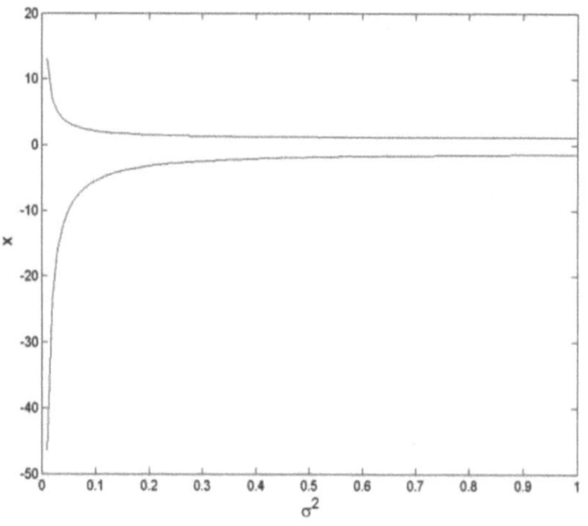

Figure 3.3. x changes with σ^2 if the fluctuations occur in λ (where $\zeta = 1$, $\lambda = 1$).

$$\delta_1' - \delta_2' = \frac{\bar{\zeta}^3}{108\sigma^6\lambda^3} + \frac{\bar{\zeta}}{12\sigma^4\lambda^3} - \frac{\bar{\zeta}}{3\sigma^2\lambda^2} \qquad (3.21)$$

Considering (2.1), (3.21) can be zero by the same way as (3.11) (3.21) change into a classic bistable structure equation similar to (3.12):

$$\delta_1' y - y^3 = 0 \qquad (3.22)$$

There is an unstable bifurcation $y = 0$ and two stable bifurcations in formula (3.22):

$$x^{\pm} = \frac{-\bar{\zeta} \pm \sqrt{3\bar{\zeta}^2 + 18\sigma^2 + 36\sigma^4\lambda}}{6\sigma^2\lambda} \qquad (3.23)$$

Meanwhile:

$$\bar{\zeta}^2 = 36\sigma^4\lambda - 9\sigma^2 \qquad (3.24)$$

Above (3.24) is obtained if (3.21) is zero. If the fluctuations occur in ζ, Figure 3.3 show that x changes.

It can be seen that the system dynamics' behaviour is totally different from the solution of the deterministic equation (3.2) if the network system has a fluctuation. If the $\sigma^2 \neq 0$, the new branched structure may be caused by this fluctuation in the system. Furthermore, from formula (3.15), (3.16), and (3.24) we also might control the dissemination dynamics behaviour of the branched structure of the CN by way of (3.11).

The denominator in (3.24) is not zero, then:

$$\bar{\zeta}^2 = 36\sigma^4\lambda - 9\sigma^2 > 0 \qquad (3.25)$$

The end result is

$$\lambda > \frac{1}{4\sigma^2} \qquad (3.26)$$

The very important result is in the Figure 3.4. In this situation, this result is totally different from

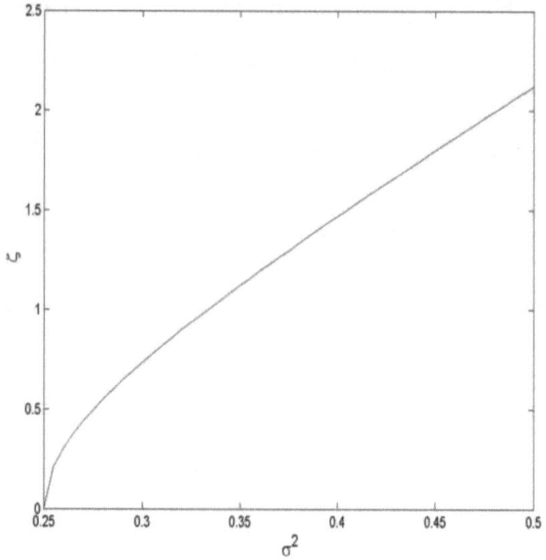

Figure 3.4. x changes with σ^2 if the fluctuations occur in ζ(where $\zeta = 1$, $\lambda = 1$).

Figure 3.2, the length scale average $\bar{\zeta}$ rapid growth with the fluctuation σ^2 is shown in Figures 3 and 4.

Obviously, the network degree k will be rapidly reduced and form the formula (2.1) if the connection probability of complex network p is invariable.

4 CONCLUSION

Firstly, if there is no fluctuation in the complex network system, that is $\sigma^2 = 0$, the system dynamics behaviour is exactly the same as the solution of the deterministic equation (3.2) by mentioning the above (3.7), (3.17).

Secondly, if the fluctuation occurs in the node interaction coefficient λ, the complex system may transit between two stable solutions caused by the fluctuation. From Figure 3.2, the NW length scale ζ has a downward tendency. If the reconnected probability p is invariable, the complex network degree k can increase and it doesn't have disadvantageous influence on the complex network. That is to say the interaction coefficient λ isn't the primary factor causing the complex network macroscopic structure mutation.

Thirdly, if the fluctuation occurs in the NW length scale ζ which is a characteristic parameter, Figure 3 .4 shows that the NW length scale average $\bar{\zeta}$ is a rapid growth with the fluctuation σ^2. It may be known by (2.1), the reconnection probability p is invariable, and the network degree k will be a rapid reduction. When k reduces to a certainly value (for instance $k < 1$), the complex network system connected with the structure will be in a fierce change. In terms of the NW length scale ζ definition, the complex network node should still exist, but the connection side between the nodes is deleted basically and the collapse which has appeared is reflected in the entire network system. This sufficiently indicates that ζ is the primary factor to change the dissemination behaviour. This result is in agreement with Yang in the literature [7].

ACKNOWLEDGEMENTS

This work was supported in part by the National Natural Science Foundation of China under Grant No. 10647005, and Science and Technology Foundation of Gui Zhou province of China, Contract No. [2010]2001, [2013]05, [2013]18, and Natural Science Foundation of the Education Department of Gui Zhou Province, No. [2012]025.

REFERENCES

[1] D.J. Watts and S.H. Strogatz, Collective dynamics of small-world' Networks, *Nature*, **393** (1998), 440–442.

[2] M.E.J. Newman and S.H. Strogatz and D.J. Watts, Random graph with arbitrary degree distribution and their applications, *Phys. Rev. E*, **64** (2001), 02618.

[3] S.H. Strogatz, Exploring complex networks, *Nature*, **410** (2001), 268–276.

[4] M.E.J. Newman and D.J. Wats, Scaling and percolation in the small-world network model, *Phys. Rev. E*, Vol. 60, 1999.

[5] C.F. Moukarzel, Spreading and shortest paths in systems with sparse long-range connections, *Phys. Rev. E*, 60, pp. 6263–6266, 1999.

[6] X.S. Yang, Chaos in small-world networks, *Phys. Rev. E*, Vol. 63, 046206, 2001.

[7] X. S. Yang, Fractals in small-world networks with time–delay, *Chaos, Solitons and Fractals,* Vol. 13, 2002.

[8] B.H. Guo and S.H. Cai and Zhu Jiaqiang, The Small-world Network Bifurcation Driven by Non-equilibrium Fluctuation, *Journal of Sichuan Normal University (Natural Science)*, Sept. 2008, Vol. 3, No. 5.

[9] B.H. Guo and S.H. Cai, The Birth and Death equation of virus spreading, *International Journal of Biomathematics*, Vol. 1, No. 2, 2008.

[10] B.H. Guo and S.H. Cai, The influence of complex network communication by Interaction coefficient, *IEEE AIMSEC,* 2011 Proceedings.

Future Communication Technology and Engineering – Chan (Ed.)
© 2015 Taylor & Francis Group, London, ISBN: 978-1-138-02777-0

Comparative study of online tools in preparation for future teachers for use in cooperative learning methods

V. Maněna, M. Maněnová, K. Myška & K. Rybenská
Faculty of Education, University of Hradec Králové, Czech Republic

ABSTRACT: One of the most popular forms of teaching today is organized group work, or cooperative education, which is also called cooperative learning. It has strong links to information and communication technologies that can be applied directly to teaching and learning. Nowadays, that is characterized by the use of modern technology, which is almost inconceivable for individual users, including teachers and pupils, who do not use any online tools for their own benefit or for education. This article compares several similar tools that teachers and their students can use to manage projects in the field of cooperative education.

Keywords: Modern technologies, smart devices, education, cooperative learning, Trello, Asana, All Things, Wunderlist.

1 INTRODUCTION

Organized group work, in which students collaborate in small groups to achieve social or other objectives, is one of the most popular forms of teaching. Students face challenges through collaborating, learn to divide roles, plan their work, divide tasks, resolve disputes, and assess the work of each other. This activity is an important aspect of the constant supervision of the teacher, who is in the role of a manager planning a teaching process. Cooperative teaching has great links to information and communication technologies. It can be applied to the teaching of ICT and the tasks which by their nature require a cooperative approach directly. These include:

- information retrieval and subsequent processing on the Web,
- creating graphics, multimedia, and databases.
- publications on the internet and creating websites,
- application development on various platforms (Dvořák, 2007).

There are many ways to use information technology as a teaching aid in cooperative teaching. ICT can greatly simplify the job of cooperative teaching mainly in terms of tools that can be used for this class. These include tools designed to:

- organize email or web conferences,
- share access to databases,
- share documents within a local or global network (cloud),
- share information through the web interface,
- support for multi-user work within development environments,
- groupware,
- applications for project management.

The article presents an application for project management Trello, which we use in teaching students (future teachers) to use for cooperative learning. Students also use these tools to work on their own projects.

2 THEORETICAL BACKGROUND

2.1 *Cooperative learning*

Cooperative learning which is different from individual learning is based on the principle of mutual cooperation in the educational process. The activity of the entire group is based on the partial activities of individuals. The basis of cooperative learning is sharing, cooperation and support (Kasíková, 2001).

The cooperation of students makes sense even in the cases where there are major differences in the abilities and skills of the students. Maybe the other reason is that they are aware of these differences and learn to communicate with each other (Rohlíková, Vejvodová, 2012).

For cooperative learning there are typical methods such as discussion, problem-based instruction, work on the product, snowballing, and buzz groups, etc.

3 THE COMPARATIVE ANALYSIS OF TOOLS SUITABLE FOR COOPERATIVE LEARNING

3.1 *Research*

There are many tools which vary in quality, appearance and type of license. When choosing the right tools, we therefore will establish certain criteria that any instrument suitable for use in cooperative learning should meet:

- free license for five or more users.
- the possibility to create unlimited projects or sets of tasks.
- A detailed descriptions of the tasks.
- The possibility of commenting on tasks.
- A user-friendly environment.
- The possibility of using tools on smart devices.

As mentioned above, there are many tools that are suitable for task sharing and collaborative work. We chose to test the following:

- Asana. A tool that offers the possibility of cooperation of many colleagues in a work environment provides unlimited number of projects and tasks, allows you to delegate tasks and is supported on devices running Android.
- Allthings (similar to Asana).
- Wunderlist (very sophisticated and user-friendly tool).
- Trello. Similar tool as the others mentioned above (Zeman, 2013).

We tested all the selected tools in a real educational environment, with the help of students from the Faculty of Education at University of Hradec Králové. We focused mainly on functions and the ease of use in student's work.

3.2 *Asana*

One of the tools for collaboration is Asana. This application runs in a web browser and on smart devices (tested on Android OS, version, but also for iOS) that let you create, assign and also comment on tasks within a work team or an entire company (Reklamní inženýři, 2014). The basic version of the tool is free for up to fifteen users in the same team, which is enough for use in collaborative teaching teams, If necessary, you can pay extra fees for the extended version (and thus increase the number of users (Asana, 2014). Asana allows you to add tasks, subtasks, comments, tags, and even upload files (from PC, Dropbox or Google Drive). To use it, registration is required, either with a new account or use the Google account (Jesisem, 2013).

3.3 *AllThings*

Like Asana, an AllThings tool makes teamwork possible. It is possible to schedule tasks, generate projects or schedules and rosters. All data can be shared and commented on. It is possible to determine the priority tasks or add terms and repetitive tasks. All is in a clear and easily editable form. It provides the ability to create a multi-coloured border to tasks, change the display and grouping. Users can search and store the searched results. Each user can invite an unlimited number of collaborators, but can only create five lists in the basic free version. Other versions are paid by monthly fee. A version of this tool is available both for computer (via a web browser), and operating system Android and iOS (AllThings, 2014).

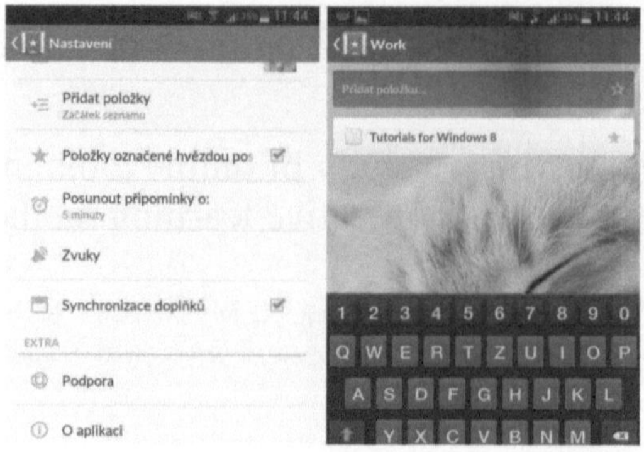

Figure 1. Wunderlist. Sample environment in Wunderlist on a smartphone. Left: settings. Right: bulletin board with an unfulfilled task.

3.4 *Wunderlist*

Another effective tool for managing tasks is wunderlist. This application is able to organize personal and business affairs. It allows users to log in via a social networks like Facebook or Google+, or by creating a new account. On the home screen are some pre-prepared lists, but users can adjust them completely by himself. After clicking on the noticeboard a list appears, showing all the uncompleted tasks (Figure 1).

Lists can be shared with other people; it is possible to organize teamwork, add comments, or insert attachments. Wunderlist is available in two versions – free and the pro version in which users can also get an option to insert attachments of unlimited size, unlimited assignments, or subtasks (Myslivеček, 2014). Since Wunderlist is a cross-platform tool that exists both in a form for web browsers and operating systems, Android or iOS, or Windows Phone, and Windows 8, BlackBerry, or Linux, the surprise is its ability to be run on the electronic book reader Kindle Fire. It also has a few specific versions for desktop, web, mobile and tablet. Also, there is a supplement for the web browser Mozilla Firefox. These devices can be synchronized with each course. Wunderlist is also partially localized into Czech (Wunderlist, 2014).

3.5 *Trello*

This tool offers advanced options for managing tasks. Users can share and collaborate on them with other users. It allows users to create lists and checklists. Each team member can add comments, photos, and videos. It also shows the information of those who works on that particular card. Trello does not have

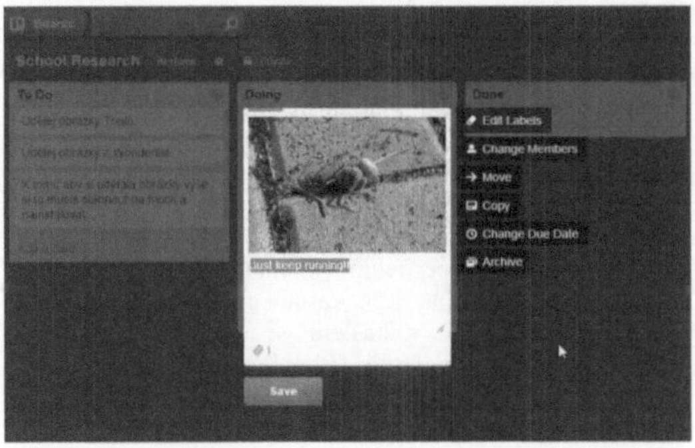

Figure 2. Trello displayed in web browser on PC.

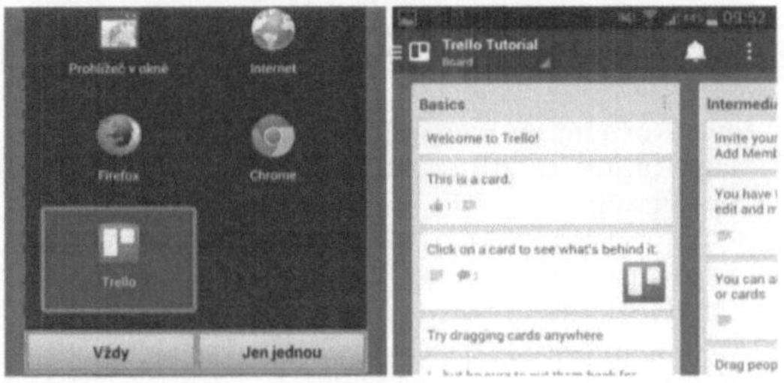

Figure 3. Using Trello, we can use a site that can be found on the internet, directly open it in this tool and save it into Trello.

advanced functions like Wunderlist. We consider this fact as a plus, for it allows cooperative learning. It is usually more valuable to have user-friendly and intuitive tools. We do not need to offer as many options to students as possible; we only need tools that make working with them easy. Trello is a very transparent tool. It is primarily a web application, which is accessible anywhere and from any place with an internet connection. It behaves a bit like a real bulletin board. It is a sophisticated tool full of different options. In the basic version for small businesses and non-profit organizations, it is free. Otherwise you have to pay an annual fee. Adding and editing individual cards is very simple and anyone can do it, even if he/she does not speak English very well (Figure 2).

Moreover, this tool is so popular. On the internet, a lot of amateur as well as professional guides on how to deal with Trello are available (Hovorka, 2014). As a project management tool, it provides great opportunities, a set of individual notes (tasks), and allows users to upload files from the computer, Dropbox or Google Drive. It is possible to work on the project with several people at once and no data is lost and everything is interconnected in real time. It is necessary to register, but users can choose to connect through a Google Account right at the beginning. Trello also allows the possibility of, for example, the aging of the notes (inactive notes fade over time).

It has its alternatives; iOS, as well as Windows 8, Windows Phone, BlackBerry, and Android.

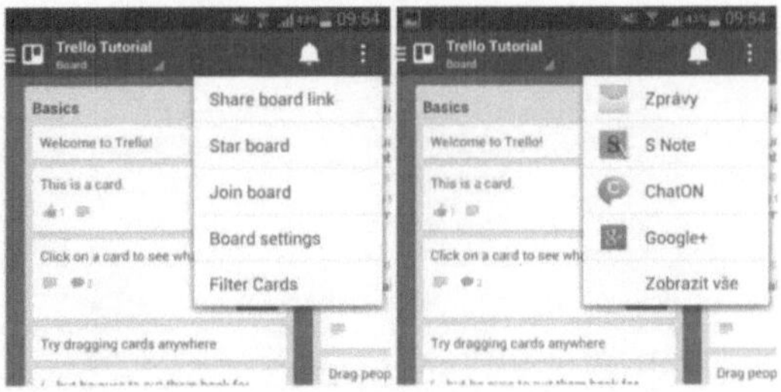

Figure 4. Instructions which we opened in Trello can be shared either using the classic menu Trello (left) or using the menu on the smart device (in our case a smartphone from Samsung, right).

Table 1. Comparison of basic features.

	Trello	Wunderlist	AllThings	Asana
Free licence	+	+	+	+
More than 5 users for free	+	+	+	+
The possibility of Inserting unlimited projects or sets of tasks for free	+	+	-	+
Ability to insert detailed descriptions of tasks	+	+	+	+
The possibility of commenting tasks	+	+	+	+
Compatibility with smart devices	+	+	+	+
Uploading a file from cloud	+	+	+	+
Signing via Google	+	+	+	+
Signing via Facebook	-	+	+	-
Ability to send email task	+	+	+	+
Tools for Vote	+	-	+	-
Shared calendar	+	+	-	+
Uploading a file from computer	+	+	+	+

3.6 *Comparison*

Each of the applications has its strengths and weaknesses. For a better overview, we compared various applications in the tables.

The table and graph above clearly shows that all applications meet most of our specified requirements. Asana also includes a feature enabling users to vote, even though it is not the classical voting when user is choosing his preferred choice. Likewise, it should be noted that the all things tool has its own calendar, but we did not know how to control it in a similar way like the other tools. For example, how to use it to schedule individual tasks and when it is appropriate to share with other users.

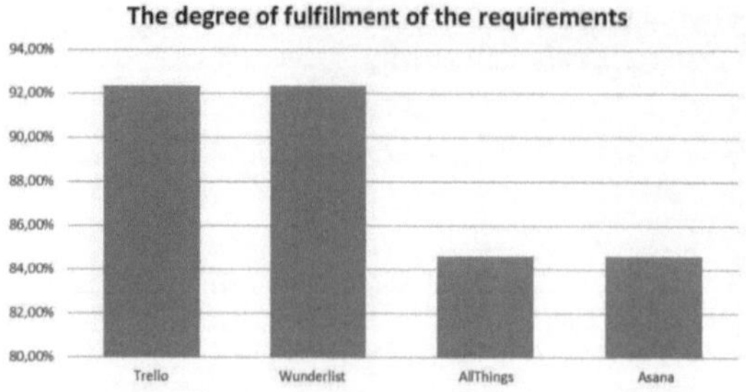

Figure 5. Graph which illustrates the compliance with the requirements of individual tools.

4 CONCLUSION

There are many tools on the internet that could be used in cooperative education. Almost all the instruments we tested met all our requirements. However, the majority of test instruments are primarily intended for managers and the work environment. All tested tools present opportunities for students to plan their tasks well and to orient in their duties. To achieve our goal which is learning to work with the tool that future teachers and students will use to plan their project, we can recommend only one. When testing the real problems that could meet the teachers and their pupils, we work best with Trello. It is a highly intuitive tool that allows easy and quick operation on smart devices. For our purposes, it is the most suitable tool. Moreover, it is fully free, so it can benefit not only students but also teachers and the whole school. Although it is only available in English, it is easy to understand the principles of its operation which make its use easy and quick.

The second tool we would recommend is Wunderlist, for it can synchronize with a lot of different smart devices (e.g. iPads, iPhones, Kindle, smart devices with Android), though it is more suitable for individual users, Besides, this tool is partially located in Czech and offers many additional language packs. This makes it possible to overcome the problem of the language barrier.

REFERENCES

Asana – Nástroj pro týmovou práci. 2014. Reklamní inženýři. Online. http://www.reklamni-inzenyri.cz/2014/08/asana-nastroj-tymovou-praci/

Dvořák. 2014. Kooperativní vyučování. Karlova Univerzita: Pedagogická fakulta. Online: http://it.pedf.cuni.cz/strstud/edutech/2003_Kooper_Dvorak/edte_filipdvorak.htm

Get more power for your team with an Asana Premium plan. 2014. Asana. Online. https://asana.com/pricing

Hovorka, M. 2014. Vykašlete se na interní firemní stránky, používejte Trello. Podnikatel: Průvodce vaším podnikáním. Online. http://www.podnikatel.cz/clanky/vykaslete-se-na-interni-firemni-stranky-pouzivejte-trello/

Jesisem. 2014. Asana. Nastroje. knihovna. Online. http://nastroje.knihovna.cz/nastroje/socialni-site-a-spoluprace/357-asana.html

Mysliveček, D. 2014. Wunderlist TO-DO List & Tasks – spravujte úkoly efektivně. Svět Androida. Online. http://www.svetandroida.cz/wunderlist-to-do-list-tasks-201408/comment-page-1

Visibility of projects, people, tasks and timescales within your business. 2014. AllThings. Online.http://www.allthings.io/

Your life in sync. 2014. Wunderlist. Online. https://www.wunderlist.com/

Zeman, J. 2013. Kooperativní nástroje. Jan Zeman3. Online. http://janzeman3.cz/index.php?a=novinky/kooperativni-nastroje&addpost&page=1

Kasíková, H. 2001. Kooperativní učení, kooperativní škola. Praha: Portál.

Rohlíková, L., Vejvodová, J. 2012. Vyčovací metody navysoké škole. Praha: Grada.

This paper is published thanks to the financial support of the project SPEV 2014 named: Entering research data in the field using mobile devices.

Future Communication Technology and Engineering – Chan (Ed.)
© 2015 Taylor & Francis Group, London, ISBN: 978-1-138-02777-0

An optimal search model for wireless sensor networks

T. Lu & Z.B. Cai
School of National Defence Science and Technology, Southwest University of Science and Technology, Mianyang, Sichuan, China

ABSTRACT: In view of the operating mode of wireless sensor networks, the node initial probability, and channel listening, an optimal search model is proposed, along with a corresponding mathematical optimization model. The goal of model optimization is to find a feasible allocation scheme for a search resource so that the probability of successful communication can reach its maximum under the constrained condition for search resources.

1 INTRODUCTION

In Wireless Sensor Networks (WSN), the nodes may have very limited energy. Once the battery runs out, either to charge the nodes or replace the battery is generally unrealistic (Heinzelman et al. 2000). Therefore, the problem is to find a way to improve energy efficiency, balance node energy consumption, and prolong the network's lifetime, avoid network splitting, and other issues that have drawn great attention in the research work on WSN.

The theory of optimal search involves the study of how to use a 'best' way to find a pre-identified object (usually called the 'search target'). In short, it is important to find the best allocation method for a search resource (eg time or energy) to either maximize the likelihood of successfully detecting the target, or to minimize the expected value for detecting the target (ie the resources consumed). The optimal search theory has been developed rapidly in recent years, with a large number of accumulated results, and is being widely used in many trades and professions: for example, geological exploration, prospecting, tracking and navigation, disease prevention and control, market research, criminalistics, economic management, and network scheduling optimization, information retrieval, mobile computing, and many other fields. (Zhu & Oommen, 1997, Zhu & Peng, 2006).

2 MATERIALS AND METHODS

2.1 *Running mode of node*

To reduce the energy consumption of the working nodes, the nodes should be in a sleep mode for most of the time. Only when there is a sensing event, should a node be involved in data communications. The cluster head should, after the incident has occurred, assist the sensing node in forwarding the data. Nodes, through

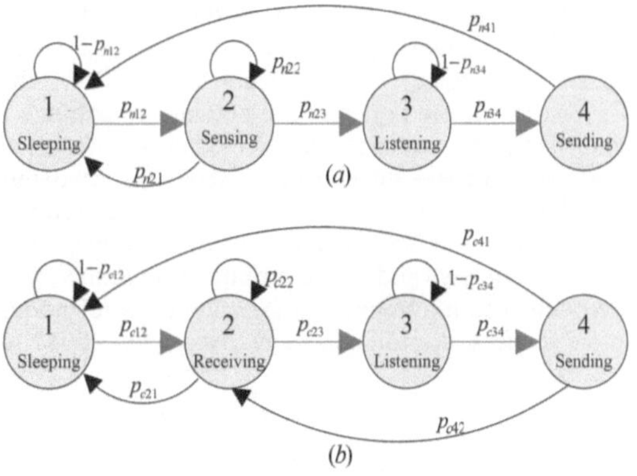

Figure 1. Operation mode of sensor nodes and a cluster head. (a) Operation Mode 1; (b) Operation Mode 2.

transitions to the work state, enable the communication module to go to sleep as much as possible in the absence of tasks, thereby reducing the energy consumption of the nodes. The conversion of the working state of nodes is both essential and effective to the sensor network. The mode of operation can be presented through different states of nodes. The sensing node and the cluster head, because of different responsibilities at different times, may work in different modes, as shown in Figure 1.

Sensor nodes run in the following way: a sensor node runs the Mode 1 in the first STAGE in each round, to the probability of p_{n12}, and is then converted from a state of hibernation to a sensing state. After the sensor's data are obtained, the node is switched to the listening state, where the node listens into a channel. If the channel is busy, then listening continues, if the channel is idle, then the node transmits the data and goes back to sleep. Sensor node ends the operation in this STAGE. Sensor node begins to run the second STAGE by way of Mode 2, mainly to collect information on adjacent nodes. Such information is the basis for selecting the operating mode for the third STAGE.

Cluster head runs in the following way: cluster head starts to run the Mode 2 in the first STAGE of each round, and the node, to the probability of p_{c12} switches from hibernation into a reception state. At this time, the head node of the cluster is responsible for receiving sensor data and the information on adjacent nodes, the cluster head in the receiving state, after staying for some time, may switch to the listening state to the probability of p_{c23}. The listening state runs in the same way as the sensor node, where the cluster head is in the idle channel to send packets, after that, the cluster head returns to the receiving state, indicating that this STAGE operation is finished, if the cluster head does not comply with the agreed terms, then in the next STAGE begins the running in the Mode 1.

2.2 *Initial probability of node*

Assuming that node's communication radius r_c is fixed, and $r_c < \frac{1}{2}d_0$ (d_0 is threshold for a free-space model or a multi-path fading model). The distance d_{toNext} between two adjacent nodes can satisfy the relationship $r_c < d_{toNext} < d_0$. This constraint allows the distribution of the nodes in the network to achieve a better state. As the node has a fixed communication range and satisfies the constraint above, in the determined scene X, covering the entire area may require the number of the desired nodes that is directly associated with the area and the communication radius r_c of the node. The minimum required number n of nodes in the scene X is -(Slijepcevic & Potkonjak, 2001):

$$\frac{X}{n\pi r_c^2} = \frac{2\pi}{\sqrt{27}} \tag{1}$$

Considering the residual energy and the number of neighbour nodes as an important parameter for the selection of cluster head, the probability for each node can be defined as the cluster head is:

$$p_{c12} = \frac{\sqrt{27}X}{2r_c^2\pi^2 N} \cdot (\alpha \frac{s(i)}{N-1} + (1-\alpha)\frac{E_P}{E_{init}}) \tag{2}$$

N is the total number of nodes in the scene X, $s(i)$ is the number of neighbouring nodes of node i, α is the weighting value to balance the energy and the number of nodes, E_p is the residual energy of the node, and E_{init} is the initial energy of the node.

2.3 *Channel listening*

Multiple nodes may communicate with each other being covered. In the use of the same channel to send data, competition will cause the node to enter the sending state at different timing states. When a node is ready to send data, it first listens on the channel, if the channel is busy, it insists listening to the next slot, if the channel is idle, then it transmits data to the probability q, and with probability $1-q$, the listening is postponed to the next slot, if the next slot channel remains idle, then it is still sending data to the probability q, with the listening being postponed to the next slot to the probability $1-q$, This is sustained until the data is sent. In order to ensure a low latency, q ranges from $[0.3, 0.5]$ (Bouabdallah et al. 2009). Considering the worst case, where there is a certain time when $k = p_{n12} \cdot N$ sensor nodes may be distributed within a single cluster, and simultaneously in the listening state, transmitting data is desired. The node state is defined as (k, m), where $0 \leq m \leq k$. Then $P(k, m)$ defined by:

$$P(k,m) = \binom{k}{m} q^m (1-q)^{k-m}, \quad \sum_{m=0}^{k} P(k,m) = 1 \tag{3}$$

$(k, 0)$ indicates that no node sends data, network being in an idle state, $(k, 1)$ represents a particular node successfully transmitted the message, when $m > 1$, it means that plurality of nodes are contenting a channel, where a conflict has occurred to the networks. One of the k nodes has the probability of successful transmission as:

$$P_s(k) = P(k,1) = kq(1-q)^{k-1} \tag{4}$$

The network with k nodes has the probability of being in an idle state as:

$$P_i(k) = P(k,0) = (1-q)^k \tag{5}$$

A contention period may appear to the probability of $1 - P_s - P_i$. Therefore, k nodes may have m contention periods to the probability of :

$$P_m(k) = P_s(k)(1 - P_s(k) - P_i(k))^m \tag{6}$$

Thus in the state (k, m) the expected value of the contention period is:

$$E[(k,m)] = \sum_{m=1}^{k} mP_m(k) = \frac{P(k,m)}{P_s(k)} \tag{7}$$

If the time to transmit a packet data comprises the channel contention time, transmission time and the end-to-end delay, assuming that each contention period is 2τ, the end-to-end delay is τ, the k nodes may send a packet of data in an average time:

$$T_{send}(k) = 2\tau \frac{P(k,m)}{P_s(k)} + \frac{l}{s} + \tau \tag{8}$$

where l is the packet length, and s is for the channel rate. $P_s(k)$ is the probability of the node being switched from listening to sending, T_{send} is the average time for the node to be switched from the beginning of listening to the end of sending, with the upper bound to be determined by the number of nodes expected to send data.

3 OPTIMAL SEARCH MODEL

To ensure the inter-node communication, an optimal search model may be established to solve it. At some point of the operation of the network, nodes that are expected to be the sending and receiving states may appear in pairs, with minimal interference between other nodes. If the data sent out happens to be received by the node in the receiving state, and also relying on this principle, the data is transmitted in the network. Then the node in the receiving state (cluster head) is assumed to be the searcher, and the node in the sending state (sensor nodes) to be the target. The problem with the node in successful communication comes to the searcher, at a minimum price (wait time for reception), successfully finds out to the target. According to the theory of optimal search, inter-node communication may be attributed to an optimal search for the discrete search space and discrete resources.

J is defined as the set of all the communication nodes covered by a certain searcher in the scene, then J is a positive integer subset. The probability of target J to be in the searcher is denoted by $p(j)$, $j = 1, 2 \ldots J$. Under normal circumstances:

$$\sum_{j \in J} p(j) = 1 \tag{9}$$

If the position of the target is not fully distributed, that is, the target might not be within communication range covered by the searcher, then:

$$\sum_{j \in J} p(j) < 1 \tag{10}$$

Assuming that the target is located in the searcher j, then the probe function $b(j, r)$ indicates the probability of successful communication within the waiting time not exceeding r times between the target j and the searcher, i.e. the searcher, with the wait time of r times, may have the probability of transmitting data to the target j. Assuming each waiting time is fixed, comprising the data transmission time and the end-to-end time delay, that is:

$$t^* = \frac{l}{s} + \tau + \beta\tau \qquad \beta = 1,2,\ldots \tag{11}$$

β is the adjustment factor of t^*, determined by the lower bound of listening time. If the communication is always successful, that means searchers can always receive data from the target, then there is $b(j,r) \to 1, r \to \infty$. Nonetheless, the interference may be actually present with the blind area from the network coverage, usually $b(j,r) \to \alpha < 1$. According to the nature of the probe function, the longer the searcher in the receiving state, the greater the probability of receiving packets is. This principle enables the probe function $b(j,r)$ to be a regular one, that is, $b(j,r)$ may have continuous derivative to each j, and the derivative function $b'(j,r)$ is a monotonically decreasing function satisfying $b'(j,r) > 0$, and assume:

$$b(j,r) = 1 - e^{-r}, (j = 1,2,\ldots J) \tag{12}$$

The detection function is derived from the theory of optimal search using the Koopman random search equation-(Stone, 1975). Let $c(j,r)$ represent the consideration for the searcher to wait for r times. The following talks about the case of the total number of wait times with a fixed upper bound. Let R be a permitted total wait times, and $\xi(r) = (\xi_1, \xi_2 \ldots, \xi_r)$ means the program for r wait times. For each $\xi(r)$, it is associated with a resource allocation function $f_{\xi(r)} : (1, 2, \ldots, J) \to (1, 2, \ldots, r)$, representing the wait times of the searcher in the implementation of the search program $\xi(r)$, that is:

$$f_{\xi(r)}(j) = \sum_{i=1}^{r} I_{\xi_i = j}, j = 1,2,\ldots J \tag{13}$$

Here $I_{\xi_i = j}$ is an instruction function for the set $\{\xi_i = j\}$. If $\xi_i = j$ is established, then $I_{\xi_i = j} = 1$. Otherwise, $I_{\xi_i = j} = 0$. In addition, the total waiting times is limited less than or equal to R, which means:

$$f_{\xi(J)}(j) = \sum_{i=1}^{J} I_{\xi_i = j} \le R \tag{14}$$

The total consideration for the search:

$$C[f] = \sum_{j \in J} c(j, f_{\xi(r)}(j)) = \sum_{j \in J} f_{\xi(r)}(j) = r \sum_{i=1}^{J} (i+1) \cdot T_{send}(i) \cdot E_{idle} \tag{15}$$

E_{idle} is the node energy consumption when idle listening. According to the above model, the communication issue is to find a workable plan for the distribution of resources and the related search programs, so that the probability of successful communication is able to achieve maximum subject to the resource constraints $C[f] \le R$. Using the Lagrange multiplier method to calculate the optimal search strategy f^*, the Lagrangian function can be defined as $l(j, \lambda, r) = p(j)b(j,r) - \lambda r$, where $1 \le j \le J, \lambda > 0, r \ge 1$.

By seeking the extreme Lagrangian function, the allowable optimal allocation of resources on the parameter r is given by:

$$\frac{\partial l}{\partial r} = p(j)e^{-r} - \lambda = 0 \tag{16}$$

Thus,

$$r = \ln \frac{p(j)}{\lambda} = r_j \tag{17}$$

The constraints lead to:

$$\sum_{j=1}^{J} \ln \frac{p(j)}{\lambda} \le R \tag{18}$$

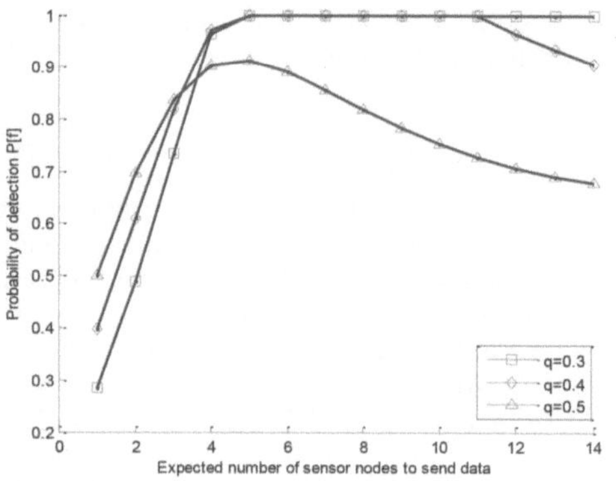

(a)

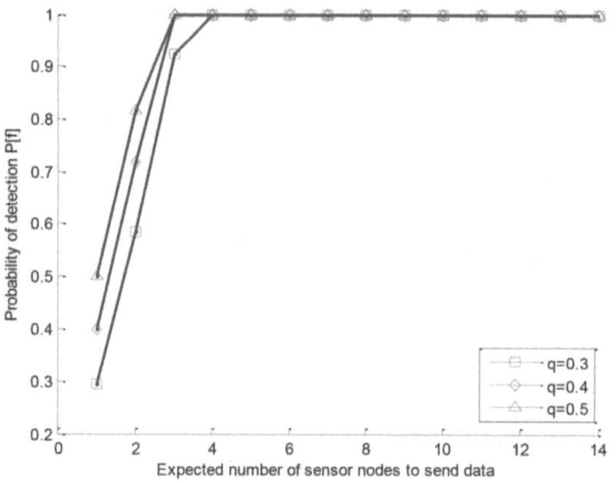

(b)

Figure 2. Statistics on Detection Probability of Nodes in Different Listening Probabilities (a)$r_j^* = 1$ (b) $r_j^* = 2$.

The inequality is taken with the equal sign at the upper bound of resource constraints. The optimal search program has a detection probability $P[f^*]$ and is given by:

$$P[f^*] = \sum_{j=1}^{J} p(j)(1 - e^{-r_j}) = \sum_{j=1}^{J} p(j)(1 - e^{-\ln\frac{p(j)}{\lambda}}) \tag{19}$$

It is easy to get the upper limit of resource allocation r_j^*. From (18):

$$\lambda \geq [p(1)\ldots p(J)]^{1/J} e^{-R/J} \tag{20}$$

Then:

$$r_j^* = \ln\frac{p(j)}{\lambda} \leq \ln\frac{p(j)}{[\prod_{j=1}^{J} p(j)]^{1/J}} + \frac{R}{J} \tag{21}$$

where, $\frac{R}{J}$ is the average waiting time assigned to each target, whereas the waiting time is given r_j^* as the upper boundary. This conclusion can be understood as follows: If the data can always be successfully received, i.e. assuming that all sensor nodes have $p(j) = 1$, and then there must be $b(j, 1) = 1$ and $r_j^* = 1$. In normal circumstances $p(j)$ is determined by sensor nodes and probability of listening, so the cluster head needs to wait for a number of times.

4 RESULTS

Figure 2 shows two cases as $r_j^* = 1$ and $r_j^* = 2$, respectively, under which the sensing node at different listening probability presents the statistics on detection probability. The analysis says, at $r_j^* = 1, q = 0.3$, when the expected nodes in the cluster exceed 4 ($J > 4$), then there must be $P[f^*] = 1$, At $r_j^* = 2r_j^* = 2, q = 0.3$, when the expected nodes in the cluster exceed 3, then $P[f^*] = 1$, At $q = 0.4$ or $q = 0.5$, when the node number is more than 2, then $P[f^*] = 1$.

ACKNOWLEDGEMENTS

This research was supported by the Doctor Foundation of Southwest University of Science and Technology (Grant No. 14zx7104) and The National Key Technology R&D Program of China (Grant No. 2013BAH32F00).

REFERENCES

Bouabdallah, N., M.E. Rivero-Angeles and B. Sericola, 2009. Continuous Monitoring Using Event-Driven Reporting for Cluster-Based Wireless Sensor Networks. IEEE Transactions on Vehicular Technology, 58(7):3460–3479.

Stone, L.D., 1975. Search Model. Theory of Optimal Search. Kettelle, J.D. (Ed.). Institute for Operations Research and the Management Sciences, pp: 1–34. USA

Slijepcevic, S. and M. Potkonjak, 2001. Power efficient organization of wireless sensor networks. Proceedings of IEEE International Conference on Communications. Jun 11–14. IEEE Computer Society, pp: 472–476.

Heinzelman, W.R., A. Chandrakasan and H. Balakrishnan, 2000. Energy efficient communication protocol for wireless microsensor networks. Proceedings of the Hawaii International Conference on System Sciences. Jan 4–7. IEEE Press, pp: 3005–3014.

Zhu, Q. and J. Oommen, 1997. On the optimal search problem: The case when the target distribution is unknown. Proceedings of the 17th International Conference of the Chilean Computer Science Society. Nov 10–15 pp: 268–277.

Zhu, Q. and B. Peng, 2006. Optimal search strategy for web-based 3D model retrieval. Proceedings of the International Conference on Advanced Web and Network Technologies, and Applications, Jan 16–18. Springer Verlag, pp: 811–818, 2006.

Future Communication Technology and Engineering – Chan (Ed.)
© 2015 Taylor & Francis Group, London, ISBN: 978-1-138-02777-0

A forwarding scheme of a publish/subscribe system based on OpenFlow

T.Y. Chen & Y. Zhang
Institute of Network Technology, Beijing University of Posts and Telecommunication, China

ABSTRACT: Nowadays, internet protocol only provides 'best effort' data forwarding services. So, the existing publish/subscribe systems' data forwarding schemes which are based on traditional networks have some inevitable problems, such as excessive delay, and unreliable data forwarding services. Because of the reasons above, we use OpenFlow to redesign a publish/subscribe system's data forwarding scheme. We design a new forwarding scheme according to OpenFlow's match methods of flow tables. In addition, we propose a new network traffic management algorithm to make the system work more stably and efficiently. At last, we build the system on Open vSwitch [8] platform and do some tests on the system's packets' processing speed and traffic management algorithm's working conditions. The results show that the system works well.

Keywords: publish/subscribe system, OpenFlow, data forwarding scheme, network traffic management algorithm.

1 INTRODUCTION

Along with the development of the network, traditional router and switch functions are expanding. Some functions like quality of service are added to switches. So when various types of data are in the network, the previous publish/subscribe systems could not ensure the prompt reliability of forwarding data.

After extensive research on the next generation of the internet, SDN (Software-defined Networking) becomes the most famous one. As the core technology of SDN, OpenFlow separates the switch into control parts and data processing parts which provides us with a programed environment [2] in which to develop new applications.

In the paper *PLEROMA: A SDN-based High Performance Publish/Subscribe Middleware*, the author proposed a new publish/subscribe system based on SDN, which was called PLEROMA [3]. The paper focused on the overall design of the publish/subscribe system but paid little attention to data forwarding schemes. So, this paper will introduce how to use OpenFlow to realize a publish/subscribe system's data forwarding scheme.

2 RELATED WORK

OpenFlow was proposed by Stanford University. As the implementation of SDN, the OpenFlow network mainly consists of an OpenFlow switch, FlowVisor and a switch controller. OpenFlow switches forward the packets according to flow tables; while FlowVisor is responsible for the virtualization of network; the controller manages the switch via the OpenFlow protocol. Using this protocol, the controller can add, update, and delete flow entries.

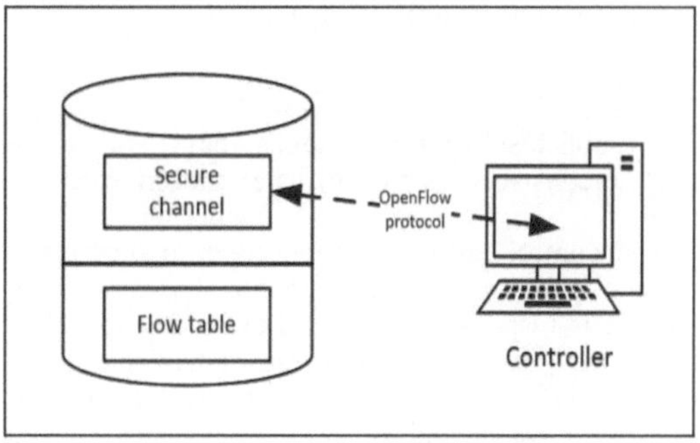

Figure 1. The OpenFlow switch and controller [6].

Match Field	Counters	Instructions

Figure 2. Main components of a flow entry in a flow table.

An OpenFlow switch consists of one or more flow tables, which perform packets lookup and forwarding. Each flow table in the switch contains a set of flow entries; each flow entry consists of match fields, counters, and a set of instructions to operate matching packets.

Match field is used for matching the coming packets. It includes an ingress port, packet headers, and optional metadata specified by previous table. As the OpenFlow standard is constantly updating, VLAN,

MPLS and IPv6 protocols has been gradually extended to the standard. Counters are used to update matching packets' number. And instructions are used to modify the action set or pipeline processing. The OpenFlow channel is the interface that connects each OpenFlow switching to a controller. Through this interface, the controller can configure and manage the switch, receive events from the switch, and send packets to the switch. Therefore, compared with a traditional switch's single and non-controllable forwarding rules, we could specify the priority and the corresponding operations when packets come to OpenFlow switches.

In addition to defining new data forwarding rules, we will design an algorithm to manage the Network traffic. Because of the limited resources and operations on an OpenFlow switch, some classic network traffic shaping algorithms [5] cannot be used here. After summarizing the advantages of token bucket mechanism [1] and RED algorithm [4], this paper presents a new feedback network traffic management algorithm. In the algorithm, the system will regularly detect current network traffic and determine the current state of the network. When the state of the network changes, the system will implement corresponding operations on packets.

3 DATA FORWARDING SCHEME

3.1 *The overall design of the system*

The publish/subscribe system's data forwarding scheme consists of a control layer and an execution layer.

The control layer is composed of a monitoring module and a management module. The monitoring module is responsible for monitoring and recording the whole system and the network's state. The management module is responsible for analysing the current state of the system and adjusting the system according to the network's state. Apart from this, the management module can also manually configured by its administrators.

The execution layer consists of command execution module and an OpenFlow switch. This layer is responsible for executing the command to detect the network's state and configure the OpenFlow switch. At the same time, it will send back the execution results to the management module.

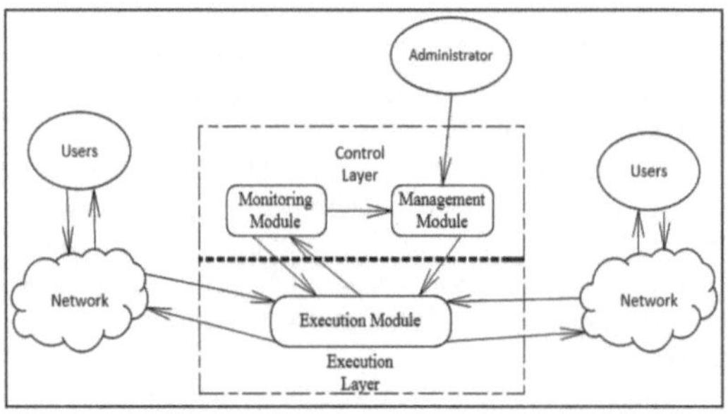

Figure 3. The overall design of the system.

3.2 *The forwarding rules*

In the publish/subscribe system, packets can be divided into three categories: general-priority, high-priority and low-priority. Apart from those packets in publish/subscribe systems, other packets are defined as being of the lowest priority. After receiving a packet, a publish/subscribe system will identify its priority. If it's a high-priority packet, it will be forwarded immediately. When the high-priority packet processing is completed, the system will forward other priorities' packets. Here we use a pipeline process to realize the method mentioned above.

The flow tables of an OpenFlow switch are sequentially numbered, starting from 0. Pipeline processing always starts at the first flow table: the packet is first matched against entries of flow table 0. Other tables may be used depending on the outcome of the match in the first table.

If the packet matches a flow entry in a flow table, the corresponding instruction set is executed, or it will go to the next table. According to the characteristics of packets in the publish/subscribe system, we use a transport source port to match every packet. The matching process of packets is shown as below.

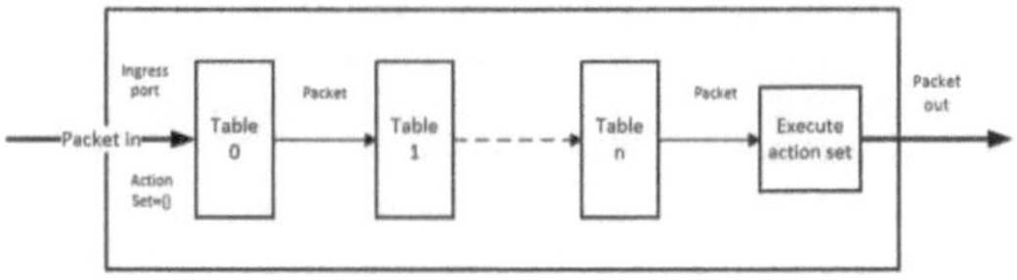

Figure 4. Packets are matched against multiple tables in pipeline [6].

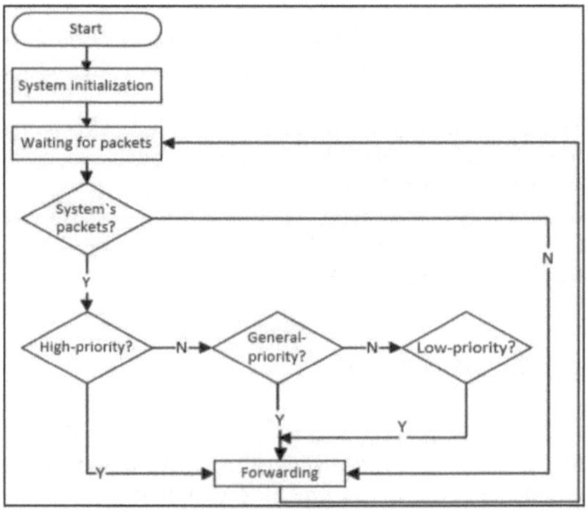

Figure 5. Packets' matching process.

3.3 *The network traffic management algorithm*

In order to ensure the stability and reliability of the publish/subscribe system, we designed a network traffic management algorithm.

We define several parameters about network traffic: alert rate/upper limit rate; detective interval; the rate of forwarding packets (per second); network state. First we set an alert value and an upper limit value through a current network environment and hardware configuration. The rate of forwarding packets can be directly collected from the OpenFlow switch, and then we can determine the network state. If the current forwarding packets' rate is less than the alert rate, we define the network state as being in a healthy state; if the rate is larger than the alert rate but less than the upper limit rate, then we define that network state as being slightly blocked; if the rate is larger than the upper limit rate, we define that network state as being heavily blocked.

When the network state is determined, the system will automatically adjust the forwarding rules according to its state. When the state is healthy, the system will do nothing. If the number of packets being processed increases and the state has changed to the state of being slightly blocked, the system will allocate more resources for forwarding packets. The lowest priority messages will be dropped in order to ensure that the remaining priority data can be totally sent. At the same time, the publishers of the system will receive a warning message to slow down the rate of dispatching the packets. If the number of packets being processed continues to increase and the state has changed to the state of being heavily blocked, only high-priority packets will be sent and lower priority packets will be dropped in the operation, and the number of dropped packets will depend on their priority level. In addition, the system's detection interval will be reduced to ensure the accuracy of the system's operation. The publishers of the system will receive a serious warning message to slow down the rate of sending the packets too. When the network state changes from being blocked to being healthy, the drop operation will be stopped and the system will return to its original state.

4 EXPERIMENTS

After finishing the whole design, we did a lot of experiments to test our system.

The test environment included three computers: two of them are the publisher and the subscriber. And the third one we build as an OpenFlow switch with three ports. Open vSwitch is used to simulate OpenFlow switch.

Firstly we tested the forwarding rules and the priority classification. The result is shown in Figure 6.

The publisher sent 30,000 packets to the subscriber, including 10,000 packets per priority. We can see that every priority packet has been received and the high-priority packets were received faster than those of other priorities'.

```
configuration finished!
server start... tcp port is 30001
server start... tcp port is 30002
server start... tcp port is 30003
number of general-priority packet received is: 1000
number of low-priority packet received is: 1000
number of high-priority packet received is: 1000
number of general-priority packet received is: 2000
number of high-priority packet received is: 2000
number of low-priority packet received is: 2000
number of general-priority packet received is: 3000
number of low-priority packet received is: 3000
number of high-priority packet received is: 3000
number of low-priority packet received is: 4000
number of general-priority packet received is: 4000
number of high-priority packet received is: 4000
number of general-priority packet received is: 5000
number of high-priority packet received is: 5000
number of low-priority packet received is: 5000
number of general-priority packet received is: 6000
number of high-priority packet received is: 6000
number of general-priority packet received is: 7000
number of high-priority packet received is: 7000
number of high-priority packet received is: 8000
number of high-priority packet received is: 9000
number of general-priority packet received is: 8000
number of low-priority packet received is: 6000
number of general-priority packet received is: 9000
number of low-priority packet received is: 7000
number of low-priority packet received is: 8000
number of low-priority packet received is: 9000
number of low-priority packet received is: 10000
number of general-priority packet received is: 10000
number of high-priority packet received is: 10000
```

Figure 6. The test result of packets' processing.

```
configuration finished!
server start... tcp port is 30001
server start... tcp port is 30002
server start... tcp port is 30003

start to receive message!
It costs 5633ms to receive 100000 messages

start to receive message!
It costs 5432ms to receive 100000 messages

start to receive message!
It costs 5549ms to receive 100000 messages

start to receive message!
It costs 5809ms to receive 100000 messages

start to receive message!
It costs 5491ms to receive 100000 messages

start to receive message!
It costs 5432ms to receive 100000 messages

start to receive message!
It costs 5302ms to receive 100000 messages
```

Figure 7. The test result of the packets' processing speed.

Then we tested the system's speed of processing the packets and the result is shown in Figure 7.

We have tested this system about seven times and every time we send 100,000 packets. When all the packets have been received, we recorded the time it took. After calculating the average number, we got the system's processing speed. It was about 18,000 packets (600 bytes per packet) per second.

Finally, we tested the traffic management algorithm. The result is shown in Figures 8 and 9.

```
root@ubuntu12:~#java -jar processPacket.jar
init finished!
###### state changed to slightly blocked ######
------- state changed to heavily blocked ------
------- state changed to heavily blocked ------
------- state changed to heavily blocked ------
------- state changed to heavily blocked ------
------- state changed to heavily blocked ------
------- state changed to heavily blocked ------
------- state changed to heavily blocked ------
...... state is changed to healthy ......|
```

Figure 8. The state changed when processing the packets.

```
configuration finished!
server start... tcp port is 30001
server start... tcp port is 30002
server start... tcp port is 30003

start to receive message!
It costs 5633ms to receive 100000 messages

start to receive message!
It costs 5432ms to receive 100000 messages

start to receive message!
It costs 5549ms to receive 100000 messages

start to receive message!
It costs 5809ms to receive 100000 messages

start to receive message!
It costs 5491ms to receive 100000 messages
|
start to receive message!
It costs 5432ms to receive 100000 messages

start to receive message!
It costs 5302ms to receive 100000 messages
```

Figure 9. The information printed by subscriber.

The publisher sends 50,000 packets per priority to the subscriber. We can see that after changing the network's state, the drop operation would start to work. High-priority packets are all received; about 400 general-priority packets are dropped, and about 1,000 low-priority packets are dropped.

5 CONCLUSION

In this paper, we designed a forwarding scheme of a publish/subscribe system based on OpenFlow. We used OpenFlow to make this publish/subscribe system work more efficiently and flexibly in forwarding packets. We also carried out experiments to test the performance and the results show that the new publish/subscribe system works well.

ACKNOWLEDGEMENTS

This research is supported by the National Grand Fundamental Research 973 Program of China under Grant No. 2011CB302506, 2012CB315802; National Key Technology Research and Development Program of China (Grant No. 2012BAH94F02); National High-tech R&D Program of China (863 Program) under Grant No. 2013AA102301; National Natural Science Foundation of China under Grant No. 61132001, 61372115); Program for New Century Excellent Talents in University (Grant No. NCET-11-0592); Project of New Generation Broad band Wireless Network under Grant No. 2014ZX03006003,2011ZX03002-002-01, 2012ZX03005008-001; The technology development and experiment of innovative network architecture (CNGI-12-03-007).

REFERENCES

[1] Daisuke Ikegami, Yasuto Nakanishi, Toshiaki Tsuchiya et al. Performance Evaluation and Throughput Formulation of TCP under Token Bucket Policer [J]. IEICE Transactions on communications, 2011, E94-B (7): 1914–1923.

[2] Egilmez, Hilmi E., Dane, S. Tahsin, Bagci, K. Tolga et al. OpenQoS: An OpenFlow controller design for multimedia delivery with end-to-end Quality of Service over Software-Defined Networks [C].//Proceedings of the 2012 Asia Pacific Signal and Information Processing Association Annual Summit and Conference. [v.1]. 2012: 1–8.

[3] Muhammad Adnan Tariq, Boris Koldehofe, Sukanya Bhowmik, Kurt Rothermel et al. PLEROMA: A SDN-based High Performance Publish/Subscribe Middleware

[4] Li Feng-juan, Yang Xiao-ping, Liu Xiao-jiao, Research of Priority Distinguished RED Algorithm, Journal of Jilin University (Information Science Edition), 2013, 31(5): 449–455.

[5] Nair, SK, Novak, DC. A traffic shaping model for optimizing network operations [J]. European Journal of Operational Research, 2007, 180 (3):1358–1380. DOI:10.1016/j.ejor.2006.04.036.

[6] OpenFlow, http://archive.openflow.org/documents/openflow-spec-v1.1.0.pdf, 2014.07.30

[7] Open vSwitch, http://openvswitch.github.io/, 2014.8.24

[8] Installing and configuring Open vSwitch on Ubuntu 12.04, http://networkstatic.net/installing-and-configuring-openvswitch-on-ubuntu-12-04-precise-pangolin, 2014.8.25

Future Communication Technology and Engineering – Chan (Ed.)
© 2015 Taylor & Francis Group, London, ISBN: 978-1-138-02777-0

An adaptive virtual machine management system for cloud computing

B. Li & Y.L. Shang
The State Key Lab of Networking and Switching, Beijing University of Posts and Telecommunications, China

ABSTRACT: Because of its high scalability and availability, cloud computing has been more and more widely used in recent years. In the process of using cloud computing platforms, we find that with the passage of time, the work load of virtual machines change constantly, but the allocated resources for virtual machines remain the same. To avoid the mismatch between the requirement and supply of resources, we designed and implemented an adaptive virtual machine management system based on OpenStack. The system will monitor the performance changes of each virtual machine, and take actions, based on a pre-configured policy, so that every virtual machine will just have the resources it needs at runtime. And the adjustment happens automatically, without extra manual operation. With its help, the scalability of the platform is enhanced.

Keywords: cloud computing, adaptive resource management, OpenStack.

1 INTRODUCTION

With the flourish of cloud computing, it is more and more widely used by business organizations, schools and even individuals nowadays. By integrating hardware and software resources, providers can offer services based on consumers demands. However, in the process of using a cloud computing platform, we find that the virtual machine's resource allocation action is done just at the moment of the creation of new instances. After that, the virtual machine's resource capacity doesn't change at all. However, as the time changes, it will inevitably happen that the size of pre-allocated resources doesn't match the size of resources that the virtual machine needs at a certain time. We can overcome this mismatched situation by manually changing the resource size, but it will involve a lot of heavy and complicated work if there are hundreds of machines that need to be handled. So we hope to design a system which can respond to the change of virtual machines, and adjust the resources allocation automatically.

The system aims at providing a way to take action at a time when a virtual machine's load is abnormal and will avoid manual operations at the same time. To do that, firstly we need a monitoring system to collect the performance data of physical hosts and virtual machines. When a monitoring system shows some unusual conditions, such as the load being too high, or a physical host crashes, the system should send out an alarm. In addition, when we decide to migrate a virtual machine or create a new instance, we should select a target host for it, using different strategies which correspond to different virtual machine placement algorithms. After that, we get the job done relying on the existing cloud computing platform.

The system reflects the advantage of cloud computing such as scalability and availability, since it ensures the balance between the resources a virtual machine needs and the resources a virtual machine allocates. By using different strategies, we can easily switch to the virtual machine placement principles we want, without modifying other parts.

This paper is organized as follows: Section 2 discusses some related works. Section 3 presents the architecture of the adaptive virtual machine management system. In Section 4, we display the implementation of the adaptive virtual machine management system in detail. Some experiments and results are displayed in Section 5. We conclude in Section 6.

2 RELATED WORK

In recent years, the energy optimization of the data centre issue has attracted many researchers' attention. Literature 1 [1] studies the virtual machine's dynamic adjustment effects on the data centre's energy consumption, and verifies that the live migration technique can reduce the energy cost by experiments. Literature 2 [2] presents a resource load forecasting technology which can reduce the number of active physical machines. Literature 3 [3] displays an energy-aware resource management system, which uses different heuristic algorithms to select the virtual machine for migrating. Literature 4–6 [4–6] represents different energy-saving placement algorithms for virtual machines. From those papers we realize that detecting resource utilization and how to allocate a virtual machine to a suitable host are two key points.

Table 1. Nova components.

Component	Function
nova-api	provides an endpoint for all API queries
nova-schedule	takes a virtual machine instance request from the queue and determines where it should run
nova-compute	a worker daemon that creates and terminates virtual machine instances
nova-volume	manages the creation, attaching and detaching of persistent volumes to compute instances
nova-network	accepts networking tasks from the queue and then performs tasks to manipulate the network

The cloud computing platform we use is OpenStack [7]. OpenStack is an open source IaaS (Infrastructure-as-a-Service) platform proposed by Rackspace and NASA (National Aeronautics and Space Administration). In the past few years, it has been attracting more and more companies and organizations to join its ecosystem, including Intel, Citrix, Dell, AMD, Canonical, HP, Cisco, Red Hat, AT&T and IBM. And in a very short period of time, it has become a large project, consisting of nine core subprojects, from two basic modules at the beginning. OpenStack aims to create abstracted pools of computing, storage, and network resources that can be used to create virtual machines on top of standard server hardware. For the users of OpenStack, they don't need to worry about the details of the underlying hardware, because the virtualization layer makes physical resources available at a higher level of abstraction.

The components of OpenStack include Compute, Object Storage, Identity, Dashboard, Block Storage, Network and Image Service. We mainly use and discuss the compute service, with a code name 'Nova'. Nova is a fabric controller with some core components shown in Table 1.

Nova component is responsible for the management of instance's lifecycle and compute resources. In fact, it supports several operations such as creating a server, pausing a server, suspending a server, and terminating a server, etc. We can easily use those actions to manipulate compute resources, without understanding the technology details of the underlying hypervisors such as KVM, XEN, Hyper-V, etc. Because of the maturity and reliability of OpenStack, we choose it as the target platform.

3 ARCHITECTURE OF ADAPTIVE VIRTUAL MACHINE MANAGEMENT SYSTEM

The primary goal of the adaptive virtual machine management system is to realize the dynamic adjustment of virtual machines at runtime, so as to achieve the flexibility of the platform. Based on that, the architecture of the system we design is shown in Figure 1.

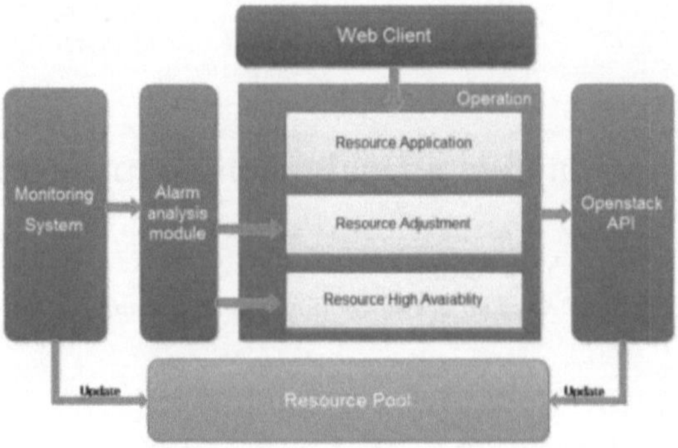

Figure 1. The architecture of adaptive virtual machine management system.

3.1 *Resource pool*

At the bottom of the system is resource pool, which contains resources of physical hosts and virtual machines. A main function of it is to maintain the state of resources. Take one virtual machine instance for example, the resource pool should not only keep some static properties such as the instance name, id, IP address, quotas, but also should keep resource usage status and load condition updated. That includes whether we can access the instance, how much the Central Processing Unit (CPU) usage is, the speed of disk Input/Output (IO), etc. When time changes or a certain client request causes a change of resource state, the resource pool should immediately collect enough information it needs and update the state of resources.

3.2 *Monitor and alarm analysis module*

The monitoring system on the left is responsible for collecting all the performance data of target machines, and saving the performance data in one place. It usually applies to C/S models, which means plugins on the client side to collect information about the running state of the machine it belongs to, and the server processes all the data reported from clients. Besides, the server also implements data persistence and data visualization functionality

Users can define at least one alarm rule for an instance, then if the monitoring system detects that workload of the instance is higher than threshold, the information should be delivered to the alarm analysis module. The alarm analysis module will look back for data in a certain time range, compute the ratio of high load time and determine whether an alarm should be generated. The reason why we do this is to avoid the jitter situation, which means the load of virtual machine goes up and down quickly.

3.3 *Operations and cloud API*

Three types of operation are supported in the system. The first is resource application. Usually the operation is requested by web client, and then the system will create a new instance and assign it to one appropriate physical host. The second supported operation is

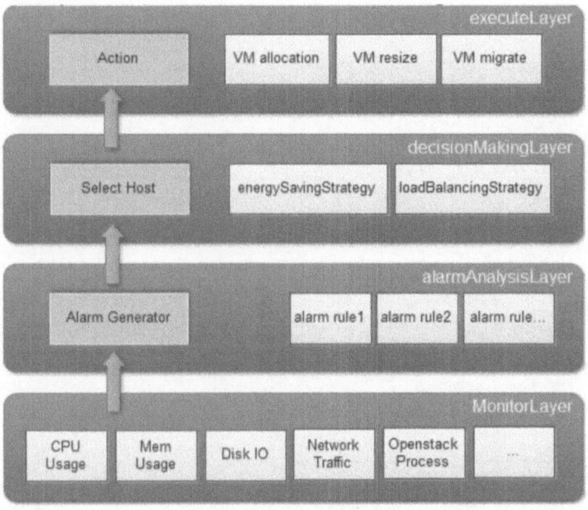

Figure 2. The framework of the adaptive virtual machine management system.

resource adjustment. When the system find that virtual machine is overloaded, an adjustment alarm is generated and takes this operation which usually means turn up or turn down the quote size. The last operation is resource high availability (HA). It is triggered when a physical host is unreachable, and the system will try to migrate all the instances on it in time, so that the availability is not affected.

The OpenStack API invokes the libvirt interface, which is a toolkit to interact with several virtualization capabilities. In libvirt, a virtual machine is named the domain and it has two properties: hypervisor type and a unique integer identifier. The interfaces libvirt library provides these including creating a domain from an xml file, rebooting a domain, attaching a device to a domain, etc. The OpenStack API encapsulates the libvirt API. The interface it provides is very abstract and descriptive, and they can be easily called using http protocol.

4 THE IMPLEMENTATION OF ADAPTIVE VIRTUAL MACHINE MANAGEMENT SYSTEM

4.1 *Framework of an Adaptive VM Management System*

The framework of the Adaptive Virtual Machine Management System is shown in Figure 2.

The system is divided into four layers: Monitor Layer, Alarm Analysis Layer, Decision Making Layer and Execute Layer. The Monitor Layer is located at the bottom of the system: it collects real-time performance data of physical hosts and virtual machines. Then it reports the data to the Alarm Analysis Layer, which evaluates historical monitoring data, and decides whether or not an alarm should be triggered. If the situation matches one alarm rule, it will tell the Decision-Making Layer to select an ideal host for the instance, depending on which strategy we use. The energy saving strategy will try to place instances into one physical host as much as possible, until the remaining resource is not capable for example, a second host will be used. While the load balancing strategy will place instances into every host equally, so each host will have very similar system load. After a host is selected, the Execute Layer will call OpenStack API to take actions including creating a new instance, resizing an instance and migrating an instance.

4.2 *Key technologies*

4.2.1 *Resource monitoring*

In order to monitor different data source flexibly, we choose Nagios [8] as our monitoring software. Nagios is an open source monitoring system and it supports a very wide range of monitoring indicators, from host resources to network services. We will benefit a lot from its flexible plugin management mechanism. Every plugin is an executive script which collects data from a specific data source. If we want to add another data source, all we need to do is to tell Nagios which plugin to use. To meet the monitoring requirements in our system, we monitor physical hosts and virtual machines separately. The performance data we collect include CPU utilization, CPU load, memory usage, disk IO and network traffic. Besides, we add the monitor for OpenStack services, such as Keystone, Glance, Neutron, so we can easily find whether a service or a public port is active or not.

We describe the monitor content in a separate configuration file, using the Nagios supporting syntax. The work consists of three parts. One is to define the check command for a service, so that Nagios will find which plugin to use and how many parameters will be passed. Another work is to define host or host group, which are Nagios objects representing the real world machine and machine group. To reduce some repeating operations, we inherit configuration files from a generic template. For a host, the host name and its IP address are the two important properties. The last work is to define service or service group. Each service represents a type of data, and a host can have multiple services related. Important properties for a service include the host or host group it belongs to, the service description and the check command. For every property, the name and value is separated by at least one whitespace. The sample configuration for a host is presented in Table 2, and Table 3 shows the configuration for some services.

4.2.2 *Alarm analysis*

User of the system will set a threshold value for a specific indicator at first, then the threshold will be passed to a Nagios plugin as a parameter, the plugin will use it to determine if the monitored machine is overloaded or not. In Nagios, once the data is beyond the threshold, the service state will be changed from OK to CRITICAL. Besides the threshold parameter, a time scale field is also needed, as it shows how long the state of

Table 2. Host definition.

```
define host{
        host_namecontroller
        address192.168.100.6
use generic-host
process_perf_data 1
check_commandcheck-host-alive
        max_check_attempts10
        notification_interval0
        notification_period24x7
        notification_optionsd,u,r
        contact_groupsadmins
        register1
        }
```

Table 3. Service definition.

```
define service {
        hostgroup_nameperf-servers
        service_descriptionCPU Usage
check_commandcheck_nrpe_1arg!check_cpu_usage
   use generic-service
        service groups perf
        notification_interval0
   }
   define service {
        hostgroup_name perf-servers
        service_descriptionMem Usage
check_commandcheck_nrpe_1arg!check_mem_usage
        usegeneric-service
        service groups perf
        notification_interval0
   }
              ...
```

overload lasts before we send out an alarm. For example, if we set the time range to one minute, nothing will happen within one minute, since not enough data is collected. Once the alarm analysis module collects enough data, it will look back into the performance history data, compute how many percentages of sample data shows that the instance is overloaded. If the value matches the alarm rule, say, greater that the threshold, then an alarm is generated and saved into database. If the sample data doesn't match the alarm rule, nothing would occur. The alarm analysis module gets service status by invoking REST API we implemented. An example of the service status API returned is shown in Figure 3.

4.2.3 *Decision Making*

The goal of the Decision-Making Layer is to select the suitable physical host for an instance. As we said before, two strategies are supported here: the energy saving strategy and the load balancing strategy. The former strategy focuses on reducing the number of active physical hosts and power consumptions as much as possible. While the latter strategy focuses on the balance of all physical hosts. It will avoid the situation where some physical hosts have a very high system load while others are left idle. If we use an energy saving strategy, the physical host, whose remaining resource capability is larger than the demand and its size is the least will be selected. If we use the load balancing strategy, the physical host whose remaining resource capability is the most will be selected. The whole selection process is shown as follows:

- filter out the unreachable hosts from the resource pool
- filter out the hosts whose remaining resources are insufficient to meet the demand
- get the current strategy we are using
- sort the hosts filtered according to the strategy, and select the first item as target host
- complete the operation by invoking the execute layer interface

localhost:8085/api/monitor/perf/compute01

```
{
  "CPU Usage": {
    "host_name": "compute01",
    "service_description": "CPU Usage",
    "check_command": "check_nrpe_1arg!check_cpu_usage",
    "current_state": "0",
    "plugin_output": "CPU OK : idle 83.90%",
    "performance_data": "user=2.25% system=13.86% iowait=0.00% idle=83.90%;80;90",
    "last_update": "1399271165"
  },
  "Load average": {
    "host_name": "compute01",
    "service_description": "Load average",
    "check_command": "check_nrpe_1arg!check_load",
    "current_state": "0",
    "plugin_output": "LOAD AVERAGE OK : 1.25,1.20,1.20",
    "performance_data": "load1=1.25;10;20;0 load5=1.20;8;18;0 load15=1.20;5;15;0",
    "last_update": "1399271165"
  },
  "Mem Usage": {
    "host_name": "compute01",
    "service_description": "Mem Usage",
    "check_command": "check_nrpe_1arg!check_mem_usage",
    "current_state": "0",
    "plugin_output": "Memory: OK Total: 16003 MB - Used: 1254 MB - 7% used",
    "performance_data": "TOTAL=16780959744;;;; USED=1313636352;;;; CACHE=10033438720
    "last_update": "1399271165"
  },
  "Disk Usage": {
    "host_name": "compute01",
    "service_description": "Disk Usage",
    "check_command": "check_nrpe_1arg!check_disk_usage",
    "current_state": "0",
    "plugin_output": "DISK OK used :  / 92.35% free",
    "performance_data": "/=14349596MB",
    "last_update": "1399271165"
  },
```


Figure 3. The return of service status API.

4.2.4 *Execution*

Relaying on the simplicity of OpenStack API provides, we can easily manipulate instances. Firstly, we get a token from the identity service, and put it into the request header of following request. When we call the API, we specify the URL, the HTTP verb, and the parameters. We will use the instance creating, resizing and migrating interfaces in the system. Table IV shows the API of resizing an instance. It is worth noting that we can set the 'resize_confirm_window' option to an appropriate number in the nova configuration file in advance, so that nova will answer the resize confirmation automatically

5 EXPERIMENTS

We have already deployed an OpenStack Havana release in a cluster with six IBM X3650 servers. The

Table 4. The resize API details.

Description	Resize the specified server
URL	/v2/{tenant_id}/servers/{server_id}/action
Method	POST
Parameter	"resize" : { "flavorRef" : flavorId }
Return	statusCode

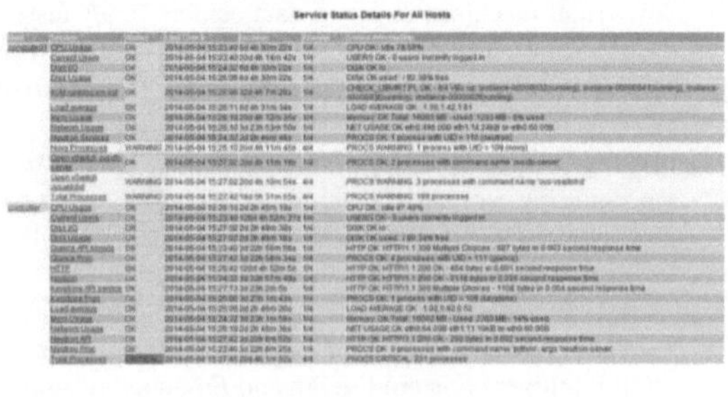

Figure 4. The service status details for all hosts at some point.

controller node has a hostname controller, the network node with a hostname network, and the others are deployed as compute nodes. One of them has a hostname compute01. After that, we create a new instance with name 'vmToAdjust'. We mainly test two functions: the monitor function and the instance adjustment function.

5.1 *Experiment of resource monitoring*

The monitor data source consists of two types: one is the performance data, the other is the OpenStack service status. The performance data are collected from both the physical hosts and virtual machines, while the OpenStack service status is just collected from the corresponding node. The performance data include: CPU utilization, CPU load, disk IO, disk usage, memory usage, and network traffic. The OpenStack service data include: Nova process, Open vSwitch process, Keystone process, Glance process, Neutron-server process and there API endpoint. We can check the monitor details from the web interface. Figure 4 shows the service status details for all hosts at some point.

5.2 *Experiment of resource adjustment*

In this experiment, we will test the resource adjustment functionality: the virtual machine instance can adapt its quota or flavour in OpenStack to the runtime load status. Before we increase the load of the instance, we define an alarm rule for vmToAdjust instance at first: if the CPU utilization is larger than 80%, and the situation lasts more than two minutes, the instance should scale up automatically. In the production environment, the duration of overload status can be longer, and we

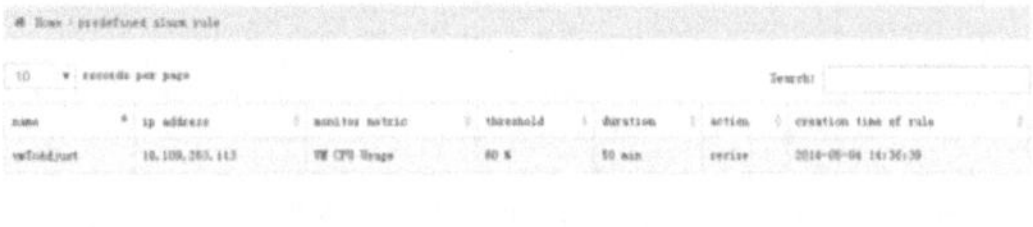

Figure 5. The alarm rule we pre-defined.

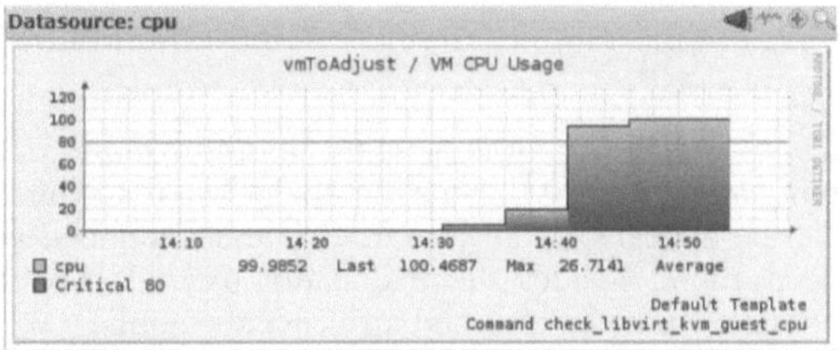

Figure 6. Change of the CPU load.

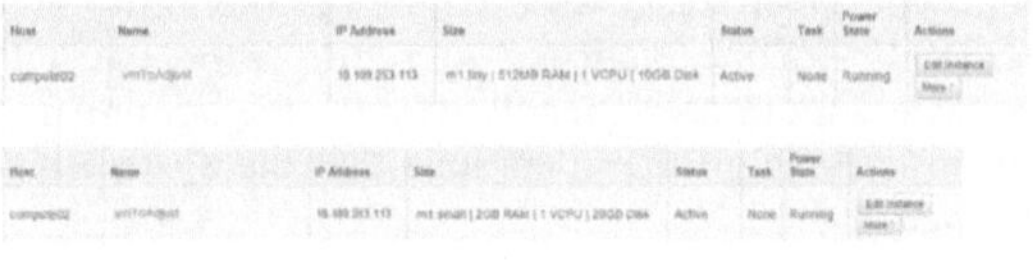

Figure 7. The instance status before and after the action.

use two minutes for simplicity. The rule we defined is displayed in Figure 5. The following steps are as follows:

step1: login to the instance command line; add the instance's pressure by using stress command. Here we create twenty processes which are running full load to simulate the condition that the load raises up very quickly;

step2: the Nagios plugins get the performance data and report to the Nagios server. The server will find the CPU Usage service's status change to CRITICAL, and then it sends the service data to the alarm analysis module;

step3: the alarm analysis module collects the service data for some time, and finds that most of the data in the past two minutes are overloaded, so it generates a new alarm, saves it to a database, and tells the decision-making layer to scale up;

step4: the decision-making layer transfers the request to the execute layer, as the adjustment operation doesn't involve migrating the instance. In the end, the execute layer gets all the flavours available in OpenStack, and gets the next flavour. Then it calls the OpenStack resize API to finish the work;

step5: if the operation is successful, the process of the action will be saved as a record on the history table in database

Figure 6 shows the change of the CPU load Figure 7 shows the instance status before the action and after the action. Figure 8 shows the saved history of this operation.

From the results we can see that instance's flavour has changed from m1.tiny to m1.small, after the load raised very high and lasted for several minutes. The instance is still active, so the adjustment operation is successful.

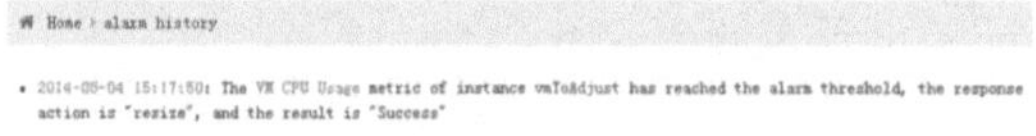

Figure 8. The history of the action is saved.

6 CONCLUSION

As we use the cloud computing platform more and more often, users will hope that their virtual machine can be allocated more resources if their application needs more. Besides, the reallocation of work should not take too much time and influence the application. In this article, we design and implement a hierarchical self-adjustment management system, then verify some functions by experiments. By integrating the system with OpenStack, we enhance the flexibility and scalability of the platform. In future work, we will provide a more user-friendly way for the administrator to execute all the operations that the system has supported.

ACKNOWLEDGEMENT

The work presented in this paper is supported by the National Grand Fundamental Research 973 Program of China (2011CB302506), the National Key Technology Research and Development Program of China "Research on the mobile community cultural service aggregation supporting technology" (2012BAH94F02).

REFERENCES

[1] Verma A, Ahuja P, Neogi A. P. Mapper: power and migration cost aware application placement in virtualized systems [M]//Middleware 2008. Springer Berlin Heidelberg, 2008: 243–264.

[2] Bobroff N, Kochut A, Beaty K. Dynamic placement of virtual machines for managing sla violations [C]//Integrated Network Management, 2007. IM'07. 10th IFIP/IEEE International Symposium on. IEEE, 2007: 119–128.

[3] Beloglazov A, Buyya R. Energy efficient allocation of virtual machines in cloud data centers [C]//Cluster, Cloud and Grid Computing (CCGrid), 2010 10th IEEE/ACM International Conference on. IEEE, 2010: 577–578.

[4] Huang D, Yang D, Zhang H, et al. Energy-aware virtual machine placement in data centers [C]//Global Communications Conference (GLOBECOM), 2012 IEEE. IEEE, 2012: 3243–3249.

[5] Xiaoli W, Zhanghui L. An energy-aware VMs placement algorithm in Cloud Computing environment [C]//Intelligent System Design and Engineering Application (ISDEA), 2012 Second International Conference on. IEEE, 2012: 627–630.

[6] Yang J S, Liu P, Wu J J. Workload characteristics-aware virtual machine consolidation algorithms [C]//Cloud Computing Technology and Science (CloudCom), 2012 IEEE 4th International Conference on. IEEE, 2012: 42–49.

[7] "OpenStack," Webpage, [Online], Available: http://OpenStack.org/

[8] "Nagios," Webpage, [Online], Available: http://www.nagios.org

Future Communication Technology and Engineering – Chan (Ed.)
© 2015 Taylor & Francis Group, London, ISBN: 978-1-138-02777-0

A comparison of free and open source technologies suitable for future authoring tools in education

V. Maněna, R. Špilka & K. Rybenská
Faculty of Education, University of Hradec Králové, Czech Republic

ABSTRACT: Smart devices, such as tablets and smartphones, are used by many people in everyday life. For this reason, it is important that they try to rank well in school. This article is based on a comparative study that compares the different technologies that can be used as a basis for the development of tools for creating training materials. Comparisons are made of the technologies that can be used through a web browser on smartphones and similar devices. In the paper, we present some options that offer test tools and we try to determine whether it would be appropriate to include similar tools for teaching in schools.

Keywords: modern technologies, open source, smart devices, education, smartphones, tablets, HTML5.

1 INTRODUCTION

The present time is characterized by a plentiful use of modern technology. Tablets and smartphones have been already owned by almost everyone and become an integral part of our daily life. However, the current generation of teachers is rarely able to keep pace with the times, having the knowledge and skills to use modern technologies as their pupils can. Therefore, the time in which we live directly encourages us to oversimplify life through individual technologies. It is necessary to keep abreast with the apparent enemy and keep it as a good friend. However, despite the fact that a confident enthusiasm inspires individual teachers to use today's modern equipment or at least they have a knowledge of these technologies, many do not even know how they could reasonably use smart devices in the classroom or especially what to do for the preparation of a lesson. It is precisely these technologies that encourage us to include them in school because they can make learning more attractive and motivate students. They make learning more interesting. A teacher who has a good command of the equipment uses these kinds of softwares to help them in the classroom and facilitate training.

In a young population, the number of owners of smart devices constantly increases, especially in terms of smartphones (mobile phones). This situation is very similar in Europe and the United States of America (eMarketer, 2014). We expect that sales of these devices will continue to rise. Thus, the introduction of technology in teaching and learning in information and communication technologies are necessary to support and seek ways to achieve them.

The aim of this article is to introduce some tools of smart devices that could be used in teaching and learning. We compare the programs with their advantages and disadvantages cons and introduce some possibilities that these tools offer. You need to understand that shyness from the use of smart devices in teaching and learning is in place

2 RESEARCH PROBLEM

2.1 *Smart devices in education*

The number of electronic devices continues to grow. Devices such as tablets, book readers, and especially the smartphones, are usually used mainly for consumption of media content, which attracts more and more people. In the first half of 2014, the number of these smart devices increased by 61% compared with the first half of last year. According to the estimates from research, for example, smartphones have already been owned by 2.8 billion people, which is 31% of the population. According to a demographic point of view, the owners of smart electronic devices over-represented people aged 20 to 24 years, mostly in higher education (MediaGuru, 2014). The research focuses mainly on smartphones because these devices usually prevail in society. Although the companies are still quite popular, the sale of tablets and perhaps electronic books has fallen considerably. According to current research, the tablet market this year only grew by 6.5%. In doing so, earlier estimates were around 12.1% (Stančík, 2014). In this trend, fewer tablets are sold, but more smartphones now are sold from a larger display than in previous years. They compete mostly 7″–7.9″ Tablet PC (Stančík, 2014). Other surveys show that about 70% of mobile phones in America are used daily in everyday situations (monitoring reports, finding a way to a certain place to find public transport, getting advice or help in some situations, etc. (Pew Research Journalism Project, 2014). According to our research, the number of mobile devices (smartphones, tablets) in the Czech population is still increasing. Especially the mobile phones where different operating systems are concerned. The most favoured devices in The Czech Republic are those with Android and iOS operating systems (Maněna, Rybenská & Špilka, 2014).

This article aims at finding out what tools could be effectively used as smart devices in educational fields.

2.2 *Methodology*

In previous research (Maněna, Rybenská, & Špilka, 2014), we used a special questionnaire and found that modern technologies were used in large numbers even in primary schools and played an important role in everyday life. We slowly started to get them into education. Now we know what types of equipment is most often found in schools. So, our task is to look for tools that can enable teachers to effectively integrate smart devices into the learning process. However, these instruments, of course, impose some requirements. It is necessary for them to be usable on smartphones, to be free (open source or free license). They also should be easy to use and user-friendly.

As a research method, we used comparative analysis. Through the comparison of selected tools and analysis of their functions, we found a suitable software for use in the educational field. We used all the tools to test the laptop and the phablet Samsung Galaxy Note N7000. We used the web browser Google Chrome.

For the selection of technologies and tools we chose the following criteria:

- Available for free (at least the basic version)
- Can be used on a computer, smartphone and tablet
- Availability of documentation
- Support for SVG,
- Manageability with your fingers (no stylus needed).

3 COMPARISON OF SELECTED TECHNOLOGIES AND TOOLS

3.1 *JointJS*

This is a JavaScript library designed to create and interact with charts and graphs. It includes a library Vectorizer, which is designed for easier handling with SVG (Scalable Vector Graphics), but it can be used separately. It also includes a Geometry library, which includes geometric operations. It was expanded and empowered. JointJS allows you to create interactive diagrams or diagrammatic tools directly in the browser. The generated diagram can be fully interactive. This tool facilitates the creation of visual tools and building applications (JointJS, 2014). The smartphone as well as the notebook is a fully functional tool.

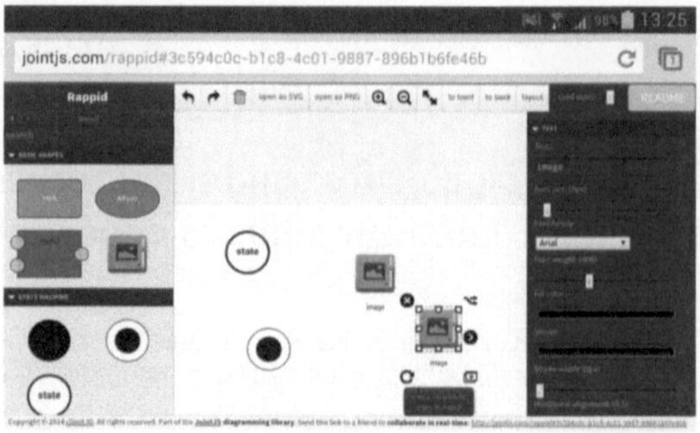

Figure 1. Demonstration of the tool JointJS on phablet Samsung Galaxy Note N7000.

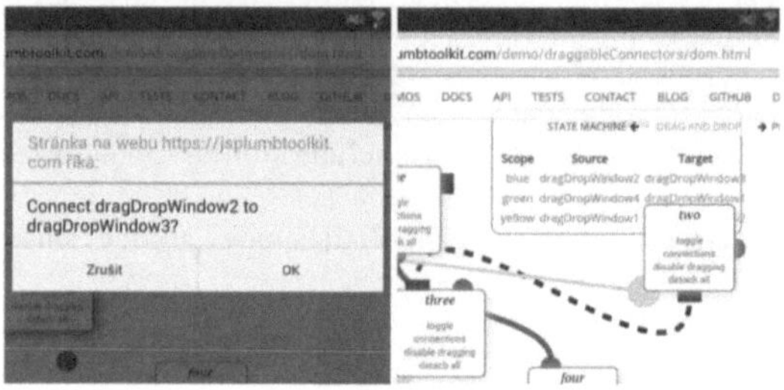

Figure 2. The picture on the left connects the two windows to the window right's third sample generated diagram.

It is certain that the laptop is a slightly better handling tool and its formation diagram than the smartphone. However, due to its screen size phablet (5.3″) and a stylus, which makes working with this software relatively easy, even if it is necessary to calculate with a smaller display of individual elements. Although, if you set the phone to 'landscape I' there is the ability to create accurate diagrams as well as the notebook (Figure 1). This very positively evaluates the fact that the tool is not necessary to download and install, and you can work with it immediately via a computer or smartphone. As for the license, the tool is free. It is licensed by Mozilla Public License version 2.0 (JointJS, 2014).

3.2 *jsPlumb*

This tool, which is intended mostly for developers, provides a simple visual element to display network diagrams. It is a tool with open source licenses (jsPlumb, 2014), fully functional, even on smartphones (Figure 2).

jsPlumb can be used in teaching, for example in networks (which are more suitable for older students or teachers).

3.3 *Draw2D touch*

This is a modern HTML5 JavaScript library that is designed for visualization and interaction with the diagram and graphs. With this tool it is possible to create tools and visual languages of all kinds very easily. It

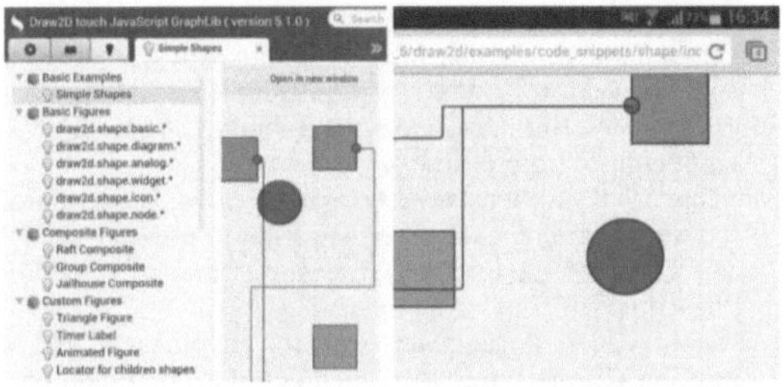

Figure 3. On the left sample is an overall view of the formation of the diagram on your smartphone. The right view is of the sample diagram of the smartphone landscape. In both cases it is possible to work with the program. When using the stylus it is easier to work.

Table 1. Comparison tools to create diagrams.

	JointJS	Draw2D touch	jsPlumb
SVG	+	+	+
Working in Google Chrome	+	+	+
Mozilla Firefox	+	+	+
Safari	+	+	+
Internet Explorer 9	+	+	+
Free for use	+	–	+
HTML5	+	+	+
JavaScript	+	+	–
Responsive web design	+	+	+
Open source or free	+	–	+
Official Tutorial or Support	+	+	+

works quickly when rendering SVG diagrams. It works in web browsers such as Google Chrome, Mozilla Firefox, Safari (so I can use it on iPad, iPhone), IE8, IE9, and IE10. The operation is simple. You can create a diagram by hand using individual elements and when it automatically writes the code. The smartphone is again quite easy to use inmanipulating individual fields, but everything is also fully functional as on the computer (Figure 3).

The disadvantage of this tool is that has to be paid for. Community Version for 4.99 Euro and Commercial Version for 499 Euro (Draw2D touch, 2014). Although the first variant is not too expensive and includes, for example, technical support for the whole year for some schools, it is almost impossible to purchase a similar software.

3.4 *SVG – edit*

This is an online vector graphics editor that works in almost any web browser. It uses JavaScript, HTML5, CSS and SVG. It draws freehand drawing curves, geometric patterns, and shapes to be coloured; it can work with layers; zoom in and out of preview, and pick a

Figure 4. Example of applications on smartphone Samsung Galaxy Note N7000. Drawing, as well as other kinds of modifications can be made relatively easily. The disadvantage may be the size of the telephone or similar device. The drawing is done by hand.

colour dropper bottom from the window. It is possible to preset the colour of the line, its thickness, and the opacity or coverage. There are also more advanced properties such as various types of lines or corner shapes. It is possible, of course, to set the shape of pen drawings blur and many others. Scripture can choose from a wide range or search for it on the computer. SVG drawings can also contain bitmaps that are embedded with the URL to the source file (Jelinek, 2011). It allows you to export the generated image formats PNG, JPEG, BMP and WebP. The big advantage is that there is no installation required, this tool can be used directly in a web browser. It is supported by the browser Google Chrome, Opera, IE version 6 (even if it is necessary to install plugin) and also Safari (it can be used on devices with iOS (SVG-edit, 2014). can be used on devices with iOS (SVG-edit, 2014). The big advantage is its localization in different languages, and it includes full Czech language support. The smartphone version works without any problems, but again there is a struggle with a certain sizes and a possible discomfort during operation of the instrument (Figure 4).

The great advantage of this tool is the fact that apart from a web browser and an internet connection, it really does not need anything. This tool can be considered almost as a standard for vector graphics on the Internet. It is an open source project.

4 CONCLUSION

The technology, based on a responsive site that works on a computer, notebook, and at the same time also on the tablet or smartphone, is available, as pointed out in our research. It is not only about open source or free technologies, there are also many professional tools (such as JavaScript library for modern web Snap.svg (Snap.svg, 2014). These professional technologies are designed for smartphones but they are often more expensive and are not affordable for schools. Instruments that are available for free are usually functional, well maintained, and user friendly. Still we do not have such a tool available that fully meets the requirements of the teacher. Such an instrument includes the promotion of responsive web, and also the option to edit vector and bitmap graphics. Through the instrument we can create a diagram and graphs, but it can also create mental maps. Utility would require all our work to be in the cloud (continue to synchronize with mobile devices and account for it). A single program similar to that would not be necessary to install and operate via the web browser. The license of such a software tool will be limited to schools. Instruments of this type would facilitate the work of many teachers and also students.

REFERENCES

Draw2D touch. 2014. Draw2D touch. Online. http://www.draw2d.org/draw2d/index.html

eMarketer. 2014. Smartphone Users Worldwide Will Total 1.75 Billion in 2014. Online. http://www.emarketer.com/Article/Smartphone-Users-Worldwide-Will-Total-175-Billion-2014/1010536

Introduction. 2014. Svg-edit. Online. https://code.google.com/p/svg-edit/

Jelínek, L. 2011. SVG-edit: Vektorový editor v prohlížeči. Linux Expres. Online. http://www.linuxexpres.cz/software/svg-edit-vektorovy-editor-v-prohlizeci

JointJS: JavaScript Diagramming Library. 2014. JointJS. Online. http://jointjs.com/

Maněna, V., K. Rybenská & R. Špilka. 2014. Research of mobile technologies used by students in primary school as a basis for the creation of teaching materials. AETIE 2014, In print.

Mediaguru. 2014. Počet tabletů se za rok zdvojnásobil na 1, 4 milionu. Online. http://www.mediaguru.cz/2014/08/pocet-tabletu-se-v-cesku-za-rok-zdvojnasobil-na-14-milionu/#.VBhsqBarYhM

Mobile Technology Fact Sheet. 2014. Pew Research Journalism Project. Online. http://www.pewinternet.org/fact-sheets/mobile-technology-fact-sheet/

JsPlumb. 2014. JsPlumb. Online. https://jsplumbtoolkit.com/demo/home/dom.html

Snap.svg. 2014. Snap.svg. Online. http://snapsvg.io/

Stančík, M. 2014. IDC: tabletů se v tomto roce prodá méně. ITBiz. Online. http://www.itbiz.cz/zpravicky/idc-tabletu-se-v-tomto-roce-proda-mene

Stančík, M. 2014. NPD: růst poptávky po tabletech kvůli větším smartphonům zpomaluje. ITBiz. Online. http://www.itbiz.cz/zpravicky/npd-rust-poptavky-po-tabletech-kvuli-vetsim-smartphonum-zpomaluje

Future Communication Technology and Engineering – Chan (Ed.)

Key technology research into the resource access framework of the Internet of Things (IoT)

C. Li & C.C. Liu
Institute of Network Technology, BUPT, Beijing, China

ABSTRACT: Nowadays a variety of sensors appear in all aspects of our real life. The data that these sensors obtain is collected by the Internet of Things (IoT) system. However, the current IoT systems work for some IoT specified resources but do not have the ability to extend themselves. In this paper, a resource access framework based on OSGi is designed to access all kinds of IoT resources. And this framework has strong extendibility.

1 INTRODUCTION

The first appearance of the Internet of Things was in the late 1990s (Gershenfeld et al. 2004). As a total information technology domain, it attracted the attention of governments, innovators and related enterprises, when it came to the world.

Especially World Summit on the Information Society held in Tunis by International Telecommunication Union in 2005 gave the 'Internet of Things' a clear definition. At the same time, it also released an 'ITU Internet report 2005 – the Internet of Things' to the scientific community, this report details the characteristics of the IoT, the key technologies, problems to be solved, and future developments.

With the rapid development of science and technology, especially the progress of sensor researches, a large number of intelligent sensing devices are used in all aspects of production and life. As an important part of the new generation of information technology, most countries regard the development of the IoT industry as a national strategy. In today's economy, IoT plays an important role in the IoT industry.

However, the scale of the IoT is becoming larger and larger, more and more sensors are being used by the IoT system (Akyildiz et al. 2002). Most of the time the sensors use their own built-in message format for data exchange with the IoT system. And there are various data transfer protocols (TCP, Modbs, H-700 and so on) and transmission (serial or network port) used by sensors. How to efficiently and easily manage sensors, get sensors' data, and parse data become urgent problems to be solved.

2 ANALYSIS AND DESIGN

2.1 *IoT resource classifications*

Under normal circumstances, IoT resources refer to an IoT sensor. However, in certain scenarios, the sub-system used to collect data from sensors will be treated as an IoT resource too.

There are so many kinds of IoT resources. And they have different messaging protocols, and different access styles. So it is difficult to model these IoT resources (Dobrev et al., 2002). But we have classified the device as follows.

2.1.1 *Classification according to working layers*

Physical layers: normally, most IoT sensors work under physical layers and deliver messages through serial ports such as common Modbus protocol. In this protocol, a master corresponds to one or more slaves. During message interaction, the master will assign addresses to slaves and send broadcast information to all slaves.

Network layers: sensors working on network layers will send data to IoT systems by access to a LAN or the internet. There are a lot of protocols working on network layers. The most common is the IP proto-col, TCP protocol (Forouzan et al., 2010).

2.1.2 *Classification according to response styles*

Active requests: this kind of IoT resource will send data to an IoT system regularly when its parameter is set. All the IoT system needs to do is to make sure its receiver is in a listening state and the link with the IoT resource is in a smooth state.

Passive responses: some IoT resources will not send data to an IoT system until the IoT system sends a data request message to it. Usually, we can find this interactive mode in a subsystem. Because subsystem is a middleware system, in a subsystem, sensors will not connect to IoT systems directly, but sends data to a subsystem instead. So, if an IoT system wants data, it must send a data request message to the subsystem.

2.1.3 *Classification according to message packaging times*

Package once: when an IoT resource packages its data to one protocol, we can say that that package is in

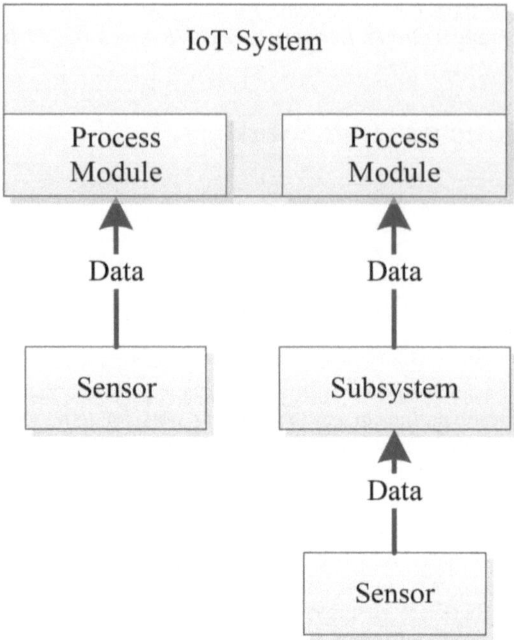

Figure 1. The current IoT system structure.

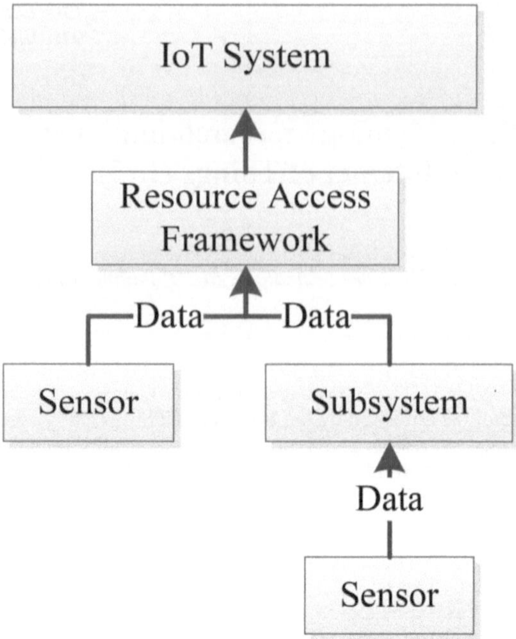

Figure 2. The position of a resource access framework.

once mode. In this mode, data is only packaged by a protocol, and the IoT system just needs to resolve the message packet once to get the final data.

Multiple Package: this is a complicated data transmission mode. IoT resources will package its data more than once. For example, IoT resources use protocol A to encapsulate data and get a middle message, and then packaged the middle message to protocol B and get a final message. Then resources will send a final message to the IoT system. When the IoT system receives a message, it will use protocol B to analyse the message and get the middle data; then it will use protocol A to analyse the middle data. By doing this step, the IoT system can obtain the final data.

According to the analysis above, we need a new framework. It can extend its protocol collection dynamically. Then the framework can resolve the data sent from the sensor by using this new protocol. And the framework can select a correct protocol from the protocol collection to resolve the message when a new sensor sends the message to our framework.

2.2 *Framework detail*

2.2.1 *Technology background*

The current IoT system uses a customization of the system according to how the sensor will access it. In this kind of system, whenever a new IoT resource needs to be added, the relevant technical staff have to write special processing modules, including data receiving and data resolving.

In Figure 1, when a new IoT resource (sensor or subsystem) is arrives, the main work of the IoT system is: 1, to finish the processing module for receiving and re-solving data from this resource. At the same time, a data convert module is needed to be done if necessary. 2, stop this system and load the new module to system. 3, modify the system to recognize the new module. 4, start the system again.

Obviously, here exist many deficiencies. First, the reusability of the module is not good. A respective module is needed when a new resource is needed to add to the system, even though this kind of resource has existed in the system. The duplication of work wastes a lot of effort. Second, it is hard to manage all modules in the system. We must stop the system if we want to add or remove a module from it. In a real-time system, stopping the system is forbidden. Third, because each module is only responsible for its own data, collected data can't be processed together. Once an IoT system wants to change its data format, all the modules in the system should be modified to adapt to a new data format. Last, the maintenance of the system is poor. When we need to update a module, we have to stop the system and modify a lot of codes.

2.2.2 *Framework structure*

In summary, we can pull the resource access module out of the IoT system (Eggins et al., 1997). Then this module can be regarded as a middleware system and we call it a resource access framework.

In Figure 2, the IoT system will focus on the processing of data, rather than spending time in managing resources or receiving data from resources. In this way, the coupling of the system is reduced greatly. And the resource access framework will focus on how to manager resources effectively and how to improve the reusability of protocols.

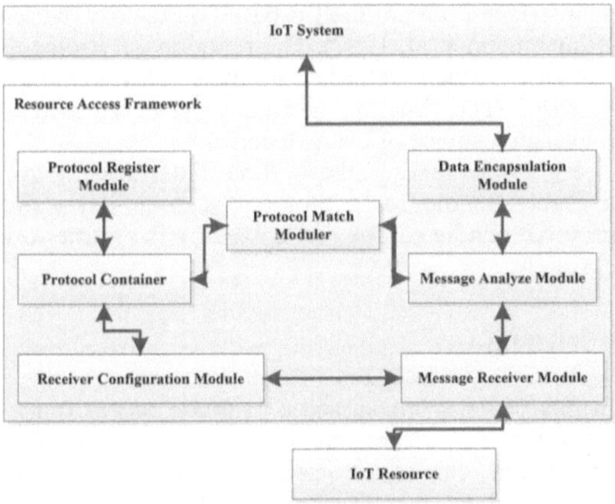

Figure 3. The structure of the resource access framework.

In order to modify a resource access framework when it is running, the framework should be based on OSGi.

Open Service Gateway Initiative (OSGi), or known as dynamic module system based on Java, is a basic framework for module development. By using OSGi, we can install, update, or delete a module without stopping the resource access framework. It is an important feature for implementing our resource access framework. The OSGi frameworks used most are Equinox, Felix, Spering-DM, etc. Equinox is suggested for use, because it has a close relationship with the Eclipse development environment, and most of Java software is developed under Eclipse. Furthermore, by using Eclipse and Equinox, we can configure the environment variables easily and accelerate the development of the resource access framework.

The figure above is a resource access framework based on Equinox.

According to the features of Equinox, we need to transfer a protocol to a bundle, so that protocol can be added to the protocol container in a resource access framework dynamically.

When a new IoT resource is using new protocol to package data, we carry out the steps below:

1. install this protocol bundle to the framework. When this bundle has been installed, the framework can access it in a protocol container.
2. initialize and receive the configuration module according to the connected attribute of the IoT resource. Then the message receive module begins to receive data from the IoT resource.
3. message receive module sends a message collected from the IoT resource to the message analysis module.
4. the message analysis module resolves the message by using a corresponding protocol stack from a protocol match module. It then transfers data resolved from the message to a data format module.
5. data format modules will format data from message analysis modules into the right format and send the data to the IoT system.

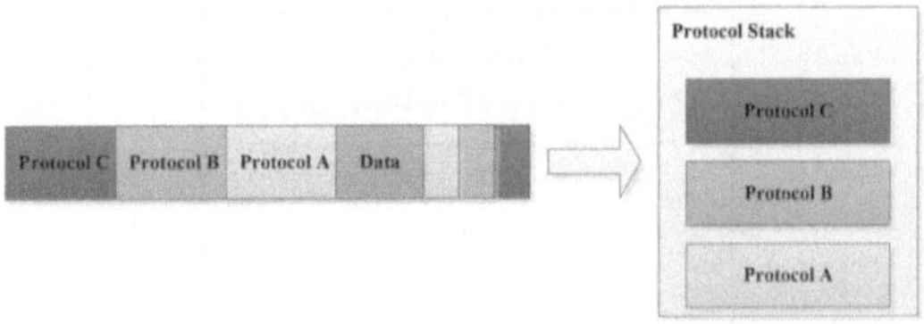

Figure 4. Structure of a protocol stack.

```
Auto adaption (Message, protocol stack)
for protocol in all protocols
   if protocol is used
      continue;
   else if protocol is right protocol
      Message = use protocol to analyse Message;
      Add protocol to protocol stack;
      Set protocol status to used;
      if this is final level
         return protocol stack;
      else
         return auto adapter (Message, protocol stack);
protocol stack = NULL;
return protocol stack;
```

Algorithm 1. Auto-Adaptation algorithm.

2.2.3 *Protocol stack*

In order to better handle a case of data, packaged with multi-protocols by IoT resources, we use protocol stack concepts to replace protocol concepts (Braden et al., 2003). A protocol stack has not just one protocol, but also one or more protocols. So a protocol stack can resolve a message that is packaged with multi-protocols.

From Figure 4, we can see the structure of a protocol stack. If the data is packaged by three protocols (Protocol A, Protocol B, Protocol C), the protocol stack will contain three of them. When resolving this data message, the protocol stack will use protocol C, protocol B, protocol A in that order.

In many cases, the protocols that a new IoT resource used exist in a protocol container. We just need to configure connect attributes simply if we know the names of the protocols. On the other hand, a re-source access framework can make a protocol stack by using an auto-adaptation algorithm in protocol match module if we cannot determine which protocols are used by this IoT resource.

2.2.4 *Encapsulate protocol on a bundle*

To allow the resource access framework recognize and use the protocol correctly, we set some rules to encapsulate the protocol. Namely, when a new protocol needs to be added to our framework, it must be transferred to a bundle first.

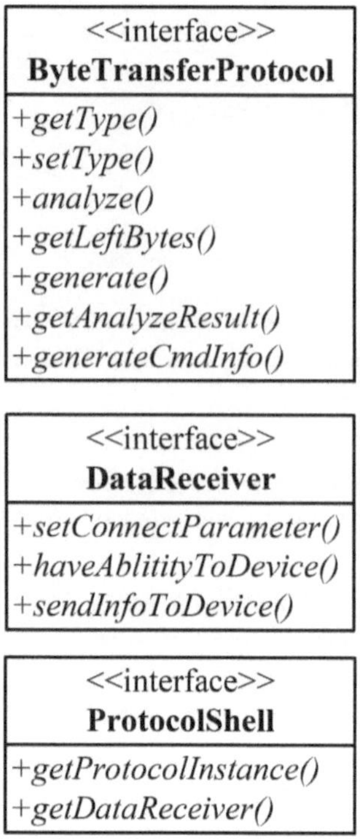

Figure 5. Interface of the protocol bundle.

A protocol bundle can be defined as in Figure 5.

Interface byte transfer protocol is used to encapsulate the protocol and our framework will use it to resolve messages. Because every protocol has its own name, the function set type/get type can be used to set and get the protocol's name. Function analysis will resolve the message passed from the framework, and return the re-solved result (success or fail). The function get LeftBytes will return the left raw bytes to be resolved by other protocols. This function generates is used to generate a feedback message, because some protocols need an ACK message when the server has received their message. The function get analysed result will return the real data in a message. The function generateCmdInfo will generate a control message according to the parameter passed from the framework. Thus, our framework can receive the data from the IoT resource and has the ability to control it.

An interface data receiver can encapsulate all kinds of connection types. By using this interface, a resource access framework can connect with IoT resources conveniently. The function Set Connect Parameter will set specific parameters for a connection, such as baud rate, stop bits, check bits in serial port or TCP port and address. The function haveAbilityToDevice will notify the framework that whether this protocol can connect to the IoT resource directly or not. This is because the protocol in the bottom has rights to connect with the IoT resource. The function sendInfoToDevice is used to send a message passed from the framework to the IoT resource.

Interface ProtocolShell is based on factory pattern (Hannemann et al. 1997). The function get Protocol-Instance is used to return an instance of Byte Transfer Protocol. The function getDataReceiver is used to return an instance of Data-Receiver.

By implementing these three interfaces above, a usable bundle for a protocol is done. Then this protocol bundle can be recognized by the framework correctly.

3 SUMMARY

In this paper, we designed a resource access frame-work that is independent of the IoT system. This frame-work can adjust itself to adapt to a new IoT resource when it is running. And with data format module, we can send easily data to different IoT systems by a setting format rule according to its requirement. This greatly increases the scope of the use of our framework.

ACKNOWLEDGEMENTS

This research is supported by the National Grand Fundamental Research 973 Program of China under Grant No. 2011CB302506, 2012CB315802; National Key Technology Research and Development Program of China (Grant No. 2012BAH94F02); National High-tech R&D Program of China (863 Program) under Grant No. 2013AA102301; National Natural Science Foundation of China under Grant No. 61132001, 61372115); Program for New Century Excellent Talents in University (Grant No. NCET-11-0592); Project of New Generation Broad band Wireless Network under Grant No. 2014ZX03006003, 2011ZX03002-002-01, 2012ZX03005008-001; The technology development and experiment of innovative network architecture (CNGI-12-03-007).

REFERENCES

Akyildiz I F, Su W, Sankarasubramaniam Y, et al. 2002. A survey on sensor networks. *Communications magazine*40 (8): 102–114.

Braden R, Faber T, Handley M. 2003. From protocol stack to protocol heap: role-based architecture. *ACM SIGCOMM Computer Communication Review* 33(1): 17–22.

Dobrev P, Famolari D, Kurzke C, et al. 2002. Device and service discovery in home networks with OSGi. *Communications Magazine, IEEE*40 (8): 86–92.

Eggins S, Slade D. 1997. *Analysing casual conversation*. London: Cassell.

Forouzan, Behrouz A. 2010. *TCP/IP protocol suite*. Boston: McGraw-Hill Higher Education.

Gershenfeld N, Krikorian R, Cohen D. 2004. *The Internet of things*. Scientific American291 (4): 76.

Hannemann J, Kiczales G. 2002. Design pattern implementation in Java and AspectJ. *ACM Sigplan Notices*37 (11): 161–173.

Future Communication Technology and Engineering – Chan (Ed.)
© 2015 Taylor & Francis Group, London, ISBN: 978-1-138-02777-0

A cucumber leaf disease classification system based on an embedded system

Y.W. Tian & X. Chen
College of Information and Electrical Engineering, Shenyang Agricultural University, Liaoning Shenyang, China

ABSTRACT: This paper describes the design of an embedded image acquisition and processing system for the classification of a cucumber leaf disease. The system will integrate image acquisition, processing, and display, and it has the advantages of having a small volume, low cost, and good portability. The other advantages are that it can truly achieve nondestructive testing. The system used the processor S3C2440A ARM9 system and a ported Linux operating system. Then the USB external contact scanning device was employed to achieve the cucumber leaf image acquisition results. The cucumber's diseased leaves image processing algorithm was studied by using an embedded system. The final result of the disease classification in the cucumber leaves was displayed through an LCD display.

Keywords: embedded, Linux, CIS scanning, image processing.

1 INTRODUCTION

In the process of growth and development, crops are often affected by various adverse factors, resulting in the occurrence of disease. Any study of the disease occurrence and development, crop disease epidemics and forecasts, and the effects of disease prevention, requires an understanding of the degree of occurrence of any plant diseases. Visibly, fast accurate measurements of degrees in crop diseases are very important.

Many domestic and foreign researchers in image processing applications in agricultural fields have made a thorough research, and put forward a lot of design schemes. At present, some foreign companies have developed a crop leaf analyser. It can be used t determine the extent of diseases in crops, but the analyser is expensive and difficult to repair. Therefore, it is not suitable for the conditions in our country, and is not conducive to local use [1]. With the rapid development of embedded technology, the purpose of making a construction of image acquisition and processing a system based on embedded technology can come true. The embedded system will overcome the image acquisition and processing system of the traditional structure, which is complex and expensive. Besides it has large volume, a big power consumption, and other shortcomings.

This paper presents a Samsung embedded USB based on a S3C2440A chip external scanning device of image acquisition and processing scheme. The use of this system can realize real-time detection in field crops.

2 SYSTEM DESIGN

2.1 *The working principle of the system*

The whole system consists of image acquisition, image processing, output, and the display of three modules. The system adopts CIS (Contact Image Sensor) scanning device for image acquisition, and then the image information can be transmitted to the ARM (Advanced RISC Machines) chip through the USB (Universal Serial Bus) interface. The ARM chip completes the image processing work, and finally displays it on an LCD (Liquid Crystal Display) based on a QT graphical interface.

2.2 *The hardware of the system*

The system uses a Samsung Corp production of the S3C2440A microprocessor as the master control chip. It is based on an ARM920T kernel 16/32 RISC embedded processor [2], and the highest working frequency can reach 533 MHz. It has an internal integration of the general LCD controller and a rich interface. The system also has the 64 MB FLASH and the SDRAM to meet the requirements of the storage operating system, and data [3]. The core board, as the hardware platform, realizes the image processing and display workings. Figure 1 is a hierarchical structure chart of the system's hardware.

2.3 *Systems software*

The software part includes the system's initialization program, the operating system kernel, the file system,

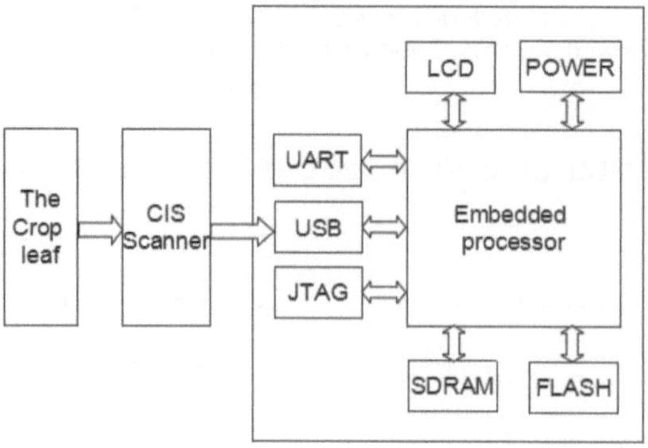

Figure 1. A hardware hierarchy diagram.

device driver, Qt image, and an application program interface. The system uses uboot as the system boot process [4], and chooses Linux as the operating system. The kernel of Linux is stable, powerful, and it supports a variety of hardware platforms [5]. At the same time adding the drivers in the driver directory is necessary. The system adopts yaffs2 file system, which includes a Qt interface, an application, relevant configuration files, and libraries etc.

3 THE IMAGE ACQUISITION MODULE

Image acquisition is the basis of the whole system which directly affects the stability of the overall system performance. The traditional image acquisition mostly uses camera acquisition. Leaves of some crops are long and wide, so the camera may not collect complete information from the image. It is also vulnerable to angle, light, and other factors. Some researchers use DSP technology, but the system is more complex and real-time image processing is difficult. Some researchers also pick crop leaves for laboratory research, but this will cause damage to crops. In view of this situation, this paper proposed a new acquisition module, a CIS scan module, which can complete the image acquisition function in high quality.

A CIS scanner, a contact image sensor, is a way to closely scan the object's surface. It has a USB interface on the bottom that is used for data transmission, and it can be connected through the interface with embedded devices. The highest optical resolution of the scanning device is 900 dpi. The scanning width is up to 21CM that can meet the requirements of broad leaf crop scanning. The crop leaf information was collected by using a CIS scanning device, and was transmitted via the USB interface to embedded devices for the corresponding image processing.

The scanning device uses contact scanning mode, so it needs to cling to the plant leaves to collect the information. Most of the blade surface is not smooth, so when you collect the crop, the surface needs to be covered with a layer of plastic plate. Figure 2 illustrates

Figure 2. Scanning image.

cucumber leaf image information that was collected by a scanning device. We can see that both the blade part and the lesion part are very clearly contrasted. The background and the leaf are clear enough to be separated easily.

4 THE IMAGE PROCESSING MODULE

The preparation of image processing software should first understand the basic principles and process of image processing. Image processing is a complex technology. It relates to many aspects of image preprocessing and image analysis. The processing flow and an algorithm using a different application background are not the same. The procedure of cucumber leaf image processing is as follows.

4.1 *Image analysis*

After image acquisition, a normal step is required to preprocess the image. Its purpose is to reduce the image noise and to enhance the contrast between the target and its background. The traditional methods of image collecting are often unsatisfactory, for example the crop lesion edge is too vague. With the scanning device close to the leaf, this would not be the case. So you can skip the image pretreatment process [6].

Because the acquisition of the image is in colour, the image processing algorithm of colour image is more complicated. However, basing it on the characteristics of the image, we can use a simpler algorithm for the processing. Figure 2 shows the collected images of a cucumber leaf, and you can see mostly green. However, the colour of these spots are yellow, and the background colour is white. Each part of the image statistics of the RGB value are shown in Table 1.

From the statistics you can see that the background and the blade part of the RGB value are different and do not overlap. Three channels can easily separate the background and the cucumber leaf. G value and B value of the blade and the lesion of the cucumber have many overlapping parts. However R value of the blade and the lesion can easily be separated. This analysis can also be seen through the following three pictures Figures 3 and 5.

Table 1. Cucumber leaf image RGB value statistics.

	Background	Between background and the blade edge	Leaf blade	Between leaf blade and the lesion edge	Lesion
R	180–255	80–180	30–80	80–100	100–150
G	180–255	80–180	60–100	60–120	50–110
B	180–255	80–180	20–50	40–80	30–80

Figure 3. R channel image.

Figure 4. G channel image.

Figure 5. B channel image.

4.2 *Image segmentation*

Image segmentation divides an image into the characteristics of the region and then extracts the targets of interest. There are many image segmentation algorithms, but they have two character images, based on

Figure 6. Separate images of diseased leaves.

pixel level: discontinuity and similarity. The specific methods include the threshold segmentation method, edge detection, and a region growing method. This design uses the threshold segmentation method.

Thresholding is a region segmentation technique, where the gray level is divided into two or a plurality of gradation range. It mainly uses the different image of the target object and the background image to select an appropriate threshold. Judging by the pixels in an image, we can see whether it can meet the requirements of the characteristics of the threshold to determine that the pixels in the image belong to the target area or belong to the background region. Then the two value images are generated [7].

It is very important to work out the threshold if you use the threshold segmentation method. The threshold can be determined according to the histogram. According to statistics, the gray level histogram of the image has three peaks and there are two valleys between the peaks. We can take the low point of gray value as a threshold to separate the diseased leaves and spots respectively. The mathematical expression is: $f(i, j) >= T$, $f(i, j) = 255$, $f(i, j) <= T$, $f(i, j) = 255$.

We can separate the diseased leaves by using the threshold T1 and separate the lesion by using the threshold T2. Images after separation were shown in Figures 6 and 7.

5 DISEASE CLASSIFICATION

Classification of crop disease hazard classes can be used to represent the lesion area percentage of the total leaf area. The number of pixels in the area of lesion is proportional to the lesion area, and the number of pixels in the area of the diseased leaf is proportional

Figure 7. Separate images of lesions.

to the diseased area of the leaf. So we can use the ratio of the number of pixels in the leaf's area and the diseased lesion area to represent the degree of harm in the plant's disease. The formula is as follows: $K = A1/A = N1/N$. A is the total area of diseased leaves and A1 is a lesion area. N is the number of pixels of the diseased leaf area and N1 is the number of pixels in the lesion area.

From the segmented image we can know that the number of diseased leaf pixels is the number of black pixels in Figure 6 and the number of lesion pixels is the difference between the number of black pixels in Figures 6 and 7. So we can calculate the ratio K. Crop disease harm degree can be calculated to grade standard according to the traditional classification methods concerning disease records. Each kind of disease can be divided into five levels. We can determine the diseases' harm grades according to the value K. What has been collected in this paper is the image of cucumber downy mildew. According to the classification standard of cucumber downy mildew, without spots, for Class 0, grade 1 for $k < 5\%$, 5.1%–10% for grade 2, grade 3 for 10.1%–25%, 25.1%–50% for grade 4, and 50.1% above for grade 5.

The image size is 750*550, a total of 412,500 pixels. The number of black pixels in the separate images of diseased leaves is 165,893 and the number of black pixels in the separate images of the lesion is 158,992. The difference between them is 6,901, namely the lesion pixel number. The above formula can be used to calculated K values, $k = A1/A = N1/N$, $k = 0.0416$. Its value is below 5%. According to disease classification this should be grade 1.

6 CONCLUSION

This design is based on embedded technology as the key to realizing crop leaf image acquisition and processing. It provides convenient, fast, and practical crop foliar information for cultivation and management for farmers. It is important for developing high quality agriculture and accelerating agricultural modernization. At the same time, it can replace the traditional crop image processing system with an embedded technique, which is cheap; has good portability, and is suitable for use in the field. The organic combination of embedded technology and image processing technology in the field of agriculture can promote the development of embedded technology in the field of agriculture and realize the innovation of agricultural technological information.

REFERENCES

[1] Tian You-Wen. Modern image recognition technique in the diagnosis of crop disease [M]. China Agriculture Press, 2010(07): 130–131

[2] Zhao Zhi-Ya, Wang Ze-Yong. Design of the Embedded System Framework for Image Collection [J]. Application of computer system, 2009, 18(12)

[3] Zheng Kai-Hua, Zhang Xu-Dong. Image Acquisition and Display System Based on S3C2440 [J]. Application of computer system, 2011, 20(8)

[4] Gan Ke-Liang. Image Acquisition and Display Based on ARM9 and Embedded System [J]. The industrial control computer, 2011, 24(8)

[5] Wu Cai-Zhang. ARM-Based Image Acquisition and Display System Design [J]. Technology and application of automation, 2010, 29(3)

[6] Zuo Fei. Technology and practice of Visual C++ digital image processing [M]. Electronic Industry Press, 2014 (03): 273–274

[7] Yang Shu-Ying, Bian Dian-Ying.The program design of VC++ image processing [M]. Tsinghua Press, 2005 (01): 143–14

Future Communication Technology and Engineering – Chan (Ed.)
© 2015 Taylor & Francis Group, London, ISBN: 978-1-138-02777-0

A study of the humanistic education of Chinese college teaching from the aspect of computer science and technology

S.Z. Liu
School of Humanities, Baoshan University, Baoshan Yunnan, China

ABSTRACT: With the application of computer science and technology in class teaching, the education methods of Chinese college teaching has entered an era of the Internet, which brings a new development opportunity to study the humanistic education of these educational institutions. Chinese is an important communication tool for human cultures. Applying humanity to Chinese teaching, and building and improving the inner world of the students are the eternal subject of Chinese college teaching. However, as computer science and technology is introduced into class, it seems that it is not so simple. Computer science and technology provide new teaching means and tools for modern teaching. Based on computer science and technology, the integration of these tools and humanity into Chinese college teaching in order to serve quality-oriented education better, are the important points that this paper will explore.

Keywords: computer, computer technology aspect, Chinese college teaching, humanity, education study.

1 INTRODUCTION

With the rapid development and popularization of Chinese computer technology, how to take advantage of modern and new teaching media to give full play to the humanistic factors in Chinese education is the important direction of our study. What is humanity education? The nature of humanity education is the education of human nature with humanistic spirits at its core, by emphasizing self-realization and awareness. It teaches people how to accept themselves happily, accept others gladly, adjust their relationship with others, the family, the group, the country and with nature, attach importance to cultivation from the outside to the inside, and undertake one's social responsibility. As a result, humanity education plays a fundamental role in college education. It needs a gradual influence of traditional Chinese culture in the class, nutrition of the culture contents, like honesty, diligence, insistence and politeness, etc. and influence of the class culture atmosphere. Chinese college teachers are assumed to undertake these important tasks to cultivate the humanistic spirit of students.

2 APPLICATION OF COMPUTER SCIENCE AND TECHNOLOGY IN HUMANITY EDUCATION IN CHINESE COLLEGE TEACHING

2.1 *Explore humanity education deeper, by taking advantage of information technology*

Dig deeper for the humanity connotation of textbooks in Chinese colleges by using information technology. With multimedia means, and present knowledge with new teaching science and technology such as video, audio, powerpoint (PPT), and so on, teachers can use a courseware to explain in detail the literature works in current Chinese textbooks, enrich class content, as well as overcome the space and time limitation of inner class teaching A modern class like this will not only enhance the learning confidence of students but can also make students understand and remember more quickly. Based on rich and substantial courseware contents, students can convert indirect knowledge into direct knowledge. Listening to background music and reading records of texts, or watching colourful documentaries enables students to put themselves in the writer's shoes, and communicate with great minds. These different emotional experiences will influence students by humanistic spirits which emphasizes the human value, and improve moral sentiments and humanity qualities. Cultivate students' humanity and humanism unnoticeably in vivid and interesting class. This helps students to pay more attention to their pursuit of the true, the good and the beautiful.

2.2 *Highlight important points and difficult points utilizing modern multimedia teaching methods*

Introducing humanity education by using multimedia appropriately in Chinese college teaching can embody abstract and general words into specific ones, which not only helps the students understand the deeper contents of texts, but also makes students relaxed when learning. For example, when learning a text, Injustice of Dou'e by Guan Hanqing, we can play some grave background music to accompany Dou'e's monologue, which will affect the students ability to grasp the feelings better in the text, and understand the helplessness of Dou'e, her hatred of that time when the rich oppressed the poor, her bravery to puisue the happy life and her struggle against the evil. In this way, we can animate the class atmosphere, relate knowledge to students' thinking, interact with them, and arouse their enthusiasm to learn.

2.3 *Help students understand figures with the help of multimedia to improve their morality*

By presenting the words and acts of excellent figures in the text by adopting information technology, the colours, pictures, and videos will draw the student's attention. In a class like this, the learning interests of students can be easily triggered. With the application of information technology and multimedia, the students can be guided to open books to understand characters in texts better, think more, and feel the advantages of learning more strongly. They can find the contents about figures at home and abroad, in modern times or in ancient times with the help of strong internet technology. Then students can throw themselves into learning actively, grasping knowledge, understanding figures, feeling times, and getting inspiration of humanistic spirits which can make them compare the past and the present and have a correct attitude toward them under the guidance of contents on the multimedia screen.

3 PROVIDE COLOURFUL AND RICH KNOWLEDGE THROUGH MODERN INFORMATION TECHNOLOGY

3.1 *Find carrier of connotation of humanistic spirits*

We need to feel Chinese literature, language, or culture, with humanity and freedom, with our heart. The writer has heard that when reading works of Tao Yuanming, you feel like a world without utilitarianism; when reading works of Li Bai characterized by freedom and easiness, you feel free and relaxed; when reading works of Wen Tianxiang, you feel uplifted. We can see that literature plays an important role in raising people's sprit and cultivating one's soul. We need to accept the fact that literature can enrich our mind, and we can communicate with others mentally in Chinese. From the aspect of computer technology, Chinese teachers should pay more attention to the selection of literature works and collection of knowledge points. College teachers can guide students to communicate with excellent cultures of human's with the help of information technology, which will provide support to approval of humanistic connotation. It is worthwhile to think deeply the influence of literature works on students. From literature works, the teachers can teach students the correct way to deal with personal relationships and noble life ideal and life wisdom.

3.2 *Take advantage of multimedia to cultivate the ability to observe*

Chinese textbooks are full of sentiments, and Chinese teachers should pay more attention to cultivate the ability to feel things, know things through close observation, and convert appearance into substantial contents. That is to say, you can only feel humanistic connotation through observation. One advantage of computer multimedia teaching is visual presentation, which can provide a foundation for the students to apply their observing ability, enable them to acquire more knowledge in a limited time, inspire them, trigger their activity to learn, help students to grasp knowledge, form skills, develop ability, and improve moral quality. Without any doubt, it brings vitality and vigour into Chinese college teaching.

3.3 *Nurture the many abilities of students by applying multimedia teaching*

In Chinese teaching, teachers must train students' ability to feel culture connotation from the aspect of visualization and emotion features of literature works. Chinese teachers undertake not only the tasks to teach books but also the tasks to cultivate people. They have to pay more attention to cultivation of the inner quality of the students and training of healthy personality besides knowledge in the book. Grasp correct teaching method, provide and create scenes for the students, nurture independence and activity of students, improve students' understanding of humanity and improve students' interests in Chinese and feeling of humanity with true feeling method by taking advantage of information technology. Then you can find honesty, kindness, and beauty in the text world. Thinking in terms of images is crucial to understand the spirits of text contents, which needs emotional experience. Nurture presentation skills and develop thinking ability through multimedia teaching, help the students understand the humanistic ideas expressed by the writer, and deepen student's understanding and feeling combination with modern politics, pay more attention to training of thinking method of students, give more chances to the students to express their opinions, ideas and raise their problems and doubts. Only in this way, can we cultivate sound and healthy personalities of students, who play important roles in the progress and development of the whole society.

4 THE HELP OF MODERN NETWORK INFORMATION TECHNOLOGY TO ENABLE STUDENTS TO CONDUCT A VARIETY OF CHINESE ACTIVITIES INDEPENDENTLY

4.1 *Advocating students found reading associations independently by the use of network*

Chinese is the carrier of national culture, while, Chinese college education faces with college intellectuals who are about to enter into the society, therefore, it is very important to carry forward the essence of traditional national culture and modern national culture from generation to generation. In addition, one of the important points is students' independent thoughts and accumulation of abilities and skills. The first step for students toward humanistic spirit independently is encouraging students to start reading groups, recording their meditation while reading and conducting experience, sharing conferences, and sharing resources with each other. Teachers could be like listeners, allowing students to interpret works in a plain way or in a deep-going way; encouraging students to express themselves freely and share their happiness in discovery.

4.2 *Encouraging students to use the network platform for Chinese composing*

Humanistic research of Chinese college education exists to improve the level of the humanistic quality. Not only paying attention to the conducting of humanistic knowledge, but also paying attention to the real improvement of students' humanistic quality, it focuses on the cultivation of the humanistic spirit. In Chinese teaching, teachers must be people-oriented, emphasizing humanistic education in Chinese teaching, and bringing students' subjective initiative into play. In the computer information age, we cannot refuse to take a ride on network technology. Blog, microblog, WeChat, etc. are rich and colourful social platforms, and they are also rich and colourful display platform for literary grace. Under the guidance of teachers, students can create Chinese-related tribes. Teachers should fully respect students' personalities, providing free creating space for students. Teachers must be good at building a democratic, equal, loose, and free study atmosphere and environment, letting students fully express their individuality, allowing humanistic spirit to permeate fully into Chinese teaching. The wide spread of the atmosphere of youth and progress is a reflection of both the 5,000 years of China's increasingly prosperous national tradition, and the humanistic spirit.

5 APPLICATION OF INFORMATION TECHNOLOGY SHOULD ADAPT TO CHINESE COLLEGE TEACHING RULES

5.1 *Pay attention to the imaginary training, by which students can form the image that is described in the text in their minds*

As a discipline research, Chinese teaching has its own laws. At the end of the last century, schools had introduced new types of teaching methods, such as the projector, etc., which can be classified as the precursor of multimedia teaching. In recent years, with the popularity of computers and the rapid development of the information highway in China, the innovation of the teaching class in keeping up with the pace of the times is taken for granted. In the classrooms of language and culture, teachers' display of a variety of information by intuitional teaching aids has a negative effect on students' independent thinking to some extent. In addition, a Chinese lesson is a palace of knowledge and emotional intelligence. It carries an inheritance of national culture, and recognition and understanding for international culture from one generation of students to the next. At the outward-looking age, it is necessary for us to realize the significance of humanity for humanistic emancipation. Humanistic education in Chinese college courses requires students' thinking and development of their self-imagination. Therefore, in the use of information technology, teachers must pay attention to the cultivation of learners' reproductive imagination.

5.2 *Pay attention to students' feelings*

Various education resources can be integrated in a very short time through the use of the multimedia computer. After all, multimedia is a kind of novel, interesting, lively, and funny high-tech. Therefore, teachers should try to avoid the students' attention being scattered and interfered, and ensure that students at the back of the classroom can hear clearly, without ignoring students' feelings.

5.3 *Stimulate students interest in learning*

The application of modern information technology should be moderate, the precondition of which is that the use of information technology is in accordance with students' psychological development regulation, and the application of humanistic being ensured. It is not that each lesson is suitable for using of an information website for teaching. Teachers should know how and what to select and judge. It is not denying the fact that information technology plugs in wings for Chinese teaching. However, the excessive use of modern teaching means, such as PPT works, focusing on the supply of informational knowledge points such as time feature for the creation and publication of works, the creation intention of the author, the social influence of works, people's evaluation, the background of the text etc., could easily overlook the embodiment of humanity. Modern information technology as an auxiliary means for Chinese teaching; on the one hand, it enriches the teaching content and presents the knowledge in a limited time for students as much as possible, on the other hand, it provides a possibility for students to be lazy. Often, students lose their enthusiasm and initiative for self-thought.

6 IMPORTANCE OF HUMANITY EDUCATION IN CHINESE LANGUAGE TEACHING IN COLLEGES AND UNIVERSITIES UNDER THE PERSPECTIVE OF INFORMATION TECHNOLOGY

6.1 *Establishing a harmonious relationship between teachers and students*

In the new relationship between teachers and students, teachers are not only student's mentors, but also their friends. In the age of change, teachers should keep pace with the times; change themselves from a class dominator to an equal communicator, from a knowledge indoctrinator to a learning guider, and from a one-way propagator to an interaction partner. An harmonious relationship between teachers and students needs teachers' meticulous love and respect for students all the time. To establish a democratic, equal, and harmonious relationship between teachers and students, Chinese teachers should listen to the heart-felt wishes of students, be tolerant to students, respect students' individual characters, enter into the inner world of students, and pay attention to their inner communication with students. The humanity characteristics of Chinese education require teachers' updating of educational concepts and improvement of education quality.

6.2 *Innovating teaching form; enriching humanistic education*

Chinese education and teaching under an information technology background, should not only instil students with subject knowledge, but also allow students getting humanistic education in study activities, by cultivating them in an invisible way. The traditional education apparently cannot follow the step of age. Therefore, teachers should reform teaching organizing patterns, and enhance students' activeness, using the advantages of information technology multimedia. Teachers should cultivate students' ability for self-study and corporative research by encouraging students to do self-study on the base of Chinese textbooks and to corporate and research to complete tasks. By enriching humanistic education patterns from all aspects and teaching through lively activities, the humanistic education of the student can be achieved subtly.

7 CONCLUSIONS

With the development of information technology and the popularity of multimedia teaching, former teaching methods appear monotonous and outdated. Therefore, with a background of information technology, the application of multimedia in teaching classes is an inevitable outcome of the age and development trend. Chinese education is one part of humanistic subject and is a humanism-applied discipline. One main aim of Chinese education is teaching student how to be an upright person. The responsibility of Chinese education is to transmit knowledge, cultivate people, and promote students' further and more comprehensive understanding about the humanistic spirit in the textbooks. From the perspective of teaching with information technology, college Chinese education should keep pace with times. Meanwhile, follow the rules of Chinese teaching and ensure the humanistic spiritual education and help students establish right world outlook, outlook on life and values.

REFERENCES

Chi Yanjie. How humanism was implemented in the Chinese teaching—taking the class teaching of "the father and son in the earthquake" for example [J]. Journal of The Chinese Society of Education. 2012(07)

Zhao Zibo, Han Tianyong. Modern education technology and the education teaching reform in colleges and universities [J]. Science & Technology Information (academic). 2006(04)

Huang Cui. With the life-oriented Chinese teaching to stimulate students' inspiration of wisdom—talking about the effective application of life-oriented teaching activities in Chinese teaching classes [J]. Modern Reading (education). 2013(02)

Chen, Hsiang; Rolf T. Wigand and Michael Nilan (2000), "Exploring Web users' optimal flow experiences", Information Technology 13 (4), pp. 263–281.

Zhang Jianwei. From Cognitivism to Constructivism [J]. Journal of Beijing Normal University (Social Science), 1996 (4), pp. 75–80.

Future Communication Technology and Engineering – Chan (Ed.)
© 2015 Taylor & Francis Group, London, ISBN: 978-1-138-02777-0

Cryptanalysis and improvement of a directed digital signature scheme using a non-interactive public-key distribution system

J.R. Sun & S.C. Lin
Department of Computer Science and Information Engineering, Asia University, Taiwan, ROC

M.S. Hwang
Department of Computer Science and Information Engineering, Asia University, Taiwan, ROC
Department of Medical Research, China Medical University Hospital, China Medical University

ABSTRACT: Tseng et al. modified a non-interactive public-key distribution system and also proposed several relevant applications based on the Maurer and Yacobi scheme. One of these applications is a directed digital signature scheme, which only uses a signer's identity to verify the signature. No pair of users can either interact with each other, or access the public files of the system. However, their scheme cannot prevent a forgery attack. Any user can forge a signature to get any message successfully and declare that it is generated by other users. In this paper, a securer directed digital signature scheme operated by using a non-interactive public-key is proposed.

1 INTRODUCTION

In 1984, Shamir (Shamir, 1984) proposed the idea that let each user's identification information (eg name, address, and E-mail address) be his/her public-key in a public-key cryptosystem. Any pair of users have the ability to verify the signature without exchanging secret and public keys, without keeping key directories, and without using the services of a third party (Hwang et al., 2014, Hwang et al., 2006). Based on Shamir's identity-based concept, many systems have been proposed which named identity-based public-key systems (Hwang & Chen, 1994, Koyama & Ohta, 1987, Tsujii & Itoh, 1989).

Before using most of the proposed schemes, the receivers must identify themselves to the trusted authority centre and then obtain their secret keys from the trusted authority centre. In other words, the users' IDs are not really their identification information. Furthermore, the trusted authority centre determines the secret keys of all users.

In 1991, Maurer and Yacobi proposed an identity-based non-interactive public-key distribution system based on a novel trapdoor one-way function of exponentiation modulo of a composite number. The latest version of the system was released in 1996 (Maurer & Yacobi, 1996). In using this system, any user's signature is verified by his/her identity information without executing an interactive public-key authentication. They adopted the squaring and the Jacobi symbol methods to guarantee each identity number corresponds to his/her identity information. In their scheme, there were no public keys, certificates or other information except the identity of each user. Many ID-based schemes have been proposed (Awasthi & Lal, 2007, Choo, 2005, Farash & Attari, 2014, He et al., 2013, Kar, 2013, Li et al., 2006, Li et al., 2007, Mohaisen et al., 2009, Quan et al., 2013, Xu et al., 2009, Zhao & Zhang, 2011, Zhao, 2008).

There are many types of Digital Signature Schemes (DSS): Directed DSS (Lu & Cao 2006, Tseng & Jan, 1998), Ring DSS (Awasthi & Lal 2007, Liu & Wong 2008, Qin, et al., 2014, Xiong, et al., 2011, Zheng, et al., 2007), Threshold DSS (Chang, et al., 2004, Chang, et al., 2005, Hwang & Chang 2005, Mashhadi 2013, Tzeng, et al., 2004, Xiong, et al., 2010, Yang, et al., 2004), Blind DSS (Chakraborty & Mehta 2012, Hwang, et al., 2002, Hwang, et al., 2003, Lee, et al., 2003, Moldovyan 2011, Moldovyan & Moldovyan 2010), Proxy DSS (Ma, et al., 2013, Mashhadi 2012, Tian & Huang 2012), Identity-based DSS (Kar 2014, Wang, et al., 2012), and Certificateless DSS (He 2009, Wang, et al., 2009, Zhou, et al., 2014).

Tseng et al., (Tseng & Jan, 1998) improved the squaring method based on the Maurer and Yacobi scheme and also proposed a directed signature scheme, a user identification scheme, and a conference key distribution system. In this article, the security of their directed signature schemes will be discussed and an improved scheme will be proposed.

The remainder of the paper is organized as follows: Section 2 is a brief review on Tseng et al.,'s directed digital signature scheme; an attack on Tseng et al.,'s directed digital signature scheme is proposed in Section 3; in Section 4, a secure directed digital signature scheme is proposed; finally, Section 5 is a brief conclusion of the paper.

2 REVIEW OF TSENG ET AL.'S SCHEME

Tseng et al., proposed a directed digital signature scheme in which only the specified receiver can verify the signature. The scheme consists of three phases: the system initialization phase, the user registration phase, and the directed digital signature phase. The three phases are described as follows.

System Initialization Phase: The trusted authority generates system parameters as follows: N denotes the product of four primes pj which are between 60 and 70 decimal digits, where the numbers (pj-1) are odd and pair-wise relatively prime; e denotes an integer in $Z^*_{\varphi(n)}$ and the secret value d is satisfied by the equation $e \cdot d \equiv 1(\bmod \varphi(N))$.

User Registration Phase: When a user U_A who wants to join the system presents his/her unique identity ID_A, the trusted authority randomly chooses a secret number t from $Z^*_{\varphi(n)}$, where φ is the Euler's totient function. Then, the trusted authority computes s_A to U_A in secret, where g is a primitive element in GF(pj), $v \equiv t^{-1}(\bmod \varphi(N))$. A one-way function, $h(\cdot)$, is published by the trusted authority:

$$s_A \equiv e \cdot t \cdot \log_g(ID_A{}^2)(\bmod \varphi(N)) \tag{1}$$

After having generated the parameters mentioned above, the trusted authority publishes $\{N, g, e, h(\cdot)\}$ and keeps $\{p_1, p_2, p_3, p_4, t, v, d\}$ secret. The legal user, U_A publishes $\{ID_A\}$ and keeps $\{s_A\}$ secret.

The Directed Digital Signature Phase: Assume the legal user U_A wants to sign a message M and send it to the specific receiver U_B. The signature generating procedure is as follows. First of all, U_A chooses a random integer k in Z_N^* and computes the signature (r, s) for message M, where:

$$r = g^{e \cdot k}(\bmod N) \tag{2}$$

$$s = (ID_B{}^2)^{s_A \cdot h(M,r)} \cdot g^k(\bmod N) \tag{3}$$

Secondly, U_A sends (M, r, s) to U_B. Upon receiving (M, r, s), U_B can verify the signature of message M by the following equation:

$$s^e = (ID_A{}^2)^{e \cdot s_B \cdot h(M,r)} \cdot r(\bmod N) \tag{4}$$

If the above verified equation holds, the signature for message M by U_A is verified. The specific verifier only verifies the validity of the signature for message M according to the unique identity of the signer.

3 ATTACK ON TSENG et al.,'S SCHEME

In this section, it will be proved that Tseng et al.,'s directed digital signature scheme cannot withstand a forgery attack. Assume U_A generates the signature (r, s) for the message M and sends it to a specified receiver U_B. In general, the values $\{M, r, s\}$ are only generated by U_A. However, in their scheme, the user U_B can forge the signature (r', s') for the message M' and declare that the signature (r', s') is generated by the U_A. The U_B forges the signature (r', s') as follows:

Step 1. Choose a random integer k' in Z_N^*
Step 2. Compute the signature (r', s') for message M' as follows:

$$r' = g^{e \cdot k'}(\bmod N)$$
$$s' = (ID_A{}^{s_B})^{2 \cdot h(M',r')} \cdot g^{k'}(\bmod N)$$

The signature (r', s') for message M' is verified correctly as follows:

$$(s')^e = (ID_A{}^2)^{e \cdot s_B \cdot h(M',r')} \cdot g^{k'}(\bmod N)$$
$$= ((ID_A{}^{s_B})^{2 \cdot h(M',r')} \cdot g^{k'})^e(\bmod N)$$

Therefore, U_B can forge the signature for any message and declare that the signature is generated by U_A. In other words, the user can forge the signature and declare that it is generated by the other user. The problem is that any user U_A or U_B, can obtain $(ID_B^2)^{s_A}$ and $(ID_A^2)^{s_B}$ respectively by the following equation:

$$(ID_A{}^2)^{s_B} = g^{d \cdot v \cdot s_A \cdot s_B} = (ID_B{}^2)^{s_A} \tag{5}$$

4 THE PROPOSED SCHEME

There are also three phases in the proposed scheme: the system initialization phase, the user registration phase, and the directed digital signature phase. The system initialization phase and the user registration phase of the proposed scheme are the same as in Tseng et al.'s Scheme (Tseng & Jan, 1998).

The Directed Digital Signature Phase: Assume the legal user U_A wants to sign a message$M$and send it to the specific receiver U_B. The signature generating procedure is as follows. First of all, U_A chooses a random integer k in Z_N^* and computes the signature (r, s) for message M, where:

$$r = g^k(\bmod N) \tag{6}$$

$$s = (ID_B{}^2)^{s_A \cdot h(M,r)} \cdot g^{d \cdot k}(\bmod N) \tag{7}$$

Secondly, U_A sends (M, r, s) to U_B. Upon receiving (M, r, s), U_B can verify the signature of the message M by the following equation:

$$s^e = (ID_A{}^2)^{e \cdot s_B \cdot h(M,r)} \cdot r(\bmod N) \tag{8}$$

If the above verified equation holds, the signature for message M by U_A is verified. The specific verifier only verifies the validity of the signature for message M by the unique identity of the signer.

The proposed scheme is securer for resisting a forgery attack. U_B cannot forge the signature for any message or declare the signature is generated by U_A. The key point is that any user, U_B, cannot obtain the signatures in Equation (7) because U_B does not know the U_A's secret value d.

5 CONCLUSION

It is shown that there is a leak in Tseng et al.'s directed digital signature scheme. Their scheme cannot withstand a forgery attack because anyone can forge the signature for any message and declare that it has been sent from another specific user. Hence, a securer directed digital signature scheme is proposed in the paper, which is operated by a using non-interactive public-key distribution system.

ACKNOWLEDGEMENTS

This study was supported by the National Science Council of Taiwan under grant 103-2221-E-468-026, 103-2622-E-468-001-CC2, and 103-2622-H-468-001-CC2.

REFERENCES

Awasthi, A.K. & Lal, S. (2007). ID-based ring signature and proxy ring signature schemes from bilinear pairings, *International Journal of Network Security* 4(2): 187–192.

Chakraborty, K. & Mehta, J. (2012). A stamped blind signature scheme based on elliptic curve discrete logarithm problem. *International Journal of Network Security* 14(6): 316–319.

Chang, T.Y., Hwang, M.S., Yang, W.P. (2005). An improvement on the lin-wu (t, n) threshold verifiable multi-secret sharing scheme. *Applied Mathematics and Computation* 163(1): 169–178.

Chang, T.Y., Yang, C.C., Hwang, M.S. (2004). A threshold signature scheme for group communications without a shared distribution center. *Future Generation Computer Systems* 20(6): 1013–1021.

Choo, K.K.R. (2005). Revisit of McCullagh-Barreto two-party ID-based authenticated key agreement protocols, *International Journal of Network Security* 1(3): 154–160.

Farash, M.S. & Attari, M. A. (2014). A pairing-free ID-based key agreement protocol with different PKGs, *International Journal of Network Security* 16(3): 168–173.

He, D., Khan, M.K., Wu, S. (2014). On the security of a RSA-based certificateless signature scheme. *International Journal of Network Security* 16(1): 78–80.

He D, Zhao W, and Wu S. (2013). Security analysis of a dynamic ID-based authentication scheme for multi-server environment using smart cards. *International Journal of Network Security* 15(5): 282–292.

Hwang M.S. & Chang, T.Y. (2005). Threshold signatures: current status and key issues. *International Journal of Network Security* 1(3): 123–137.

Hwang M.S., Chong S.K., and Chen T.Y. (2010). Dos-resistant ID-based password authentication scheme using smart cards, *Journal of Systems and Software* 83: 163–172.

Hwang, M.S., Lee, C.C., Lai, Y.C. (2002). Traceability on low-computation partially blind signatures for electronic cash. *IEICE Fundamentals on Electronics, Communications and Computer Sciences* E85-A(5): 1181–1182.

Hwang, M.S., Lee, C.C., Lai, Y.C. (2003). An untraceable blind signature scheme. *IEICE Transactions on Foundations* E86-A(7): 1902–1906.

Hwang M.S., Lo J.W., and Lin S.C. (2004). An efficient user identification scheme based on ID-based cryptosystem, *Computer Standards & Interfaces* 26(6): 565–569.

Hwang T and Chen J. L. (1994). Identity-based conference key broadcast system. *IEE Proceedings – Computer Digital Technology* 141(1): 57–60.

Kar, J. (2013). ID-based deniable authentication protocol based on Diffie-Hellman problem on elliptic curve. *International Journal of Network Security* 15(5): 357–364.

Kar, J. (2014). Provably secure online/off-line identity-based signature scheme for wireless sensor network. *International Journal of Network Security* 16(1): 29–39.

Koyama, K. & Ohta, K. (1987). Identity-based conference key distribution system, in *Advances in Cryptology, CRYPTO'87*, Lecture Notes in Computer Science, LNCS 293: 194–202.

Lee, C.C., Hwang, M.S., Yang, W.P. (2003). Untractable blind signature schemes based on discrete logarithm problem. *Fundamenta Informaticae* 55(3–4): 307–320.

Li F., Xin X., and Hu Y. (2006). ID-based signcryption scheme with (t,n) shared unsigncryption, *International Journal of Network Security* 3(2): 155–159.

Li Z., Chong C.F., Hui L.C.K., Yiu S.M., Chow K. P., Tsang W.W., Chan H.W., Pun K.K.H. (2007). An attack on libert et al.,'s ID-based undeniable signature scheme. *International Journal of Network Security* 5(2): 220–223.

Liu, J.K. & Wong, D.S. (2008). Solutions to key exposure problem in ring signature. *International Journal of Network Security* 6(2): 170–180.

Lu, R. & Cao, Z. (2006). A directed signature scheme based on RSA assumption. *International Journal of Network Security* 2(3): 182–186.

Ma, C., Xue, K., Hong, P. (2013). A proxy signature based re-authentication scheme for secure fast handoff in wireless mesh networks. *International Journal of Network Security* 15(2): 122–132.

Mashhadi, S. (2013). A novel non-repudiable threshold proxy signature scheme with known signers. *International Journal of Network Security* 15(4): 274–279.

Mashhadi, S. (2012). A novel secure self proxy signature scheme. *International Journal of Network Security* 14(1): 22–26.

Maurer U.M. & Yacobi Y. (1996). A non-interactive public-key distribution system. *Designs, Codes and Cryptography* 9(3): 305–316.

Mohaisen A., Nyang D.H., Lee K.H. (2009). Hierarchical grid-based pairwise key pre-distribution in wireless sensor networks, *International Journal of Network Security* 8(3): 282–292.

Moldovyan, N.A. (2011). Blind signature protocols from digital signature standards. *International Journal of Network Security* 13(1): 22–30.

Moldovyan, N.A. & Moldovyan, A.A. (2010). Blind collective signature protocol based on discrete logarithm problem. *International Journal of Network Security* 11(2): 106–113.

Qin, Z., Xiong, H., Li, F. (2014). A provably secure certificate based ring signature without pairing. *International Journal of Network Security* 16(4): 278–285.

Quan Q., Xiao C.J., and Zhang R. (2013). Grid-based data stream clustering for intrusion detection. *International Journal of Network Security* 15(1): 1–8.

Shamir A. (1984). Identity based cryptosystems & signature schemes, in *Advances in Cryptology, CRYPTO'84*, Lecture Notes in Computer Science, LNCS 196: 47–53.

Tian, M. & Huang, L. (2012). Breaking a proxy signature scheme from lattices. *International Journal of Network Security* 14(6): 320–323.

Tseng Y.M. & Jan J.K. (1998). ID-based cryptographic schemes using a non-interactive public-key distribution system. In *Proceedings of the 14th Annual Computer Security Applications Conference (IEEE ACSAC98)*: pp. 237–243.

Tsujii S & Itoh T. (1989). An ID-based cryptosystem based on the discrete logarithm problem. *Journal of Selected Areas in Communications* 7(4): 467–473.

Tzeng, S.F., Yang, C.Y., Hwang, M.S. (2004). A nonrepudiable threshold multi-proxy multi-signature scheme with shared verification. *Future Generation Computer Systems* 20(5): 887–893.

Wang, C., Long, D., Tang, Y. (2009). An efficient certificateless signature from pairings. *International Journal of Network Security* 8(1): 96–100.

Wang, Z., Wang, L., Zheng, S., Yang, Y., Hu, Z. (2012). Provably secure and efficient identity-based signature scheme based on cubic residues. *International Journal of Network Security* 14(1): 33–38.

Xiong, H., Qin, Z., Li, F. (2010). Identity-based threshold signature secure in the standard model. *International Journal of Network Security* 10(1): 75–80.

Xiong, H., Qin, Z., Li, F. (2011). A certificateless proxy ring signature scheme with provable security. *International Journal of Network Security* 12(2): 92–106.

Xu C., Zhou J., Xiao G. (2008). General group oriented ID-based cryptosystems with chosen plaintext security. *International Journal of Network Security* 6(1): 1–5.

Yang, C.Y., Tzeng, S.F., Hwang, M.S. (2004). On the efficiency of nonrepudiable threshold proxy signature scheme with known signers. *Journal of Systems and Software* 73(3): 507–514.

Zhao X. & Zhang F. (2011). A new type of ID-based encryption system and its application to pay-TV systems. *International Journal of Network Security* 13(3): 161–166.

Zhao Z.M. (2008). ID-based weak blind signature from bilinear pairings. *International Journal of Network Security* 7(2): 265–268.

Zheng, D., Li, X., Chen, K. (2007). Code-based ring signature scheme. *International Journal of Network Security* 5(2): 154–157.

Zhou, M., Zhang, M., Wang, C., Yang, B. (2014). CCLAS: A practical and compact certificateless aggregate signature with share extraction. *International Journal of Network Security* 16(3): 174–181.

Future Communication Technology and Engineering – Chan (Ed.)
© 2015 Taylor & Francis Group, London, ISBN: 978-1-138-02777-0

Cryptanalysis and improvement of the Li-Liu-Wu user authentication scheme

T.H. Feng
Department of Computer Science and Information Engineering, Asia University, Taiwan, ROC

M.S. Hwang
Department of Computer Science and Information Engineering, Asia University, Taiwan, ROC
Department of Medical Research, China Medical University Hospital, China Medical

W.Y. Chao
Department of Management Information Systems, National Chung Hsing University, Taiwan, ROC

ABSTRACT: It is important to authenticate the legitimacy of remote users over the public Internet. A Password-based Authentication Scheme (PBAS) is one of the schemes used to authenticate the legitimacy of remote users. PBASs have been widely deployed to verify the legitimacy of remote users. Recently, Chen et al. proposed a user authentication scheme using smart cards. However, Li, et al., have proved that their scheme is vulnerable to a forgery attack, a server spoofing attack, and a password guessing attack. Li, et al., also proposed a modified scheme to eliminate the security vulnerability. Unfortunately, we found that insecurity in their scheme also exists. In this article, we will prove their scheme is vulnerable to a password guessing attack. Finally, we will propose an improved scheme to eliminate the security vulnerability.

1 INTRODUCTION

It is important to authenticate the legitimacy of remote users over the public Internet. A Password-Based Authentication Scheme (PBAS) is one of the schemes used to authenticate the legitimacy of remote users. PBASs have been widely deployed to verify the legitimacy of remote users. In 2006, Tsai et al., (Tsai et al., 2006) classified password authentication schemes into three classes: RSA-based password authentication schemes (Hwang 1999, Shen, et al., 2003, & Yang, et al., 2004), El Gamal-based password authentication schemes (Hwang & Li 2000, Kumar 2004, & Yang, et al., 2003), and hash-based password authentication schemes (Kim & Koç 2005, Lee, et al., 2002, Lee, et al., 2013, Lin, et al., 2006, Mangipudi & Katti 2006, & Zhuang, et al., 2014).

There are many user authentication schemes that have been proposed (Sood, et al., 2011, & Tao & Adams 2008). Some of these schemes are using smart cards for storing user's security information (He, et al., 2011, Hwang et al., 2010, Hwang et al., 2005, Kumar, et al., 2011, Lee, et al., 2002, Ramasamy & Muniyandi 2012, Shen, et al., 2003-1, Tang, et al., 2013, Wang & Yang 2006, & Yang, et al., 2012). Some of these schemes are applied to multi-server environments (Feng, et al., 2014, He, et al., 2013, & Lin, et al., 2003). Some of these schemes are based on biometrics (Li, et al., 2010, Li & Hwang 2010, & Prakash 2014). Some of these schemes are based on neural networks (Li, et al 2001 & Lin et al., 2005). Some of these schemes are applied to mobile environments (Hwang, et al., 2002, Lee, et al., 2006, Li & Chu 2009, Liao, et al., 2006, & Wu, et al., 2005).

Recently Chen et al proposed a user authentication scheme using smart cards (Chen & Lee 2008). However, Li, et al., have proved that their scheme is vulnerable to a forgery attack, a server spoofing attack, and a password guessing attack (Li, et al., 2012). Li, et al., also proposed a modified scheme to eliminate the security vulnerability. Unfortunately, we found that insecurity in their scheme also exists. In this article, we will prove their scheme is vulnerable to a password guessing attack. Finally, we will propose an improved scheme to eliminate the security vulnerability.

The remainder of the paper is organized as follows. In Section 2, we briefly review Li-Liu-Wu's user authentication scheme. An attack on Li-Liu-Wu's user authentication scheme is proposed in Section 3. In Section 4, we propose an improved Li-Liu-Wu's user authentication scheme. Finally, we give a brief conclusion in Section 5.

2 REVIEW OF LI-LIU-WU'S SCHEME

There are three participants in Li-Liu-Wu's user authentication scheme (Li, et al., 2012): a Key Information Centre (KIC for short), a Server (S for short), and a User (U for short). The scheme involves three phases, namely the registration phase, the login phase, and the authentication phase (Li, et al., 2012).

Registration Phase: In this phase, the user U initially registers with the KIC over a secure communication channel. The main purpose of this phase is that the KIC generates and sends a smart card with the secure information (n, e, g ID_U, CID_U, S_U, h_U) to the user U. We denote these parameters in the smart card as follows:

$n = p \times q$, where p and q are two large prime numbers which are generated by the KIC.
e: a public key chosen by the KIC.
g: a primitive element in both F_p and F_q.
ID_U: an identity of user U which is generated by the KIC.
$CID_U = h(PW_U) \oplus h(ID_U \oplus d)$.
PW_U: the user's password which is chosen by the user U.
d: the private key, such that ed mod $\Phi(n) = 1$, where $\Phi()$ is the Euler's totient function and $\Phi(n) = (p - 1)(q - 1)$.
$S_U = h(ID_U \oplus d)^d$.
$h_U = g^d$.

Login Phase: In this phase, the user U inserts his smart card into a smart card reader in the computer and then inputs his password PWU. Next, the user's smart card generates and sends the login request message, $M1 = \{ID_U, X_U, Y_U, n, e, g, T_U\}$, to the server S. We denote these parameters in the login request message as follows:

$X_U = g^r$, where *r* is a random number which is generated by the smart card.
$Y_U = S_U h_U^{rh(h(ID_u \oplus d), T_u)}$, where T_U is the current time stamp.

Authentication Phase: In this phase, the server S verifies the authenticity of the login message M1 requested by the user U as follows:

1) The server S checks whether T_U is in the valid time interval of the current time or not.
2) The server S checks whether the following equation holds or not:

$$(Y_U)^e = h(ID_U \oplus d)\, X_U^{h(h(ID_u \oplus d),\, T_u)}.$$

3 CRYPTANALYSIS OF LI-LIU-WU'S SCHEME

In this section, we will show that Li-Liu-Wu's user authentication scheme (Li, et al., 2012) cannot withstand a password guessing attack when the user U loses his/her smart card. If an attacker steals or finds out a user's smart card details, and extracts the valuable stored information (n, e, g, ID_U, CID_U, S_U, h_U) in some ways (Messerges, et al., 2002), we show that Li-Liu-Wu's scheme cannot withstand a password guessing attack as follows:

Step 1. Obtain the login request message $M_1 = \{ID_U, X_U, Y_U, n, e, g, T_U\}$ by interception from the Internet between the user U and the server S.

Step 2. Guess a password PW'_U and obtain CID′ as follows:

$$CID'_U = CID_U \oplus h(PW_U)$$
$$= [h(PW_U) \oplus h(ID_U \oplus d)] \oplus h(PW'_U)$$

Step 3. The server S checks whether the following equation holds or not:

$$(Y_U)^e = CID'_U\, X_U^{h(CID'_u,\, T_u)} \quad (1)$$

If the above equation holds, the attacker guesses the correct password PWU. Otherwise, the attacker does not yet guess the correct password. The attacker repeats Steps 2 and 3 until the correct password is found.

We show that Equation (1) holds, implying that the attacker guesses the correct password as follows. The left side of Equation (1) is:

$$(Y_U)^e = h(ID_U \oplus d)\, X_U^{h(h(ID_u \oplus d),\, T_u)} \quad (2)$$

If the correct password is guessed, $h(PW'_U) = h(PW_U)$:

$$CID'_U = CID_U \oplus h(PW_U)$$
$$= [h(PW_U) \oplus h(ID_U \oplus d)] \oplus h(PW'_U)$$
$$= h(ID_U \oplus d)$$

The right side of Equation (1) is:

$$CID'_U = CID_U \oplus h(PW_U)$$
$$= [h(PW_U) \oplus h(ID_U \oplus d)] \oplus h(PW'_U).$$

The above equation is equal to Equation (2). Therefore, Equation (1) holds. However, if the attacker does not guess the correct password, $h(PW'_U) \neq h(PW_U)$:

$$CID'_U = CID_U \oplus h(PW_U)$$
$$= [h(PW_U) \oplus h(ID_U \oplus d)] \oplus h(PW'_U).$$

The right side of Equation (1) is:

$$CID'_U\, X_U^{h(CID'_u,\, T_u)}$$
$$= [h(PW_U) \oplus h(ID_U \oplus d)] \oplus h(PW'_U)\; X_U^{h([h(PWU) \oplus h(IDU \oplus d)] \oplus h(PW'U),\, T_u)}$$

The above equation is not equal to Equation (2). Therefore, Equation (1) does not hold. The attacker knows that it is an incorrect password and he/she needs to guess other passwords.

4 THE PROPOSED SCHEME

In order to eliminate the security vulnerability of Li-Liu-Wu's user authentication scheme, we will propose an improved user authentication scheme in this section. Like Li-Liu-Wu's scheme, there are also three participants: the KIC, a server S, and a user U; and three phases in the proposed scheme: the registration phase, the login phase, and the authentication phase.

Registration Phase: In this phase, the user U initially registers with the KIC over a secure communication channel. The main purpose of this phase is that the KIC generates and sends a smart card with the secure information (n, e, g, ID_U, CID_U, h_U) to the user U. These parameters in the smart card are the same as those in the registration phase of Li-Liu-Wu's user authentication scheme. The difference between the proposed scheme and Li-Liu-Wu's user authentication scheme is that the proposed scheme removes the parameter SU from the smart card in Li-Liu-Wu's user authentication scheme. The proposed scheme does not need the parameter SU for authentication.

Login Phase: In this phase, the user U sends a login request message to the server S whenever the user U wants to access some resources upon the server S. The login request message is produced by the following steps:

Step 1. The user U inserts his/her smart card into a smart card reader in their computer and then inputs his/her password PW_U.

Step 2. The user's smart card computes CID'_U as follows:

$$\begin{aligned} CID'_U &= CID_U \oplus h(PW_U) \\ &= [h(PW_U) \oplus h(ID_U \oplus d)] \oplus h(PW_U) \\ &= h(ID_U \oplus d) \end{aligned}$$

Step 3. The user's smart card generates a random number r and computes X_U as follows:

$$Y_U = CID'_U \, h_U^{rh(CID'_u, T_u)},$$

Step 4. The user's smart card computes Y_U as follows:

$$Y_U = CID'_U \, h_U^{rh(CID'_u, T_u)},$$

where T_U is the current time stamp.

Step 5. The user's smart card sends the login request message, $M_1 = \{ID_U, X_U, Y_U, n, e, g, T_U\}$, to the server S.

Authentication Phase: In this phase, the server S verifies the authenticity of the login message M1 requested by the user U as follows:

Step 1. The server S checks whether T_U is in the valid time interval of the current time or not.

Step 2. The server S computes CID'_U as follows:

$$CID'_U = h(ID_U \oplus d).$$

Step 3. The server S checks whether the following equation holds or not:

$$(Y_U)^e = (CID'_U)^e \, X_U^{h(CID'_u), T_u)}. \quad (3)$$

If so, the server S accepts the user login request, otherwise, the server S rejects the user U's login request. The remaining steps of the authentication phase are the same as those of the authentication phase in Li-Liu-Wu's scheme.

We show that Equation (3) holds implying that the legal user inputs a correct password as follows. The left side of Equation (3) is:

$$\begin{aligned} (Y_U)^e &= (CID'_U)^e \, h_U^{rh(CID'_u, T_u)}, \\ &= (CID'_U)^e \, X_U^{h(CID'_u, T_u)} \end{aligned}$$

The above equation is thus equal to the right side of Equation (3). Therefore, the server S verifies the legal user U.

The proposed scheme can avoid a password guessing attack when the user U loses his/her smart card. If an attacker steals or finds out a user's smart card details, and extracts the valuable stored information (n, e, g, ID_U, CID_U, h_U) in some ways (Messerges, et al., 2002), we show that the attacker cannot guess the correct password by Equation (3). In spite of the guessing password, whether correct or not, the verification Equation (3) always holds. Therefore, the attacker cannot obtain the correct password.

5 CONCLUSION

We have shown that there is a leak in Li-Liu-Wu's user authentication scheme. This scheme cannot withstand a password guessing attack when a user U loses his/her smart card. We also proposed an improved and secure user authentication scheme.

ACKNOWLEDGEMENTS

This study was supported by the National Science Council of Taiwan under grant 103-2221-E-468 -026, 103-2622-E-468-001-CC2, and 103-2622-H-468-001-CC2.

REFERENCES

Chen, T.H. & Lee, W.B. (2008). A new method for using hash functions to solve remote user authentication. *Computers and Electrical Engineering* 34: 53–62.

Feng, T.H., Ling, C.H., Hwang, M.S. (2014). Cryptanalysis of Tan's improvement on a password authentication scheme for multi-server environments. *International Journal of Network Security* 16(4): 318–321.

He D, Zhao W, and Wu S. (2013). Security analysis of a dynamic ID-based authentication scheme for multi-server environment using smart cards. *International Journal of Network Security* 15(5): 282–292.

He, D., Chen, J. Hu, J. (2011). Weaknesses of a remote user password authentication scheme using smart card. *International Journal of Network Security* 13(1): 58–60.

Hwang, M.S. (1999). A remote password authentication scheme based on the digital signature method. *International Journal of Computer Mathematics* 70(4): 657–666.

Hwang M.S., Chong S.K., and Chen T.Y. 2010. Dos-resistant ID-based password authentication scheme using smart cards. *Journal of Systems and Software* 83: 163–172.

Hwang, M.S., Lee, C.C., Yang, W.P. (2002). An improvement of mobile users authentication in the integration environments. *International Journal of Electronics and Communications* 56(5): 293–297.

Hwang, M.S. & Li, L.H. (2000). A new remote user authentication scheme using smart cards. *IEEE Transactions on Consumer Electronics* 46(1): 28–30.

Hwang, M.S., Lo, J.W., Liu, C.Y., Lin, S.C. (2005). Cryptanalysis of a user-friendly remote authentication scheme

with smart card. *Pakistan Journal of Applied Sciences* 5(1): 99–100.
Kumar, M. (2004). New remote user authentication scheme using smart cards. *IEEE Transactions on Consumer Electronics* 50(2): 597–600.
Kim, M. & Koç, Ç.K. (2005). A simple attack on a recently introduced hash-based strong-password authentication scheme. *International Journal of Network Security* 1(2): 77–80.
Kumar, M., Gupta, M.K., Kumari, S. (2011). An improved efficient remote password authentication scheme with smart card over insecure networks. *International Journal of Network Security* 13(3): 167–177.
Lee, C.C., Li, L.H., Hwang, M.S. (2002). A remote user authentication scheme using hash functions. *ACM Operating Systems Review* 36(4): 23–29.
Lee, C.C., Liu, C.H., Hwang, M.S. (2013). Guessing attacks on strong-password authentication protocol. *International Journal of Network Security* 15(1): 64–67.
Lee, C.C., Hwang, M.S., Liao, I.E. (2006). Security enhancement on a new authentication scheme with anonymity for wireless environments. *IEEE Transactions on Industrial Electronics* 53(5): 1683–1687.
Lee, C.C., Hwang, M.S., Yang, W.P. (2002). A flexible remote user authentication scheme using smart cards. *ACM Operating Systems Review* 36(3): 46–52.
Li, C.T. & Chu, Y.P. (2009). Cryptanalysis of threshold password authentication against guessing attacks in ad hoc networks. *International Journal of Network Security* 8(2): 166–168.
Li, C.T. & Hwang, M.S. (2010). An online biometrics-based secret sharing scheme for multiparty cryptosystem using smart cards. *International Journal of Innovative Computing, Information and Control* 6(5): 2181–2188.
Li, C.T. & Hwang, M.S. (2010). An efficient biometrics-based remote user authentication scheme using smart cards. *Journal of Network and Computer Applications* 33: 1–5.
Li, J., Liu, S., Wu, S. (2012). Cryptanalysis and improvement of a YS-like user authentication scheme. *International Journal of Digital Content Technology and its Applications* 7(1): 828–836.
Li, L.H., Lin, I.C., Hwang, M.S. (2001). A remote password authentication scheme for multi-server architecture using neural networks. *IEEE Transactions on Neural Networks* 12(6): 1498–1504.
Liao, I.E., Lee, C.C., Hwang, M.S. (2006). A password authentication scheme over insecure networks. *Journal of Computer and System Sciences* 72(4): 727–740.
Lin, I.C., Hwang, M.S., Li, L.H. (2003). A new remote user authentication scheme for multi-server architecture. *Future Generation Computer Systems* 19(1): 13–22.
Lin, I.C., Ou, H.H., Hwang, M.S. (2005). A user authentication system using back-propagation network. *Neural Computing & Applications* 14(3): 243–249.
Lin, C.W., Tsai, C.S., Hwang, M.S. (2006). A new strong-password authentication scheme using one-way hash functions. *International Journal of Computer and Systems Sciences* 45(4): 623–626.
Mangipudi, K.V. & Katti, R.S. (2006). A hash-based strong password authentication protocol with user anonymity. *International Journal of Network Security* 2(3): 205–209.
Messerges, T.S., Dabbish, E.A., Sloan R.H. (2002). Examining smart-card security under the threat of power analysis attacks. *IEEE Transactions on Computers* 51(5): 541–552.
Prakash, A. (2014). A biometric approach for continuous user authentication by fusing hard and soft traits. *International Journal of Network Security* 16(1): 65–70.
Ramasamy, R. & Muniyandi, A.P. (2012). An efficient password authentication scheme for smart card. *International Journal of Network Security* 14(3): 180–186.
Shen, J.J., Lin, C.W., Hwang, M.S. (2003). Security enhancement for the timestamp-based password authentication scheme using smart cards. *Computers & Security* 22(7): 591–595.
Shen, J.J., Lin, C.W., Hwang, M.S. (2003-1). A modified remote user authentication scheme using smart cards. *IEEE Transactions on Consumer Electronics* 49(2): 414–416.
Sood, S.K., Sarje, A.K., Singh, K. (2011). Inverse cookie-based virtual password authentication protocol. *International Journal of Network Security* 13(2): 98–108.
Tang, H., Liu, X., Jiang, L. (2013). A robust and efficient timestamp-based remote user authentication scheme with smart card lost attack resistance. *International Journal of Network Security* 15(6): 446–454.
Tao, H. & Adams, C. (2008). Pass-Go: A proposal to improve the usability of graphical passwords. *International Journal of Network Security* 7(2): 273–292.
Tsai, C.S., Lee, C.C., and Hwang, M.S. 2006. Password authentication schemes: Current status and key issues, *International Journal of Network Security* 3(2): 101–115.
Wang, R.C. & Yang, C.C. (2006). Cryptanalysis of two improved password authentication schemes using smart cards. *International Journal of Network Security* 3(3): 283–285.
Wu, H.C., Liu, C.Y., Chiou, S.F. (2005). Cryptanalysis of a secure one-time password authentication scheme with low-communication for mobile communications. *International Journal of Network Security* 1(2): 74–76.
Yang, C.C., Chang, T.Y., Hwang, M.S. (2003). The security of the improvement on the methods for protecting password transmission. *Informatica* 14(4): 551–558.
Yang, L., Ma, J.F., Jiang, Q. (2012). Mutual authentication scheme with smart cards and password under trusted computing. *International Journal of Network Security* 14(3): 156–163.
Yang, C.C., Yang, H.W., Wang, R.C. (2004). Cryptanalysis of security enhancement for the timestamp-based password authentication scheme using smart cards. *IEEE Transactions on Consumer Electronics* 50(2): 578–579.
Zhuang, X., Chang, C.C., Wang, Z.H., Zhu, Y. (2014). A simple password authentication scheme based on geometric hashing function. *International Journal of Network Security* 16(4): 271–277.

Future Communication Technology and Engineering – Chan (Ed.)
© 2015 Taylor & Francis Group, London, ISBN: 978-1-138-02777-0

Evaluation of publish/subscribe-based routing over Software-Defined Networking (SDN)

S.J. Wang & Y. Zhang
State Key Laboratory of Networking and Switching Technology, Beijing University of Posts and Telecommunications, Beijing, China

ABSTRACT: The objective of this work is to give an outline on the architecture and routing protocol of publish/subscribe models deployed under SDN. Examining the performance of SDN transporting publish/subscribe-based traffic under OpenFlow protocol, gives us a new way to apply publish/subscribe architecture. Our evaluation work focuses on studying the routing protocol according to the topology detected by floodlight and constructed by ourselves. Open Shortest Path First (OSPF) and Jackson's Theorem are included in the design of routing protocol. The main advantages and disadvantages of applying publish/subscribe mechanisms are shown under SDN.

1 INTRODUCTION

The main guiding objective in designing current Internet architecture is the end-to-end principle. The exchange of information within a publish/subscribe context is done by distributing available data through the set of publications. End points, which are interested in receiving such information, obtain it by a corresponding subscription. The transfer of data from the publisher to the subscriber is triggered upon the relevant match between the publication and the subscription [1].

In the last revision of the publish/subscribe system based on WSN (Web Services Notification), we compute the routing path by the topology which forms in the present Internet Protocol (IP) based Internet using OSPF, and transfer publish messages by socket knowing the IP address of the next receiver which has been calculated, ignoring the underlying link. It performs well in common situations, but it cannot avoid a traffic jam caused by itself or other transmissions. Therefore, we pay attention to a content-centric network.

The optimization of data delivery paths in topology management is a critical operation since it directly influences the efficiency of the entire system. The path from the publisher to the subscriber can be optimized with respect to a wide range of network parameters, and thus affects the performance of the system in different ways [3]. One of the most important performance measures of a data network is the average delay required to deliver a packet from origin to destination. Furthermore, delay considerations strongly influence the choice and performance of network algorithms, such as routing and flow control [4].

Queuing theory is the primary methodological framework for analysing network delay. Its usage often requires simplifying assumptions. Unfortunately, more realistic assumptions make meaningful analysis extremely difficult. For this reason, it is sometimes impossible to obtain accurate quantitative delay predictions based on queuing models. Nevertheless, these models often provide a basis for adequate delay approximations, as well as valuable qualitative results and worthwhile insights [4].

We consider that Jackson's Theorem is suitable for the publish/subscribe system. Jackson's Theorem assumes that: 1. Arrival processed from outside the network is Poisson; 2. At each queue, streams have the same exponential service time distribution and one single server; 3. Inter-arrival times and service times are independent. With this theorem, we no longer need to look at the short-term behaviours, but concentrate on the derivative of the long-term behaviours [5].

We collect information about the network by Floodlight [6], which is one of the implementations of OpenFlow [7], and we deliver our routing methods, which according to Jackson's Theorem, are calculated to each node and each node will invoke actions when comparison satisfies the received packet topic and its calculated next stops.

2 BACKGROUND AND RELATED WORK

The architecture of the present Internet was designed with a strong focus on the sender-oriented way of communication. However, over time, end-user requirements changed and the data content has become of central importance. This initiates the work towards content-centred communication [3]. We use Floodlight as the information collector and topology maintainer inside a group, and WSN as messaging architecture. The current work concerning these is as follows.

2.1 *Floodlight*

Floodlight implements OpenFlow protocol which is an open standard managed by Open Networking Foundation. The OpenFlow protocol connects controller software with network devices, thus enabling the Floodlight controller to command the switches.

The Floodlight Open SDN Controller is an enterprise-class, Apache-licensed, Java-based Controller [6]. It provides means to manage or direct the network that lies beneath based on source, destination, topic, or other features of a packet.

Floodlight can handle mixed OpenFlow and non-OpenFlow networks, which can manage a multiple "island" of OpenFlow hardware switches [6], and this feature gives us the chance to deploy it on the real networks.

2.2 *WSN*

The purpose of the Web Services Notification (WSN) National Standardization Technical Committees (TC) is to define a set of specifications that standardize the way Web services interact using "Notifications" or "Events." They form the foundation for Event Driven Architectures by using Web services. These specifications provide a standardized way for Web services, or other entities, to disseminate information to a set of other Web services, without having the prior knowledge to these other Web Services. They can be considered as defining "Publish/Subscribe for Web services." These specifications have many applications, for example, in the arena of system or device management, or in commercial applications such as electronic trading [8].

3 DESIGN AND IMPLEMENTATION

We highlight our architectural design choices and routing protocol of the publish/subscribe system under SDN in the following section, as a brief overview to the system implementation.

3.1 *Architecture design*

Our simulation scenarios are based on mixed circumstances of an OpenFlow network and a traditional IP-based network. The scenarios are composed of administrators and some groups. Each group contains a representative and some agents that all are implemented by a Floodlight Controller. Figure 1 illustrates the topology architecture and main components. Floodlight provides essential modules for performing publish-subscribe functionality, e.g. maintaining topology inside a group of the publish/subscribe system, gathering linkage information. These are prerequisites for creating publish/subscribe-based systems

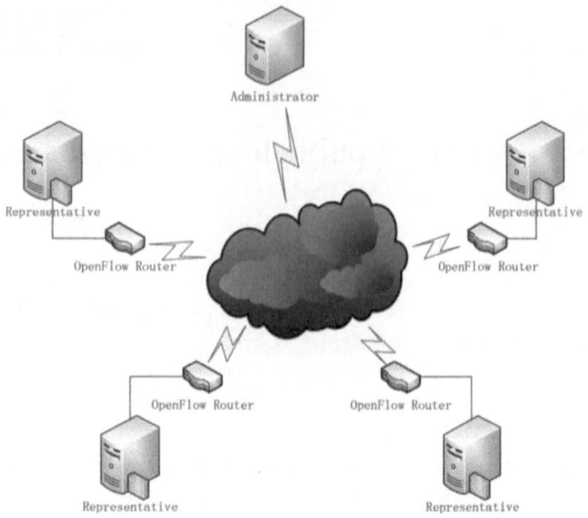

Figure 1. Topology architecture.

and computed routes. We also build neighbours for every group for topology maintenance.

The administrator maintains fundamental information about every group, e.g. the IP address of the representative, the User Datagram Protocol (UDP) port, and the subnet mask. It provides the information to new entrants and receives updates from groups. It can also give an overview of the topic tree that is retrieved from Lightweight Directory Access Protocol (LDAP) and change the tree in LDAP when a command is performed through its interface. A backup coexists with the administrator which stores the same repository, and will be transformed to the one which really functions when the former is down. The administrator and its backup detect each other by Hello message.

Representatives collect all subscribe and unsubscribe messages of its group, and merge and spread them all over the network. It periodically updates subscription of its group by publishing a Link State Advertisement (LSA) that contains all the link information and subscriptions. At the same time, it stores subscriptions from other groups as well as its own, and works out the forwarding method based on subscription and link states. It also determines the forwarding tree for every topic after computing routes.

Agents can appear in a group simultaneously and replace the representative by election when the representative is down or cannot be reached. Agents also subscribe and publish, but it can only broadcast its messages beyond the group information through the representative. Agents maintain the same subscription form and store neighbours of this group as the representative, but they cannot compute routes or forward messages to other groups.

Every client belongs to a representative or an agent and publishes/receives messages through it. Clients retrieve topics from a topic tree, which is now kept in LDAP, and subscribe topics which are needed and can be made use of, or for creating topics. We utilize OpenFlow routers within a group and common IP-based networks, thus easing the burden of controllers, and reducing the cost of buying the devices. The OpenFlow router, which connects one group with the outside network, is under control of its controller and collects information of nodes, which can reach directly to their links.

LSAs are transmitted through neighbours who are selected when a group joins the network according to link states that are gathered by its OpenFlow router and their existing neighbours. Representatives can also adjust their neighbours when one of them cannot be reached. LSA is used for transmission of new subscriptions and the synchronization of them.

A topic tree consists of several layers that can be defined by users through the administrator, and each father topic contains all the topics of its branch. Clients publish their subscriptions by following the topic tree or creating topics by themselves (which is not recommended).

Neighbours are first selected from directly connected groups, then associated with the distance between groups, and the number of existing neighbours other representatives have. When a node becomes representative, it will start the examining process to find neighbours as in Algorithm 1. Tnbr is set by the user which means the most neighbours a group could has. It will select Rnbr (to make Snbr which stands for the number of its neighbours in effect approach Tnbr/2) as potential neighbours from the whole network (expressed by Nnetwork) while Snbr is less than Tnbr, or none of the existing groups could be selected. Algorithm 1 illustrates the selecting process as described above.

Algorithm 1 Found neighbours

```
Select directly connected groups as neighbours
while S_nbr < T_nbr do
   for N_group in N_network do
      Calculate the promising score of selecting the
      group as a neighbour.
   end for
   Select R_nbr nodes as its potential neighbours.
   Start the neighbour building process.
end while
```

3.2 *Routing protocol*

We can get information about network states, link properties, and traffic conditions through Floodlight, and representatives. Incorporating this information into the path creation algorithm leads to more intelligent routing decisions, which could increase overall performance and save network resources.

For calculating the shortest paths, Jackson's Theorem is applied; we expect less average delay of the whole network with our routing decision. In this system, we mostly focus on packet delay between groups. This delay is the sum of delays on each link traversed by the packet. Each link delay in turn consists of processing delay, queuing delay, transmission delay, and propagation delay [4]. This accounting neglects the possibility that a packet may require re-transmission to a link due to transmission errors or various other causes, as we do not guarantee if delivery is absolutely unmistakable.

The main difficulty in the analysis of networks of transmission lines is that the packet inter-arrival times after traversing the first queue are correlated with their length. It turns out that if somehow, this correlation is eliminated and randomization is used to divide traffic among different routes, then the average number of packets in the system can be derived as if each queue in the network is M/M/1 (Memoryless arrival/Memoryless service time/Single server) [4].

We assume that our system consists of K first-come first-served, single-server queues in which packets from each group arrive at each queue i in accordance with independent Poisson processed at rate r_i. Once packets from a group are served at queue i, they proceed to join each queue j with probabity P_{ij}, or to exit the network with probability $1-\sum_{j=1}^{K} P_{ij}$. The routing probabilities P_{ij} together with the external input rates r_j can be used to determine the total arrival rate of groups λ_j at each queue j as shown below:

$$\lambda_j = r_j + \sum_{i=1}^{K} \lambda_i P_{ij}, \quad j = 1, \dots K \ [4] \tag{1}$$

We get the link distance between groups through a periodical distance detecting message and applying it for route computing. We calculate the forwarding tree for each topic and suppose that the node which holds the lowest λ_j. of the subscribers of this topic is its root. We add λ_j. to relative link distance when computing with OSPF and re-compute for every topic when the arrival rate of groups or the whole topology changes.

Most messages in a queue that a representative could efficiently deal with q_j, affect the result of route selecting. We add λ_j/q_j to route length r_{ij} while calculating and restricting the number of children of a group to a defined value. Then the process is as below. In Algorithm 2, s_k stands for sum of subscribers of topic k; n_k stands for the number of nodes already in the tree. It starts from 0 and ends at s_k. G stands for groups that are to be selected. M stands for groups that are already in the tree.

Algorithm 2 Route computing of a topic

```
Select root of the forwarding tree
while n_k < s_k
   select node closest to root
   for g in G
Refresh route length if g is connected to new node
      and make g closer to root
   end for
end while
```

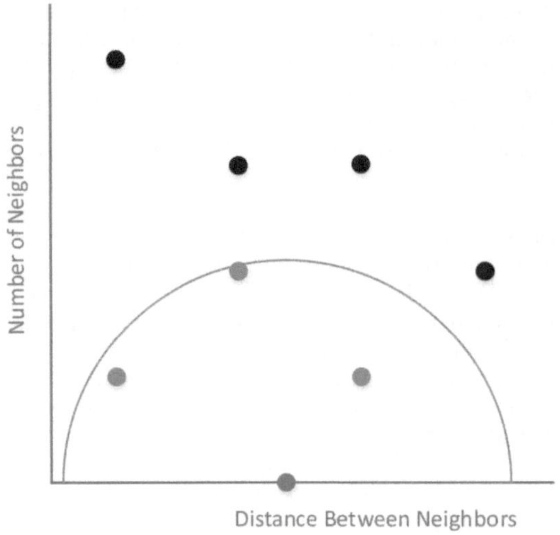

Figure 2. Neighbours' selection.

4 IMPLEMENTATION DETAILS

A representative or agent of a group will register with administrator for messages about this group and other groups such as the UDP port, the physical address of a group's OpenFlow router/switch, and the TCP (Transmission Control Protocol) port of other groups. The representative will then ask any other groups for the subscription table that contains all the groups and publish the subscriptions itself, and then the agent will get the form from its representative and subscribe messages through it.

We need neighbours and Hello messages between groups to know the others' states. If the router of this group detects that some routers of groups are directly connected, it will select them as neighbours. If the sum of neighbours does not meet the expected value, the following process will be performed. We take the distance between groups as the X-line; the number of neighbours as the Y-line. A group is eliminated when its neighbours are up to the limit. Figure 2 illustrates a new group stand by the node on X-line selecting potential neighbour nodes within the circle while other nodes are outside.

An Hello message is sent between neighbours every 30 seconds (can be adjusted). The representative will wait for a message or send TCP detection when the Hello message of its neighbour is timed-out, and tell the others about the disconnection between them if no response is heard, which will lead to re-computing the route of all topics.

We sort groups by their names when we calculate for the forwarding tree of a topic and select the first group as its root. Then Dijkstra are used when computing with the arrival rate of a group adding to the route length as a factor. They will store their next stops from the same tree for message transmission as they have the same input and computing method, avoiding a transmission loop.

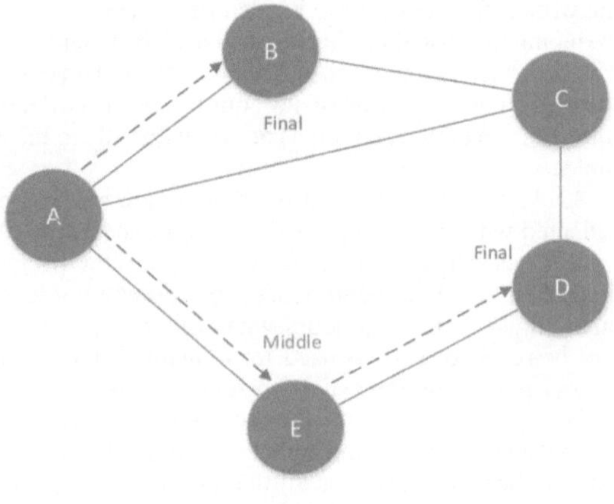

Figure 3. Forwarding paths.

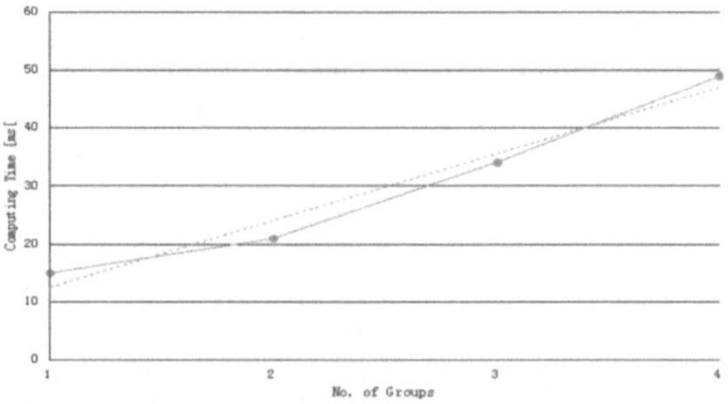

Figure 4. Route computing time.

We take the distance between routers that are directly connected as 1, and we assume the maximum of efficiently processed message as 200, according to current experimenting results (the number adjusted to experimental environment). We add arrival rate/200 to route length when we compute routes using Dijkstra.

Figure 3 illustrates a forwarding process of a topic. Group A, B, and D subscribe this topic, and A is the root of the tree. We add arrival rate to route length as a factor. We get two paths: 1. A to B; 2. A to D (E as a middle stop). When a group publishes messages about this topic (not in the forwarding tree), it will forward these messages to A first and A will transmit it later. If a group in this tree publishes such kinds of messages, say E, it will forward the messages to A and D, and A will forward to B later.

Figure 4 illustrates a sample of the time cost for route computing in the case of 1, 2, 3, and 4 groups consisting of the network. This result under SDN is similar to the result under a traditional network, indicating that SDN does not slow down the speed of computing paths. The trend of time consuming is like the dashed line of Figure 4. However, we just did our experiment with a small quantity of groups; a long-term trend is still to be sought.

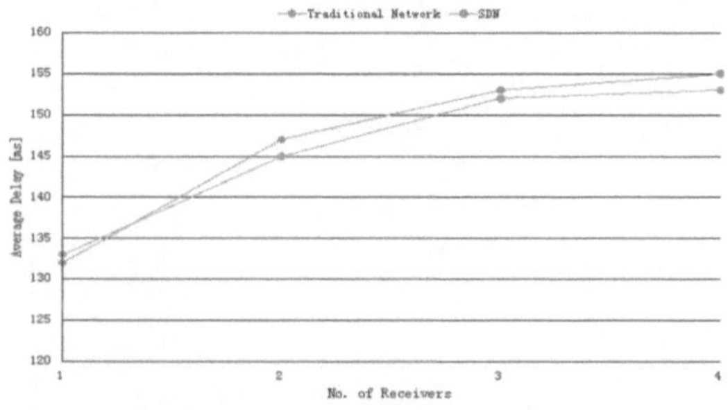

Figure 5. Performance evaluation.

We also found that topology converges when connections of the network changes, and selects neighbours if the number of their neighbours is less than the figure that we set.

Figure 5 illustrates a sample result of our work. We publish messages about a topic through a group under a traditional network and SDN with speed of 200 items/second by using the same forwarding method that the subscribers are receiving. We can see that average delay under SDN is smaller than under a traditional network.

5 CONCLUSION AND FUTURE WORK

The publish/subscribe model is increasingly popular due to its flexibility, asynchrony, and data-oriented nature. Its application under SDN is also analysed to be effective according to the work we have described above. There are still many potential and interesting network designs to be investigated in the future.

We can do some research with the newest OpenFlow protocol 1.4 under other open sources (such as OpenDaylight), as Floodlight just serves OpenFlow 1.0 for now. We just deployed OpenFlow routers inside a group for the concern of limitation of application and budget, but we surely got good results with the research. New routing protocol will be needed when more OpenFlow routers/switches are applied in our system.

ACKNOWLEDGEMENT

This research is supported by the National Grand Fundamental Research 973 Program of China under Grant No. 2011CB302506, 2012CB315802; the National Key Technology Research and Development Program of China (Grant No. 2012BAH94F02); the National High-tech R&D Program of China (863 Program) under Grant No. 2013AA102301; the National Natural Science Foundation of China under Grant No. 61132001, 61372115); the Program for New Century Excellent Talents in University (Grant No. NCET-11-0592); the Project of New Generation Broad band Wireless Network under Grant No. 2014ZX03006003,2011ZX03002-002-01, 2012ZX03005008-001, and The Technology Development and Experiment of Innovative Network Architecture (CNGI-12-03-007).

REFERENCES

[1] B., Gajic; J., Riihijarvi; P., Mahonen. *Intra-Domain Topology Manager for Publish-Subscribe Networks.*

[2] V., Jacobson; M., Mosko; D. Smetters; J., Garcia-Luna-Aceves. *Content-centric networking. Whitepaper describing future assurable global networks.* Response to DARPA RFI SN07-12, 2007.

[3] B., Gajic; Jesus, Palenzuela; J., Riihijarvi; Petri, Mahonen. *Mobility-aware Topology Manager for Publish-Subscribe Networks.*

[4] Delay Models in Data Networks. [online] Available: *http://web.mit.edu/dimitrib/www/Queueing_Data_Nets.pdf*

[5] Delay Models and Queueing. [online] Available: *http://www.mit.edu/people/medard/6.02s/6.02sday2.pdf*

[6] Floodlight Introduction. [online] Available: *http://www.projectfloodlight.org/floodlight/*

[7] OpenFlow Protocol Specification. [online] Available: *https://www.opennetworking.org/sdn-resources/onfspecifications/openflow, OpenFlow Switch Specification 1.0.1, Open Networking Foundation, Sept. 6, 2012*

[8] WSN Introduction. [online] Available: *https://www.oasis-open.org/committees/tc_home.php?wg_abbrev=wsn*

Future Communication Technology and Engineering – Chan (Ed.)
© 2015 Taylor & Francis Group, London, ISBN: 978-1-138-02777-0

An adaptive web-based learning system based on ontology

Q.L. Diao & F.H. Qiu
School of Information Science Technology, East China Normal University, China

ABSTRACT: Effective web-based learning should be able to adapt the learning process to students' personalized needs. This paper proposes an ontology-based model for web-based learning systems, where dynamic adaption is achieved through the analysis of students' objectives and learning activities. Then the paper presents the implementation of an adaptive web-based learning system based on ontology, and suggests related technologies.

Keywords: Ontology, web-based learning, adaptability.

1 INTRODUCTION

With the rapid development of information technology and computer network technology, more and more institutions have delivered education and training on a web platform. Different from teaching in classrooms where teachers can perceive students' learning behaviour directly and solve students' problems in real time, can a web-based learning system find its way to provide an effective learning process? Learning theorists justify that in order to reach an objective, such as acquiring a skill or learning something, a learner must be actively involved through practice to incorporate it cognitively into long-term memory (Dick, et al. 2005). Therefore, an effective e-learning system should provide a comprehensive idea of content organization, system functionalities, and advanced personalization of contents. However, to accomplish all these characteristics is a difficult task (Dan-El et al. 2013).

Ontology can provide a representation of a shared conceptualization of a specific domain that can be communicated between people and applications (Chandrasekaran et al. 1999). The use of ontologies in education brings many benefits: ontologies render education systems smart and reflective, deliver explicit knowledge, standardize the vocabulary of a knowledge domain, and make communication easier and knowledge reusable (Mizoguchi et al. 2000).

In this paper, we propose an ontology-based model for the design and implementation of adaptive web-based learning systems.

2 ONTOLOGY-BASED MODEL FOR ADAPTIVE WEB-BASED LEARNING SYSTEMS

2.1 *About ontology*

Originating in philosophy, ontology has been applied to the field of knowledge engineering, artificial intelligence, and information systems, to solve the problem of knowledge sharing and reusing. There exist several definitions of ontology. One of the most recognized definitions, given by Studer et al. 1998 is that it is a formal explicit specification of a shared conceptualization of a domain of interest. This definition captures the core characteristics of ontology as one aspect of formality, explicitness, consensus, conceptuality, and domain specificity. Ontology used in an information system provides a tool for machines to "understand" knowledge by means of knowledge representation techniques and, therefore, can be used as a base for making decisions on reasoning about specific knowledge (domain ontology).

2.2 *Proposed ontology model*

To provide a personalized learning process, it is necessary to acquire users' real-time learning activities at the beginning. Due to the complexity of the learning process, we introduce Fuzzy Logic (Barnabas Bede 2013) into our model. Therefore, the formal ontology model is as follows:

$$\Omega = (C, \mathfrak{R}, I, A) \tag{1}$$

An ontology Ω is a quad tuple of a set of concepts C, a set of relations $\mathfrak{R}$, a set of instances I, and a set of axioms A.

where, C = CNormal ∪ CFuzzy (CFuzzy is a set of concepts which includes fuzzy properties, while CNormal includes no fuzzy properties),

$\mathfrak{R}$ = $\mathfrak{R}$ Normal ∪ $\mathfrak{R}$ Fuzzy ($\mathfrak{R}$ Fuzzy is a set of relations which includes fuzzy properties, while $\mathfrak{R}$ Norma includes no fuzzy properties).

I = INormal ∪ IFuzzy (IFuzzy is a set of instances which includes fuzzy values, while INormal includes no fuzzy values).

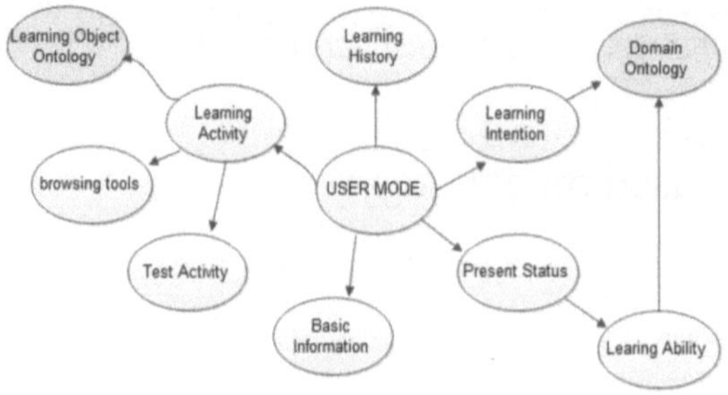

Figure 1. User ontology.

Our ontology model consists of *Domain Ontology*, *User Ontology*, *Learning Object Ontology*, and *Strategy Ontology*.

2.2.1 *Domain ontology*

Domain ontology can help users locate and learn related information more effectively. Domain ontology describes specific knowledge concepts and their relations in the system. In our model, domain ontology mainly represents specific curriculum knowledge. Thus, in our domain ontology, relations between concepts include "*is SubclassOf*" and "*isSuperclassOf*" which are used to declare a parent-child relation between two concepts. "*is PreRequsite*" is used to declare one concept (knowledge) required to learn prior to the related concept(knowledge). "*is Synonymous*" is used to match related concepts in domain ontology when students use synonymous words or expressions (terminologies),

In addition to relations between concepts, we apply some *datatype properties*, which link an individual to a particular value, such as "*optional*" (not for all students but for capable ones to learn).

2.2.2 *User ontology*

User Ontology describes the characteristics of a student. Our user ontology consists of each student's Basic Information (including personal information such as name, ID, birthday, etc.), Learning Intention, Learning Activity, Test Activity, Present Status (including learning ability, learning habit, etc.), and Learning History as shown in Figure 1.

2.2.3 *Learning object ontology*

Learning Object Ontology describes all learning resources. It consists of two parts; one is "resource documents," serving as integrated learning objects, the other is "resource atom unit," i.e. a minimum unit of learning resources, which is used to meet the special needs of users by appropriating the integration of these units.

2.2.4 *Strategy ontology*

Strategy Ontology describes strategies used in the whole system, which includes *Instruction Strategy*, *Test Strategy*, *Context Processing Strategy*, and *Security Strategy*.

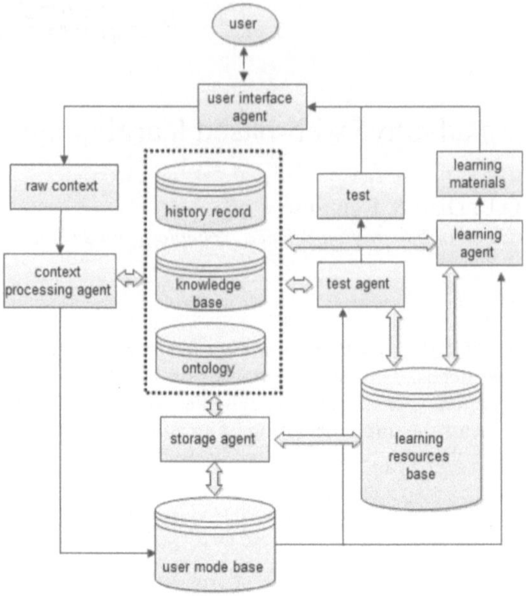

Figure 2. System framework.

3 DESIGN AND IMPLEMENTATION OF AN ADAPTIVE WEB-BASED LEARNING SYSTEM BASED ON ONTOLOGY

3.1 *System framework*

In order to provide an adaptive learning process, the system should detect all the real-time activities with high performance at first. Therefore, we take the context concept into consideration. There are many definitions of context (Schilit, et al. 1994, Dey et al. 2000). In the application of an adaptive web-based learning system, we consider context as all the information of user activities in a web-based system. On this basis, we present the system framework as shown in Figure 2.

3.2 *Related technologies*

3.2.1 *User interface agent*

The main function of *User InterfaceAgent* is to provide users with a friendly interface, acquire users' learning activities, hand in raw context information to the *Context Processing Agent,* and present the contents made by the *Learning agent* and *Test Agent* to users with proper form.

3.2.2 *Context processing agent*

The main function of *Context Processing Agent* is to process raw context handed in by the *User Interface Agent* and provide high-level context (user mode).

We use Jena (Hewlett Packard 2005), used for Semantic Web application based on Java open source, to refine raw context (such as users browsing activities using browser, test results, etc.) to get the user's present status.

Following are two exemplary rules used to process raw context by Jena:

Rule 1:
[(?x rdf:type :browsing_record)
(?x:browsing_object ?t1)
(?x :browsing_quantity ?u)
(?y rdf:type :browsing_judge)
(?y : browsing_upperLevel ?n1)
(?y : browsing_lowerLevel ?n2)
(?y :degeree_of_learning ?v)
(?t1 : domain_knowledge ?t2)] –> [greatThan(?u ?n1) lessThan(?u ?n2)
(?t2 :degree_of_learning ?v)]

Rule 2:
[(?x rdf:type :test_record)
(?x : domain_knowledge ?knowledge)
(?x : test_result ?score)
(?y rdf:type :teaching_strategy)
(?y:threshold_for_master_of knwoledge ?threshold)
greatThan(?score ?threshold)]
–>[(?z rdf:type : user_knowledge_mode)
(?z :domain_knowledge ?knowledge) (?knowledge :master_ of_ knowledge
'true'^^http://www.w3.org/2001/XMLSchema# boolean)]

Rule 1 is used to detect the user's learning status (degree of hard work) on the basis of the user's browsing activities.

Rule 2 is used to detect the user's learning ability on the basis of the user's test result.

3.2.3 *Test agent*

The main function of *Test Agent* is to produce proper test materials for users to test their learning abilities based on users' present status.

3.2.4 *Learning agent*

Based on users' present status, *Learning Agent* provides related learning materials suitable for users' learning intention and present learning ability.

Learning Agent can also provide the function of expanding query.

Given the domain ontology (curriculum ontology), $\Omega = (C, \Re, I, A)$

let *qExpand* be expanded query, then

$$qExpand : Query(qC) \rightarrow Query(q'C') \quad (2)$$

where $qC = \{qC_1 \ldots qC_n\}$ is a set of concepts to query by users in curriculum ontology,

$$q'C' = \{x \in C \mid \forall c \in qC, \exists r \in \Re, < c, r, x >\}$$

is a set of concepts after expanded by the relation r.

In our system, the query is expanded by the relations of "*isSubclassOf*," "*isSuperclassOf*," and "*isSynonymous*" (see *2.2.2 User Ontology*).

3.2.5 *Storage agent*

The main function of *Storage Agent* is to work cooperatively with *"history record," "knowledge base," "learning resources base,"* and *"user mode base"* storage bases, and provide convenient storage and retrieval services for *Context Processing Agent*, *Test Agent*, and *Learning Agent*.

In view of its flexibility and extensibility, *Storage Agent* also provides an interface of ontology file operations for the communication with other systems.

4 CONCLUSION AND FUTURE WORK

In this paper, we proposed an ontology-based model to support an adaptive learning system on web platforms. With the benefits of using ontology models, fuzzy logic, and context-aware technologies, the proposed web-based learning system is suitable for students to enjoy a personalized learning process based on their learning objectives and activities.

Future work will focus on the improvement of approaches to detect and analyse students' learning abilities with proper evaluation measures.

REFERENCES

Barnabas Bede 2013. Mathematics of Fuzzy Sets and Fuzzy Logic, *Studies in Fuzziness and Soft Computing* 295

Chandrasekaran, B., Josephson, J.R., Benjamins, V.R 1999. What Are Ontologies, and Why Do We Need Them, pp. 20–25 (January/ February)

Dan-El Neil Vila Rosado, Margarita Esponda-Argüero, 2013. Raúl Rojas. Power Chalk: An Adaptive e-Learning Application. Multimedia and Internet Systems: *Theory and Practice Advances in Intelligent Systems and Computing* Volume 183, pp 179–188

Dey A K. Providing 2000. *Architectural Support for Building Context-Aware Applications: [PhD Thesis]*. Georgia USA: Georgia Institute of Technology, Dec.

Dick, W., Carey, L., Carey, J.O. 2005. The Systematic Design of Instruction, 6th edn. pp. 1–12. Allyn & Bacon ISBN 0205412742

Mizoguchi, R., & Bourdeau, J. 2000. Using ontological engineering to overcome common AI-ED problems. *International Journal of Artificial Intelligence in Education*, 11 (2), 107–121.

Hewlett Packard Laboratories B. 2005. Jena. Asemantic Web framework for java. http://Jena.sourceforge.net

Schilit, B., Adams, N. Want, R. 1994. Context-aware Computing Applications. In the Proceedings of the 1st International Workshop on Mobile Computing Systems and Applications, Santa Cruz, CA, IEEE,. 85–90

Studer, R., Benjamins, V.R., Fensel, D. 1998. Knowledge engineering: principles and methods. IEEE Trans. Knowl. Data Eng. 25 (1–2), 161–197

Future Communication Technology and Engineering – Chan (Ed.)

Solving of an erythrometry problem using a specialized Hough transform

A.S. Potapov, O.V. Shcherbakov & I.N. Zhdanov
University ITMO, Saint Petersburg, Russia

ABSTRACT: The problem of erythrometry is considered. A novel method based on a modified Hough transform is proposed to solve this problem. The method includes the steps of both counting red blood cells and producing histograms of their radii construction. The achieved precision of erythrometry with the use of developed methods is acceptable for medical applications.

1 INTRODUCTION

The processing of medical images of red blood cells is considered in this work. Due to the complex nature of blood slide microscopic images, it is important to solve the task of recognizing and counting blood cells (Maitra et al., 2012). Microscopic images of blood slides provide a lot of valuable information about the health condition of a person, but still today such information, including red blood cells' quantity and the size of each cell, is collected by special human operators. Therefore, the problem of automation of these processes is the main goal of the present research. There are different known approaches to solving the problem; however, most of them only consider the counting of cells. In research papers (Veluchamy et al., 2012, Poomcokrak et al., 2008, & Kimbahume et al., 2011), the application of artificial neural networks is discussed, and the methods based on morphological processing (Nasution et al., 2008, & Priyankara et al., 2006), and feature extraction (Hamouda et al., 2012) are also being investigated.

In this paper, an algorithm based on modified Hough transform is presented. There are different modifications and optimizations of Hough transform applied to ellipse and circle detection (Nguyen et al., 2009, Ballard, 1981). The application of ellipse detection Hough transform to red blood cell counting is also described by Maitra et al., 2012. We not only give a solution to blood cell counting problems, but also consider the erythrometry problem that consists in retrieving histograms with the amount of blood cells of each radius.

2 OVERVIEW OF THE DEVELOPED METHOD

2.1 *Modification of Hough transform for counting red blood cells*

The developed image-processing algorithm for blood cell counting consists of several steps:

1. Auto-resizing the input image.
2. Readjusting the image dynamic range of brightness.
3. Smoothing the image using a Gaussian filter.
4. Applying the Canny filter for edge detection.
5. Numerating the contours.
6. Voting for centres of the contour chains (erythrocytes) yielding the image, in which brightness corresponds to a chance of a pixel to be the centre of the blood cell.
7. Accepting votes that exceed threshold.
8. Smoothing the image of centres using a Gaussian filter.
9. Selecting connected regions in the resultant image.
10. Counting the number of the resulting regions, in each of which one cell centre can be situated.

2.2 *The problem of erythrometry*

Erythrometry is one of the most important problems in blood analysis. The main element of erythrometry consists in evaluating the Price-Jones curve. It consists of several steps:

1. Calculating the radii of red blood cells.
2. Counting the number of red blood cells that have a similar radius.

3. Drawing a plot N(R), where N is the number of blood cells, which have radius R.

The obtained plot is used afterwards for analysis of human health.

3 COUNTING OF RED BLOOD CELLS

3.1 *Image preprocessing*

The proposed method is tested on a red blood cell image base, several samples of which can be seen in Figure 1. It can be seen that this image has a complex structure, especially for counting red blood cells.

The developed methods include auto-resizing, readjusting, smoothing, and Canny filtering. Obviously, the steps of smoothing and Canny filtering require no explanations, thus auto-resizing and readjusting steps will be considered.

At first, the user should choose an appropriate red cell on a picture in order to mark interactively its diameter using a mouse. After the diameter is indicated, the whole picture is resized to such scale, in which the diameter will be equal to 100 pixels. This step can be performed automatically, but interactive measurement of the typical diameter increases reliability when arbitrary images are used. The default size of blood cells can be known from imaging parameters (focal length, etc.), since the normal physical size of blood cells is known.

Image brightness adjusting results in the extension of the dynamic range of brightness from minimum to maximum of intensity of image pixels, as can be seen in Figure 2.

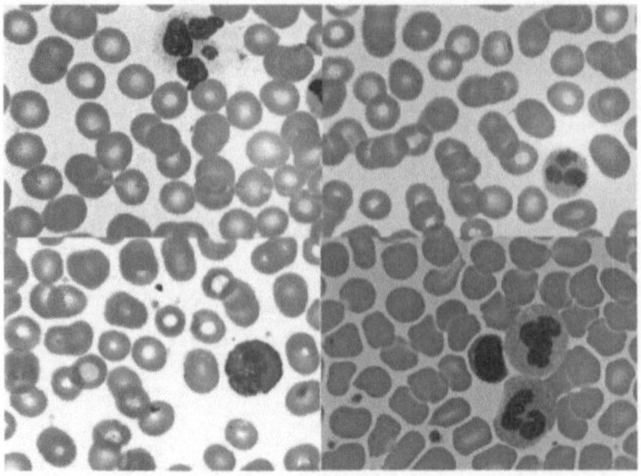

Figure 1. Samples of red blood cells image bases.

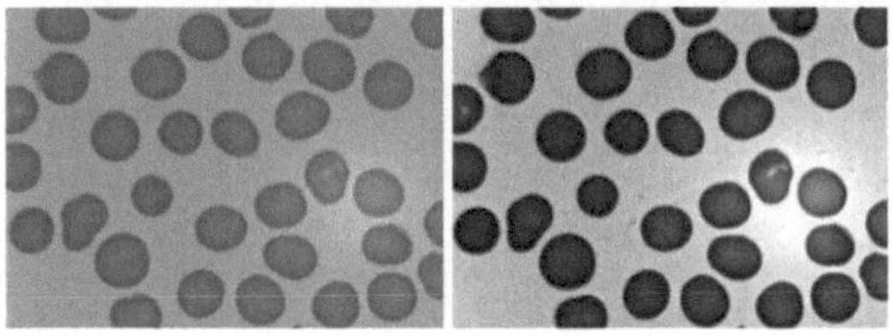

Figure 2. Sample of blood containing red cells (left) and adjusted image (right).

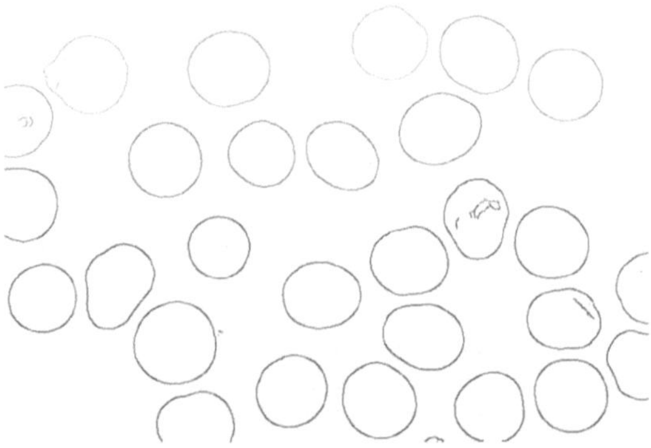

Figure 3. Inverted image of numerated contours.

3.2 *Numerating of contours*

Numerating contours and resultant centres are carried out after image normalization. The essence of these operations can be described by the following steps:

1. Searching line-by-line for non-zero elements.
2. If this element is a part of an existing contour, its colour is set equal to the index of this contour. Otherwise, a new contour is created, and this element becomes the first element of this contour.
3. Steps (1)–(3) are repeated until the end of evaluating the image.

When this algorithm is finished, the resulting image corresponds to a set of contours with intensity equal to its index (see Fig. 3).

3.3 *Voting and counting of erythrocytes*

At this step, a modification of Hough transform is used. After detecting and numerating the contours, it is necessary to understand which of them do not belong to erythrocytes. For this purpose, the following steps are required:

1. Three points are randomly chosen for every contour (each point belongs to the contour under consideration, and all contours are processed separately).
2. A circle is constructed using these three points.
3. If the radius of the circle belongs to the necessary range and the circle centre lies in the image, the value of the accumulator corresponding to the circle centre is increased by 1.
4. Steps (1)–(3) are repeated for a sufficient number of times for the current contour to accumulate votes.
5. Steps (1)–(4) are executed for each contour.
6. Accumulator elements that do not have enough votes are excluded from further consideration.
7. The image, in which the brightness of the pixels corresponds to the number of votes for centres, is smoothed.
8. Regions of connected pixels with non-zero numbers of votes are detected and numerated.

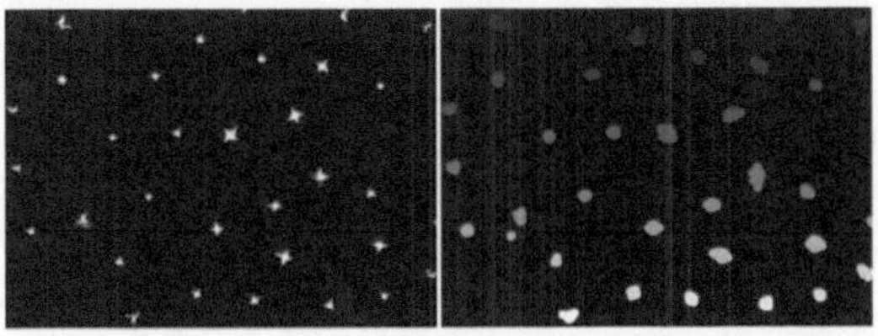

Figure 4. Accumulator image (left) and cells central areas (right).

The number of erythrocytes can be easily counted from the image consisting of regions numerated with the use of the algorithm described above.

3.4 *Results of counting*

Examples of an accumulator image and an image with numerated regions that represent the cells central areas are shown in Figure 4.

The suggested basic method for counting blood cells allows for an achievement of 85% recognition rate. Additional heuristics for contour enhancing on noisy images and merging close regions belonging to the same cells result in an increase in the counting precision up to 96–98%, which is compatible or better than in existing methods.

4 SOLVING THE ERYTHROMETRY PROBLEM

4.1 *Erythrometry*

As mentioned above, it is necessary to calculate the radii of all found erythrocytes in order to solve the erythrometry problem.

Consider all contours found at the step of blood cell counting. To calculate the radii of cells, firstly it is necessary to determine which contours correspond to erythrocytes. Contours, which do not vote for regions recognized as cell centres, are removed. Then, the following steps are carried out:

1. Using the vote system described in 2.3, appropriate circles take part in voting for centres of erythrocytes.
2. The centre that gets the maximum number of votes becomes the centre of the current contour.
3. Lines are drawn from the centre-winner point to all points of the contour.
4. All lengths of lines are accumulated and averaged. The averaged length is considered as the radius of the erythrocyte.
5. Information about erythrocyte radius for each contour is accumulated in the histogram.
6. Using information from this histogram, plot N(R) is drawn.

4.2 *Results of solving the erythrometry problem.*

An example of results is represented as the plot N(R) (see 1.2) shown in Figure 5.

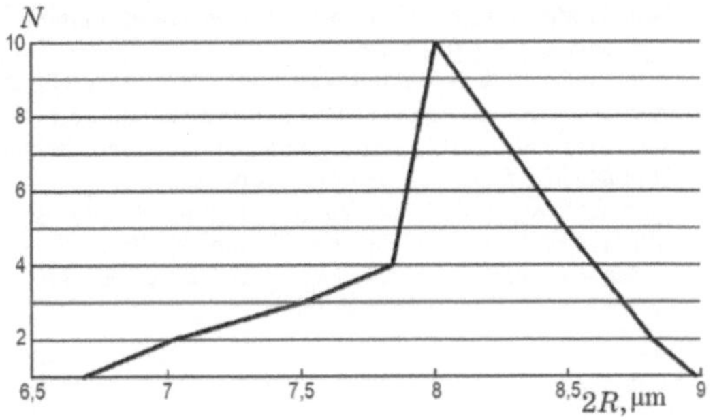

Figure 5. Plot N(R).

One of the main parameters of this histogram is its width (or dispersion of radii), which can be used to give diagnoses of blood illnesses. Automatic evaluation of this parameter is much faster, but still less (yet sufficiently) precise than its evaluation by human experts. However, additional experiments are necessary, which also require novel medical image databases with ground truth information appropriate for automatic evaluation of image analysis methods.

5 CONCLUSION

This paper presents a solution to the erythrometry problem in the case of automatic detection and counting of red blood cells in microscopic images. Results of the presented method indicate the possibility of achieving sufficient accuracy in automated image-based erythrometry.

Further research can be focused on gathering databases of blood cell images containing erythrometry ground truth data, which is necessary for the well-directed development of robust automatic erythrometry methods.

ACKNOWLEDGEMENTS

This work was financially supported by the Ministry of Education and Science of the Russian Federation, and by the Government of the Russian Federation, Grant 074-U01.

REFERENCES

Ballard D.H. 1981. Generalizing the Hough transform to detect arbitrary shapes. *Pattern recognition* 13(2): 111–122.

Hamouda A. & Khedr A.Y. & Ramadan R.A. 2012. Automated red blood cell counting. *International journal of computing science* 1(2): 13–16.

Kimbahune V.V. & Ukepp N.J. 2011. Blood cell image segmentation and counting. *International journal of engineering science and technology* 3(3): 2448–2453.

Maitra M. & Gupta R.K. & Mukherjee M. 2012. Detection and counting of red blood cells in blood cell images

using Hough transform. *International journal of computer applications* 53(16): 18–22.
Nasution A.M.T. & Suryaningtyas E.K. 2008. Automated morphological processing for counting the number of red blood cell. *2008 International Joint Conference in Engineering; Proc.*, 4–5 Jakarta: Indonesia.
Nguyen T.M. & Ahuja S. & Wu Q.M.J. 2009. A real-time ellipse detection based on edge grouping. *IEEE International Conference on Systems, Man and Cybernetics; Proc.*, 3280–3286, 11–14 October.
Poomcokrak J. & Neatpisarnvanit C. 2008. Red blood cells extraction and counting. *The 3rd International Symposium on Biomedical Engineering*: 199–203.
Priyankara G.P.M. & Silva R.K.O.H. & De Silva C.R. 2006. An extensible computer vision application for blood cell recognition and analysis.
Veluchamy M. & Perumal K. & Ponuchamy T. 2012. Feature extraction and classification of blood cells using artificial neural network. *American journal of applied sciences* 9(5): 615–619.

Future Communication Technology and Engineering – Chan (Ed.)
© 2015 Taylor & Francis Group, London, ISBN: 978-1-138-02777-0

Fast reconstruction of Go board grids using the modified Hough transform

A.S. Potapov, O.V. Shcherbakov & I.N. Zhdanov
University ITMO, Saint Petersburg, Russia

ABSTRACT: The task of the detection of straight lines arranged in some regular ways on images is considered in the example of the reconstruction of the lined grid on a Go board. The modified Hough transform is developed by introducing additional constraints and decomposing the task of parameter evaluation. Experimental results are given, which validate that even low-contrast regular structures can be detected with high robustness and precision.

1 INTRODUCTION

In this paper, we consider the task of the recognition of positions in the Go game, which is interesting, because artificial players are still weaker than human players here. Automatic recognition of positions can be useful for games played in the real world being transferred into electronic databases, for online translation of tournaments on Go servers, for amateurs to save their real game records using smart phones, and so on.

A Go board (goban) is marked with a grid (usually 19 lines × 19 lines grid), and black and white stones are placed on the line intersections. The main difficulty of the task of Go game position recognition consists of the detection of the goban and its lined grid. After the grid is reconstructed it is relatively easy to analyse which intersections are occupied by stones of what colour.

In some cases, camera position can be known as a priori. However, in general, the goban can be observed from an arbitrary aspect and angle, and sometimes it is necessary to analyse only one current position independently. Thus, we consider the most difficult case of reconstructing the grid on the goban using one single image.

It should be pointed out that not many works on this topic exist. For example, a method for reconstructing game positions on the base of key points on images is presented by Seewald, 2010. This method uses the well-known SIFT (Scale Invariant Feature Transform) algorithm for feature detection. However, since SIFT is patented, a demo version is used in the mentioned paper. Thus, the problem of keypoint detection and construction of the descriptors should still be solved in this approach in order to use it in practice.

Work of Hirsimäki 2005 should be mentioned as well. This method deals with the same problem and also uses Hough transform, which lacks speed because of dealing with classical Hough transform implemented on Matlab.

Since the task of the lined grid reconstruction is connected with the task of detection of regular structures on images, and these structures are partially occluded by stones, usage of the Hough transform seems to be the most appropriate approach. The task of detection of geometric elements (and straight lines, in particular) on images is very common (Duda et al., 1972, Olson 1999, Galambos et al., 1999, Rau et al., 2003, & Huijun et al., 2003). For example, this task is solved (Duda et al., 1972) using Hough transform. Constrained Hough transform for lines and curves detection is introduced (Olson, 1999) for minimizing line positioning errors. The usual approach of increasing computation time is based on selecting some fraction of input points, eg Galambos et al., 1999. However, this leads to a decrease of probability for detecting good candidates for lines. Modified Hough transform with Principal Axis Analysis is proposed by Rau et al., 2003. This method gives a faster detection of straight lines arranged in a regular structure.

Most of the existing modifications of the Hough transform focus on robust detection of separate low-contrast or short lines. The task of reconstruction of the grid on gobans is specific in several respects. The grid to be recognized is plotted on the wooden board whose texture elements can be linear and have similar contrasts. The main difference of grid lines is their regularity. Thus, the methods of improved Hough transform for robust detection of low-contrast or short lines will tend to extract more texture elements instead of separating them from the grid. Instead, some specialized modification of Hough transform is needed which makes use of prior information about regularities in the grid lines, which allows for precise discrimination of these lines from all other arbitrary lines.

2 DEVELOPED METHOD

Regularities in the line arrangement can be accounted for in different ways. For example, one can perform a direct fitting of the grid to the image considering all possible cell sizes and grid orientations. However, this would be computationally expensive. Another possible approach is to introduce a complete parameter space of lined grids and calculate the Hough transform in this space. However, the dimension of this space will be greater than that of the separate lines, which will lead to a decrease in robustness.

In this paper, we propose to factorize the task of the estimation of lined grid parameters separating pairs (coordinates of the vanishing point), and (grid position, cell sizes) for acceleration of computation. Also, modified Hough transform is being used in this method that speeds up execution of the program. Thus, the developed method consists of two steps.

The first step consists of determining the vanishing point, in which all parallel lines after projective transformation intersect each other. The grayscale image $I(x, y)$ is given, and its gradient

$$\nabla I(x,y) = (\partial I(x,y)/\partial x, \partial I(x,y)/\partial y) \tag{1}$$

is estimated.

Pixels with high value of the gradient amplitude (selected using Otsu thresholding) are considered. Then, direction angle α perpendicular to the gradient direction at every point (x_0, y_0) is calculated. It defines a line that contains this point and has this direction:

$$\begin{aligned} x(t) &= t\cos(\alpha) + x_0, \\ y(t) &= t\sin(\alpha) + y_0. \end{aligned} \tag{2}$$

Each such line specifies a set of possible vanishing points, which can be parameterized by the distance and direction from the image centre to the certain vanishing point. Direction angles are taken uniformly, while $1/d$ is taken as the distance parameter (the bin #0 corresponds to infinity, for which α coincides with the direction from the centre of the image to the vanishing point). It appears that for most images the projective transformation is weak, and only bin #0 for distances can be considered.

For the sake of robustness, voting weights are estimated on the base of the surroundings of the pixel. All pixels situated on line (1) within certain radius R are considered, and pixels with a high value of the gradient amplitude are counted. The number of these pixels is taken as the weight for voting for the corresponding vanishing points. The vanishing point with the highest number of votes is chosen.

The recovered information is used in order to make the image rectified in such a way that the grid lines become horizontal and vertical.

The next step consists of the detection of this rectified grid. The distance between lines and their positions are unknown. Again, the gradient field of the image is considered. Only the vertical component of the gradient is used while detecting horizontal grid lines, and vice versa. At first, some rows of the image are selected as salient lines. In order to contain a salient line row, it should contain enough long chains of pixels with a high value of the vertical component of the brightness gradient. Then, for each pair of salient rows, which are situated in a certain range of distance, it votes for the corresponding grid size. The permissible range is [0.5 MinSize/D + ε, MinSize/D − ε], where MinSize is the minimal image size (width or height), D is the number of lines in the grid (eg 19 for usual gobans), and ε is a small additional offset. This restriction on distance between adjacent lines implies that the goban is expected to be entirely visible, and should not be photographed from a great distance. After all votes are calculated, the accumulator with the maximum number of votes is chosen. This accumulator corresponds to the most probable grid step. The same sequence of operations is carried out for image columns in order to recover vertical lines of the grid.

The next unknown parameter to be evaluated is the beginning position of the grid. This parameter can be estimated by searching for the leftmost equidistant vertical lines (or the topmost horizontal lines), the distance between these lines will correspond to the sought grid step. Starting from these lines, all other lines of the grid are detected until their total number equals D.

3 IMPLEMENTATION DETAILS

Now consider some details of the implementation of the above-described algorithm. The first minor issue is the robustness of the gradient field estimation. In our task, successful results were achieved with the usage of Gaussian smoothing and Sobel masks.

Then, while marking rows (or columns) of the rectified image as horizontal (or vertical) lines, many adjacent rows can be marked. One can reduce the number of such lines using more strict thresholds on the average gradient module or line length, but there are gobans with low-contrast grids, which will be missed in this case. Instead, we used less strict thresholds, but performed post-processing of the detected lines, which consisted of the grouping of the adjacent lines concentrated in the surroundings of the most salient lines. Lines with averaged parameters were constructed on the base of such groups.

Since rectification was carried out before detecting the specific lines on the base of integral Hough transform, the projective transformation was compensated not too precisely. Thus, deviations from the perfect equidistant arrangement of the lines in the grid should be taken into account. Therefore, the voting by pairs of lines for grid steps should not be exact. In our implementation, pairs of lines voted for several nearby accumulators with different weights, as shown in Figure 1.

As can be seen from Figure 1, not only the accumulator of the grid steps corresponding to the precise distance between lines (N) was affected, but other accumulators of steps in the range [N−3, N+3] also

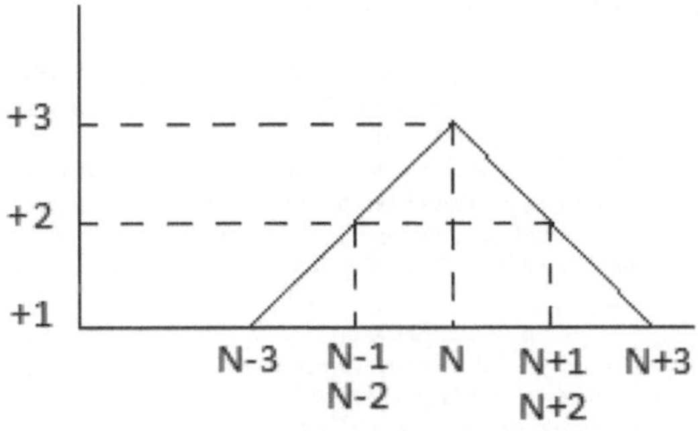

Figure 1. Voting for several grid steps.

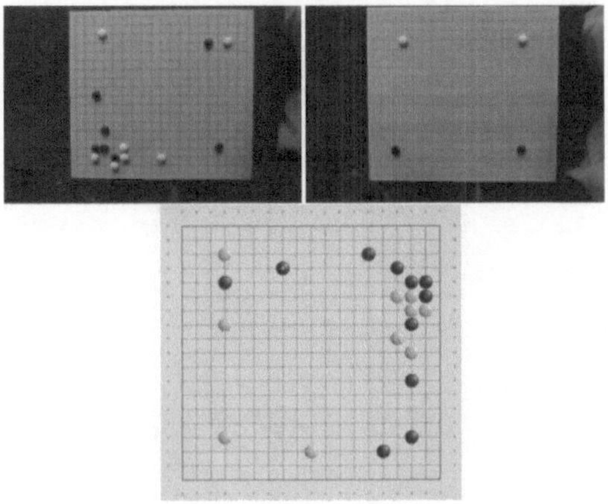

Figure 2. Image example of type A (left), type B (right), and type C (bottom).

increased by less number of votes, decreasing with the distance.

Identification of the next grid line on the stage of the whole grid reconstruction using the evaluated grid step was also carried out in the window of the same size [N−3, N+3].

4 EXPERIMENTS

Experimental validation of the developed algorithm is carried out using three types of images including real images with the clearly visible grid (type A), real images with the low-contrast grid mixed with the texture of the goban (type B), and synthetic images (type C), which are shown in Figure 2.

The test set contained forty images, among which there were two images of type C, sixteen images of type B, and the remaining images were type A. It appeared to be necessary to set a special value for the parameter for contrast thresholding for the images of type B. Except for this adjustment, the grid for all images

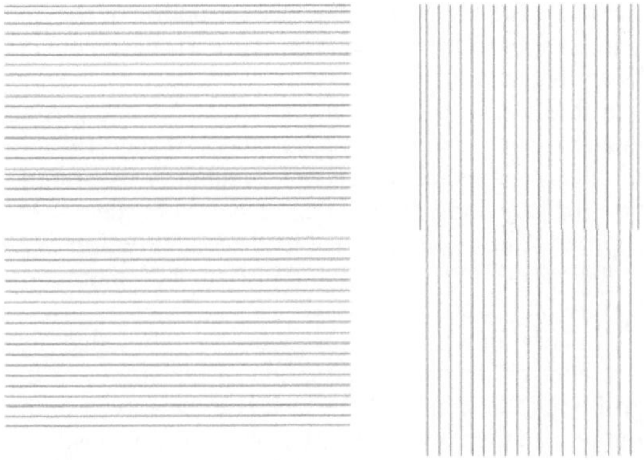

Figure 3. Inverted image of intermediate steps. Horizontal lines without post procedure (top left), vertical lines with post procedure (top right), horizontal lines with post procedure (bottom left) and vertical lines with post procedure (bottom right).

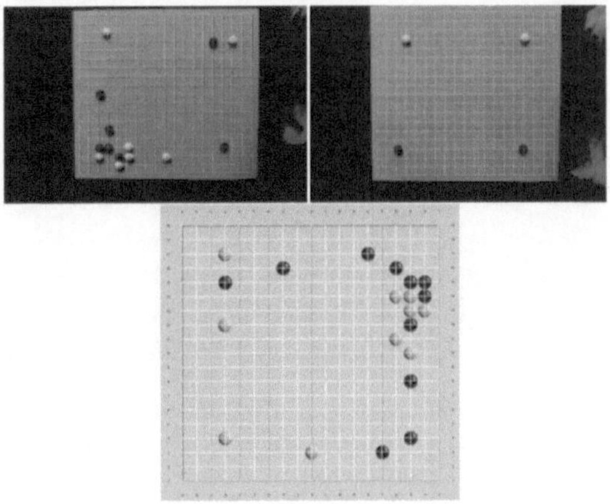

Figure 4. Example of result of the grid detection on the image of type A (left), type B (right), and type C (bottom).

of the test set was successfully found, automatically. Inaccuracy of the node positions of the recovered grid was less than 10 pixels for type A images (resolution 1024×683), 10 pixels for type B images (resolution 1024×683), and 2 pixels for type C images (resolution 1049×1049). Lower precision for the images of types A and B was caused by imprecise rectification (small errors in the recovered orientation angle, presence of weak projective component, and probable distortion). Examples of the intermediate steps are shown in Figure 3. The grid detection results can be found in Figure 4.

5 CONCLUSION

The method proposed in the presented work is capable of detecting regular structures on images that are shown on the example of the reconstruction of the line grid on a Go playing board (goban). This is achieved by applying additional restrictions imposed by the regular structure on the line parameters, and decomposing the task of evaluating the parameters of the grid into relatively independent tasks of evaluating its orientation (or vanishing point), step size, and borders. This approach allows for a robust grid reconstruction even in the case of low contrast and presence of distracting texture on the board material.

It is not only a robust reconstruction, but also precise enough for further detection of Go stones in the grid line intersections. This task is to be solved in the future, and then it will be possible to convert positions on the Go board into a digital form.

ACKNOWLEDGEMENTS

This work was supported by the Russian Federation President's grant Council (MD-1072.2013.9) and the Ministry of Education and Science of the Russian Federation.

REFERENCES

Duda, R.O. & Hart, P.E. 1972. Use of the Hough Transformation to Detect Lines and Curves in Pictures. *Communications of the ACM* 15(1): 11–15.

Galambos, C. & Matas, J. & Kittler, J. 1999. Progressive Probabilistic Hough Transform for Line Detection. *Computer Vision and Pattern Recognition*: 554–560.

Hirsimäki T. 2005. Extracting Go Game Positions from Photographs.

Huijun Di & Lei Wang & Guangyou Xu. 2003. A Three-step Technique of Robust Line Detection with Modified Hough Transform. *Third International Symposium on Multispectral Image Processing and Pattern Recognition; Proc. SPIE 5286, 20–22 October 2003.*

Olson, C.F. 1999. Constrained Hough Transforms for Curve Detection. *Computer Vision and Image Understanding* 73(3): 329–345.

Rau, J.Y. & Chen, L.C. 2003. Fast Straight Lines Detection Using Hough Transform with Principal Axis Analysis. 8(1): 15–34.

Seewald, A.K. 2010. Automatic Extraction of Go Game Positions From Images: A Multi-Strategical Approach to Constrained Multi-Object Recognition. *Applied Artificial Intelligence: An International Journal* 24(3): 233–252.

Future Communication Technology and Engineering – Chan (Ed.)
© 2015 Taylor & Francis Group, London, ISBN: 978-1-138-02777-0

Strategies and modeling of rumors defense and privacy protection on complex networks

Z.L. Sun, M.M. Tang & X.C. Fu
Shanghai University Department of Mathematics, Shanghai, China

ABSTRACT: In this paper, we present a kind of information spreading model on complex networks, and design two strategies to defend rumours and protect privacy. We consider the variation of information and the behaviour of nodes to send message so that multiple versions of information will evolve in the network. The main innovation of this paper is that we combine rumors defense and privacy protection in one class of strategies with different goals. Finally, we give some numerical simulations to lay a solid basis for further research.

1 INTRODUCTION

Nowadays, with the popularity of the Internet and social platforms, the way of information spreading is quite different from before. Messages exchanged between individuals can be regarded as a dynamical system on complex networks, so a series of dynamic behaviours for information spreading on complex networks have attracted more and more attention.

Traditional models for information spreading are widely used, such as SI, SIS, SIR, which study the spreading of single information[1],[2]. Daley and Kendall (DK)[3] introduced the original model for rumor spreading. In the DK model, a closed and homogeneously mixed population can be classified into three distinct classes. These classes are called ignorants, spreaders, and stiflers. M. Nekovee and his colleagues study the theory of rumors spreading in complex social networks[4], and the paper introduces a general stochastic model for the spread of rumor, and derives mean-field equations that describe the dynamics of the model on complex social networks. It shows that scale-free social networks are prone to the spreading of rumors, just as they are to spreading infections. D Trpevski and his colleagues[5] introduce a model for rumor spreading over networks, in which two rumors, termed rumor 1 and rumor 2 with different probabilities of acceptance, may propagate among nodes. Y Zhang and his colleagues[6] study the rumor evolution and spreading on social networks, where the content of the rumor is modified during the spreading by the individuals with a certain probability. They have two choices, to forward or to modify. As a forwarder, an individual disseminates the rumor directly to their neighbours. As a modifier, conversely, an individual revises the rumor before spreading it out. Yaghoob Naimi and his colleagues[7] introduce the generalized model to investigate the impact of the interaction rules on the efficiency and reliability of the rumor process.

2 STRATEGIES OF RUMORS DEFENSE AND PRIVACY PROTECTION

We face two major problems after analysing the process of information spreading on a network:

1) If the information has been tampered with in the spreading process, with the increasing complexity of networks, the tampered information (such as rumors) will be trusted by more and more individuals, and the original information will be ignored.
2) If some sensitive, private information regarding the individual is exposed once in the network, it will be spread widely and quickly. Then, immediately, the individual will face security risks.

The first problem needs a solution to ensure that information can resist all kinds of noises in the network to be spread effectively. The second problem needs to work out that the information cannot be easily spread in the network, thus decreasing the security risk for individuals. In fact, if we measure the effect of information spreading on complex networks based on whether the individuals trust the information or not, the two problems above can be considered in one system:

1) The better the anti-noise capacity of the information, the more the original information can be trusted.
2) The better the privacy protection effect, the less the original information can be trusted.

Based on the above two goals, we give the following corresponding strategies.

2.1 *Majority vote strategy*

In order to make more information that is correct to be trusted, the original information will be broadcasted several times on the network. The distribution of the broadcasts node in the network and the frequency of broadcasts will be controlled to improve the anti-noise capacity.

2.2 *Smoke bomb strategy*

In order to make the correct information be less trusted, the rumors will be broadcasted several times on the network. The distribution of the broadcasts node, the frequency of broadcasts, and the versions of rumors will be controlled to improve the efficiency of the privacy protection.

Due to better quantification of the strategies and observation of the results, the compartment model is not enough. Therefore, in the following we will present node modelling.

3 NODE MODELLING FOR INFORMATION SPREADING

Many works are based on epidemic modelling. The information is always considered as a disease, but in fact, it has some obvious characteristics. In the node modelling, we have summed up the characteristics of the information as the basic assumptions.

3.1 *Assumptions*

Each node is an individual in a complex network and,

1) The same original information will be changed to different versions of information by the modifier of the rumor monger. Each node will receive many versions of information but will only trust one version or trust none, due to some reasons each time.
2) The node which is not a rumor monger, trusts a version of information once. The behaviour of its sending information is a stochastic process.
3) The node which is a rumor monger, never pays attention to the information received. The behaviour of its sending information is planned and can be quantified.

In some cases, a few rumor mongers can immediately affect the entire network. It depends on the topological properties of the rumor monger node j in the complex network such as node degree k_i. Also, the behavioural characteristics of these nodes play an important role. Different from the diseases, information is a kind of signal, which needs to be understood so the information can be encoded. It can always be expressed as binary code. Tampering and variation of information will be considered as code transformation.

We use stochastic variable M to denote the finite information:

$$M = \{a_0 a_1 \cdots a_m\} \in S^m \tag{1}$$

where $S = \{0, 1\}$ and $m \in N$. In the case of non rumor, the variation of information can be represented as a

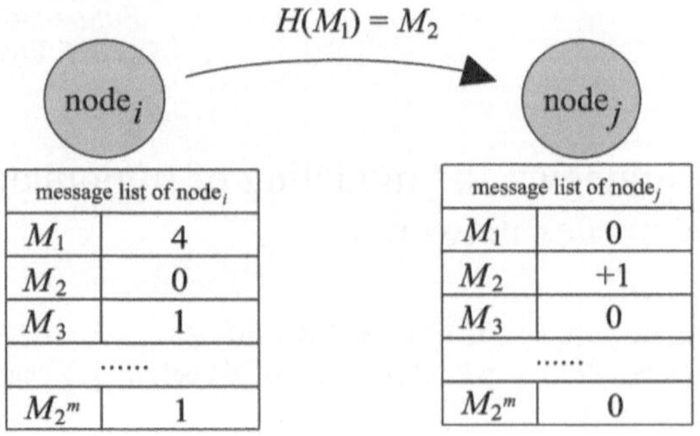

Figure 1. The message propagation mechanism between nodes.

stochastic process with transition probability matrix P_m :

$$P = \begin{pmatrix} p_{1,1} & \cdots & p_{1,2^m} \\ \vdots & \ddots & \vdots \\ p_{2^m,1} & \cdots & p_{2^m,2^m} \end{pmatrix} \tag{2}$$

where $p_{i,j}$ denotes the transition probability from M_i to M_j.

The following figure shows the message propagation mechanism between nodes.

whether the node i trusts the message M may depend on many factors such as personal values, public media guides, personal experience, and so on. One of the major factors is the messages it receives from its neighbours.

Therefore, without rumor mongers the spreading function should be:

$$\dot{\mathbf{M}} = f \circ \mathbf{M} + g \circ \mathbf{A}H(\mathbf{M}) \tag{3}$$

where $\boldsymbol{M}$ is the message vectors of all nodes, and $\boldsymbol{A}$ is the adjacency matrix of the complex network. The mapping $H: \boldsymbol{M} \to \boldsymbol{M}$ contains the process of variation and behaviour of the sending message. The operator f and g determine the basis of whether the node trusts message M or not.

Furthermore, the statistics on the received messages can also contain final τ historical information, which means:

$$\dot{\mathbf{M}} = f \circ \mathbf{M} + g \circ \mathbf{A}\int_{t-\tau}^{t} W(s)H(\mathbf{M})ds \tag{4}$$

where W is the weight of final τ historical information.

In the case of rumor, the rumor mongers will broadcast a rumor M_r with plans, and the rumor is independent of its received messages. The modelling will be:

$$\dot{\mathbf{M}} = f \circ \mathbf{M} + g \circ (\mathbf{A}_{\text{no rumor}}H(\mathbf{M}) + \mathbf{A}_{\text{rumor}}H_{\text{rumor}}(\mathbf{M}_r)) \tag{5}$$

where $\boldsymbol{A}_{\text{no rumor}}$ is the adjacency matrix without rumor mongers, and $\boldsymbol{A}_{\text{rumor}}$ is the one with only rumor mongers. The $H_{\text{no rumor}}$ is the plan of rumor mongers.

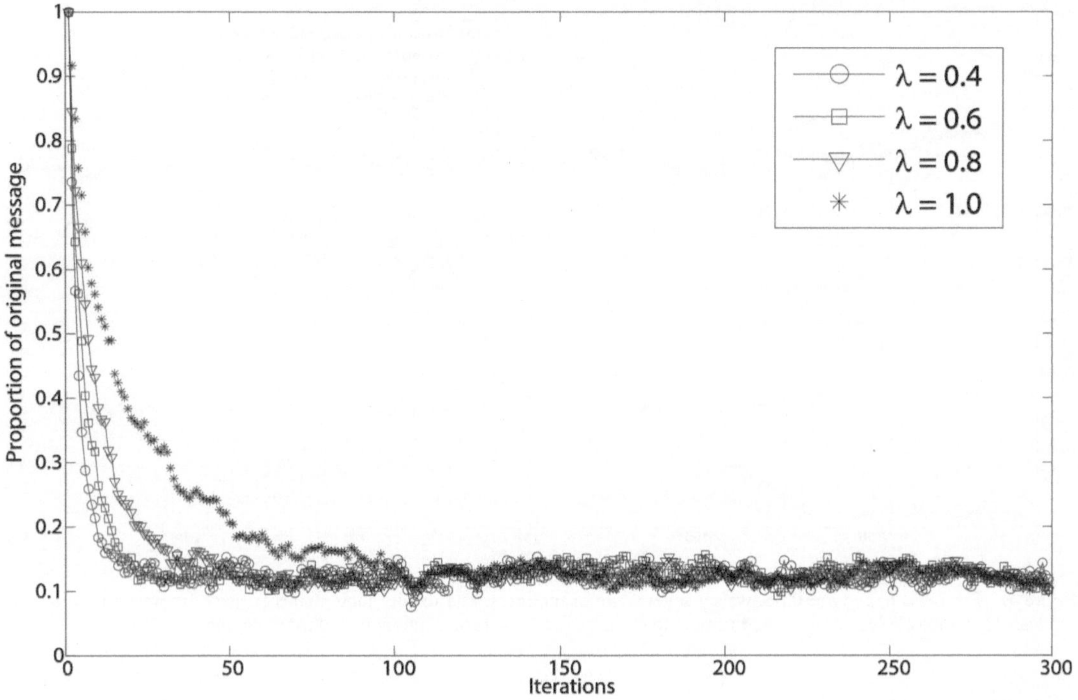

Figure 2. On BA network with 500 nodes and $<k> = 2$, the proportion of original messages tends to 12.5% with variation probability $p = 0.15$. It means that we cannot distinguish the authenticity of messages in the network. λ affects the convergence rate of progressive distribution.

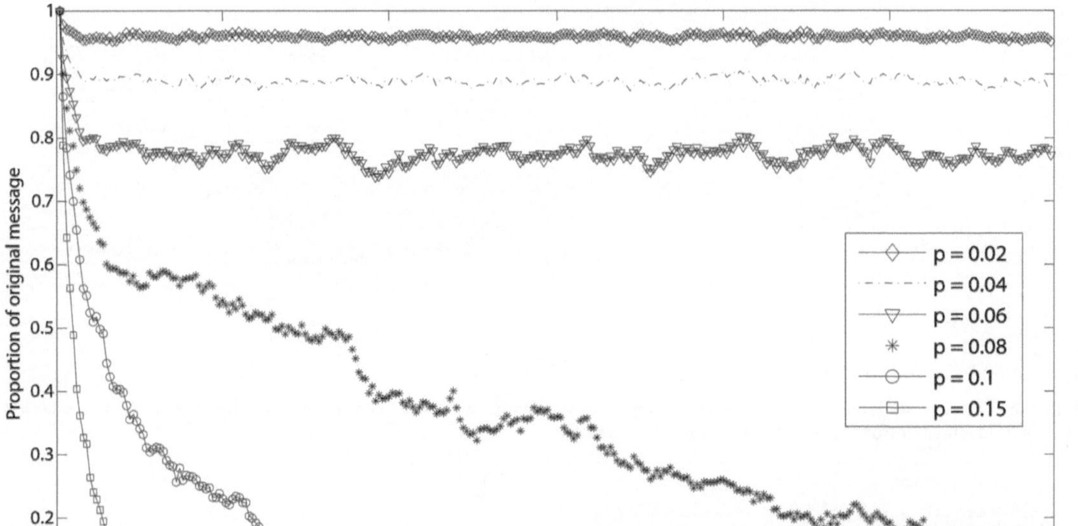

Figure 3. Fix $\lambda = 0.6$, we will see that the distribution of original messages is related to variation probability p. It seems that there exists a threshold of p to determine whether the original messages are still the major information or not.

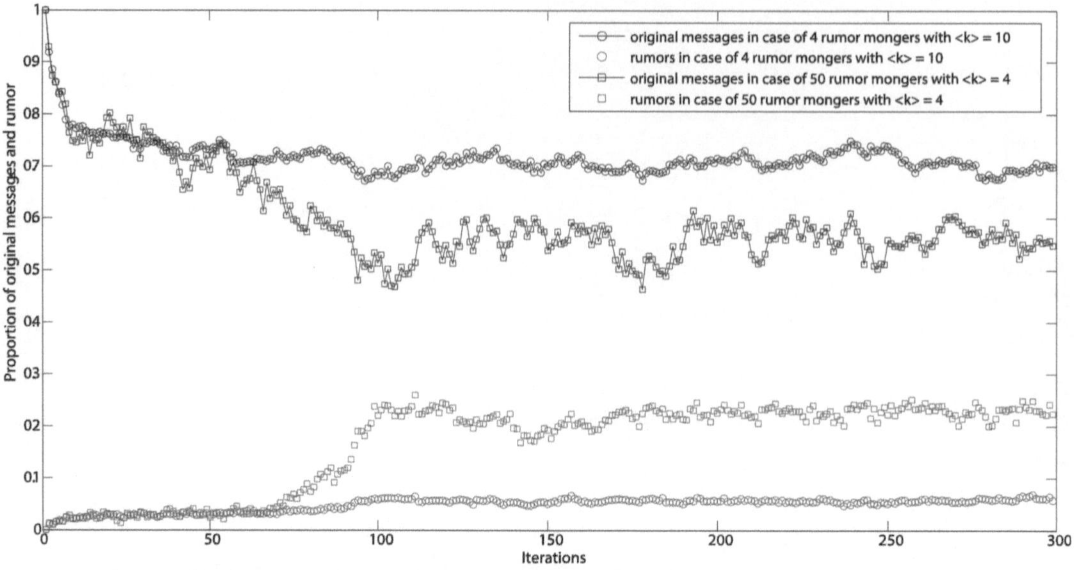

Figure 4. Fix $\lambda = 0.6$, and $p = 0.06$, we set several rumor mongers. The results show that the case of more rumor mongers and less degree may be easier to spread rumors than the case of less rumor mongers and more degree.

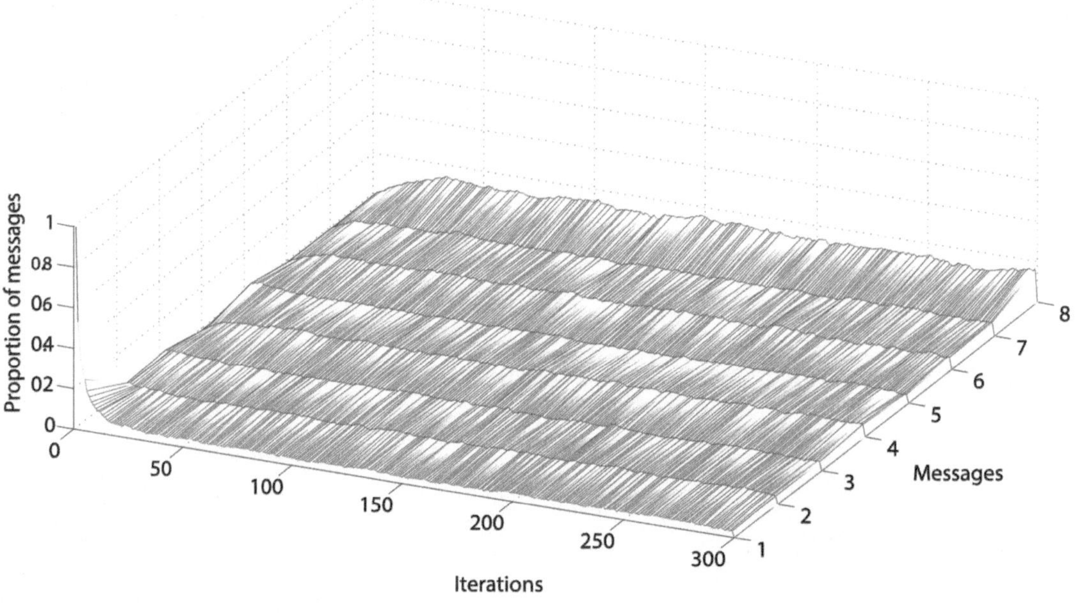

Figure 5. Fix $\lambda = 40.6$, and $p = 0.10$, we set 4 rumor mongers with $<k> = 10$. The result shows that the network in which it is easy to invalidate information may feed rumors effectively.

Finally, we are concerned with the distribution of trusted information in networks and the correlation between the distribution and some parameters. These parameters include, but are not limited to, the structure of the complex network, the stochastic process parameters of sending the message, the transition probability matrix P_M, the parameters W and τ about historical information, and the parameters about rumor mongers.

4 NUMERICAL SIMULATIONS

In this paper, we assume that the operator g has the following rules:

1) The node i trusts the message M_k which is the most one of its received messages. In the historical information case, it trusts the message M_k which has the maximal weight.

2) If the message which has the maximal weight is not the only one, the node i trusts none of them.

We also assume that the stochastic process of sending the message is the Bernoulli process with parameter λ.

Without loss of generality, we suppose $m = 3$ such that $M \in \{000,001,010,011,100,101,110,111\}$. The variation probability of each bit is p. It means that the transition probability matrix P_M is:

$$P = \begin{pmatrix} (1-p)^3 & (1-p)^2 p & \cdots & p^3 \\ (1-p)^2 p & (1-p)^3 & \ddots & (1-p)\ p^2 \\ \vdots & \ddots & \ddots & \vdots \\ p^3 & \cdots & \cdots & (1-p)^3 \end{pmatrix} \quad (6)$$

Without rumor mongers, the entire network is all trust M_{000} initially. The results are shown, after 300 iterations, in Figures 2–5.

5 RESEARCH PROSPECTS

In this paper, we give a finer model to describe information spreading in a network of nodes. The idea of multiple versions of information from the original information after tampering or variation meets the actual situation better. In fact, it is helpful to study rumor modelling in different cases by changing the information trust mechanism and sending behavioural mechanisms. Moreover, we can use this model to study how to improve the efficiency of resource utilization, such as less broadcasted information and fewer rumor mongers.

REFERENCES

[1] Fu X C, Small M, Walker D M, Zhang H F. Epidemic dynamics on scale-free networks with piecewise linear infectivity and immunization [J]. Physical Review E, 77 (2008) 036113.

[2] Zhou T, Fu Z Q, Wang B H. Epidemic dynamics on complex networks [J]. Progress in Natural Science, 2006, 16(5): 452–457.

[3] Xu X J, Peng H O, Wang X M, et al. Epidemic spreading with time delay in complex networks [J]. Physica A: Statistical Mechanics and its Applications, 2006, 367: 525–530.

[4] Fu X C, Small M and Chen G R. Propagation Dynamics on Complex Networks–Models, Methods and Stability Analysis [M]. John Wiley \& Sons, Inc., Chichester, England, 2014.

[5] Daley D J, Kendall D G. Stochastic rumours [J]. IMA Journal of Applied Mathematics, 1965, 1(1): 42–55.

[6] Nekovee M, Moreno Y, Bianconi G, et al. Theory of rumour spreading in complex social networks [J]. Physica A: Statistical Mechanics and its Applications, 2007, 374(1): 457–470.

[7] Trpevski D, Tang W K S, Kocarev L. Model for rumor spreading over networks [J]. Physical Review E, 2010, 81(5): 056102.

[8] Zhang Y, Zhou S, Zhang Z, et al. Rumor evolution in social networks [J]. Physical Review E, 2013, 87(3): 032133.

[9] Naimi Y, Naimi M. Reliability and efficiency of generalized rumor spreading model on complex social networks [J]. Communications in Theoretical Physics, 2013, 60(1): 139.

Future Communication Technology and Engineering – Chan (Ed.)
© 2015 Taylor & Francis Group, London, ISBN: 978-1-138-02777-0

Container multimodal transport system model study based on the RFID technology

C.H. Pan & H.M. Fan
Dalian Maritime University Transportation Management College, Liaoning, China

ABSTRACT: In recent years, with the rapid development of import and export trade, the requirement for the logistics to speed up the construction of container multimodal transport system has become more and more urgent. Like most advanced automatic identification technology, RFID technology now is widely used in manufacturing, retail, logistics, transportation and other industries. Based on the urgent demand for constructing the container multimodal transport system, the paper studied the applications of RFID technology in container multimodal transport system which provides a reference for a wide range of RFID technology applications in the future.

1 INTRODUCTION

As the most advanced automatic identification technology, RFID technology now is widely used in manufacturing, retail, logistics, transportation and other industries. FID technology can identify fast-moving objects and can identify more than one label, and the operation is quick and easy. It can greatly improve the efficiency of the port cargo acceptance. Besides, RFID technology's real-time information access technology enables the goods to be all-the-way tracked to ensure cargo's safety and avoid accidents in a timely manages.

Based on the urgent demand for constructing the container multimodal transport system, the paper studied the applications of RFID technology in container multimodal transport system. The paper analyzed the applications and problems of RFID technology in container multimodal transport system at first, and then based on RFID technology, there is an analysis for container multimodal transport system's overall architecture and the establishment of functional modules.

2 THE STATUS OF CONTAINER AUTOMATIC DIAGNOSIS TECHNIQUE RESEARCH

When it comes to the theory research of container tag system, in foreign countries, Doo-Jin Park, Young-Bok Choi and others presented real-time positioning systems based on RFID will be applied into the container management, and positioning algorithm will still be the yard assignment algorithm. But using RFID for data collection can greatly improve the efficiency of container management system[1]. E.W.T.Ngai and others proposed to apply RFID technology combined with mobile business in container warehouses and the system's goal is to track containers and to provide transparent data management and to improve the container management processes[2]. Francesco Rizzo and Marcello Barboni proposed a novel RFID system based on active identification to improve container transport and supply chain security issues and Detailed designed and introduced the logic and the main characters of the system[3]. Magalie Laniel and Jean-Pierre Emond studied using RFID reader/writer of three frequency bands within the closed container to measure the signal strength received by tags for temperature monitoring in real time in order to achieve better management of the perishable goods in the container store. They also applied this technique in cold chain logistics to monitor the internal temperature of reefer containers[4–5].

Among domestic studies, Zhengwu Yuan and Dongli Huang designed a new architecture of container positioning and recognition based on the RFID in a detailed way. Wei Wang and Shidong Fan introduced the RFID technology application in container transport from the management of port container transport equipment. Zhang Jia from Nanjing Aeronautics and Astronautics University created a detailed container transportation management model based on RFID and analyzed the concrete process of applying RFID in container yards in regards to yards. XUE Xiaoping, MEI Su-ping and others proposed a research of railway container transport management system combined with RFID information acquisition device with ATIS information system.

3 APPLICATION OF RFID TECHNOLOGY IN CONTAINER PORTS

As technology continues to improve, many container automatic diagnosis systems are put into use both at home and abroad, and automatic identification system based on RFID technology is one of them. At present, examples of RFID application in containers mainly include the following:

(1) Container Security Initiative (CSI) advocated by the United States

Since "9·11" terrorist attacks, the United States took the lead in proposing concepts like "The smart containers", "The green channel" and etc. Besides, its cooperated well-known enterprises do related studies. CSI is a new action took by the United States to ensure the safety of containers entering the ports, and to reduce threats of terrorist attacks. Electronic seals will be installed on all imported containers. The reading instruments get data from electronic seals, and then the data will be transmitted to the specially designed information platform. When the container is damaged or the route changes or the delay and other incidents happen, managers can quickly receive automatic alarms from the system by computers, cell phones or PDAs.

(2) "Smart Containers Plan" implemented by the South Korea Government

The South Korea Government planned "Smart Containers" project in 2004, and in December of this year the "RFID Maritime Logistics" project will be officially launched and funded. The goal of this project is to enhance the South Korea's international trades' efficiency and safety by applying RFID technology scheme. The "Smart Containers" with intelligent labels and sensing devices are shipped from Pusan port to the west coast of the United States and major ports of Europe under the motoring of transport security system (TSS) software. This project will use more than one active RFID products, including an electronic seal, a sensor seal supporting RFID. They can be attached to the container, and transit some key information such as location, container security, changes in temperature and humidity to the reader network. People will be able to monitor containers within the supply chain.

(3) Application of container tag system for two ports and one shipping line in Shanghai port

December 3rd 2005, the "Zhejiang Shipping No. 325" officially set sail heading for Shanghai from Yantai port. The liner ship was loaded with more than 300 containers with electronic tag attached for the first time. This industrial test of automatically recognition technology system applied in containers of two ports and one shipping line, opened up the first route of containers with electronic labels, combined transportation features container in domestic trade. It used integrated technology of double frequency electronic label system, and the label stored the information of container and the loaded goods, and can record the related information such as the switch-on and off of the door automatically, which completely realized the automatically recognition of information during container transport links, and achieved container transport links and real-time information exchange. Prior to this, there had not been successful precedents in the world that applied electronic labels during the container route between ports.

(4) RFID put into commercial operation in Hong Kong and Shenzhen ports

Ports in Hong Kong and Shenzhen have been equipped with radio frequency identification tracking equipments for commercial operation. The technology provider of this project operates a global information network which uses an effective radio frequency identification equipment and software to provide carriers with information about the location of their containers and status of goods.

4 RESEARCH ON CONTAINER MULTIMODAL TRANSPORT SYSTEM MODEL BASED ON RFID TECHNOLOGY

4.1 *Container multimodal transport system architecture*

The establishment of platform aiming for container multimodal transport system for business processes, based on the RFID Port information technology platform, can not only optimize port operations organization and improve work efficiency, but also integrate the port EDI Information for multimodal transport container system which can provide real-time information enquiry service for service object of every link of container multimodal transport system which lay the foundation for achieving the integrated management of international container's full, real-time and online monitoring.

The overall architecture of container multimodal transport system based on RFID technology, conceptually, can be divided into four layers from the bottom up: the device layer, the communication layer, the data-tier and application-tier.

(1) Device layer

Device layer, also called the automatic identification system for electronic tags, is the most basic and the most important part of the system, which mainly completes front automatic data acquisition and recognition of the container transport and cargo information systems. This layer mainly includes hardware devices ranging from RFID tags, FID reader, FID edge server, GPS, bar-code equipment, image recognition equipment, which can be used to collect the information automatically, quickly and reliably. It also includes electronic billboards, gantry cranes, trucks, containers, forklifts, pallets, bridge cranes and other infrastructures.

(2) Communication layer
Communication layer is a real-time information interactive system which mainly completes real-time information of container or goods seamless access to background information. This information is read in identification equipment in pier crossings in and out, yard gantry cranes, container ship and quayside container crane or gantry crane and other equipment. This layer mainly consists of wireless LAN devices, Ethernet, transmission equipment, RS232/485 equipment, data communication protocols, GSM/GPRS transmission networks and so on.

(3) Data layer
Data layer mainly completes the screening, classification, storage, analysis and processing of real time data access, which not only maintains real-time interaction with the communications layer, but also provides the basis for analyses and deal with problems for the application layer. The data layer plays a connecting role in the process. This layer mainly consists of docks, yard Center database and so on.

(4) Application layer
Application layer, actually, is a WAN system which integrates data of all LAN system for cross-system, and cross-platform's using. Through setting access permission, it can not only achieves each enterprise's track management on the logistic information of container and the transport vehicles, but also meets the regulatory requirements of customs, state quality inspection and other management institutions on container or on the transport vehicles. Meanwhile, it can meet simple operation of Terminal, yard and other actual operation sectors on container business process.

The whole model adopted Browser/Server (B/S) Architecture to do the real time control of container transport and provided yard management, graphical representation of shipping, query and monitor interface for customs and other government departments. Freight forwarders, shipping agents can track ships and cargos through PDA in real time.

4.2 *The function module of intermodal container system*

According to intermodal container system for business processes and characteristics of the port container assignment, and considering the core task of container port, using RF recognition technology is completed on container information of real-time collection and on container and the vehicles of dynamic monitoring management, which determines the basic function based on RFID technology of intermodal container system. These functions can both solve the node bottleneck problem in the process of container port production and meet customer's needs on timely receiving goods information and further perfect the work of customs, inspection and official sectors and other official departments in the function of real-time monitoring the import and export goods. Thus, the function modules

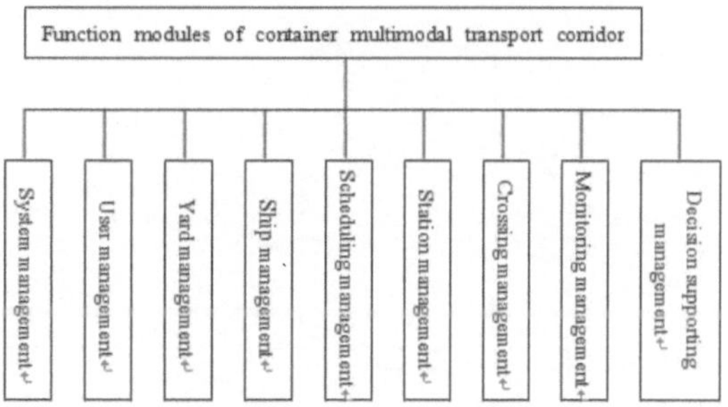

Figure 1. Function modules of container multimodal transport corridor based on RFID.

of intermodal container system include: system management, user management, yard management, ship management, job scheduling management, entrance of terminal management, monitoring management, decision support management and so on, as Figure 1 shows.

System guided by modern logistics management concepts, taking RFID as the main technical means, with GPS Intelligent terminals, EDI Digital and information technology, can monitor the transport, management and upload or download in the process of port logistics, and can realize the information collection and automatic identification in domestic and international container port enter and exit crossings, yards, boat handling and other link. While building container terminal EDI and trade enquiry system based on RFID technology, which can track the whole information of container logistics processes, have full cargo container logistics information and realize the intelligent and visual management of container port transport process in order to improve the efficiency of port operations and reduce management costs. Systems running can strengthen the management of container port for containers and truck vehicles, and can further strengthen the customs supervision and improve custom's work efficiency. Thus the container port logistics management level and service level will be upgrade and logistics costs will be reduced.

In addition, the system provides a number of functions modules designed for many different types of users. These different types of users include: import and export companies, freight forwarders, shipping agents, port managers, handling companies, government departments as well as financial institutions. The system assigns different permissions for different users to make each type of users can have different operation interface, and users can just manage and monitor the information within the scope of their duties. Of course, the system is equipped with a dedicated system administrator in order to maintain and coordinate the overall operation of the system timely. Management model of container multimodal transport corridor facing different users as shown in Figure 2 below.

Figure 2. Management model of container multimodal transport corridor facing different users.

5 CONCLUSION

With the rapid development of the multimodal transportation of container and the RFID technology, the paper studied applications of RFID technology in multimodal transportation system. First of all, it gave a conclusion of the basic requirement which meet the application of RFID technology in the container of port according to the comparative analysis for automatic identification technology of the container. In addition, it analyzed the status quo of RFID technology showing that many scholars at home and abroad have been committing to the RFID technology application and analyzed the application and problems of RFID technology in container ports. The problem mainly included: different criteria, lack of integration and the variety of applications. Finally, the paper gave the analysis of the overall architecture and the establishment of functional modules based on RFID technology for container multimodal transport system.

REFERENCES

[1] Thevissen, Patrick. lmplantation of an RFID tag into human molars to reduce hardforensic identification labor, Forensic Science International, 2006, 159(5): 33–39.

[2] L. Frisk, J. Jarvlnen, R. Ristolainen. Chip on flex attachment with thermoplastic ACF for RFID applications, Zoology, 2003, 106(3): 169–171.

[3] Doo-Jin Park, Young-Bok Choi, Ki-Chan Nam. RFID-Based RTLS for Improvement of Operation System in Container Terminals [R]. Institute of Electrical and Electronics Engineer, 2006.

[4] E.W.T. Ngai, T.C.E. Cheng, S. Au, Kee-hung Lai. Mobile commerce integrated with RFID technology in a container depot [R]. Decision Support Systems, 2001, 43: 62–76.

[5] Francesco Rizzo, Marcello Barboni, Lorenzo Faggion, Graziano Azzalin, Marco Sironi. Improved security for commercial container transports using an innovative active RFID system [C]. Journal of Network and Computer Applications, 2010.

Future Communication Technology and Engineering – Chan (Ed.)
© 2015 Taylor & Francis Group, London, ISBN: 978-1-138-02777-0

Design of an OAM radio beam generator using microstrip patch array

F.C. Mao, J.J. Yang & M. Huang
Wireless Innovation Lab of Yunnan University, School of Information Science and Engineering, Kunming, Yunnan, China

S. Yan
School of Information and Electronics, Beijing Institute of Technology, Beijing, China

Y. Feng
Radio Monitoring Center of Yunnan Province, Kunming, Yunnan, China

ABSTRACT: Orbital Angular Momentum (OAM) of Laguerre-gaussian beams can form infinite dimensional Hilbert spaces. As thus, the same carrier frequency using OAM muliplexing will have infinite transmission capacity theoretically. How to generate radio beams carrying OAM with arbitrary mode number is therefore a core issue. This paper presents the design of an OAM radio beam generator using microstrip patch array antenna at 5.3 GHz. The transmission processes of the OAM beams are simulated. The influence of array radius and the number of elements on OAM antenna performance are analyzed in detail.

1 INTRODUCTION

Since the openness of wireless channel, how to share simultaneously a finite amount of radio spectrum through multiplexing becomes a hot spot. To a great extent, the development of wireless communication stems from the development of multiplexing techniques. For instance, the first generation (1G) cellular systems like AMPS rely on FDM/FDD in the extreme; the second generation (2G) cellular networks, GSM, IS-136 and PDC employ TDM/FDD while IS-95 employs CDM/FDD; the third generation (3G) wireless networks, such as W-CDMA, cdma2000 and TD-SCDMA, introduce SDM based on smart antenna techniques; and the fourth generation (4G) mobile communication systems, like TD-LTE as well as FDD-LTE, combine the OFDM and MIMO [1–3]. The first experimental test based on OAM muliplexing is implemented by Tamburini et al [4], soon after, a terabit free-space data transmission with OAM muliplexing is realized by Wang et al [5]. OAM is a new degree of freedom of enhancing wireless spectrum efficiency.

In 1992, Allen et al recognized that well-defined OAM could be generated in beams of laser light with a Laguerre-Gaussian amplitude distribution in paraxial approximation theory [6]. Since then OAM has been used in domains such as optical physics, atomic and molecular physics, radio astronomy, atmosphere and ionosphere and plasma [7,8]. Besides, the application of OAM in radio communications has been explored [9]. At optical band, an OAM mode state can be achieved by multiple ways, such as dipole-coupled plasmonic spiral antenna [10], q-plate [11], metasurface [12], and the computer generated holography method [13]. However, at radio band, the OAM states are mainly generated by antenna techniques. For example, a helicoidal parabolic antenna with a single radial cut introduced on its surface is adopted to construct a helical phase front [4,14,15]. By selecting the surface pitch parameter a desired OAM state can be generated. Another noticeable method is to use an array antenna. In 2007, the first numerical experiments performed by Thidé et al [7] have shown that a circular dipoles array can be used to generate the radio beams with OAM characteristics. A system study on such a scenario was carried out by Mohammadi et al [16,17] in 2010. Then in 2013, Deng et al [18] presented a kind of Vivaldi antenna array to produce OAM beams. Recently, Bai et al [19] proposed a circular patch array using rectangular element to radiate radio beams with OAM properties. A patch antenna for generating OAM radio wave beam is designed by our group [20]. In this work, we extend the previous work and design a rectangular patch antenna to transmit radio beams carrying OAM at 5.3 GHz. The rest of the paper is organized as follows. Firstly, the structure of the patch array antenna is introduced; secondly, mutual coupling among the elements is discussed; thirdly, the influence of array radius and the number of elements on antenna performance are simulated and analyzed, and the transmission process of the OAM beam is shown. All the simulations are performed by using the finite-element software HFSS.

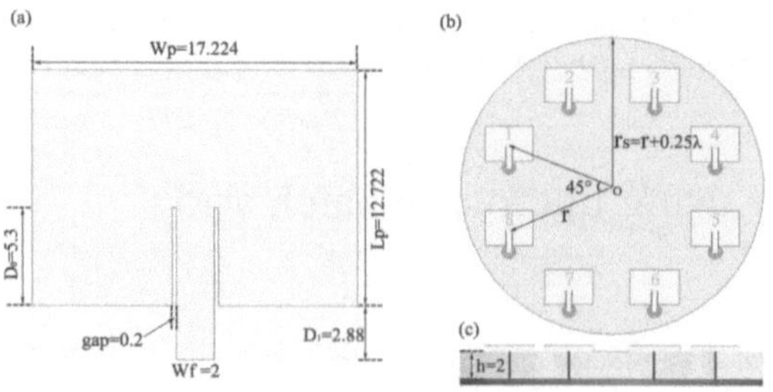

Figure 1. Antenna Configuration: (a) geometry details of the single rectangular patch; (b) the top view of the circular array antenna; (c) the front view of the circular array antenna.

2 ELEMENT ANTENNA AND THE ARRAY

In order to generate OAM at radio frequency with antenna array, we consider the configuration depicted in Figure 1. The radiation element employed in the design is a rectangular patch antenna, which is conformable to both planar and curved surfaces, as well as easy to manufacture. The inset-feed allows this antenna to be designed for a wide impedance range without using extra impedance matching circuitry [21]. As shown in Figure 1(a), the radiation element consists of a 12.722 mm by 17.224 mm patch fed by a 2 mm wide microstrip line. Details of the designing procedure of such a patch can be found in reference [21]. Figure 1(b) provides the top view of the proposed circular patch array antenna. Each patch is located equidistantly along the perimeter of a circle with radius r, radius of the substrate is $r_s = r + \lambda/4$. The patches are fed by microstrip line to retain the planar geometry, and they can be etched on a same substrate along with the feeding networks together. The azimuthal distance between each two adjacent elements is $360°/N$, where N is the number of array elements. Figure 1(c) shows the front view of the array antenna. This design uses FR4 substrate with a relative permittivity 4.4 and 2 mm thickness.

Firstly, we simulated the S_{11} and the 3D radiation pattern of a single patch, as shown in panels (a) (c) of Figure 2. It is seen that this patch antenna operates at 5.3 GHz with a -12.5 dB bandwidth from 5.1 GHz to 5.5 GHz; its radiation pattern has a unipolarity property. The calculated antenna gain is 1.21 dB with radiation efficiency of 85%. As an example, a patch array antenna with 8 elements operating in mode 1 has been simulated. The array radius is set to $r = 0.7\lambda$. Figure 2(b) and Figure 2(d) give the reflection coefficients and transmission coefficients of the phased patch array, respectively. It demonstrates that these radiation elements have identical resonant frequencies since the very weak mutual coupling among them. The strongest mutual coupling is obtained between element 1–8 and 4–5, which is less than -25 dB (Figure 2(d)). The mutual coupling can be tuned by adjusting r and N. Next section will focus on analyzing the performance of the phased patch array with different array radii and different OAM modes.

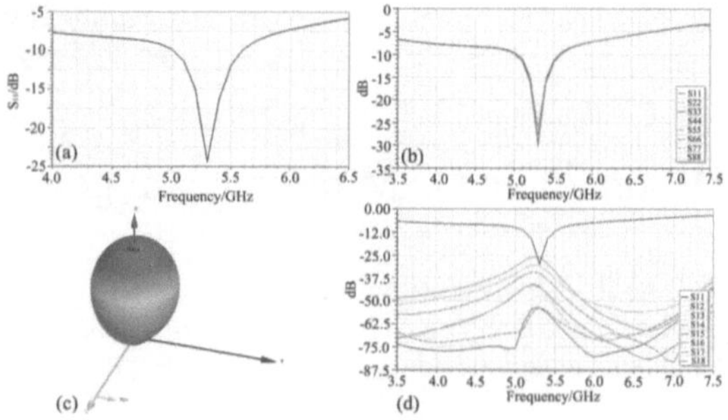

Figure 2. Antenna performance: (a) reflection coefficient of an element antenna; (b) reflection coefficients of circular patch array; (c) 3D radiation pattern of an element antenna; (d) transmission coefficients of the circular patch array.

Figure 3. Phase distribution of E_z: Generated by an 8-element circular phased patch array. Array radius is $r = 0.5\lambda$. Dimension of view window is $6\lambda \times 6\lambda$.

3 OAM GENERATED BY CIRCULAR ARRAY

It has been verified that a radio beam carrying OAM can be generated by a circular phased antenna array [7,17]. In such circular array, all the N radiation elements are excited by the same signal but successively delayed relative to each other with an increment $2\pi l/N$, where l is the mode number of OAM. The total field from the array can be calculated by a superposition of the fields from each element. Using the pattern multiplication rule the total field could be expressed as the product of the array factor and the field from single element. When fed by current source $Ie^{j\beta_n}$, where I is current amplitude and $\beta_n = 2\pi nl/N$ is current phase, the array factor F_a of such array can be wrote as

$$F_a = \sum_{n=1}^{N} \exp[jkr\sin\theta\cos(\varphi - \varphi_n) + j\beta_n] \tag{1}$$

where k is the wave number and ϕ_n is the azimuthal angle of the nth element. This array factor contains phase factor $\exp(jl\phi)$ which results in a helical phase front structure [22].

A key property of an OAM-carrying beam is the helical phase front in its field. The number of branch cuts of the helix where the phase jumps from π to $-\pi$ depends on the OAM mode of the beam. A beam with OAM mode $l = 0$ has 0 branch cut, $l = 1$ has 1 branch cut, and so on. This can be observed in Figure 3. The largest l mode an array can generate is determined

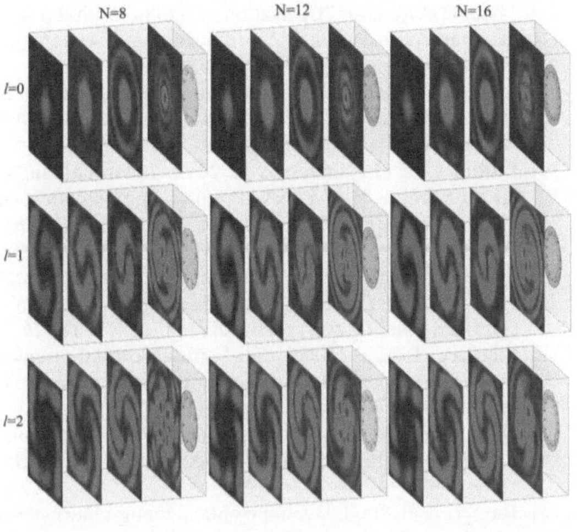

Figure 4. Intensity distribution of E_z: The magnitudes of electrical fields are drawn at planes $z = \lambda$, $z = 2.67\lambda$, $z = 4.34\lambda$, $z = 6\lambda$, from the nearest one to the farthest relative to the array position. The dimension of view window is $5\lambda \times 5\lambda$.

by its element number [7]. Theory predicts $|l| < N/2$. Beams with a large l will not produce a pure rotating phase front and hence no perfect OAM mode. These beams will be distorted and will not carry a stable OAM mode. In our 8-element array case (Figure 3(d)), the advised modes are $l = 0, \pm 1, \pm 2$.

The transmission process of OAM beams generated by the array antenna with different element number is shown in Figure 4. The OAM beams with $l = 0$, 1 and 2 for each case are simulated. It is seen that for the same l, increasing the array number will improve the continuity of the phase. Besides, we can get higher OAM modes by increasing the element number, and the antenna gain can be improved slightly by increasing the array number. For instance, an enhancement of 0.5dB can be obtained by increasing the number from 8 to 12 for the case of $l = 0$.

To study the influence of array radius on antenna performance, we simulate the far field radiation pattern of the antenna with 8 elements. The radius varies from 0.5λ to 2λ, while OAM modes are $l = 0, 1, 2$. The 2D radiation patterns in the plane of $\phi = 0°$, 45° and 90° are plotted in Figure 5. In OAM mode $l = 0$, the array mainly radiates in the boresight direction. From the subfigures in columns, it is obvious that enlarging the array radius will narrow the main lobe beam width while elevate the side lobe. For $l = 1$ and $l = 2$, a null appears at the boresight due to each opposite pair of elements are operating out of phase. The centre null scale can be reduced when changing the array radius from 0.5λ to 2λ, meanwhile, the antenna gain will be increased. In addition, it can also be viewed from each panel that the antenna gain decreases with increasing mode number and the null size increases accordingly. On the other hand, we can see that the radiation patterns in different azimuthal angles for the antenna with the same array radius and modes

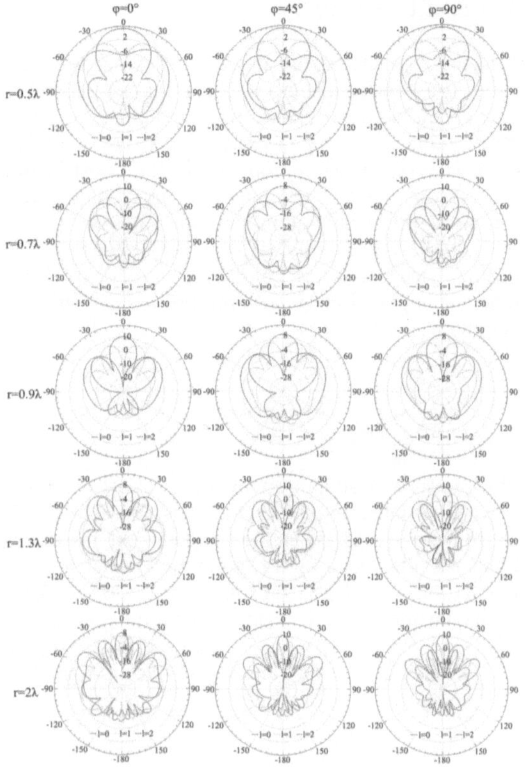

Figure 5. Radiation patterns of 8-element circular phased patch array. The array radius r varies from 0.5λ to 2λ. The azimuthal angles are $\phi = 0°, \phi = 45°, \phi = 90°$.

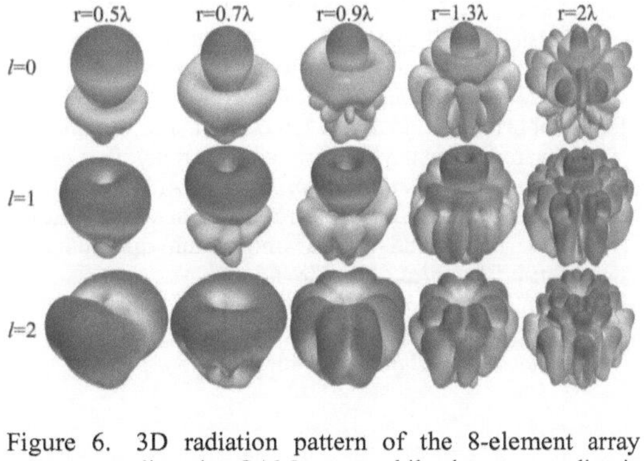

Figure 6. 3D radiation pattern of the 8-element array operates at diversity OAM states while the array radius is changing from 0.5λ to 2λ.

discriminate against each other. Such phenomenon can be further observed in Figure 6 which provides the corresponding 3D polar plots. This is due to the local effect caused by the geometry structure of element antenna, resulting in non-perfect omni bearing radiation, instead of the ideal point source or electrically short dipole whose radiation pattern possesses rotational symmetry property.

4 CONCLUSIONS

A microstrip patch array antenna is designed to generate OAM radio beam. By arranging the patches equidistantly along the perimeter of a circle, an array factor containing the phase factor $\exp(jl\phi)$ is configured. The transmission processes of the OAM beams are demonstrated. The impact of array parameters including the array radius and the number of elements on OAM antenna performance are analyzed in detail. These results will guide us to further exploit the practical application of OAM multiplexing in wireless communications.

ACKNOWLEDGMENTS

This work was supported by the National Natural Science Foundation of China (Grant Nos. 61161007, 61261002, 61461052), the Specialized Research Fund for the Doctoral Program of Higher Education (Grant No. 20135301110003, 20125301120009), China Postdoctoral Science Foundation (Grant No. 2013M531989, 2014T70890), and the Key Program of Natural Science of Yunnan Province (Grant No. 2013FA006).

REFERENCES

[1] Rappaport T S 1996. *Wireless communications: principles and practice* New Jersey: Prentice Hall PTR.

[2] Blogh J S & Hanzo L L 2002. *Third-generation systems and intelligent wireless networking: smart antennas and adaptive modulation* John Wiley & Sons.

[3] Ghosh A et al 2010. LTE-advanced: next-generation wireless broadband technology [Invited Paper]. *IEEE Wireless Communication* 17(3): 10–22.

[4] Tamburini F et al. 2012. Encoding many channels on the same frequency through radio vorticity: first experimental test. *New Journal of Physics* 14(3): 033001.

[5] Wang J et al. 2012. Terabit free-space data transmission employing orbital angular momentum multiplexing. *Nature Photonics* 6(7): 488–496

[6] Allen L et al. 1992. Orbital angular momentum of light and the transformation of Laguerre-Gaussian laser modes. *Physical Review A* 45(11): 8185–8189.

[7] Thidé B et al. 2007. Utilization of photon orbital angular momentum in the low-frequency radio domain. *Physical review letters* 99(8): 087701.

[8] Leyser T B et al. 2009. Radio pumping of ionospheric plasma with orbital angular momentum. *Physical review letters* 102(6): 065004.

[9] Edfors O & Johansson A J. 2012. Is orbital angular momentum (OAM) based radio communication an unexploited area. *Antennas and Propagation, IEEE Transactions on* 60(2): 1126–1131.

[10] Rui G et al. 2012. Beaming photons with spin and orbital angular momentum via a dipole-coupled plasmonic spiral antenna. *Optics express* 20(17): 18819–18826.

[11] Marrucci L et al. 2006. ptical spin-to-orbital angular momentum conversion in inhomogeneous anisotropic media. *Physical review letters* 96(16): 163905.

[12] Yu N et al. 2011. Light propagation with phase discontinuities: generalized laws of reflection and refraction. *Science* 334(6054): 333–337.

[13] Tao S H et al. 2003. Dynamic optical manipulation with a higher-order fractional Bessel beam generated from a spatial light modulator. *Optics letters* 28(20): 1867–1869.

[14] Tamburini F et al. 2013. Experimental demonstration of free-space information transfer using phase modulated orbital angular momentum radio. *arXiv preprint arXiv*: 1302.2990.

[15] Abhay S A et al. 2013. Simulation of the S-Band photon vorticity modulation scheme. *American Journal of Electromagnetics and Applications* 1(2): 38–43.

[16] Mohammadi S M et al. 2010. Orbital angular momentum in radio: Measurement methods. Radio Science 45(4): RS4007.

[17] Mohammadi S M et al. 2010. Orbital angular momentum in radio—a system study. *Antennas and Propagation, IEEE Transactions on* 58(2): 565–572.

[18] Deng C et al. 2013. Generation of OAM Radio Waves Using Circular Vivaldi Antenna Array. *International Journal of Antennas and Propagation* vol. 2013: 1–7.

[19] Bai Q et al. 2013. Generation of orbital angular momentum (OAM) radio beams with phased patch array. *Antennas and Propagation Conference (LAPC), 2013, IEEE, 11–12 November 2013.* 2013 Loughborough.

[20] Qin C L et al. 2014. Generation of OAM Radio Waves Using Patch Antenna. *Advanced Materials Research.* 2014, 926: 2337–2340.

[21] Matin M A & Sayeed A I. 2010. A design rule for inset-fed rectangular microstrip patch antenna. *Issue*, vol. 1: 63–72.

[22] Knudsen H L. 1953. The field radiated by a ring quasi-array of an infinite number of tangential or radial dipoles. *Proceedings of the IRE* 41(6): 781–789.

Future Communication Technology and Engineering – Chan (Ed.)
© 2015 Taylor & Francis Group, London, ISBN: 978-1-138-02777-0

Research and design of a forest management mobile service cloud platform for the natural forest protection project

F. Li, Y. Ma, X. Zhang, X.W. Yu & P.F. Feng
Institute of Forest Resource Information Technique, Chinese Academy of Forestry, Beijing, China

M.B. Zhang
Science and Technology Agency, Chinese Academy of Forestry, Beijing, China

ABSTRACT: This paper analyses the actual demand of the forestry protection project daily management, and puts forward solutions and methods for the forestry protection project and forest management service platform, based on mobile cloud computing technology. The project, equipped with intelligent terminal equipment for each ranger, is effective in monitoring the management of a forest area. The location information can be sent back to the cloud centre node at any time. The intelligent terminal equipment is designed based on mobile cloud computing technology. At the same time, images, video, and other information about fire, pests, deforestation, and so on, can be acquired and reported timely.

1 INTRODUCTION

The natural forest protection project, also known as the natural forest resource protection project, in China, is mainly implemented in the upper reaches of the Yangtze River, the upper and middle reaches of the Yellow River, the Northeast, Inner Mongolia, and other key state-owned forest regions. The engineering construction aims to achieve three major goals: to protect the existing forest resources in order to accelerate the fostering of forest resources, to shunt the placement of surplus staff and workers, and to realize the strategic transfer of forest industry enterprises and the reasonable adjustment of the industrial structure. The first phase of the project started in 1998 and ended in 2010. During this time, national investment for the project added up to 106.4 billion Yuan. The main task of the second phase of the project is to continue stopping the commercial logging of the natural forest in the Yangtze river upstream, the middle and upper reaches of the Yellow River, the forest region of the Northeast, Inner Mongolia, and other key state-owned forests, whose yield has reduced from 10.941 million cubic metres to 4.025 million cubic metres in three years (the protected forest is 1.732 billion acres, the public welfare forest is 1.16 billion acres, and the state-owned sapling forest is 2.63 billion acres.) 48.9 million acres of reserve resources continue to give subsidies to state-owned employee social insurance and social expenditure.

In such a huge natural forest protection project, management work is a very important working content; China has established a multi-level regulatory mode. The State Forestry Bureau is responsible for the organization, coordination, guidance, and supervision of the protection work of the project. The competent forestry authorities of provinces, autonomous regions, and municipalities, which are directly under the central government, allocate the forest management index, and establish and improve the forest management responsibility system. Units of the natural forest protection project (including the County Forestry Department, the state-owned forestry enterprises of state-owned forest farms, etc.) have established the forest management organization system, which consists of the county (bureau), township (forest), village (group, area), and management site, to establish the perfect management system for forest management. The forest workers or full-time rangers are the main part of the management responsible for patrols; supervisors are responsible for the implementation of management responsibility, including non-regular inspections of workplaces, recording data, and reporting the information to the management department. At present, the regulation efficiency is low; access to information is not timely and comprehensive, which cannot meet the needs of the information management of the natural forest protection project. In order to further strengthen the ranger management work, and their sense of work responsibility, as well as to prevent forest fires, pests, deforestation events, and prevent patrol inspection does not reach the designated position, we need to collect and transmit the location information of the rangers and effectively monitor the management area of the forest ranger.

With cloud computing, 3G/4G communication technology, the LBS (Location-based service) technology, mobile GIS (Geographic Information System) technology, and the popularity of smart phones' built-in GPS (Global Position System) modules, the mobile Internet has been involved in both people's personal life and work. Cloud computing, as a new mode of network applications, gets the heterogeneous and autonomous computing resources together through the Internet, which is able to handle huge amounts of equipment. Cloud computing combined with smart phones is formed by mobile cloud computing via smartphones, which give people access to services from the cloud platform in the communications network coverage within the scope at any place, any time, and in any environment.

According to the demands of the daily management of the natural forest protection project area, we construct the project's Intelligent Inspection System based on mobile cloud computing technology. In addition, through the smart mobile phone and cloud computing platform of the research institute of forest resource information techniques, we can provide a high performance computing environment as a priority project area. The key project only needs to equip patrols with smartphones equipped with GPS modules and installed with the intelligent inspection App. It can set an inspection plan, an inspection line, and a checking point for the ranger, and send information to the mobile client through a web browser. It not only reduces the work intensity of the ranger and improves working efficiency, but also reduces a lot of informatization investment.

2 ANALYSIS OF KEY TECHNOLOGIES

2.1 *Mobile cloud computing technology*

The concept of "cloud computing" was proposed in 2006 by Google. Cloud computing is a network-based technology; it provides shared resources, software, and information to the requested computer or other devices. The development of mobile cloud computing began with the development of the mobile Internet. Mobile cloud computing is a major branch in the future development of cloud computing. Users can use the interest network services via mobile cloud computing and not accept an equipment limit. The successes of mobile cloud computing companies are:

(1) The Canadian RIM (Research In Motion Ltd.) provides a BlackBerry enterprise application server project. The BlackBerry E-mail server connects the enterprise applications, wireless network, and mobile terminals together; users can access their mail account via the BlackBerry terminal, which can enable users to synchronize their mail and calendar, view attachments, and their address books;

(2) Google, as the pioneer of mobile cloud computing, actively develops the Android platform for the mobile environment and terminals; it continuously introduces new applications based on mobile terminals and cloud computing, including integration of mobile searches, voice search services, geographic searches, Google mobile maps, etc.

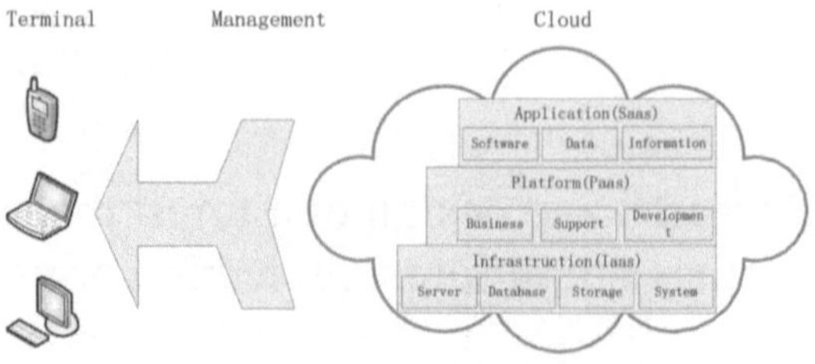

Figure 1. Mobile cloud computing "terminal", "management", "cloud" model, and service type.

Mobile cloud computing completes the main calculation in the "cloud"; the "terminal" is the smartphone. Its data and technical requirements are sent in the form of services to the cloud. An effective management platform also needs to allocate resources and request scheduling between the "cloud" and the "terminal."

2.2 *LBS*

A location-based service obtains the position information of the mobile terminal users (geographic coordinates, or the earth coordinates) through the telecom operators' radio communication network (GSM, CDMA network), or external positioning way (e.g. GPS), with the support of a geographic information system platform in order to provide a value-added service of corresponding services for the users.

The system uses Baidu map Position-SDK to provide basic positioning capability for the mobile client's LBS application. Position-SDK using GPS, base station, and Wi-Fi signal to acquire location information. When the application launches a location request to the position-SDK, SDK positioning will determine the position according to the actual situation (such as whether to open the GPS, whether to connect to the Internet, if it has any signals, etc.).

3 THE SYSTEM DESIGN

In accordance with the natural forest resource conservation project management requirements, our platform can be divided into the mobile cloud App and the cloud services platform. The topological structure of the system is shown in Figure 3.

The system work-flow is as follows. Managers set the district inspection scheme in the cloud services platform; rangers operating protection work in the forest carry the cloud terminal, which is connected with the cloud service platform, so the system can query the duties. For example, if the inspector does not patrol the specified site, the platform will not receive geographic information data of that point. By statistical analysis, the software cloud service platform can clearly reflect the ranger's work each month. With a scientific and accurate appraisal basis, the ranger can be rewarded or penalised, the key project management district overall workload quantification analysis can be made, and a reasonable allocation can be made of the maintenance

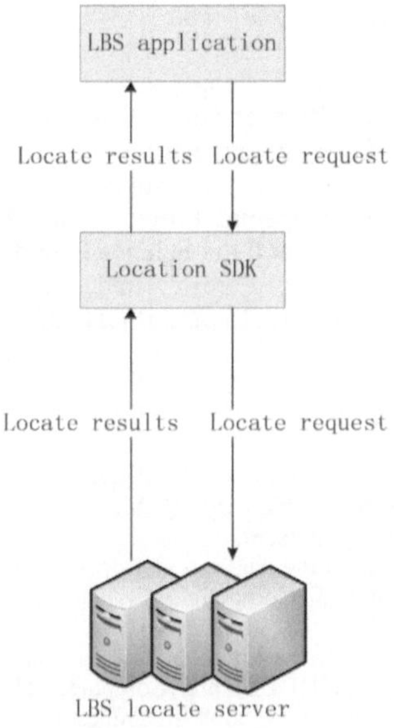

Figure 2. LBS location services.

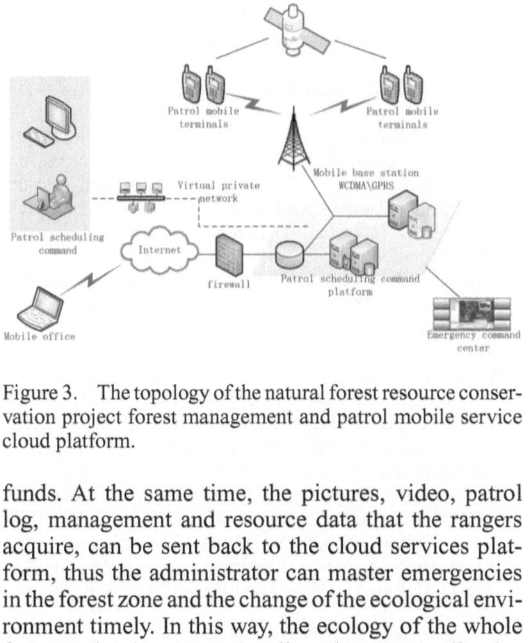

Figure 3. The topology of the natural forest resource conservation project forest management and patrol mobile service cloud platform.

funds. At the same time, the pictures, video, patrol log, management and resource data that the rangers acquire, can be sent back to the cloud services platform, thus the administrator can master emergencies in the forest zone and the change of the ecological environment timely. In this way, the ecology of the whole forest region is protected well, and we can really play the supervisory role of the forestry frontier ranger.

3.1 *The design of the mobile cloud terminal*

The natural forest protection project with the intelligent inspection cloud terminal App contains many features, including checking the patrol mission, an attendance clock, an alarm, checking location data acquisition, a key, a predefined inspection area, the route of inspection, the inspection time, the inspection personnel, the scheduling plan, a network inquiry, track playback, remote management, and long distance transmission. It can effectively monitor the management area of the forest ranger, and usefully report pictures and video of fires, insect pest, deforestation, and other information.

An intelligent mobile phone installed with a patrol management App is professional equipment based on the Android operating system. It has many features including a global positioning satellite (GPS), a data communication transmission function, and high definition video acquisition and transmission (300,0000-megapixel autofocus camera, h. 264 video compression format). It has quasi-three functions, large capacity batteries (the effective working time is 12 hours or more), and a professional grade GPS superstar search ability (the average signal strength can reach 49dB, which is five times that of an ordinary mobile phone). It is of high performance, high quality, and easy to operate as one of the specialized terminals.

A mobile terminal login system can automatically download the inspection task and line through the web service from the cloud centre system. Through the terminal's navigation system, inspection lines can be visually displayed on the electronic map and guide the inspector to carry out the work. When inspectors arrive at the inspection point, the terminal automatically judges and suggests that inspectors need to check the project. Defects in the inspection terminal recorded in the process will be uploaded to the cloud centre system.

3.2 *Protection forest management of cloud platform design*

The user access control in the cloud computing platform of the research institute of the forest resource information techniques of CAF (Chinese Academy of Forestry), constructs a patrol management portal system for each test natural forest protection area, which integrates the experimental zone of basic geographic information data, social and humanistic information, and control station and management staff information. All this data is uniformly stored and managed by the cloud storage system.

After the administrator logs into the system, all rangers' locations can be displayed to the client on the map; by comparing the pre-set patrol route, ranger's attendance of each inspection point can be clearly judged. At the same time, the history track and altitude information of the forest ranger can be viewed anytime, judging whether inspection is in accordance with the pre-set line. Through the real-time transmission of the network, administrators can usefully access the last information that the rangers found (including descriptions, pictures, and video). In an emergency, the portal can be connected to the emergency command centre to on-site scheduling, and accident treatment.

The function module and the data used in the portal exist in the form of a web service in the cloud platform. As long as the test area demands it, the cloud platform can provide personalized needs by rapid assembly of

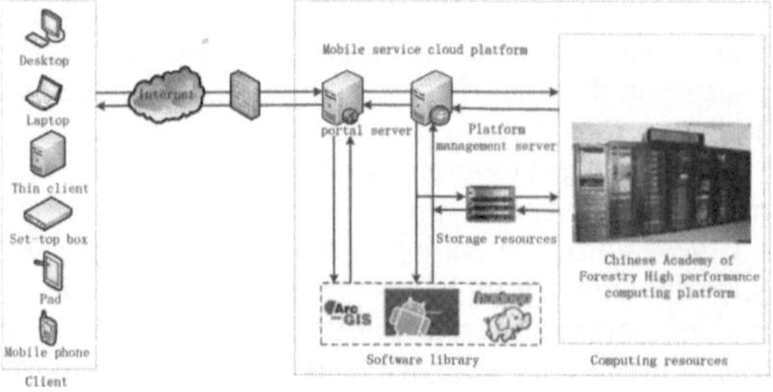

Figure 4. Cloud platform architecture.

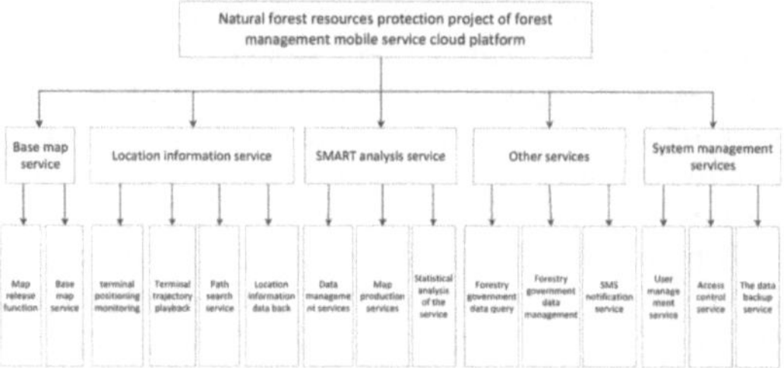

Figure 5. The service integration of forest management mobile service cloud platform for the natural forest protection project. The natural forest resource conservation project management mobile service cloud forest service integration a platform.

these computing services and data services for different test areas. The architecture of the cloud platform is shown in Figure 4.

The natural forest resources protection project with a forest management mobile service cloud platform main packaging, has a basic map service, location information services, SMART analysis services, and system management, etc. These services are in the form of a web service. This cloud platform is an open, flexible platform.

4 CONCLUSION AND PROSPECT

The informatization construction of a forestry system is relatively backward, especially in the primary forest where there is a lack of IT talents and equipment. Through the construction the forest mobile service cloud platform for the natural forest protection project, mobile cloud computing technology is introduced into the major engineering project management of the forest. Advanced information technology, combined with a traditional work mode, makes the natural forest protection project implementation management intelligent, improves the forest management level, and saves a lot of manpower and material resources.

ACKNOWLEDGEMENTS

Many thanks to Li Zeng-yuan's (Vice Superintendent of the Institute of Forest Resource Information Technique, CAF) great instruction and guidance of this project. This paper was supported in part by the National Major Special Projects and Development Program of China and the Science Foundation of The Chinese Academy of Forestry, under contract number 21-Y30B05-9001-13/15, and IFRIT201104.

REFERENCES

X. Zhou. Cloud computing challenges and opportunities for the telecoms industry [J]. Mobile Communications, 2010(11) (In Chinese).

H.J. Ni. Based on the Android platform of mobile terminal GPS research [J]. Computer technology and development, 2012 (5): 198–201 (In Chinese).

X.H. Shu. Based on the Android platform of mobile Web map service design [D]. Liaoning, Computer College of Dalian Maritime University, 2009 (In Chinese).

D.P. Yang, W.M. Lang, H.S. Li. Cloud computing service model research [J]. Telecommunications Information, 2011(11) (In Chinese).

X.C. Lan, R.X. Yin. Analysis and study of mobile cloud computing [J]. Microprocessors, 2010, 3: 114–115 (In Chinese).

M. Mowbray, S. Pearson. A Client-Based Privacy Manager for Clouding Computing [C]. Proceedings of the Fourth International ICST Conference on Communication System Software and Middleware, 2009: 1–8.

L. Nikolaos, K. Eleni, T. Konstantinos. A Semantic Interoperability Framework for Cloud Platform as a Service [C]. Third IEEE International Conference on Cloud Computing Technology and Science, 2011: 280–287.

Y. Liu. Cloud computing summarized and the application research of mobile cloud computing [J]. Journal of information and communication technology, 2010, 4(2): 14–20 (In Chinese).

R.Y. Deng, C. Qin, X.Z. Xie. Mobile cloud computing application status and existing problems of analysis [J]. Journal of Chongqing University of Posts and Telecommunications (Natural Science Edition), 2012, 24 (6): 716–2, 3, 7 (In Chinese).

Christensen J.H. Using Restful Web-services and cloud computing to create next generation mobile applications//the 24th ACM SIGPLAN Conference Companion on Object Oriented Programming Systems Languages and Applications (OOPSLA), USA: ACM Press, 2009: 627–634.

Future Communication Technology and Engineering – Chan (Ed.)
© 2015 Taylor & Francis Group, London, ISBN: 978-1-138-02777-0

Design of miniaturized microstrip antenna using metamaterials

Y. Zhang & X.C. Ma
Wireless Innovation Lab of Yunnan University, School of Information Science and Engineering, Kunming, Yunnan, China

S. Yan
School of Information and Electronics, Beijing Institute of technology, Beijing, China

J.J. Yang & M. Huang
Wireless Innovation Lab of Yunnan University, School of Information Science and Engineering, Kunming, Yunnan, China

ABSTRACT: A miniaturized antenna based on double negative (DNG) metamaterials is designed in this paper. Metamaterials using hexagonal Split-Ring Resonator (SRR) and wire strip shows DNG property at two frequency bands. By inserting the matamaterials into the substrate of the rectangular microstrip antenna, the size of rectangular patch breaks through the limitation of half wavelength. Simulation results show that the reduction of the patch's size are 85.4% and 93.6% when resonant frequencies are 8.7 GHz and 12.7 GHz compared with the conventional patch antenna, which shows the significant miniaturization of microstrip antenna. The working frequencies can be controlled by adjusting the size of hexagonal SRR. This method of antenna miniaturization has potential application in antenna engineering.

1 INTRODUCTIONS

Microstrip antenna with the characteristics of small size, low cost and good performance is widely used in wireless communications. The demand for reduced antenna size has increased tremendously over the past years [1]. Examples of microstrip antenna miniaturization such as slotting, using metamaterials and high dielectric material substrate, fractal geometry and so on, have been proven theoretically and, in part, experimentally in the recent technical literatures [2–7].

Metamaterials are ordered composites that exhibit exceptional properties not readily observed in nature [8]. They can achieve a series of strange electromagnetic properties such as negative refraction and perfect lens [9]. The electromagnetic properties of metamaterials can be represented by the effective permittivity (ε_{eff}) and effective permeability (μ_{eff}). Metamaterials exhibit ε_{eff} and μ_{eff} both negative are called the double negative (DNG) materials [10,11]. Because of these unusual properties, they offer some interesting changes in radiation characteristics of an antenna. Many studies have been done in antenna miniaturization by using metamaterials [12–15].

In this paper, we design a microstrip patch antenna with a metamaterial substrate. Firstly, the simulation model of dual-band DNG metamaterials based on a hexagonal split-ring resonator (SRR) [16] is presented. To confirm the double negative characteristic, we extract ε_{eff} and μ_{eff} from the S-parameters. Then, metamaterial patch antenna is simulated with different sizes of the patch to get a better performance. Finally, we compare the proposed metamaterial patch antenna with conventional one, and the effectiveness of the miniaturization strategy is proved.

2 SIMULATION MODEL

2.1 *DNG metamaterial design*

In the design of DNG materials, negative dielectric constant can be easily obtained by metal wires, while obtaining negative permeability has become the key issue. When the material is penetrated by electromagnetic wave from one direction, it can create magnetic resonance phenomenon which achieves negative permeability because of the electromagnetic induction. Recently, the DNG metamaterials made up of hexagonal-shaped split-ring resonator, wire strip and substrate is proposed by Ouedraogo et al. [12]. The wire strip ensures negative permittivity and the SRR ensures the negative permeability. When both structures are combined together, they simultaneously offer negative permittivity and negative permeability, hence it becomes double negative metamaterials.

Figure 1 (a) is the front view of DNG metamaterial unit. SRR consists of two regular hexagons with an outer radius a = 3.81 mm, the width of the ring c = 0.3 mm, the distance between the two rings e = 0.1 mm, the width of the gap is 0.2 mm and the

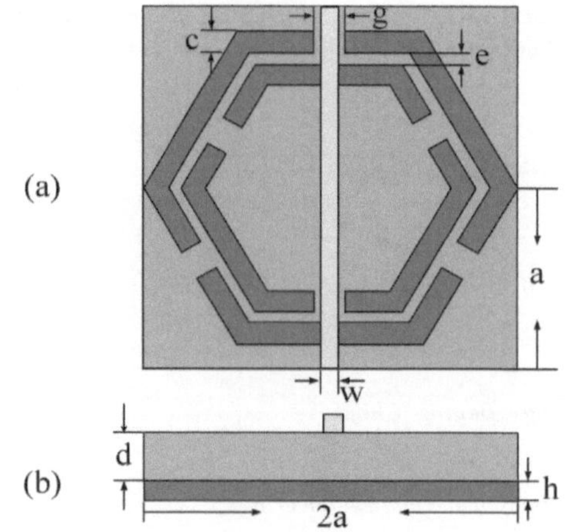

Figure 1. (a) Front view of DNG metamaterial; (b) Side view of DNG metamaterial.

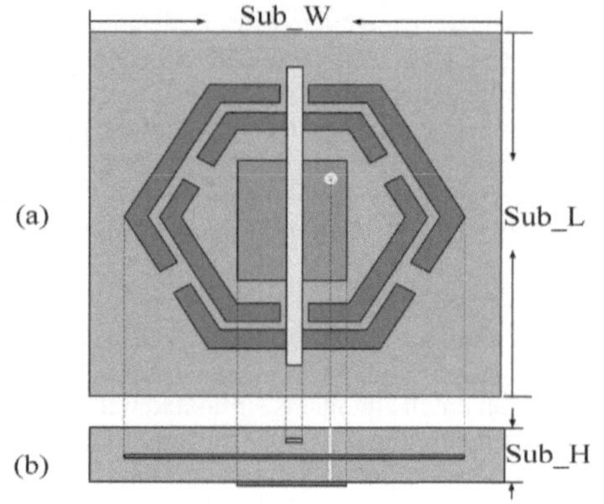

Figure 2. (a) Front view of proposed antenna; (b) Side view of proposed antennaresults and discussion.

width of the metal wire w = 0.15 mm. Figure 1(b) is the side view of the unit. The lower layer is hexagons SRR and the middle is microwave substrate material Roger RT/duroid 6006, which has a relative permittivity of $\varepsilon_r = 6.15$ and a dielectric loss tangent tan of $\delta = 0.0027$. The upper layer is wire strip. The thickness of the substrate d = 0.12 mm and that of both SRR and wire strip h = 0.04 mm.

2.2 *Antenna design*

Figure 2 shows the simulation model of the microstrip patch antenna. The antenna is mounted on Roger RT/duroid 6006 substrate and fed by a coaxial transmission line. Patch and ground plane are made of copper. DNG structure is inserted into the sub-strate with the same size as presented in Figure 1.

The patch is excited by a wave port. The width and length of the patch are denoted as Pat_W and Pat_L. The width, length and height of the substrate are denoted as Sub_W = 6.8 mm, Sub_L = 6.8 mm and Sub_H = 1.278 mm.

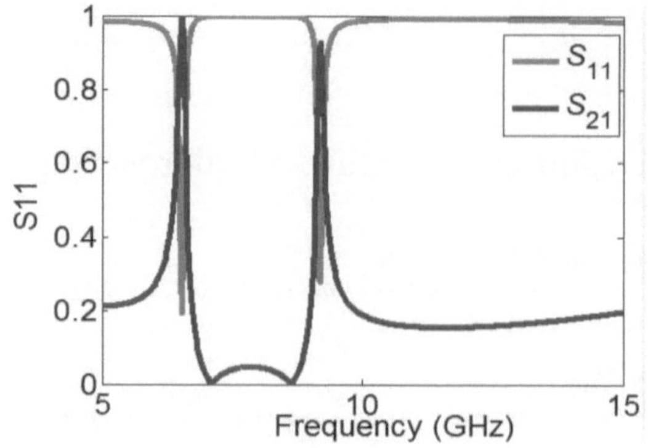

Figure 3. S-parameters of DNG metamaterial.

3 RESULTS AND DISCUSSION

3.1 *DNG characteristic of metamaterials*

The simulation is performed based on the software Anosoft HFSS, and S-parameters are shown in Figure 3. It can be seen that the S-parameters have a sudden change between 6–7 GHz and 9–9.6 GHz which shows a very obvious dual-band characteristic.

Relying solely on S-parameters can not explain that the DNG structure has achieved the negative permeability and negative permittivity. Extracting the equivalent permittivity and permeability from S-parameters is an effective method [16]. We use the inversion algorithm based on Kramers–Kronig relations [17] by firstly calculating the characteristic impedance Z (Eq. 1) and refractive index n (Eq. 2) by using S-parameters and then derive the equivalent permittivity and permeability according to Eq. (3) and Eq.(4).

$$z = \pm\sqrt{\frac{(1+S_{11})^2 - S_{21}^2}{(1-S_{11})^2 - S_{21}^2}} \tag{1}$$

$$n = \frac{\arccos\left(\frac{1-S_{11}^2+S_{21}^2}{2S_{21}}\right)}{k_0 d_{eff}} \tag{2}$$

$$\varepsilon_{eff} = n / Z \tag{3}$$

$$\mu_{eff} = n \times Z \tag{4}$$

k_o is the free-space wavenumber, d_{eff} is the effective thickness of the metamaterial cell.

Figure 4 shows the variation of μ_{eff} and ε_{eff} of DNG metamaterials with frequency. In frequency band of 8.5 GHz–9.1 GHz and 10.5 GHz–11.5 GHz, the metamaterial shows double negative property.

3.2 *Miniaturization of the antenna*

By changing the size of rectangular patch, we simulated the S11 of the antenna, as shown in Fig. 5. We find that the resonant frequencies of the antenna are always located around 8.7 GHz and 12.7 GHZ, which

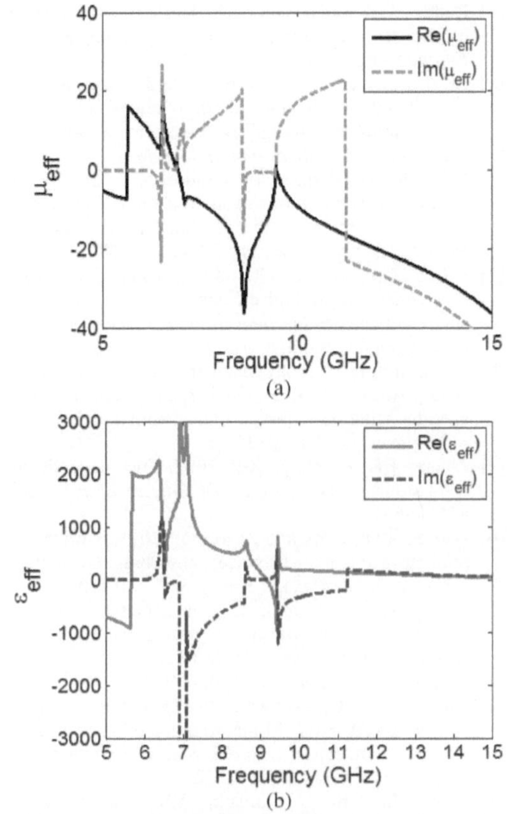

Figure 4. (a) μ_{eff} of DNG metamaterials, (b) $\varepsilon_{\mathrm{eff}}$ of DNG metamaterials.

is independent on the size of the patch. This is quite different from traditional microstrip patch antenna.

We design two single band antennas working at 8.7 GHz and 12.7 GHz respectively, the sizes of the two patches are 3.6 mm × 2.8 mm and 1.8 mm × 1.1 mm. Their S11 parameters are shown in Figure 6. We compared the performance of the proposed metamaterial antenna with the conventional patch antenna, as shown in Table 1. It is seen that working at the same frequency of 8.7 GHz (12.7 GHz) the size of the proposed antenna is only 14.65% (6.43%) of the conventional one, which shows the effectiveness of the metamaterials for antenna miniaturization. But the bandwidth of the proposed antenna is relatively narrow.

Then, we studied the influence of the hexagon size on antenna performance. The outer radius of hexagon is supposed to vary from 3 mm to 3.75 mm with an interval of 0.15 mm, while the patch size is fixed at 1.8 mm × 1.1 mm. The simulation result is shown in Figure 7. We can see the operating frequency is controlled by the size of the hexagonal structure. Moreover, the resonant frequency decrease the size of the hexagonal structure. Therefore, by changing the sizes of the hexagonal structure and the patch we can get a patch antenna working at arbitrary frequency with a satisfactory performance.

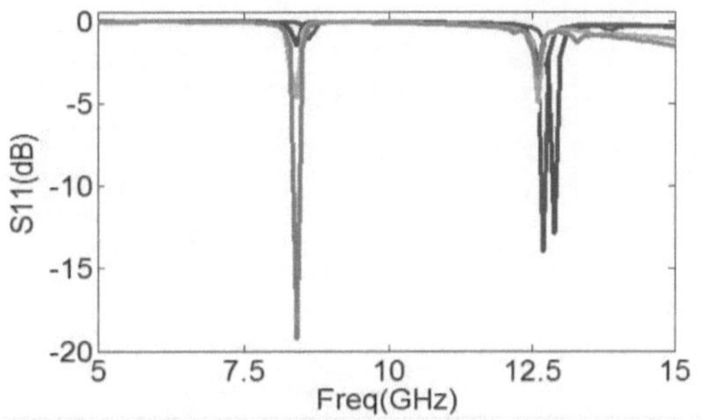

Figure 5. S11 of antenna with different sizes.

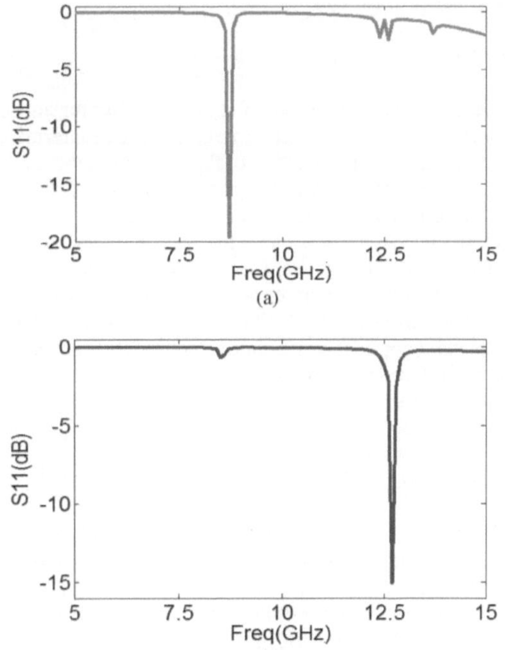

Figure 6. S11 parameters of proposed patch antennas. (a) patch size = 3.6 mm × 2.8 mm; (b) patch size = 1.8 mm × 1.1 mm.

Table 1. Analysis of antenna miniaturization.

	Patch size (mm^2)	f_0 (GHZ)	BW (MHz)	Percent
Proposed antenna 1	3.6×2.8	8.7	110	14.65%
Conventional antenna 1	9.1×7.56	8.7	570	
Proposed antenna 2	1.8×1.1	12.7	90	6.43%
Conventional antenna 2	6.2×4.97	12.7	1100	

• f_0: Resonant frequency; BW: Bandwidth.

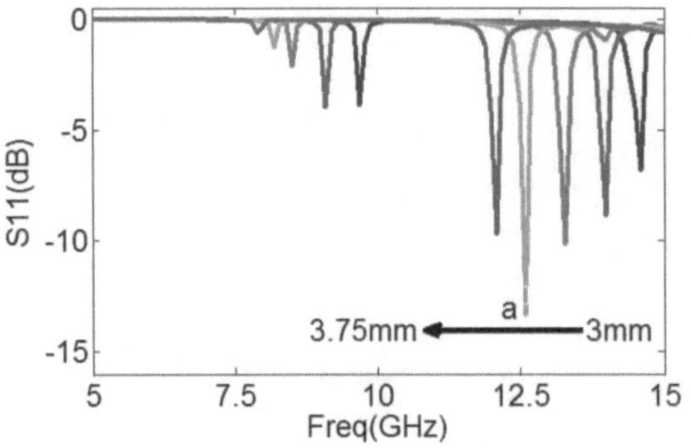

Figure 7. S11 parameters of proposed antenna with different outer radius of hexagon.

4 CONCLUSIONS

In this paper, a method of antenna miniaturization based on dual-band double negative metamaterial is presented. The physical size of the patch with the metamaterial substrate is reduced to 14.65% and 6.43% of the conventional antennas. By adjusting the size hexagonal SRR, one can change the working frequency of the antenna, and by adjusting the size of the patch, the desired performance can be obtained. The antenna based on the DNG metamaterials may have important applications in antenna engineering. Experimental demonstration of the proposed patch antenna is our future work.

ACKNOWLEDGMENTS

This work was supported by the National Natural Science Foundation of China (Grant Nos. 61161007, 61261002, 61461052), the Specialized Research Fund for the Doctoral Program of Higher Education (Grant No. 20135301110003, 20125301120009), China Postdoctoral Science Foundation (Grant No. 2013M531989, 2014T70890), and the Key Program of Natural Science of Yunnan Province (Grant No. 2013FA006).

REFERENCES

[1] Fan, X.K. 2008. Research of miniature multi-band micro-strip antennas. Jinan: Radio Physics. Shandong University.

[2] Wang, L., Liu, J., Wei, Q., and Safavi-Naeini, S. 2013. A design of miniaturized dual-band antenna. *Antennas & Propagation*, vol. 2: 734–736.

[3] Singh, I., Jain, S., Tripathi, V. S., and Tiwari, S. 2013. Microstrip patch antenna miniaturization using planar metamaterial unit cell. *Quality, Reliability, Security and Robustness in Heterogeneous Networks*. vol. 115: 358–364.

[4] Reddy V, & Sarma N. 2013. Koch fractal boundary patch over reactive impedance. *Materials Science and Engineer-ing*. 44(1): 012005.

[5] Lizzi, L., Azaro, R., Oliveri, G., and Massa, A. 2012. Mul-tiband fractal antenna for wireless communication systems for emergency management. *Journal of Electromagnetic Waves and Applications*, 26(1): 1–11.

[6] Ullah, M.H., Islam, M.T., Mandeep, J.S., and Misran, N. 2013. Design and analysis of a multi band electrically small antenna using ceramic material substrate. *Przeglad Elektrotechniczny*, 89(1): 271–274.

[7] Ullah, M.H., Islam, M.T., and Mandeep, J.S. 2013. A parametric study of high dielectric material substrate for small antenna design. *International Journal of Applied Electromagnetics and Mechanics*. 41(2): 193–198.

[8] Indrasen Singh, 1 Vijay Shanker Tripathi, 1 and Sudarshan Tiwari, 2. 2013. Dual-Band microstrip patch antenna minia-turization using metamaterial. *Journal of Engineering*. Vol. 2013: 1–5.

[9] Zhang, F.L. & Zhao, X.P. 2007. Tunable split ring resonator and its effect. *ACTA Phisica Sinica*. 56(8): 4661–4667.

[10] Bao, S., Luo, C.R., and Zhao, X.P. 2011. S-wave band microstrip antenna with perfect absorbing metamaterial substrate. *ACTA Phisica Sinica*. 60(1): 014101.

[11] Shelby, R.A., Smith, D.R., Nemat-Nasser, S.C., and Schultz, S. 2001. Microwave transmission through a two-dimensional, isotropic, left-handed metamaterial. *Applied Physics Letters*, 78(4): 489–491.

[12] Ouedraogo, R.O., Rothwell, E.J., Diaz, A.R., Fuchi, K., and Temme, A. 2012. Miniaturization of patch antennas using a metamaterial-inspired technique. *Antennas and Propagation*, 60(5): 2175–2182.

[13] Singh, I., Jain, S., Tripathi, V.S., and Tiwari, S. 2013. Microstrip Patch Antenna Miniaturization Using Planar Metamaterial Unit Cell. *Quality, Reliability, Security and Robustness in Heterogeneous Networks*. pp: 358–364.

[14] Alici, K.B., Caliskan, M.D., Bilotti, F., and Vegni, L., 2013. Experimental verification of metamaterial loaded small patch antennas. COMPEL: *The International Journal for Computation and Mathematics in Electrical and Electronic Engineering*. 32(6): 1834–1844.

[15] Yang J. J. Huang M., Mao F. C., Sun J., Antenna miniaturization based on supperscattering effect, *Radio engineering*, 2012, 21(1): 512–516.

[16] Yang, H., Wang, C.H. And Guo, XR. 2014. A novel dual -band left-handed metamaterials composed of multi-defects hexagonal structure. *ACTA Phisica Sinica*. 63(1): 014103.

[17] Ding M., Xue H. & Wu B. 2013. The comparisons between two retrieve algorithms for metamaterials. *ACTA Phisica Sinica*. 62(4): 044218.

[18] Zsolt Szabó, Gi-Ho Park & Ravi Hedge. 2010. A unique extraction of metamaterial parameters based on Kramers–Kronig relationship. *IEEE Transactions on Microwave Theory and Techniques*, 58(10): 2646–2653.

Future Communication Technology and Engineering – Chan (Ed.)
© 2015 Taylor & Francis Group, London, ISBN: 978-1-138-02777-0

A ZigBee wireless network factory monitoring system design

W.K. Huang, F.J. Wu & Z. Xiao
Guangzhou University, Guangzhou Guangdong, China

H.B. Wu
Zhanjiang Port Group Co. Ltd., Zhanjiang Guangdong, China

ABSTRACT: Because of the cable transmission wiring complexity, construction difficulties, low reliability, and a single type of sensor information being incomplete and other deficiencies, we propose a ZigBee-based wireless network technology industrial site with an automatic monitor solution. A common industrial site controls temperature, humidity, smoke, etc. To detect the amount of data processed by STM32, the ZigBee wireless network communication will integrate the information of each node to achieve distributed measurement and control. The test results show that the system can realize dynamic changes in the node and ad hoc networks because of the integration of effective information and other functions; its low power consumption, low cost, fast data transfer, extendibility, stability, and security, means that it is worthy promoting.

Keywords: ZigBee network, industrial field, wireless node, automatic monitoring.

1 INTRODUCTION

The control network of traditional industry usually uses cable to transmit information, which has some disadvantages, such as wiring complexity, instability, difficult extension, and inconvenient installation.[1] With the rapid development of measurement technology, network communication technology, and computer technology, wireless transmission technology gradually substitutes the cables. Its future is promising because it has many advantages, such as network flexibility, easy extension, easy installation, etc.

The ZigBee has advantages of low cost, scalability, low power consumption, and a self-repairing capability. The STM32 has low power consumption, great efficient data processing ability, powerful chip resources, and a rich peripheral interface. This paper proposes an efficient solution, which combines the ZigBee technology with STM32 single chip technology to design a multi-sensor ZigBee wireless network controller, which enables us to build a multicast wireless control network. It builds a real-time monitoring system by using the features of the ZigBee. This design also provides support to the Internet of Things.

2 SYSTEM DESIGN

In the industrial control field, ZigBee ad-hoc network function links each wireless sensor node to construct a wireless network monitoring system. This system consists of a PC, a coordinate node, a routing node, and a sensor node. The coordination node is responsible for setting up the wireless network; the sensor node and routing node search the existing network and join the network.[3] The routing or sensor node includes the sensor module, data processing control module, and wireless transmitting module.

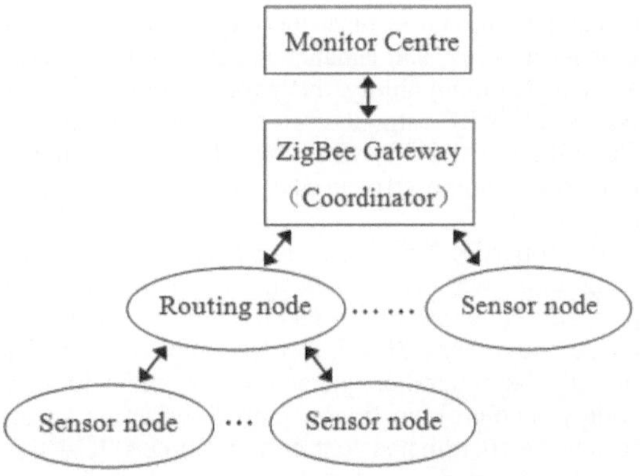

Figure 1. A wireless monitoring system structure.

STM32 is used to collect and process sensor data. It sends the data to the coordination node by wireless radio frequency. The coordinating node receives data from each node, and then sends this to the PC through the serial port. The PC provides the operating interface to users, and displays and saves the real-time data from each node. The system architecture diagram is shown in Figure 1.[4]

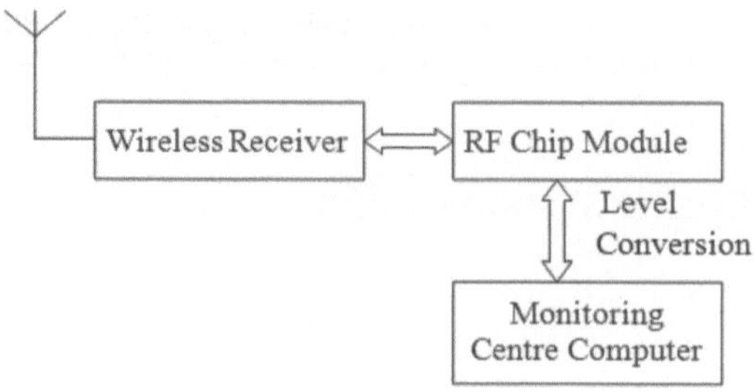

Figure 2. Block diagram of the interface coordinator node.

3 HARDWARE DESIGN OF A ZIGBEE WIRELESS NODE INTERFACE

3.1 *Hardware design of the coordinator interface*

Figure 1 shows that the coordinator is a bridge between the ZigBee wireless network and other networks; it serves as a gateway. It is used to receive various data, parameters, and commands. After unpacking, it is sent to the central computer for data analysis and treatment, and database management. The computer can also send the instructions to control the monitoring points according to monitoring requirements. The structure is as shown in Figure 2.

In order to adapt the industrial control field environment with large quantities of monitoring points and many types of sensors and complex distribution of equipment, and to meet the requirements of real-time monitoring, and enhance the system integration and anti-jamming ability, this paper chooses the ZigBee wireless RF chip CC2530 as the RF transceiver. CC2530 is a 2.4 GHz IEEE 802.15.4 standard RF transceiver composed of an RF transceiver and an 8051 composition.

Its properties include a fast calculation speed, high integration, rich on-chip resources, and simply built peripheral circuits. It also has the ability to communicate with other CPUs. It can meet the requirements of multi-point parallel data acquisition and achieve a variety of control functions and miniaturization. It can realize the coordinated control of multiple CPUs[5]. A coordinating node interface hardware circuit diagram is shown in Figure 3. In order to enhance the reliability of the control node and anti-jamming e ability, the circuits which use a monopole antenna make use of shielded inductors L1, L2, and capacitors C4, C8, C10, C5, and C3 to make up an imbalanced transformer to improve the performance of the antenna. In addition, the DCOUPL pin of CC2530 is concatenated to a decoupling capacitor C1 and then grounded to ensure that the 1.8 V digital logic on-chip regulator can provide stable operation.

3.2 *Hardware design of the routing or sensor node interface*

In the industrial control field wireless network monitoring system, each monitoring point can be used as a sensor node or routing node through the ZigBee

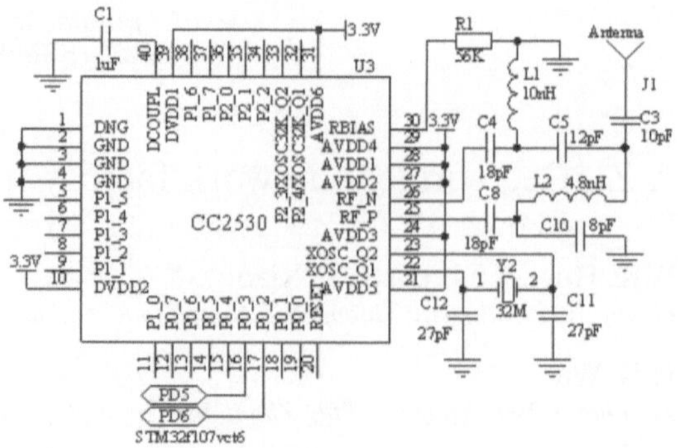

Figure 3. Coordinating node interface hardware circuit diagram.

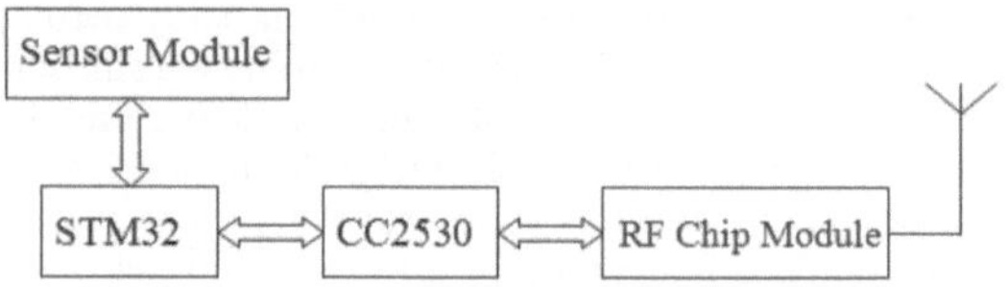

Figure 4. Routing or sensor block diagram node interface.

wireless sensor network access plant floor control. The main work is to collect and send the information about temperature, humidity, smoke, personnel in the control field, and explain the received command and performance order. Because the 8051 kernel of the CC2530 chip data has a poor processing capability, it will reduce the node's efficiency and lead to transmission delay, packet loss, and other problems when dealing with system multi-sensors. This paper chooses the STM32F107VCT6 type 32-bit ARM chip as a sensor or routing node micro controller.[6]

The sensor or routing node interface is shown in Figure 4. The STM32 MCU deals with the sensor's data collecting and processing at regular times to communicate through the CC2530 UASRT interface. The CC2530 receives the sensor data, packages the information of the node MAC address, web address, device type etc. into packets, and uses unicast to transfer the packet to the coordinator node. Because the STM32 and CC2530 can use 3.3 V logic power supply, the use of a serial port communication will not require the voltage conversion, and thus simplify the development process.

4 ZIGBEE WIRELESS NETWORK BUILDING

The ZigBee wireless network of this system used a peer-to-peer topology network structure. The wireless network building process mainly includes two steps. The first step is to form a network by the coordinator and network configuration; the second step is to access the wireless network using another node as a router or sensor node.[7]

(1) The coordinator forms a network.

(2) The coordinator is the core of the ZigBee network; a complete ZigBee network can only have one coordinator. The coordinator begins to initialize boot device after its start-up. Then, according to the setting, it will choose a proper channel and PAN ID from a ZigBee network. When the application for joining the network from a router or sensor node is detected, it will randomly assign a 16-bit network address for it. Moreover, the coordinator has to detect whether there is new data; if new data is detected, it will unpack the handle and send it to the PC. The formation process of the ZigBee network is: 1. To initialize the network equipment, and use the initialize function of the ZDO device to initialize network formation. 2. To set the network initialization event in the network formation initialization functions, and use the ZDO equipment start function when dealing with the network formation initialization event function. 3. The network layer performs network formation requests; if it was built successfully, it will receive the feedback information from the network layer; then it is sent to the application layer to form a system message event by the ZDO layer in the form of a message. The router or sensor nodes join the network. The router in the ZigBee network has the function of routing forwarding. Only when a ZigBee network exists, can the routing node and sensor node join the network and work properly. Otherwise, it will keep sending a network join request until the coordinator allows it to join and assigns it a network address. Then it can work normally. In this system, because the sensor nodes are only responsible for sending data collected by the STM32, they will remain in a dormant state when they are not working.

The process for the routing node and terminal node to join the ZigBee network is: 1. After initializing the electrical system, it calls for the ZDO device initialization function; 2. The network initialization event is then set in the network formation initialization function. Then it starts to call for the ZDO function. 3. The ZDO equipment star function calls for the network discovery function on the network layer. If a network is found at this time, the ZDO layer will receive a feedback message about finding the network from the network layer. Then it sends a request to join the network by the network layer; the coordinator will return the response of successfully joining the network after treatment, and allocate a network short address for the node.

5 DETECTION SYSTEM ANALYSIS OF EXPERIMENTAL RESULTS

After completing all the node deployment, the wireless network system runs normally, as shown in Figure 5. Each node sensor of the system is at a normal state.

In the process of detection, if the node of a sensor value exceeds the preset warning threshold, the corresponding sensor in the display box will display items in red, as shown in Figure 6, to remind the user that the sensor value exceeds the bid.

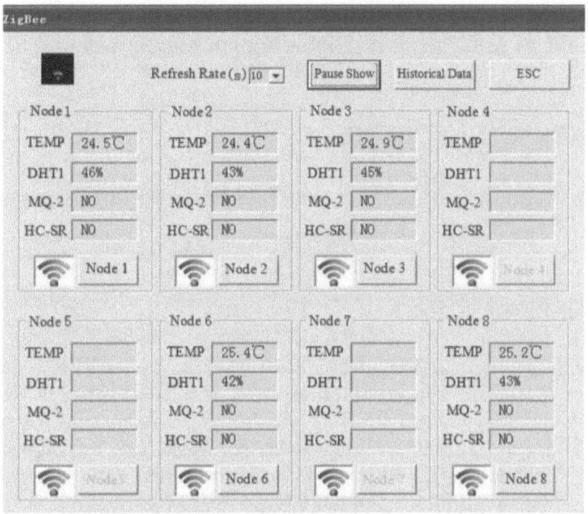

Figure 5. System uptime interface.

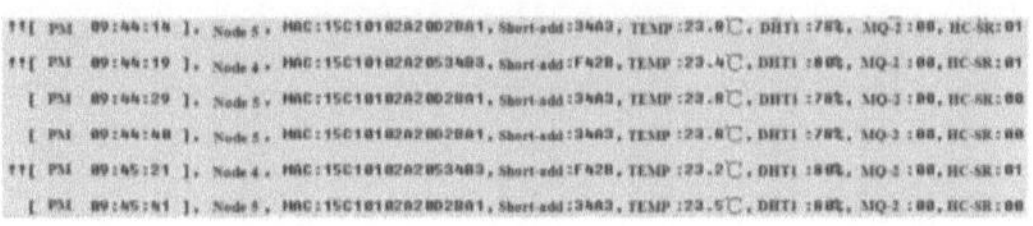

Figure 6. Historical data storage format system monitoring.

For the convenience of the user query and management of data, the system will store the testing data of each node in the local database. The data storage format is shown in Figure 7, including the specific time and node details, etc. In the historical data, if there are two exclamation marks, "!!", in the front, it indicates that the moment sensor value exceeds the preset warning threshold. Smoke and infrared data using a Boolean variable, states that if the normal situation is 00, then the alarm state is non-00.

6 CONCLUSION

The ZigBee wireless technology solves the complex wiring problem of modern factory automatic monitoring systems caused by the high plant and the large space workshop. It also avoids the complex wiring among the monitoring modules in the industrial control field, and ensures reliable transmission of real-time data of each sensor. The system has accomplished the ZigBee node hardware interface's design and wireless network building, the acquisition and processing of the analogue signal, digital signal and switch signal, and the display of temperature, humidity, gas concentration, personnel, etc., to realize the real-time status of the industrial control field automatic monitoring. The commissioning test shows that the system data transfer rates are up to 250 kBps, and it has a good reliability and anti-interference ability. It can monitor the plant-running status better and provide real-time, and an accurate and reliable sensor data environment.

ACKNOWLEDGEMENTS

Shaoguan City Science and Technology Plan Projects (2012 cxy/C13) Funding.

REFERENCES

Zhang Chengli. Machinery factory integrated wiring between wiring and wiring equipment design [J]. 2010, 4(6):75–78.

Wang Dong, Zhang Jin-rong, Wei Yan, Cao Chang-xiu, Tang Zheng. Building Wireless Sensor Networks by Zigbee Technology [J]. Journal of Chongqing University (Natural Science Edition). 2006, 29(8):95–97, 110.

Cuilu, Cai Jueping, Zhao Bochao, WangXin. Mine personnel positioning safety monitoring system based on ZigBee technology [J]. Journal of Dalian University of Technology. 2011(4):102–106.

Wu Fengjie, Huang Wenkai, Wu Hongbi. Application of ZigBee for automatic monitoring system of infusion workshop [J]. Process automation instrumentation. 2014, 35(4):49–52.

Jian Song. Wireless sensor network of wearing shop based on CC2530 [J]. 2010 International Conference on Circuit and Signal Processing: 222–224.

Sun Qifu, Sun Yunqiang, Yao Aiqin. Design and application of based on the STM32 general intelligent instrument [J]. Instrument Technique and Sensor. 2010(10): 34–36.

Jiang Jianping, Chen Hui. Design of Based on the CC2530 ZigBee wireless city street lamp control system. Measurement and control technology. 2012, 31 (9): 56–59.

Future Communication Technology and Engineering – Chan (Ed.)
© 2015 Taylor & Francis Group, London, ISBN: 978-1-138-02777-0

A VaR (Value-at-Risk) model based on GARCH approach and EVT

J. Yang
School of Economics and Management, Yunnan Normal University, Kunming, China

S.Z. Zhang
School of Mathematical Science, Yunnan Normal University, Kunming, China

ABSTRACT: In recent years, financial disasters have emphasized the importance of effective risk management for financial regulators and market practitioners who have become particularly sensitive to changes in the assets'value. The use of quantitative risk measures has become an essential management tool to be placed in parallel with the models of returns. In this paper, we introduce all kinds of GARCH models and EVT theories and apply these methods to measure the exchange rate risk of the Chinese exchange market. Firstly, we examine the heteroscedasticity of the return series of USD/RMB and EUR/RMB data, and the results suggest that there is obvious heteroscedasticity. Secondly, we choose the best GARCH model to filter the return series to i.i.d residual series, and employ extreme value theory to estimate the tails of those i.i.d series; we find the degree of fitting of GPD is very high. Finally, we calculate corresponding VaR estimates, which can help the investor to measure the exchange risk accurately.

Keywords: VaR, GARCH, EVT, exchange rate, market risk measurement.

1 INTRODUCTION

In the past decades, financial disasters have emphasized the importance of effective risk management for financial regulators and market practitioners who have become particularly sensitive to changes in the assets' value. The use of quantitative risk measures has become an essential management tool to be placed in parallel with the models of returns. These measures are used for investment decisions, supervisory decisions, risk capital allocation, and external regulation. In the fast-paced financial world, effective risk measures must be as responsive to news as other forecasts and must be easy to grasp, even in complex situations. Generally speaking, risk management consists of four complementary steps: identification, assessment, monitoring, and control (mitigation). Among them, risk assessment (measurement) is an important prerequisite for the following risk control. The method of risk measurement has undergone many changes from the traditional tool – the standard deviation to the slightly more complicated sensitive indicators of risk factors (e.g. duration and Gamma), to the now widely used but highly controversial method of Value-at-Risk (VaR).

Since Value-at-Risk received its first wide representation in July 1993 in the Group of Thirty report, the number of uses of – and uses for – Value-at-Risk have increased dramatically. However, it is important to recognize that the Value-at-Risk technique has gone through significant refinement and passed essential process changes since it originally appeared. Theoretical research that relied on the Value-at-Risk as a risk measurement was initiated by Jorion (1997) and Dowd (1998), who applied the Value-at-Risk approach based on risk management emerging as the industry standard by choice or by regulation. Value-at-Risk-based management by financial as well as non-financial institutions was researched and described by J.P. Morgan (1998). Its wide use occurs from the fact that Value-at-Risk is an easily interpretable summary measure of risk, and also has an attractive explanation, as it allows its users to focus attention on the normal market condition in their routine operations. Value-at-Risk models aggregate the several components of price risk into a single quantitative measure of the potential for losses over a specified time horizon.

Recognition of such models by financial and regulatory communities is evidence of their growing use. For instance, financial regulation, like the Basel Committee on Banking Supervision, requires US banks to use VaR to determine the minimum capital to support their trading portfolios. Since 1998, the Securities and Exchange Commission requires registrants to provide quantitative information about market risk with VaR being one of the disclosure alternatives.

The rest of paper is structured as follows: In section 2, a short overview of the general concept of VaR and ES will be presented. In section 3, we review the ARCH type model, including GARCH models. Meantime, an overview on extreme value theory (EVT) will also be presented. In section 4, we focus on an application of the VaR model based on the GARCH approach and EVT on the Chinese Exchange rate including USD/RMB and EUR/RMB data. In section 5, we will give a summary of the whole text.

2 VaR MODELS AND EXTREME VALUE THEORY

2.1 *General concept of VaR and ES (Expected Shortfall)*

In this section, we firstly give definitions of VaR and ES and discuss their uses and basic properties. Let X be a random variable with the cumulative distribution function $F_X = P\{X \leq x\}$. X may represent the loss or gain of a return or portfolio. In this paper, X has the meaning of loss.

Definition 1 (Value-at-Risk) The VaR of X with confidence level $\alpha \in (0, 1)$ is:

$$VaR_{\alpha}(X) = \inf\{x \in R, F_X(x) \geq \alpha\} \tag{1}$$

By the above definition, $VaR_{\alpha}(X)$ is a left α quantile of the random variable X. For a given time horizon t and confidence level α, the VaR of a portfolio is the potential loss in the portfolio's market value over the time horizon t that is exceeded with probability $1-\alpha$. VaR has become a popular risk measure for risk management both for the purpose of reporting and measurement of capital adequacy. Despite its wide acceptance, it is not a coherent risk measure. A coherent risk measure should satisfy the axioms of translation invariance, subaddivity, positive homogeneity, and monotonicity. Unfortunately, VaR lacks subadditivity.

Another informative measure of risk is the Expected Shortfall (ES), which is also known as Mean Excess Loss or Tail VaR. ES is a more consistent measure of risk since it is subaddivity and convex. Next, we give a formal definition of ES.

Definition 2 (Expected Shortfall) For a random variable X with a conditional distribution function, ES is the conditional expectation of the loss above VaR for the time horizon t and the confidence level α, i.e.:

$$ES_{\alpha}(X) = E[X | X > VaR] = \int_{-\infty}^{+\infty} z dF_X^{\alpha}(z) \tag{2}$$

where

$$F_X^{\alpha}(z) = \begin{cases} 0 & , \ when\ z < VaR_{\alpha}(X) \\ \dfrac{F_X(z)-\alpha}{1-\alpha} & , \ when\ z \geq VaR_{\alpha}(X) \end{cases}$$

2.2 *EVT (Extreme Value Theory) and POT (Peak Over Threshold) method*

Extreme Value Theory (EVT) is a classical topic in probability theory; many books and surveys are available on the subject. Here, we intend to give some intuition and basic results of EVT.

Generally, there are two related ways of identifying extremes in real data. Let X_i $(i = 1, 2, \dots)$ denote a random variable which may represent daily losses or returns. The first approach considers the maximum (or minimum) the variable takes in a successive period, for instance, weeks, months, or years. These selected observations constitute the extreme events, and a method that determines these selected extreme values, is called the Block Maxima Method (BMM); the second approach focuses on the realizations which exceed a given threshold, all exceed the threshold u constitute extreme events. This method is called Peak Over Thresold method(POT). The BMM is the traditional method used to analyse data with seasonality. However, POT method uses data more efficiently and seems to be a popular choice method in recent applications. In the following section, the POT method is introduced in detail.

Let X_i $(i = 1, 2, \dots)$ denote a sequence of i.i.d random variables with an unknown distribution function $F(x)$. We focus on estimating the distribution function F_u of values of x above a certain threshold u. The distribution function F_u is called the *conditional excess distribution function (cedf)*. Formally:

$$F_u(y) = P\{X - u \leq y | X > u\},\ y \geq 0 \tag{3}$$

where X is a random variable, and u is a given threshold. Based on a conditional probability formula, we have:

$$F_u(y) = P\{X - u \leq y | X > u\} = \frac{F(u+y)-F(u)}{1-F(u)} = \frac{F(x)-F(u)}{1-F(u)} \tag{4}$$

Define also the Generalized Pareto Distribution (GPD) as:

$$G_{\xi,\beta}(y) = \begin{cases} 1-(1+\xi\dfrac{y}{\beta})^{-\frac{1}{\xi}} & if\ \xi \neq 0 \\ 1-e^{-\frac{y}{\beta}} & if\ \xi = 0 \end{cases} \tag{5}$$

The following is a crucial property of GPD, due to Pickands (1975). For a large class of underlying distribution function F, the conditional excess distribution function F_u, for a certain threshold u is satisfied as $\lim_{u\to\infty} \sup_{0<x<x_F-u} |F_u(x) - G_{\xi,\beta}(x)| = 0$, where x_F is the right endpoint (which could be infinite) of the distribution F. This result indicates that GPD becomes an accurate approximation of the excess distribution function F_u.

The parameters of the GPD can be estimated by maximizing the likelihood function, once the threshold has been chosen. The log likelihood based on (5) and computed on the n upper order statistics is:

$$L(\xi,\beta|y)=\begin{cases}-n\log\beta-(\frac{1}{\xi}+1)\sum_{i=1}^{n}\log(1+\frac{\xi}{\beta}y_i) & if\ \xi\neq 0\\ -n\log\beta-\frac{1}{\xi}\sum_{i=1}^{n}y_i & if\ \xi=0\end{cases}$$

An attractive feature of GPD is that we have a close form formula for VaR. In fact, from (4) we have:

$$F(x)=[1-F(u)]F_u(y)+F(u)$$

After replacing F_u by the GPD and $F(u)$ by its estimate $(n-N_u)/n$, where n is the total number of observation and N_u is the number of observations above the threshold u, we can get the following result:

$$F(x)=1-\frac{N_u}{n}(1+\xi\frac{x-u}{\beta})^{-\frac{1}{\xi}} \tag{6}$$

From formula (6), the calculation of the VaR of confidence level q is quite an easy task. We only need to solve the equation $F(VaR)=q$. We then have:

$$VaR_q=u+\frac{\beta}{\xi}[(\frac{n}{N_u}(1-q))^{-\xi}-1] \tag{7}$$

Similarly, we can get a close formula for ES_q:

$$ES_q=VaR_q+\frac{\beta+(VaR_q-u)\xi}{1-\xi}=\frac{VaR_q}{1-\xi}+\frac{\beta-\xi u}{1-\xi} \tag{8}$$

3 ARCH AND GARCH MODELS

3.1 *GARCH models*

Since the introduction of the ARCH model introduced by Engel, GARCH models, introduced by Bollerslev, generalized Engle's earlier ARCH models to include Autoregressive (AR) as well as Moving Average (MA) terms. GARCH models can be more parsimonious (use fewer parameters), thereby increasing computational efficiency. After that, a numerous extension of GARCH models emerged to specify the conditional variance models. The class of GARCH models has become extremely popular for application to financial data over the past twenty years. In the following section, the GARCH class of models is explained as follows.

3.1.1 *GARCH (p,q)*

A typical GARCH (p,q) process is composed of two parts. The first part is the mean function:

$$y_t=c+\sum_{i=1}^{R}\phi_i y_{t-i}+\varepsilon_t+\sum_{j=1}^{M}\theta_j\varepsilon_{t-j}+\sum_{k=1}^{N_x}\beta_k X(t,k) \tag{9}$$

The above equation implies that y_t follows a $ARMA(R,M)$ process. The second part is composed of the variance equation. The GARCH model can be characterized by the assumption that the conditional heteroscedastic series satisfies:

$$\begin{cases}\varepsilon_t=\sigma_t\eta_t,\ \eta_t\sim i.i.d.\ N(0,1)\\ \sigma_t^2=k+\sum_{i=1}^{p}G_i\sigma_{t-i}^2+\sum_{j=1}^{q}A_j\varepsilon_{t-j}^2\end{cases} \tag{10}$$

where $p\geq 0, q>0, k>0,\ G_i>0\ (i=1,2,\ldots,p)$, $A_j>0\ (j=1,2,\ldots,q)$, and $\sum_{i=1}^{p}G_i+\sum_{j=1}^{q}A_j<1$.

3.1.2 *EGARCH (p,q)*

An Exponential GARCH model was proposed by Nelson in 1991[4]. The variance equation of EGARCH(p,q) is:

$$\log\sigma_t^2=k+\sum_{i=1}^{p}G_i\log\sigma_{t-i}^2+\sum_{j=1}^{q}A_j[\frac{|\varepsilon_{t-j}|}{\sigma_{t-j}}-E\{\frac{|\varepsilon_{t-j}|}{\sigma_{t-j}}\}]+\sum_{j=1}^{q}L_j(\frac{\varepsilon_{t-j}}{\sigma_{t-j}}) \tag{11}$$

Since the EGARCH model takes a log transformation, the conditional heteroscedasticity is guaranteed to be positive. No restrictions need to be imposed on coefficients. Furthermore, the conditional heteroskedasticity depends on the absolute magnitude of shocks and reflects the effect of the shock direction. If the parameter estimate is $L_j\neq 0$, it suggests the existence of asymmetry. GARCH-GJR (p,q) (TGARCH).

In addition, a model reflecting financial market asymmetric volatility is called the Threshold GARCH (TGARCH) introduced by Glosten, Jagannathan, and Runkel in 1993. Its conditional heteroskedasticity function is:

$$\sigma_t^2=k+\sum_{i=1}^{p}G_i\sigma_{t-i}^2+\sum_{j=1}^{q}A_j\varepsilon_{t-j}^2+\sum_{j=1}^{q}L_j\lambda_{t-j}\varepsilon_{t-j}^2$$

$$\text{where } \sum_{i=1}^{p}G_i+\sum_{j=1}^{q}A_j+\frac{1}{2}\sum_{j=1}^{q}L_j<1\text{, and } \lambda_{t-j}=\begin{cases}1, & \varepsilon_{t-j}<0\\ 0, & \varepsilon_{t-j}\geq 0\end{cases} \tag{12}$$

L_j is called the leverage.

4 EMPIRICAL STUDY

In this section, we will apply those theories to empirical research. First, we choose the Chinese foreign exchange rates as samples for our empirical study. Second, we investigate the statistical features and fat-tail behaviour of the samples. Third, we aim to find out a good candidate for the log return distribution and build a fitted VaR model based on it.

4.1 *Sample selection and data statistical features*

China's official exchange rates mainly involve five currencies: the US dollar (USD), the Euro (EUR), the Japanese Yen (JPY), the Hong Kong dollar (HKD), and the British Pound Sterling (GPD). In this paper, we will use the exchange rate of the USD/RMB and EUR/RMB as our main object. All the data come from

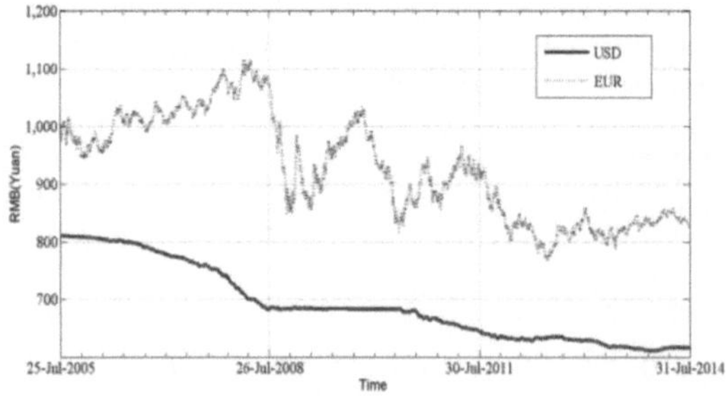

Figure 1. USD and EUR exchange rate time series.

Table 1. Statistical features of logreturn.

	Mean	Std	Skewness	Kurtosis	J-B statistic	Prob
USD	0.00012	0.0008	−0.4621	5.8791	835.8577	0.0001
EUR	0.00006	0.0064	−0.3050	10.6697	5411.6	0.0000

the middle price of the daily exchange rate of the RMB published on the official website of the State Administration of Exchange (www.safe.gov.cn). The data that we use in this paper runs from July 25, 2005 to July 31, 2014, 2,195 days' data in total. The reason for choosing this period is that we hope our study can reveal the main feature of the Chinese exchange market after the Chinese currency reform. It is known that the major change in the Chinese foreign exchange market started on July 21, 2005 when China began to carry out a managed floating exchange rate policy, which is based on the market supply and demand, and makes adjustments in accordance with the reference currency basket. Figure 1 below shows the trend of the price sequence of the USD and EUR in the chosen period.

We introduce our notation now. We will use r to denote the logreturn, ie $r_i = 100[\log(P_t) - \log(P_{t-1})]$, where P_t is the closing quotation price of the daily exchange rate. Using mathematical software, such as MATLAB2011(b), we easily get the statistical features of these two types of foreign exchange data, and the results are recorded below:

From Table 1 and Figure 2, we find some typical statistical features of the financial time series, such as leptokurtosis, fat tails, and non-normal distribution. All these significant features should be taken into consideration when we choose VaR models to estimate exchange rate risks.

4.2 *GARCH model selection about logreturn*

The return series are generally required to be stationary to build a model. Therefore, we conduct an ADF (unit root) test, and the result shows that the logreturn series of the USD and EUR time series are stationary. In addition, we also conduct Engle's ARCH test for heteroscedasticity on the residual. The results, in Table 2, show that there is obvious heteroscedasticity.

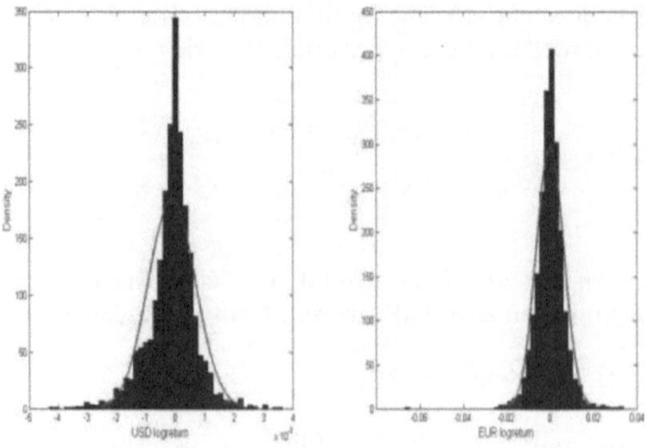

Figure 2. USD and EUR exchange rate logreturns density and normal density.

Table 2. Test for heteroscedasticity.

Order	LM-statistics For USD	pValue	LM-statistics For EUR	pValue
1	55.7321	0.000376	231.7028	0
2	71.6247	0.001232	235.4935	0
3	110.3689	0.002645	235.9931	0

Therefore, it is reasonable to build a GARCH model to fit the logreturn series.

Since two time series have obviously fat tail characteristics, we assess the performance of the GARCH-t(1,1), the EGARCH-t(1,1), and the GARCH-t-GJR(1,1) model according to AIC-BIC criterion, where t means their marginal distributions both obey general Student t distribution. The estimated results are shown in Table 3 and Table 4. For USD data, we find that the AIC and BIC of EGARCH-t(1,1) is the smallest, which indicates that the EGARCH-t(1,1) model is the best to fit logreturn's volatility. However, for EUR data, the GARCH-t(1,1) model is probably the best one. The best estimation is in Formulas (13) and (14).

$$\begin{cases} r_{USD_t} = -8.1391\times10^{-6} + \varepsilon_t \\ \quad (-2.4371) \\ \log\sigma_t^2 = -0.0141 + 1\times\log\sigma_{t-1}^2 + 0.2066\dfrac{|\varepsilon_{t-1}|}{\sigma_{t-1}} - 0.03915\dfrac{\varepsilon_{t-1}}{\sigma_{t-1}} \\ \quad (-0.31)\quad(344.5)\quad(6.0413)\qquad(-1.5986) \end{cases} \tag{13}$$

$$\begin{cases} r_{EUR_t} = 9.1636\times10^{-5} + \varepsilon_t \\ \quad (0.6245) \\ \sigma_t^2 = 2\times10^{-7} + 0.958\sigma_{t-1}^2 + 0.03547\varepsilon_{t-1}^2 \\ \quad (1.1899)\quad(-94.04)\quad(3.8613) \end{cases} \tag{14}$$

4.3 *VaR calculation*

From the GARCH models discussed above, we attain the series of standardized residuals of the logreturn. By employing EVT, we can estimate parameters by using

Table 3. AIC-BIC criterion for USD GARCH models.

	GARCH-t(1,1)	EGARCH-t(1,1)	GARCH-t-GJR(1,1)
AIC	−13443	−13812	−13446
BIC	−13418	−13781	−13416

Table 4. AIC-BIC criterion for EUR GARCH models.

	GARCH-t(1,1)	EGARCH-t (1,1)	GARCH-t-GJR (1,1)
AIC	−8367.9	−8358.5	−8367.7
BIC	−8342.8	−8328.4	−8337.6

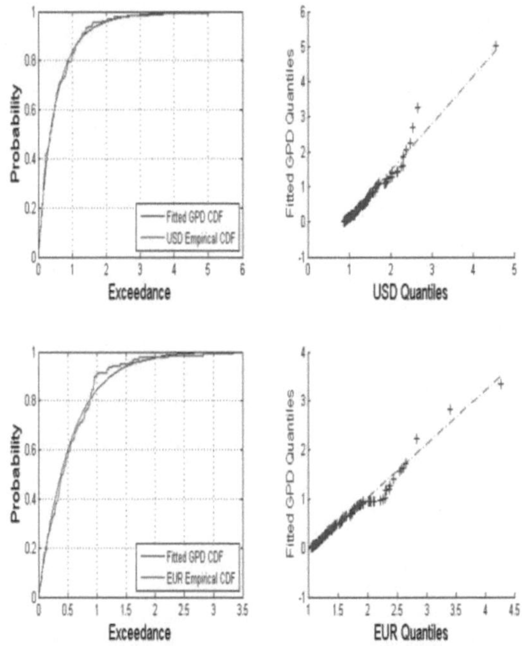

Figure 3. GPD fitted to the USD and EUR residuals (Right tail).

the maximizing the likelihood function (7). The estimated results are shown in Table 5. Figure 3 shows the effect of the residual data fitted by GPD and the corresponding Quantile-Quantile plot (Q-Q plot); both suggest that a very satisfactory fitting has been achieved.

Furthermore, we apply the conditional variance function to estimate volatility σ_{t+1}, and employ the VaR of the sequence of residuals to the conditional mean function $r_{t+1} = c + \sigma_{t+1} Z_{t+1}$, where Z_{t+1} is standardized residual. Then, the VaR estimates for returns of exchange rate can be obtained. Table 6 displays the VaR of the right tail and left tail of the USD and EUR with high (99% or 1%) and low (95% or 5%) confidence level.

Table 5. Estimated Parameters of GPD.

	u	Exceedance	ratio	$\hat{\xi}$	$se(\hat{\xi})$	$\hat{\beta}$	$se(\hat{\beta})$
Right Tail							
USD	0.8592	263	11.97%	0.1585	0.100	0.4744	0.0625
EUR	1.3179	190	8.64%	0.0161	0.0903	0.4814	0.0656
Left Tail							
USD	−1.344	219	9.99%	0.0766	0.0928	0.5433	0.0720
EUR	−1.292	219	9.99%	0.2290	0.1024	0.3944	0.0544

Table 6. VaR estimates of empirical data.

	Right Tail		Left Tail	
	$VaR_{0.99}$	$VaR_{0.95}$	$VaR_{0.01}$	$VaR_{0.05}$
USD	0.030	0.021	−0.0213	−0.0184
EUR	0.023	0.017	−0.0115	−0.0057

5 CONCLUSION

The research of risk measurement is the ubiquitous question in the field of finance and economics. Unlike traditional methods that directly apply Extreme Value Theory to the raw empirical data, in this paper we apply ARMA-GARCH models to EVT to capture the statistical features of the return series of the Chinese exchange market. First, we choose the best GARCH model to filter the return series to independent identical distribution's residual series, and select t distribution as the marginal distribution to compensate for the fat tails. Then, we employ the GPD to analyse the standard residual series and estimate the relevant parameters used by the POT method. Finally, we examine the Chinese exchange rate data USD/RMB and EUR/RMB, and calculate the corresponding VaR estimates, which may provide investors with a more accurate measurement of foreign exchange risk.

ACKNOWLEDGEMENT

This research is supported by the Scientific Research Foundation of applied basic project in Yunnan province under grand number 2010ZC063 and 2010ZC079.

REFERENCES

Javier F. Navas, "Calculation of Volatility in a Jump-Diffusion Model," The Journal of Derivatives, Oct, 2003, pp. 1–17.

Engle R, "Autoregressive Conditional Heteroskedasticity with Estimates of The Variance of United Kingdom Inflation," Journal of Finance, 1982 , of United 50(3), pp. 812–851.

Bollerslev T., "Generalized Autoregressive Conditional Heteroskdasticity," Journal of Econometrics, 1986, 31, pp. 307–327.

Nelson, "Conditional Heteroskdasticity in Asset Returns: A New Approach," Econometrics, 1991, pp. 347–370.

Gencay. R, and Selcuk. F, "Extreme Value Theory and Value-at-Risk:Relative Performance in Emerging Markets," International Journal of Forecasting, 2004, 20.

John C. Hull "Risk Management and Financial Institutions," China Machine Press, 2008, pp. 144–149.

Pan Zhibin. "Risk measurement in financial market". Shanghai: Shanghai Academy of Social Sciences Press, China, 2008, pp. 78–115.

Wang Xinyu. "Measure Methods and Empirical Analysis of Financial Market Risk," Beijing: Economic Management Press, China, 2008, pp. 34–85.

Future Communication Technology and Engineering – Chan (Ed.)
© 2015 Taylor & Francis Group, London, ISBN: 978-1-138-02777-0

Laser signal attenuation due to sand and dust particle scattering

Y.G. Wang, H.Q. Wang, & M.H. Cao
School of computer and communication, Lanzhou University of Technology, Lanzhou, Gansu, China

ABSTRACT: The scattering characteristics of laser beam transmitting in sand and dust weather conditions is analysed based on Mie theory. The influence of extinction, scattering, and absorbing on laser signals is analysed under conditions of different dust particle radii. Afterwards, the influence of single scattering and multiple scattering on laser signals in different atmospheric visibility is simulated. Moreover, the laser intensity attenuation of different optical wavelengths is calculated according to the measured data from the Helan Mountain area in sand and dust weather. The result shows that the attenuation induced by sand and dust particles is inversely proportional to atmospheric visibility, and multiple scattering plays a remarkable role in long-distance transmission.

1 INTRODUCTION

Free Space Optical (FSO) communication is considered as one of the potential ways to realize last mile high-speed data transmission, especially in natural disaster and emergency rescue communication. FSO has the advantages of wide bandwidth, security, and being immune to electromagnetic interference. Since signals are transmitted in atmospheric channels, factors like atmospheric turbulence, rain, cloud, fog, and dust will have an impact on signal quality. In recent years, there has been much research focused on the field of atmospheric turbulence, rain, and fog [1–4]. However, on the issue of sand and dust weather, only research into the microwave domain has been studied [5,6]. Unfortunately, the impact of sand and dust weather on wireless optical domains was neglected, even though particles causing scattering and absorbing would seriously affect the free space optical signals. Therefore, the laser attenuation and dust causing the scattering need to be analysed in FSO systems to ensure reliable delivery of information.

2 SIGNAL ATTENUATION CAUSED BY SINGLE SCATTERING

The semiarid region in northwest China suffers from sandstorms frequently due to its neighbouring desert. The average sand and dust particle radius there is less than 0.1 mm, and the particle size follows lognormal distribution [7], as shown in Equation (1):

$$p(r)=\frac{1}{2r\sigma\sqrt{2\pi}}\exp\left\{-\frac{1}{2}\left[\frac{\ln(2r)-\mu}{\sigma}\right]^2\right\} \quad (1)$$

where r denotes the particle radius, σ denotes the standard deviation, and μ denotes the mean value of $\ln(2r)$.

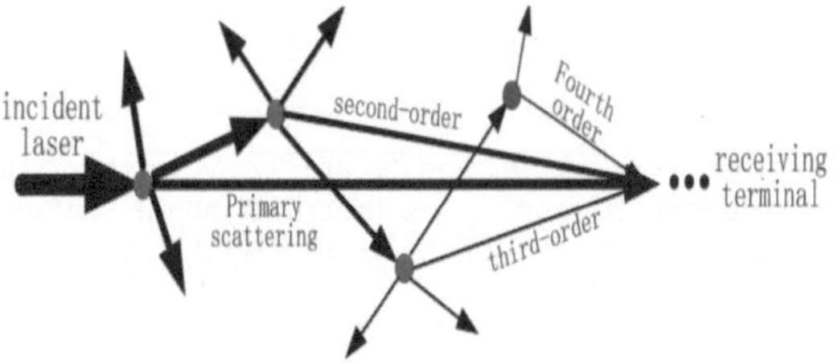

Figure 1. Schematic diagram of sand and dust particles causing scattering.

The typical value of μ and σ is -9.72 and 0.405 [7], respectively.

The sand and dust particles causing scattering are described in Figure 1. The random scattering results in signal intensity fluctuation, which affects the Bit Error Rate (BER) performance of intensity, modulated by FSO communication systems.

When the distance between the sand and dust particles is greater than three times the particles' diameter, it is considered that scattering is independent, which means only the effect of single scattering needs to be taken into account [8]. According to the Mie theory, the signal light intensity after single dust particle scattering can be expressed as [8]:

$$I=\frac{\lambda^2}{8\pi^2 d^2}I_0(i_1+i_2) \quad (2)$$

where λ denotes the wavelength of the laser signal, d denotes the distance from the dust particle to the observation point, I_0 denotes the original signal light intensity, and i_1 and i_2 denote the perpendicular and parallel components after scattering, which can be expressed respectively as:

$$i_1=S_1(m,\theta,\alpha)\Box S_1^*(m,\theta,\alpha) \quad (3)$$

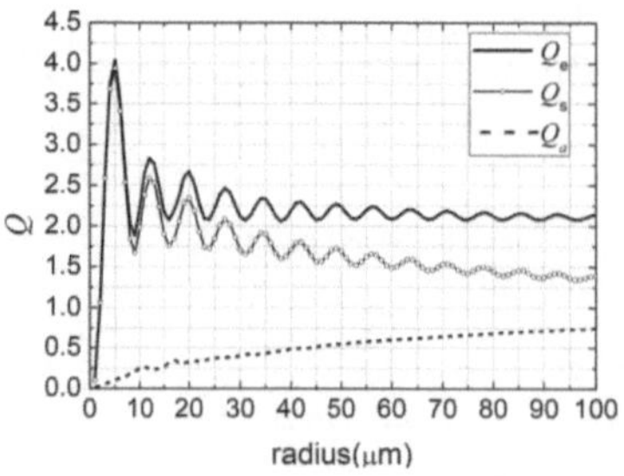

Figure 2(a). The attenuation, scattering, and absorbing efficiency factors versus the sand and dust particle radius (laser wavelength = 8 μm).

$$i_2 = S_2(m,\theta,\alpha) S_2^*(m,\theta,\alpha) \tag{4}$$

where θ is the scattering angle, α is the sand particle size, m is the sand particle refraction factor, and S_1 and S_2 are the scattering functions which can be given by:

$$\begin{cases} S_1(\theta) = \sum_{n=1}^{\infty} \frac{2n+1}{n(n+1)} [a_n \pi_n(\cos\theta) + b_n \tau_n(\cos\theta)] \\ S_2(\theta) = \sum_{n=1}^{\infty} \frac{2n+1}{n(n+1)} [a_n \tau_n(\cos\theta) + b_n \pi_n(\cos\theta)] \end{cases} \tag{5}$$

In Equation (5), a_n and b_n are the Mie scattering coefficients, π_n and τ_n are functions of the scattering angle θ, which can be expressed by the first order Legendre function:

$$\pi_n = \frac{P_n^{(1)}(\cos\theta)}{\sin\theta} \tag{6}$$

$$\tau_n = \frac{dP_n^{(1)}(\cos\theta)}{d\theta} \tag{7}$$

In a single scattering scenario, the signal attenuation and scattering can be defined by the extinction efficiency factor and scattering efficiency factor, denoted as Q_e andQ_s, respectively:

$$Q_e = 2/\alpha^2 \operatorname{Re} \sum_{n=1}^{\infty} (2n+1)(a_n + b_n) \tag{8}$$

$$Q_s = 2/\alpha^2 \sum_{n=1}^{\infty} (2n+1)\left(|a_n|^2 + |b_n|^2\right) \tag{9}$$

The difference of Q_e and Q_s is known as the absorption efficiency factor, denoted as Q_a:

$$Q_a = Q_e - Q_s \tag{10}$$

Figures 2 (a) and (b) show the relation between sand and dust particle radius with respect to the Q_e, Q_s, and Q_a when the signal light wavelength is fixed at 8 μm and 10.6 μm, respectively. We discover from Figure 2 (a) that with an increase of the particle radius, the extinction efficiency factor and the scattering efficiency factor gradually reach a constant of 2.1 and

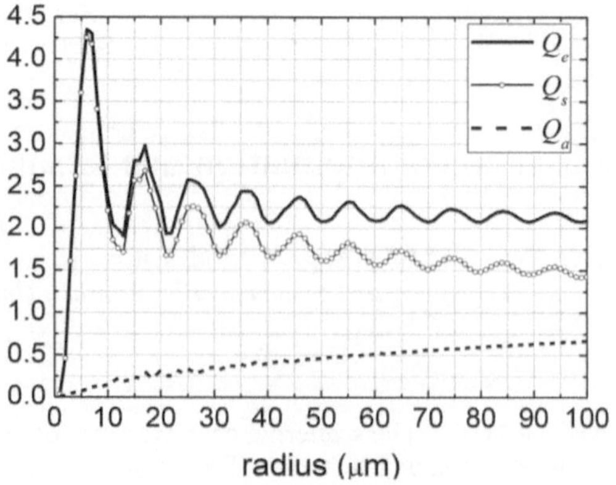

Figure 2(b). The attenuation, scattering and absorbing efficiency factors versus the sand and dust particle radius (laser wavelength = 10.6 μm).

1.35. When the laser wavelength rises to 10.6 μm, as shown in Figure 2 (b), the extinction efficiency factor and scattering efficiency factor fluctuate more, and gradually reach a larger constant of 2.15 and 1.48.

In a single scattering scenario, the extinction coefficient and the scattering coefficient can be expressed by Equations (10) and (11):

$$\mu_e = \int_0^{\infty} \pi r^2 Q_e \rho N(r) dr \tag{11}$$

$$\mu_s = \int_0^{\infty} \pi r^2 Q_s \rho N(r) dr \tag{12}$$

where ρ denotes the average sand and dust particle density (cm^3), r denotes the particle radius, and $N(r)$ denotes the distribution of particle size.

When the optical signal propagates in discrete random medium, according to Bouguer's Law, the signal intensity can be described as:

$$I = I_0 \exp(-\int_0^l \mu_e dl) \tag{13}$$

where l is the transmission distance, and I_0 is the initial signal intensity. When Equation (10) is substituted for Equation (12), the light intensity will be:

$$I = I_0 \exp(-\rho \int_0^l \int_0^{\infty} \pi r^2 Q_e N(r) dr dl) \tag{14}$$

According to Koschmieder's Law, the atmospheric visibility can be expressed as [9]:

$$V = \frac{15}{\alpha_0} \tag{15}$$

where $\alpha_0 = 8.686 \times 10^{-3} \pi \rho \int r^2 \mathrm{N}(r) dr$ [10]. When Equation (14) is substituted for Equation (13), the signal light intensity will be:

$$I = I_0 \exp\left(\frac{-15 \int_0^l \int_0^{\infty} \sigma_e N(r) dr dl}{8.686 \times 10^{-3} \pi V \int_0^{\infty} r^2 N(r) dr}\right) \tag{16}$$

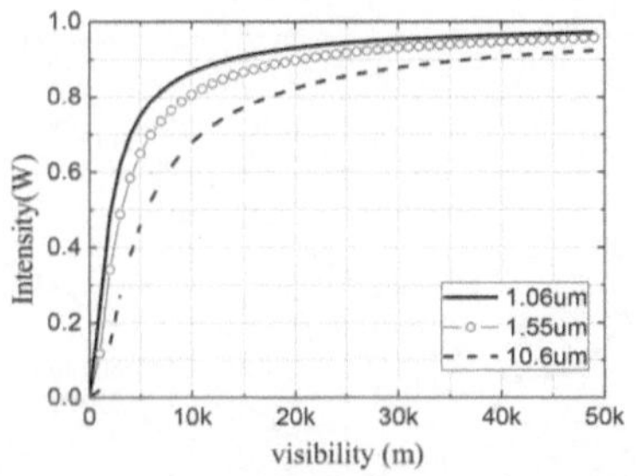

Figure 3(a). The laser signal intensity as a function of atmospheric visibility (l = 1000 m).

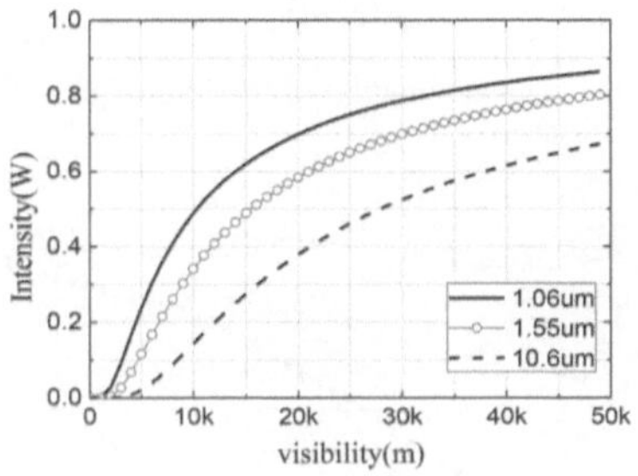

Figure 3(b). The laser signal intensity as a function of atmospheric visibility (l = 5000 m).

Figure 3 shows the relationship between the received signal light intensity and atmospheric visibility in different light wavelengths (1.06 μm, 1.55 μm and 10.6 μm), where the signal intensity is normalized to 1W. We can conclude from Figure 3(a) that the received laser intensity increases along with the increase of atmospheric visibility, and finally it reaches a stable value of 0.97 w, 0.96 w, and 0.92 w for wavelengths of 1.06 μm, 1.55 μm, and 10.6 μm, respectively. When increasing the transmit distance to 5 km, as shown in Figure 3(b), the curves grow gradually. This is because when a laser signal propagates at long-distance the dominant factor will become multiple scattering, and single scattering is insufficient to describe the scenario. Therefore, a multiple scattering scenario was required for analysis.

3 SIGNAL INTENSITY ATTENUATION CAUSED BY MULTIPLE SCATTERING

We assume the signal transmits along the z-axis direction. Collision continued until the photon escaped from, or was absorbed by, the sand and dust medium, which follows uniform distribution. The distance of one photon in two random collisions can be described as [10]:

$$L = -\frac{\ln(\xi)}{\sigma_e} \tag{17}$$

where ξ is a random number within [0,1].

When a photon travels from the n-th collision point (x_n, y_n, z_n) to the $(n+1)$-th collision point (x_n+1, y_n+1, z_n+1), the coordinates of the two points will be:

$$\begin{cases} x_{n+1} = x_n + \mu_x L \\ y_{n+1} = y_n + \mu_y L \\ z_{n+1} = z_n + \mu_z L \end{cases} \tag{18}$$

where μ_x, μ_y, μ_z denote the intensity component of the x-axis, y-axis, and z-axis, respectively. We assume μ'_x, μ'_y, μ'_z is the intensity component of the x-axis, y-axis, and z-axis after the $(n+1)$-th collision, respectively. We can express them as Equations (19) to (21):

$$\mu_x' = \frac{\sin\theta}{\sqrt{1-\mu_z^2}}(\mu_x\mu_z\cos\varphi - \mu_y\sin\varphi) + \mu_x\cos\theta \tag{19}$$

$$\mu_y' = \frac{\sin\theta}{\sqrt{1-\mu_z^2}}(\mu_x\mu_z\cos\varphi + \mu_x\sin\varphi) + \mu_y\cos\theta \tag{20}$$

$$\mu_z' = -\sin\theta\cos\varphi\sqrt{1-\mu_z^2} + \mu_x\cos\theta \tag{21}$$

where ϕ denotes the azimuthal angle, which is uniformly distributed over the interval of $[0, 2\pi]$. θ denotes the scattering angle which is defined by the Henyey-Greenstein function:

$$\theta = arc\cos\left\{\frac{1}{2g}\left[1+g^2-\left(\frac{1-g^2}{1-g+2g\xi}\right)^2\right]\right\} \tag{22}$$

The probability of the n-th photon escaping from the medium after i-th collision can be given by [7]:

$$P_n^i = \begin{cases} w_i\exp(-\mu_e\frac{H-z_i}{\mu_{zi}}), & \mu_{zi} > 0 \\ 0 & , \mu_{zi} \le 0 \end{cases} \tag{23}$$

where H is the thickness of the dielectric medium, w_i is the gain constant, and μ_{zi} is the z direction cosine component of the i-th collision. The relation of the gain constant of the two adjacent collisions can be described as:

$$w_i = w_{i-1}\exp(-\mu_a L) \tag{24}$$

when assuming that the original value of w_0 is 1 and μ_a is the absorption coefficient which can be expressed as $\mu_a = \mu_e - \mu_s$. According to Equation (23), the collision probability of the n-th photon will be:

$$p_n = \sum_{i=0}^{M-1} p_n^i \tag{25}$$

where M is the number of collision times. If the total photon number is K, the average atmospheric transmittance can be described as:

$$\tau = \frac{1}{K}\sum_{n=1}^{K} p_n \tag{26}$$

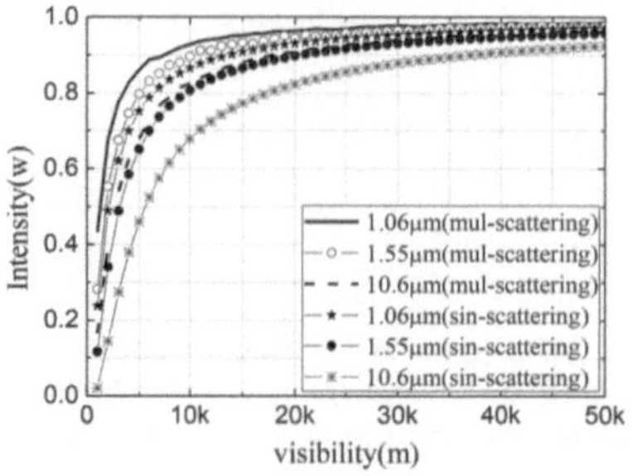

Figure 4(a). The laser signal intensity as a function of atmospheric visibility (l = 1000 m).

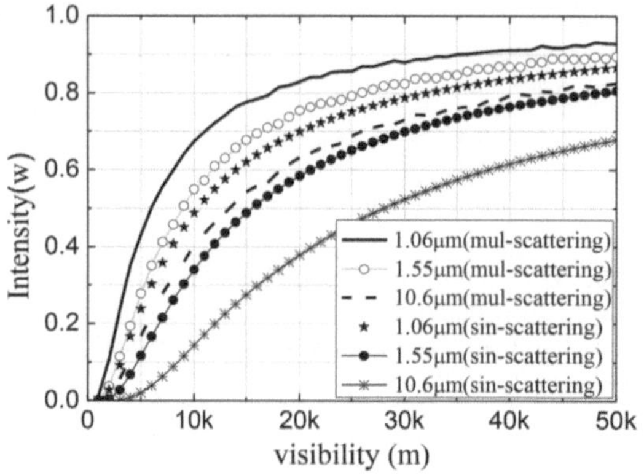

Figure 4(b). The laser signal intensity as a function of atmospheric visibility (l = 5000 m).

Therefore, laser intensity after multiple scattering can be expressed as:

$$I = I_0 \tau \tag{27}$$

Figure 4 shows the relationship of the received signal light intensity and atmospheric visibility in different light wavelengths (1.06 μm, 1.55 μm and 10.6 μm), where the signal intensity is normalized to 1W and the refractive index of sand and dust particles is set as 1.55-i0.005. We can conclude from Figure 4(a) that the laser signal intensity will increase to a peak value when the atmospheric visibility is beyond 10^4 m. Furthermore, the transmit distance has a different impact on the signal intensity of different wavelengths. For example, when the atmospheric visibility is 20 km and the transmission distance is 1000 m, the different laser intensities between single scattering and multiple scattering are 0.03w, 0.047w, and 0.08w, with wavelengths of 1.06 μm, 1.55 μm, and 10.6 μm, respectively. When the transmission distance increases from 1000 m to 5000 m, as shown in Figure 4(b), the difference rises to 0.13w, 0.17w, and 0.25w, respectively. We can conclude that multiple scattering is the dominant influencing factor on signal intensity in long-distance transmission because of the occurrence of more collisions during the signal propagation. Conversely, single scattering is the dominant influencing factor on short-distance transmission.

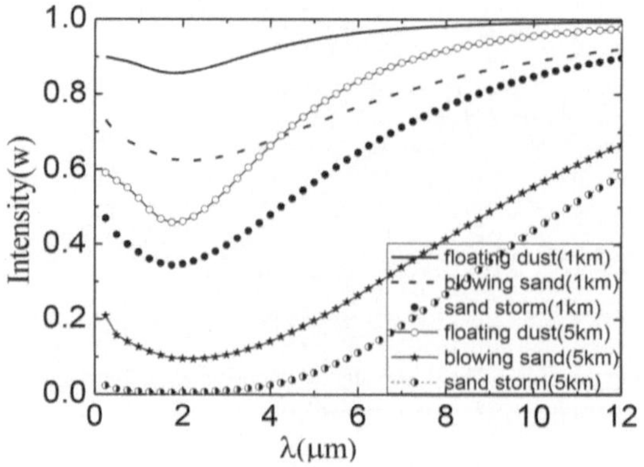

Figure 5(a). The attenuation of the laser signal in a single scattering scenario.

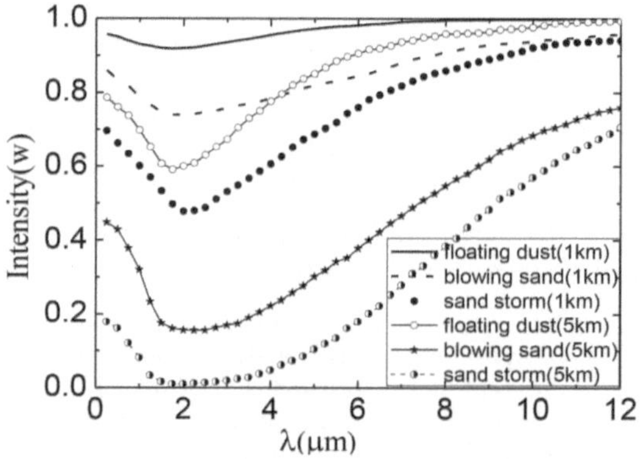

Figure 5(b). The attenuation of the laser signal in a multiple scattering scenario.

According to the measured data of sand particle size distribution from the Helan Mountain area in floating dust weather, blowing sand weather, and sandstorm weather, respectively [11], the laser intensity attenuation of different optical wavelengths is calculated under the conditions of single scattering and multiple scattering scenarios, where the transmission distance is 1 km and 5 km, respectively. It can be seen obviously from Figure 5 that ① there is an attenuation band at 2 μm; ② the sand and dust density is the most critical influence on signal attenuation; and ③ multiple scattering is the dominant influencing factor on signal intensity in long-distance transmission in all three kinds of weather in the Helan Mountain area.

4 CONCLUSIONS

We analyse the relationship between the extinction efficiency factor, the scattering efficiency factor, and the absorption efficiency factor on the sand and dust particle radii. Moreover, the relationship between laser intensity and atmospheric visibility in single and multiple scattering scenarios is analysed in sand and dust weather. Furthermore, the signal attenuation of different wavelengths is calculated according to the measured data of the Helan Mountain area. The result shows that the received laser intensity will reach a peak value with an increasing in atmospheric visibility. In addition, multiple scattering is the dominant influencing factor on signal intensity in long-distance transmission, both in single and multiple scattering.

REFERENCES

[1] Grabner, Kvicera. 2014. Multiple Scattering in Rain and Fog on Free-Space Optical Links. *Journal of Lightwave Technology* 32(3): 513–520.

[2] Ijaz, Zabih, Joaquin et al. 2013. Enhancing the Atmospheric Visibility and Fog Attenuation Using a Controlled FSO Channel. *IEEE Photonics Technology Letters* 25(13): 1262–1265.

[3] Fadhil, Al-Khafaji, Abd et al. 2012. New Priority-Based Parameter Optimization Technique for Free Space Optics under Bad Weather Conditions. *2012 IEEE 3rd International Conference on Photonics (ICP)*: 116–120.

[4] Capsoni, Luini, Nebuloni. 2012. Prediction of Cloud Attenuation on Earth-space Optical Links. *2012 6th European Conference on Antennas and Propagation (EUCAP)*: 326–329.

[5] Saleh, Abufares, Snousi. 2012. Estimation of wave attenuation due to dust and sand storms in southern Libya using Mie model. 2012 IEEE 13th Annual Wireless and Microwave Technology Conference (WAMICON): 1–5.

[6] Alhuwaimel, Mishra, Inggs. 2012. Review of Radar Signal Attenuation Due to Sand and Dust Storms. *Proceedings of the 2012 International Conference on Electromagnetics in Advanced Applications, ICEAA'12*: 1096–1099.

[7] Yang Ruike, Su Zhenling, Liu Kexiang et al. 2008. Influence of multiple scattering on attenuation of millimeter wave propagation in sand and dust storm. *Chinese Journal of Radio Science* 23(3): 530–533.

[8] Dong Qunfeng, Li Yingle, Xu Jiadong et al. 2012. Scattering of electromagnetic pulse wave by spherical sand and dust particles. *2012 10th International Symposium on Antennas, Propagation and EM Theory, ISAPE 2012*: 490–492.

[9] Muhammad, Zabih, Jiri et al. 2013. Modeling of fog and smoke attenuation in free space optical communications link under controlled laboratory conditions. *Journal of Lightwave Technology* 31(11): 1720–1726.

[10] Chicea, Turcu. 2005. A Random Walk Monte Carlo Approach to Simulate Multiple Light Scattering on Biological Suspensions. *Romanian Reports in Physics* 57(3): 418–425.

[11] Gu Yudan, Nui Shengjie. 2006. Analysis and Fitting of the Instantaneous Size Distribution of Sand-Dust Aerosol Particles in Helan Mountainous Area. *Journal of Nanjing Institute of Meteorology* 29(4): 500–506.

Future Communication Technology and Engineering – Chan (Ed.)
© 2015 Taylor & Francis Group, London, ISBN: 978-1-138-02777-0

Research on the innovation and development of electronic commerce based on centralized procurement

L.Z. Zhang
Glorious Sun School of Business, Donghua University, Shanghai, China

ABSTRACT: The article focuses on how the domestic commercial banks systematically implement the innovation and development of e-commerce by relying on procurement operations. It analyses the current main modes of the development of domestic e-commerce and investigates the status of the development of domestic commercial banks in e-commerce. It studies three stages of progressive development of electronic commerce relying on centralized procurement operations, and analyses the opportunities and challenges of this innovative way for the development of e-commerce.

Keywords: Commercial bank, Internet finance, centralized procurement, e-commerce, SMEs (Small and Medium-sized Enterprises).

1 INTRODUCTION

Internet finance is a hot topic. In the Internet financial ecosystem, commercial banks and electronic commercial enterprises are participants, whose human and financial resources are to be put into new areas of e-commerce the most. The author argues that the model is a robust development road for a commercial bank entering the field of electronic commerce, and providing Internet financial innovation services to SMEs.

2013 was the first year of the development of Internet finance, and Internet finance is one of the hottest issues. Since the birth of the concepts of Internet finance, Internet companies, banks, regulators, media, and research institutions are constantly supplemented with new interpretations and connotations, and the debate associated with it has not stopped. I believe that Internet finance and e-commerce are not only interinfiltration but also interdependence and the goal is to promote the development of both finance and e-commerce through the Internet, and constantly improve the customer experience and productivity. Through severe competition and testing in the market and the Internet, finance and the development of electronic commerce tend to integrate with each other inevitably. The article focuses on the innovative ways of commercial banks to promote e-commerce development.

2 THE PURPOSE AND STATUS OF THE COMMERCIAL BANKS TO EXPAND E-COMMERCE

Commercial banks are the main source of financing for enterprises and government, and bank loans are the main way to get valuable funds. In recent years, with the opening of domestic capital markets, and with the rapid development of shadow banking and private lending, new ways to get money are constantly emerging, thus, to some extent, alleviating the problem of financing in small and medium-sized enterprises. The development of state-owned commercial banks and national joint-stock commercial banks represents the direction of the development of electronic commerce in the financial industry.

At present, China's e-commerce models in accordance with the transaction object, mainly cover the B2B e-commerce, B2C e-commerce, C2B e-commerce, and C2C e-commerce models. Analysis finds that currently the domestic commercial banks are still in the exploratory stage, and commercial banks developing e-commerce business pursue three main objectives.

The first goal is to obtain the data. Banks facilitate e-commerce by businesses' online services, and they want to continue to accumulate SMEs or personal real data, and then use this data to provide a reliable basis for making decisions for enterprise loan and personal loan As we all know, it is difficult to get SMEs loans from commercial banks for the following important reasons: Commercial banks are unable to determine the real situation of SMEs operating, it has unfavourable conditions of monitoring business, operating costs are high, and risks and benefits of the loan cannot be accurately judged. To solve these problems, the banks offering e-commerce can alleviate the problems of missing data in small and medium business.

The second goal is to increase the revenue of the intermediary business. Since China joined the World Trade Organization (WTO), the time of the financial market has come, the domestic rates market will be gradually liberalized, and the different ways of depositing will become narrower. Financial institutions will face unprecedented challenges, and market competition will become more intense. Faced with such an environment, commercial banks must be restructured to look for new business growth in order to make up for the loss of policies to protect the upcoming dividend, which is an increase of intermediary business revenue. Business innovation is very important for the commercial banks to solve the existing problems.

The third goal is to enhance the core competitiveness to boost the share price. According to industry experts, From February 11 to 13, 2014, the share price of such banks as CITIC, Ping An, and Minsheng rose continuously, and even appeared in a number of trading board, and the Internet financial innovation subjects closely related.

Since the beginning of 2005, some commercial banks began to seek cooperation with e-commerce companies and promote the development level of e-commerce business, but ultimately they did not form a good model of cooperation. Some commercial banks began to try to build their own e-commerce platform to explore e-commerce business innovation about Internet finance, and they achieved remarkable results.

2.1 *Industrial and Commercial Bank of China*

ICBC's "melting e purchase" e-commerce platform formally launched operations in January, 2014. The "melting e purchase" shopping mall is dominated by B2C, which has advantages in financial services, such as the flexibility in banking payment and convenient financing, and that highlight such features as shopping can loan, integral can arrive at present, quality, features login very convenient.

According to reports, after being online for only two months, the mall is a collection of digital home appliances, automotive, financial products, shoes and clothing, food and beverage, jewelry gifts, tourist traffic and more than 10 industry hundreds of well-known brands, real-time online sales with nearly ten thousand pieces of salable goods.

2.2 *China construction bank*

The China Construction Bank started its own e-commerce platform "good financial business" in 2010. Because of "high" operating costs, a "big" site-scale, and "full" product categories, the industry considered that "good financial business" was taking the "higher grade" line. The platform formally launched into operation in late 2011. Currently, the platform mainly consists of two parts; one is a good financial enterprise mall, whose flagship is SME B2B, and the other is a good financial personal mall, whose flagship is B2C model that the SMEs sell to the general consumer.

2.3 *China Minsheng Bank*

China Minsheng e-commerce limited liability company was established in Qianhai, Shenzhen on August 29, 2013. Minsheng e-commerce registered capital is up to 3 billion CNYand the e-commerce platform name is "unity line." It is reported that Minsheng e-commerce plans to expand the IT technical staff to 800 or 900 people within six months, focusing on the development of IT systems and platform construction.

3 INNOVATIVE IDEAS FOR RELYING ON CENTRALIZED PROCUREMENT TO DEVELOP E-COMMERCE

Centralized purchasing by commercial banks is generated with the deepening of China's financial reform. The so-called centralized purchasing by commercial banks refers to commercial banks to determine the procurement object through a tender, inviting tender, competitive negotiation, inquiry, a single source negotiation, and in other ways, in order to get bulk items and works by centralizing non-personal funds for purchase, lease commission, employment, etc. The centralized procurement management model of the commercial banks has recently become one of the hot topics of domestic banks. Research on Commercial Bank of centralized procurement is just try to research on the centralized procurement business on the Internet, the Internet financial, e-commerce, the business of commercial bank sets a commercial bank should be used to focus on purchasing management is still a blank.

With the rapid development of economy in China, the scale and efficiency of commercial banks have been quick to upgrade, and the amount of centralized procurement by commercial banks has also shown a rapid increase in momentum. Only in 2009, the centralized procurement of a state-owned commercial bank amounted to more than 5 billion RMB. Some commercial banks have launched a centralized procurement management system, but the centralized procurement regulations is uneven; some only provide a centralized procurement system in the headquarters, and branches' procurement operations are still self-managed; some centralized procurement management part of goods procurement, in most cases, is authorized to branch management.

Through the analysis, I believe that by relying on centralized procurement, business expansion of e-commerce business is a viable way commercial banks can explore to develop e-commerce. In Internet financial and e-commerce environments, the study of the centralized procurement B2B model of commercial banks not only optimizes the traditional procurement model and the integration of the internal allocation of resources, but also highlights financial services innovation and offers an opportunity. Commercial banks can be divided into three step development of B2B electronic commerce.

3.1 *A single buyer stage of e-commerce*

This stage is the laying down of the foundation of a commercial bank engaged in e-commerce. The key point is to develop a plan to expand e-commerce, which includes a product system, a business structure, a brand strategy, and an implementation plan, and specifically the plan includes standardized vendor management on the specification of centralized procurement, the introduction of supplier access and exit mechanisms, and classification and ratings management. Implementation of the procurement management plan can either arrange for the procurement plan, or effectively guide the supply products of vendor, improving efficiency and reducing costs. Regulating the internal procurement process is also essential at this stage, and is the essential link to achieve online internal management of the centralized procurement. This phase is the construction of the initial e-commerce platform, and the buyer only has a commercial bank, and online commodity control production and operation requirements and non business demand in the range of their own, the commercial bank is the relative strength of the buyer, has a comparative advantage on carding supplier.

3.2 *The bank's internal staff to buy a stage of e-commerce*

After the foundation of the first phase of the development, supplier management has entered the track specification, price, quality and service in the process of centralized procurement of goods advantage has been shown, business cooperation between commercial banks and suppliers increasingly tacit understanding. The supply of commodities supplier capacity far more than commercial banks routinely buy demand, both to further expand in the product quantity, types of desire. It is a good solution to introduce the internal stall for procurement in commercial banks. The stage has three modes to choose from; one is that staff is in the name of another client to participate in centralized purchase needed goods; another is that employee on behalf of commercial banks to purchase needed goods with the same price; The last model is relatively complex, that is, with reference to the commercial banks B2B centralized purchasing commodity prices, internal staff through the e-commerce platform, buy the desired product online. The phase is from a single client to expand to two or more internal customers and discover the C2B model to expand in the B2B platform.

3.3 *The market operation stage of e-commerce*

This stage is where a number of suppliers, commodity types, and transaction frequency of the transaction have been on a larger scale and system platform, and the operating mechanism and brand building have reached a level which can be marketed. The construction of the system can be expanded to external public network services after addressing security mechanisms, and users can synchronize the internal migration or add external access services. An e-commerce platform can try a market-oriented operation mechanism, to explore the introduction of division, and an independent legal person or commissioned external operation uses other means to achieve independent operation of the e-commerce platform. Then it introduces a customer base of commercial banks and integrates e-commerce platforms and e-banking channel services, and it further integrates factoring commercial banks, cash management, supply chain finance, online banking, and many other products into the e-commerce business, thus forming true commercial banking e-commerce one-stop services and achieving a breakthrough in all aspects of scale, efficiency, brand, and competitiveness.

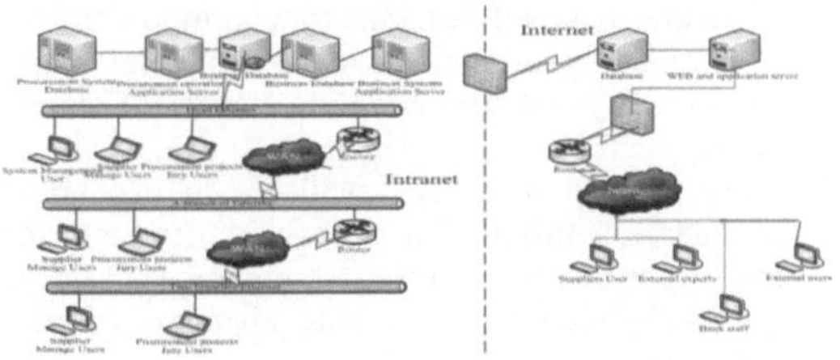

Figure 1. Centralized procurement – e-commerce structure chart.

After these three stages, the service system of commercial banks will build competitive e-commerce. Initial ideas to build a product system can be seen in Figure 1.

4 COMPARATIVE ANALYSIS OF THE ADVANTAGES AND DISADVANTAGES OF COMMERCIAL BANKS RELYING ON CENTRALIZED PROCUREMENT TO EXPAND E-COMMERCE SERVICES

Commercial banks relying on centralized procurement operations can steadily into the field of e-commerce. In the e-commerce ecosystem, each participant has a different understanding and the respective advantages of resources. From the perspective of SWOT, can be compared to the analysis of commercial bank along the centralized procurement, increase internal staff group purchase, and then realize the comprehensive customer service.

Strengths

1. Commercial banks have sufficient capital to support and promote the construction and operation of an e-commerce platform.
2. The trading platform has enough volume of business. It is understood that purchases from a national joint-stock commercial bank amount for at least 3 billion yuan or more per year. As the business grows, purchases will increase year by year.

3. There are abundant demands for commodities. The procurement requirements of commercial banks for operating and non-operating activities are very strong.
4. There are the resources available to expand e-commerce. Introduction of bank customers and employees buying into the e-commerce platform, and commercial banks can take advantage of price and service.
5. The credit of commercial banks can attract more suppliers. Banks are credit-operating financial institutions and can provide credit protection for e-commerce business.

Weaknesses

1. A self-built e-commerce platform lacks effective measures for risk control. A problem about a bank self-built e-commerce platform is how to ensure the risk operation business platform generated.
2. The banking operations of e-commerce are underpowered in the system. Bank conservative style of acting, bank dealers must exist act conservative, competitive weakness, lack of characteristics of wolf etc.
3. The e-commerce of the bank is vague on supervision.
4. E-commerce is just the supplement of the banking business. As can be seen from the analysis of the aim of e-business for commercial banks, commercial banks only consider e-commerce as a means of accessing data and middle incomes, and it is difficult to incorporate them into a highly developed strategy.

Opportunity

1. With the era of big data, commercial banks can obtain valuable data by expanding e-commerce business in addition to conventional funds information.
2. E-commerce activities can enhance existing customer cooperation.
3. The procurement activities can expand opportunities for the collaboration of commercial banks with suppliers.
4. Procurement activities can provide more business opportunities for the e-commerce platform and commercial banks.
5. The orderly development of the commercial bank supply chain finance can be promoted.
6. The benign supply of goods from the commercial bank centralized procurement can be promoted and protected.
7. The strict credit evaluation system of commercial banks, the strict risk control mechanism, and advanced technology systems are used to make e-commerce activities form a closed loop, so as to effectively reduce and control the risk of e-commerce in all activities.

Threaten

1. Commercial banks are facing challenges from the rapid growth of electronic business of Business Company.
2. Commercial banks are facing from the third party payment company business expanding threat.
3. Commercial banks are facing the threat of banks from the internal operation mechanism.

5 CONCLUSIONS

Overall, commercial banks in the Internet financial boom, whether in product system construction, or making changes in marketing ideas, are experiencing challenges and tests from various parties. I think that commercial banks rely on centralized procurement to develop e-commerce will be a new direction. I hope that the commercial bank can seek for the development of electronic commerce road as soon as possible.

REFERENCES

Guosheng He. Sustainable development of China's commercial banks to explore [M]. Beijing: China Financial Publishing House, 2012.6.

Pathak J, Lind M R. An e-business audit service model in the B2B context [J]. Information Systems Management, 2010, 27(2).

Fucui Xiao, Min Qin, Lihua Huang. Strategies transition to platform-based B2B e-commerce business model [J]. Business Economics and Management, 2011, 1 (8).

Wiersema F. The B2B Agenda: the current state of B2B marketing and a look ahead [J]. Industrial Marketing Management, 2013.

Bangtai Peng. Problems of e-commerce development in China's commercial banks and Countermeasures [J]. Audit and Finance, 2012 (3): 39–40.

Jiaying Wang, Juemin Wang. The study of commercial bank e-commerce platform development [J]. Financial Aspect, 2013 (6): 86–91.

Xuerong Lv. Grassroots People's Bank of centralized procurement analysis [J]. Jilin Financial Research, 2012 (12): 74–76.

Qingfeng Yan. B2B e-commerce platform for centralized procurement model analysis [J]. Chinese bidding, 2012 (1): 14–16.

Jing Yang. Thinking focused on the procurement of commercial banks [J]. Financial Times, 2011 (36): 132.

Jian Liu, Rui Zhu. Research modes of operation of commercial bank strategic alliance [J]. Modern Finance, 2012 (8): 30–32.

Future Communication Technology and Engineering – Chan (Ed.)
© 2015 Taylor & Francis Group, London, ISBN: 978-1-138-02777-0

The simulation of cluster bombs' damage efficiency and analysis of damage factor

S. Liu, X.J. Zhao, F. Zhao & X. Wei
Science and Technology on Complex Land Systems Simulation Laboratory, Beijing, China

W.F. Li
Chinese Defense Science and Technology Information Centre, Beijing, China

ABSTRACT: First, the paper emulated cluster bombs' random point of fall, the radius of damage of the bombs, drew the bombs' damage range, and counted the damage effect. The final damage efficiency of cluster bombs was calculated and the influence from each damage factor of the cluster bombs was studied through many random experiences. The method solved the tough problem of the precise calculation of damage efficiency, which was caused by the overlapped damage size.

1 INTRODUCTION

The damage efficiency of cluster bombs was analysed comprehensively through modeling and simulation, which may provide corresponding researchers with reference data.

At present, there are three available methods to evaluate the damage efficiency of cluster bombs, which are the practical method, the analytical method, and the simulation method. In the paper, the simulation method was used in order to solve the problem of precisely calculating the damage efficiency.

2 SHOT ERROR ANALYSIS

2.1 *Components and classification of shot error*

Shot error is the random offset between the actual shot point and the target. In order to describe the shot error, a Cartesian coordinate system was built in which O represented the target, X represented the vertical axis, and Z represented the horizontal axis, as shown in Figure 1. The aiming position was marked A, and its coordinates were assigned (x_0, z_0). The point of fall C of the average trajectory in the coordinate system was a random point, with an error which was marked (x_n, z_n) towards the point A. The error (x_n, z_n) was defined to be the shot error of various factors, as well as the scattered centre error. The actual trajectory did not accord with the average trajectory because of the scattering. The centre of the scattered bombs was the point P, with a random error which was marked (x_p, z_p). The fall point of each bomb was random too when the bombs were scattered. The deviation of the bomb's fall point Z towards the point P was defined to be the bomb's scattered error.

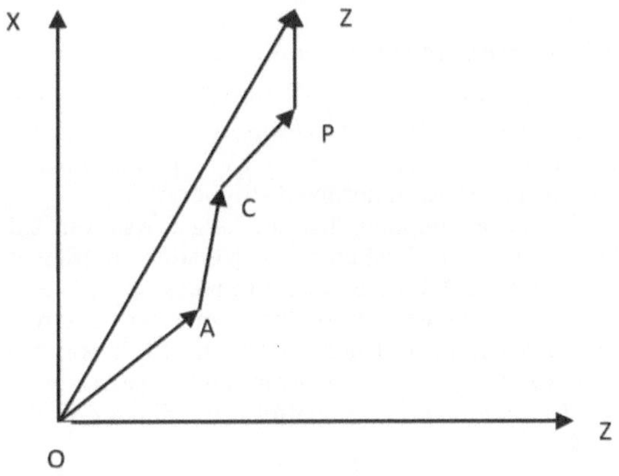

Figure 1. Scattered error of bombs.

- Analysis of the main bomb's scattered error

Generally, the vertical error and the horizontal error were independent and conformed to the normal distribution. The fall point of the main bomb conformed to the two-dimensional normal distribution $(\mu_1, \mu_2, \sigma_1, \sigma_2, \rho)$. μ_1 and μ_2 stood for the average coordinates of the main bomb's fall point, σ_1 and σ_2 stood for the shot accuracy of the cluster bombs, and ρ stood for the correlation coefficient of X and Y.

- Analysis of the cluster bombs' scattered error

The figure of the cluster bombs' fall points was usually round or elliptical, which conformed to the uniform distribution in the scattered area. However, the distribution of the direction and distance was not independent, which was not convenient for converting the error. In order to make the scattered direction and distance independent, the relevant transformation was necessary For example, the uniform distribution of the cluster bombs in the eclipse was demonstrated as follows.

When the fall points of the cluster bombs conformed to the uniform distribution, the joint density function was:

$$f(x,z)=\begin{cases}\dfrac{1}{\pi hl} & (\dfrac{x^2}{l^2}+\dfrac{z^2}{h^2}\le 1)\\ 0 & (\dfrac{x^2}{l^2}+\dfrac{z^2}{h^2}>1)\end{cases} \tag{1}$$

The rectangle of length $2L_l$ and width $2L_h$ can replace the eclipse:

$$\begin{aligned}L_l&=\sqrt{\pi}l/2\\ L_h&=\sqrt{\pi}h/2\end{aligned} \tag{2}$$

The joint density function was as follows:

$$f(x,z)=\begin{cases}\frac{1}{\pi hl} & \left(|x|\le\frac{\sqrt{\pi}l}{2},\ |z|\le\frac{\sqrt{\pi}h}{2}\right)\\ 0 & ohers\end{cases} \tag{3}$$

2.2 *Construction of the error model*

The shot error of the cluster bomb was composed of a variety of errors. In order to simplify the computation, the fall point of the main bomb (X, Y) conformed to the two-dimensional normal distribution.

On the assumption that the target was vertical against the shot direction, the correlation coefficient of the vertical and horizontal coordinates was 0. Therefore, the scattered area of the bombs was a circle with a radius of R. The centre of the circle was the expected fall point of the main bomb. The scattered bombs conformed to the uniform distribution in the circle.

- Scattering Random Simulation of the Main Bomb's Fall Point

The fall point of the main bomb (X, Y) conformed to the two-dimensional normal distribution in no consideration of the relevant errors. The random numbers of the standard two-dimensional normal distribution were generated by the following formula based on the random numbers of (0, 1) uniform distribution:

$$\begin{cases}g_1=\sqrt{-2\ln r_1}\cos(2\pi r_2)\\ g_2=\sqrt{-2\ln r_1}\sin(2\pi r_2)\end{cases} \tag{4}$$

r_1 and r_2 were random numbers of (0, 1) uniform distribution. g_1 and g_2 were independent random numbers of standard normal distribution. Suppose that g_1 and g_2 conformed to the two-dimensional normal distribution of ($\mu_1, \mu_2, \sigma_1, \sigma_2, \rho$):

$$\begin{pmatrix}g_1\\ g_2\end{pmatrix}=\begin{pmatrix}\sigma_1 g_3\\ \sigma_2(\rho g_3+g_4\sqrt{1-\rho})\end{pmatrix}+\begin{pmatrix}u_1\\ u_2\end{pmatrix} \tag{5}$$

and g_3 and g_4 were random numbers of standard two-dimensional normal distribution.

The coordinates of the main bomb's fall point, ($x+g_1$, $y+g_2$), were the shot point (x, y) plus the deviation (g_1, g_2). If the target was a point, the coordinates of the shot point were the same as the coordinates of the point; if the target was a line, the coordinates of the shot point were the coordinates of the line's midpoint; if the target was a rectangle, the coordinates of the shot point were the coordinates of the rectangle's centre.

- Scattering Random Simulation of the Cluster Bombs' Fall Point

If the coordinate of the circle' centre was (X_R, Y_R) and the radius was R, the coordinate of the i bomb's fall point which scattered uniformly in the circle was as follows:

$$\begin{cases}X_i=X_R+R\cdot Rand1\cdot\cos(2\pi\cdot Rand2)\\ Y_i=Y_R+R\cdot Rand1\cdot\sin(2\pi\cdot Rand2)\end{cases} \tag{6}$$

Rand 1 and *Rand* 2 were random numbers of (0, 1) uniform distribution.

3 REALIZATION OF SIMULATION PROGRAM OF CLUSTER BOMBS' DAMAGE EFFICIENCY

In order to study cluster bombs' damage efficiency, damage factors, such as the scattered radius of the main bomb and the number of bombs, would be initialized before the simulation. The pixel method was adopted to design the simulation programs.

3.1 *Design principles*

The damage area of each bomb was indicated by the number of pixels in the damage field.

During the simulation of m bombs, the damage area was marked as a distinct colour from the target area. The simulation result of the damage area was obtained by summing the number of pixels of the distinct colour from the target area, and multiplying the area of each pixel.

3.2 *Program realization*

A flow chart is shown as in Figure 2.

3.3 *Interface and function of the program*

The interface of the program was based on the dialog, as is shown in the following figure,

The program has realized the following functions.

- Initialization of Parameters

The parameters initialized before simulation consisted of the number of bombs, the damage radius of the bomb, the scattered radius of the main bomb, and simulation repetition times.

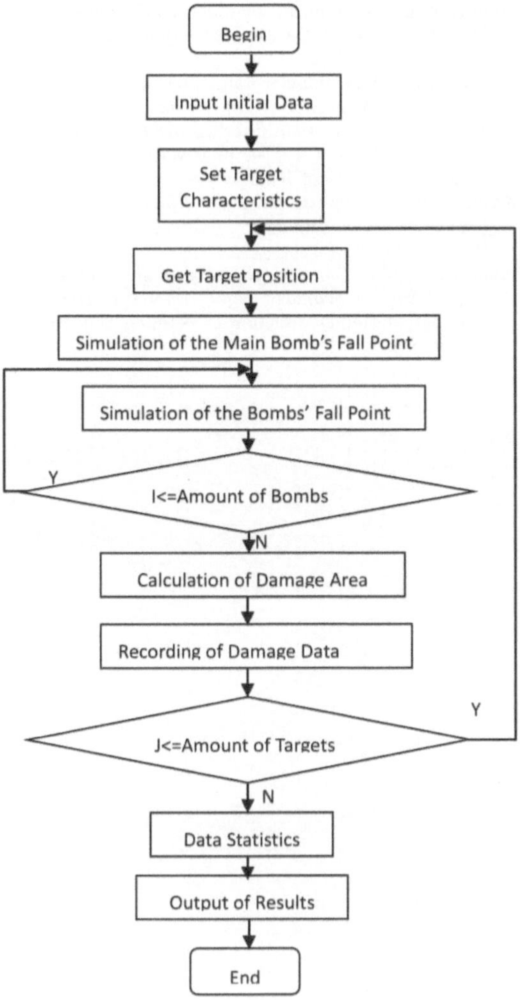

Figure 2. Flow chart of the program.

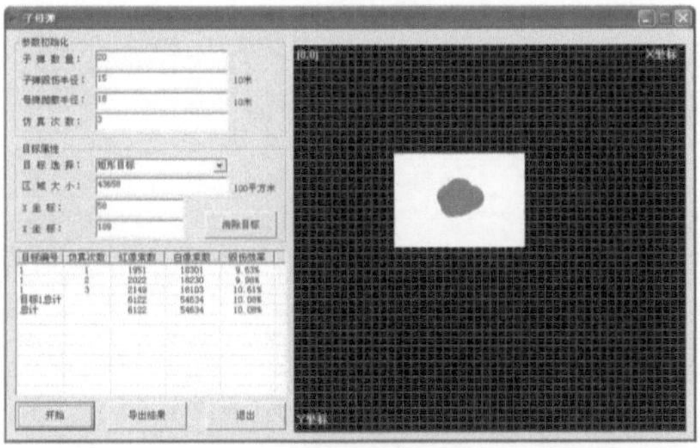

Figure 3. Interface of the program.

- Set the Kind of Target

  The target could be set to be point, line, or area.

- User-defined Simulation
- Output of Results

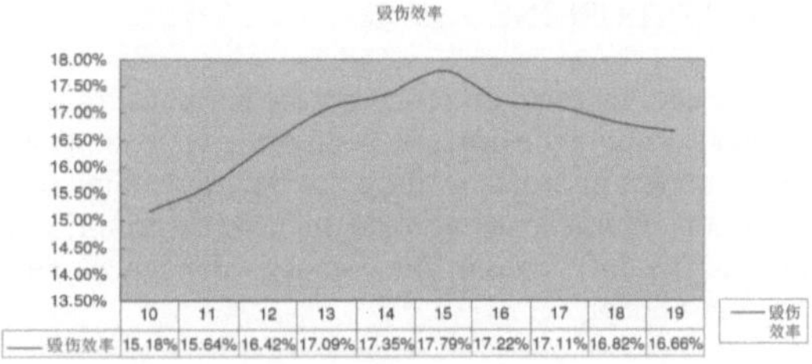

Figure 4. Effect of scattered radius on damage efficiency.

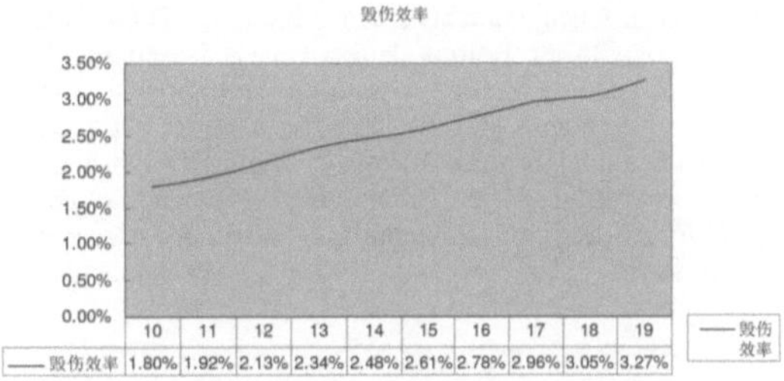

Figure 5. Number of bombs on damage efficiency.

4 ANALYSIS OF SIMULATION RESULTS

The scattered radius of the main bomb and the number of bombs were analysed comprehensively by means of simulation.

4.1 *Scattered radius of the main bomb*

As to the cluster bombs, the scattered radius of the main bomb was a very important performance parameter, which indicated the shot density of the cluster bombs. The damage efficiency of the target under the circumstances of different scattered radius is shown in Figure 4.

When the scattered radius of the main bomb increased from 10 to 12, the damage efficiency increased remarkably. However, the damage efficiency decreased when the scattered radius went beyond 15. So setting the correct scattered radius of the main bomb was significant to improve the damage efficiency of the cluster bombs.

4.2 *Number of cluster bombs*

When the number of cluster bombs increased from 10 to 19, the damage efficiency of the target was as shown in Figure 5. The damage efficiency increased linearly as the number of bombs increased, but the increasing rate was comparatively slow. It inferred that increasing the number of bombs to a limited extent was not significant in improving the damage efficiency of the cluster bombs.

Therefore, in order to enhance the damage efficiency, it was essential to reduce the size of the individual bomb and increase the number of bombs as much as possible, while ensuring the damage power of the bombs.

5 CONCLUSIONS

The factors that had an effect on the shot accuracy of cluster bombs in the process of shooting were analysed in the paper. In order to study the damage efficiency, the pixel method was applied in programming, and the factors influencing the damage efficiency were analysed deeply.

REFERENCES

Chen Lixin, Chen Wanchun, Yin Xingliang. 2004. Evaluation of Cluster Bombs' Effectiveness Based on Pixel Simulation. *Journal of System Simulation* (16): 474–476.

Lin Li. 2003. Four-step Line Drawing Algorithm Based on Bresenham Algorithm. *Journal of Jinan University*.

Que Corporation. 1995. *Connection of Open Database-ODBC2*. Beijing: Publishing House of Electronics Industry.

Song Kun, Liu Ruining, Li Weiming. 2007. *References for MFC Programming Developers*. Beijing: Future Science and Technology Press.

Sun Xiaoya. 2006. Computer Random Numbers Generation of Normal Statistic. *Journal of Dalian University*.

Yang Jianxiong. 2002. *Analysis of Cluster Bombs' Tactically Technical Indexes*: 7–12. Nanjing: Master Dissertation of Nanjing University of Technology.

Zhang Yanliang, Chen Lixin. 2000. *Efficiency Analysis of Ground-to-ground Ballistic Missile*. Beijing: National Defense Industry Press.

Zhang Chan. 2007. *Damage Efficiency Analysis of Field Missiles to Surface Scattered Light Armored Object*: 5–10. Nanjing: Master Dissertation of Nanjing University of Technology.

Future Communication Technology and Engineering – Chan (Ed.)
© 2015 Taylor & Francis Group, London, ISBN: 978-1-138-02777-0

The virtual display of a modern printing press

X.H. Wang, L.K. Luo, Y.S. Qi & Z.Y. Sun
Department of Mechatronic Engineering, Beijing Institute of Graphic Communication, Beijing, China

ABSTRACT: The virtual display of a modern press is to promote the print culture and show mechanical charm. Through functional analysis, mechanical design, mapping, and structural design, the overall design of the machine is completed, especially for the unclear part of the supplement. The machine's three-dimensional digital model was built in the assembly environment. By using virtual reality technology, we can create a virtual and interactive scene to show organizations, compositions, and printing works. It can be observed at any angle, it shows a perfect virtual printing, and provides better conditions for popularizing mechanical knowledge at the same time.

Keywords: Mechanical engineering, press, virtual prototyping.

1 INTRODUCTION

Printing is one of the four great inventions of ancient China; it has witnessed the long history of our country. From the hand-written book to the invention of movable type, the printing press was then invented. Printing vividly manifested the development and progress of human civilization. Typography makes the dissemination of culture more rapid and popular. Professor Eisenstein, the representative of the United States Environmental School of Communication Media, pointed out that the press was the motivation for the spread of the early modern European culture revolution! [1].

Most of the early printing presses consisted of feeding, inking, printing, and delivery [2]. Their structure was compact, clear, and functional, and perfectly demonstrated the mechanical charm. To mechanical learners and people who are familiar with printing, it is a good model for teaching. However, as time has gone by, these devices have mostly been scrapped; some of them became museum exhibits, and others remained in photographs The equipment in the museums cannot be operated because of its old age, a serious lack of spare parts, a lack of maintenance personnel, and other reasons. As a result, we have many difficulties and obstacles in fully understanding the printing of that time. A virtual display prototype system is based on the model of computer-based technology, and it has the function of the physical prototype. It reflects the characteristics of the machine, including appearance, spatial relationships, and kinematics and dynamics. This display technology has the immersion, interaction, and idea of the three basic features. This enables visitors to enter a realistic virtual environment, and operate virtual objects in the environment.

Therefore, using specialized technology to build a model of a modern printing press, and then display it in a virtual environment, we can fully demonstrate the principle and the printing process more clearly, with a concise display of its mechanical structure, which is the perfect complement to the real exhibits. This is the best learning environment for printing and machinery.

2 VIRTUAL DISPLAY METHODS OF A PRINTING PRESS

A virtual printer display follows three steps: design, 3D modelling, and virtual display [3,4].

The main structural design of the press is to make clear how it works, to complete the missing parts of the machine based on existing dataaccording to the printing and mechanical principles. 3D modelling is used to establish three-dimensional digital models based on the actual size of the real press and the working principle of the machine. The virtual display is to show the works, structure, and composition of the machine and its parts.

In order to clarify the mechanical movement principle, the machine is explained using typical mechanisms. The specific methods are shown in Figure 1.

In the 3D modelling phase, based on a pre-structure design, typical mechanisms are built in an assembly modelling environment to ensure the body parts fit together as shown in Figure 2.

In the virtual display phase, the three-dimensional model is placed in a virtual scene, creating a scene by setting the lights, cameras, etc., setting the script based on demonstrated needs, and creating interactive scenes, as shown in Figure 3.

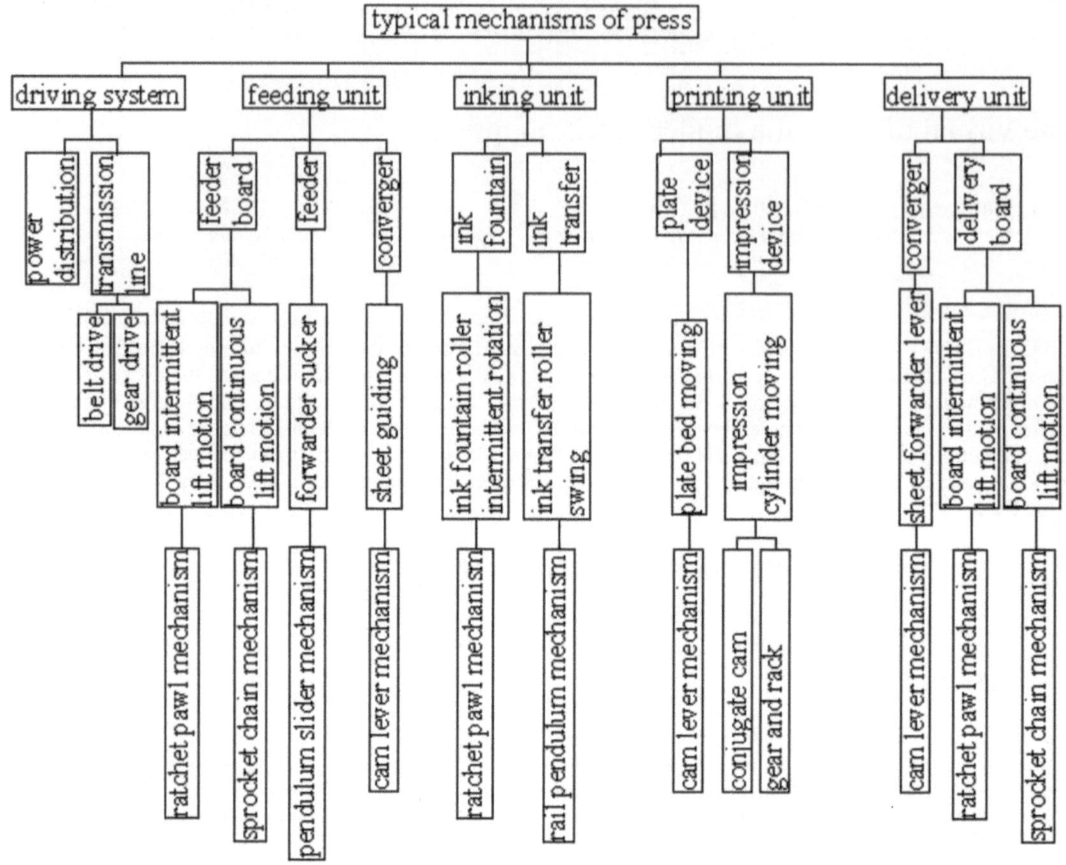

Figure 1. Typical mechanisms of a press.

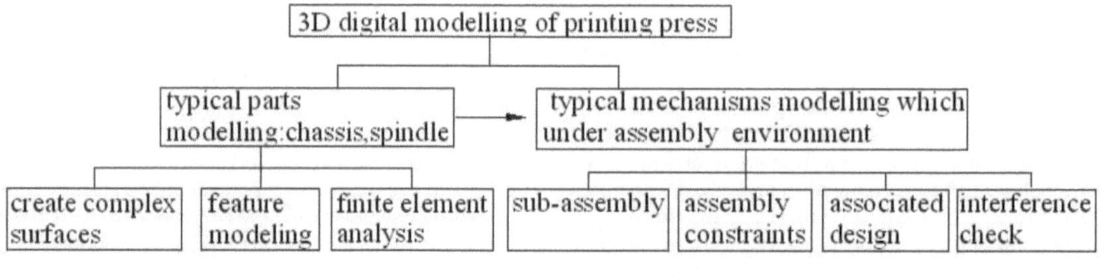

Figure 2. 3D modelling.

3 APPLICATIONS

We take the flat-bed cylinder press as an example. The printing plate is fitted onto the planar plate bed. The imprinting device is a cylindrical roller. When printing, the paper is bitten by the impression cylinder gripper and rotated with it, and impressed on the printing plate,. Print is completed. Every version of the flat plate reciprocates once, completes a cycle, and prints a product. Fitted under the flat-bed, racks meshed with the cylindrical roller gears cause the impression cylinder rotation printing machine to stop. When the platform moves forward, the impression cylinder rotates, then the platform moves to the front of the drum gear in the version rack (rack gear), and the embossing is done. At this time, the cylinder should be stopped immediately, overcoming inertia; then the platform returns back. When printing starts again, the rack should be re-kneaded and geared.

3.1 *Plate drive mechanism design*

A printing unit is driven by a cam link mechanism. In Figure 4, the main cam (8) led a group of four agencies to achieve edition sets of horizontal reciprocating, so the flat-bed has variable motions; its speed is

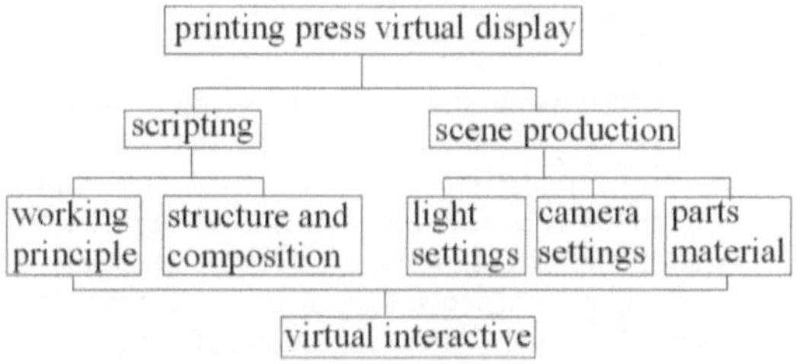

Figure 3. Virtual display.

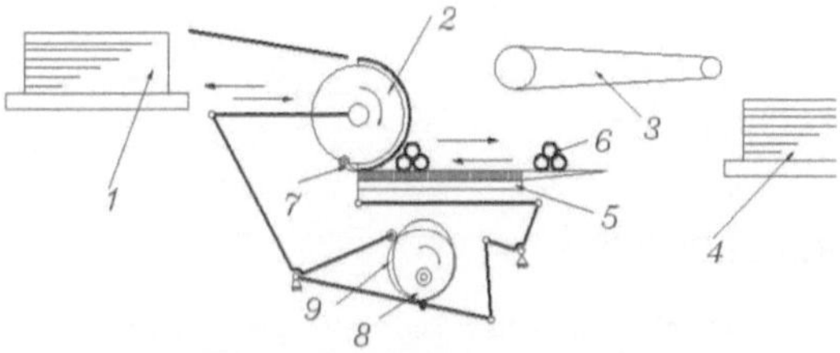

1. Feeding device 2. Impression cylinder 3. Delivery device 4. Delivery table 5. Flat plate 6. Inking unit 7. Gripper 8. Main cam 9. Vice cam

Figure 4. Flat-bed cylinder press working principle diagram.

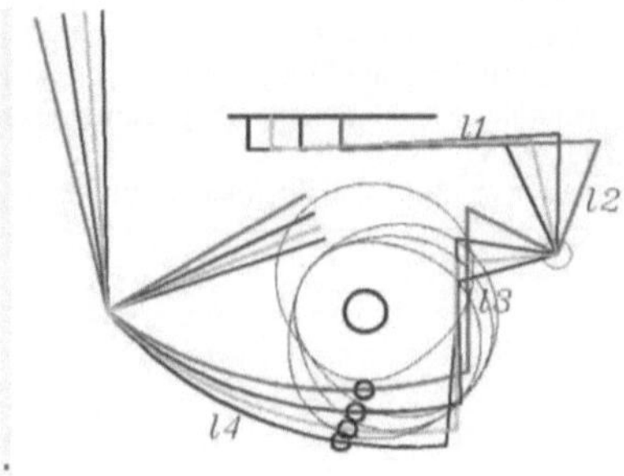

Figure 5. Kinematic loop diagram.

slow at working stroke and fast at another time. It does increase the printing time to improve the printing quality. The vice cam (9) pushes a group of rockers, and makes the cylinder reciprocating motion. This design can shorten the travel time of printing edition sets reciprocating, shorten the printing time, and increase printing efficiency.

The main cam (8) and vice cam (9) are conjugate cams; they are designed to be compact and smooth to reduce abrasion. The conjugate cams make it possible for the cam to be lifted and returned smoothly.

With initial printing speed, the limit position of the plate bed, the cam centre, and each rod length, the cam profile can be determined by using the rubbing or reverse method (the calculation process is omitted), and then the cam lever design agencies can be calculated. Figure 5 shows an institutional campaign cycle diagram.

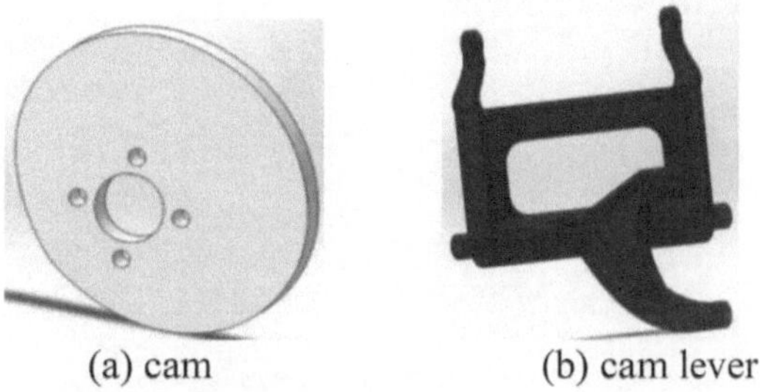

Figure 6. The modelling of the part.

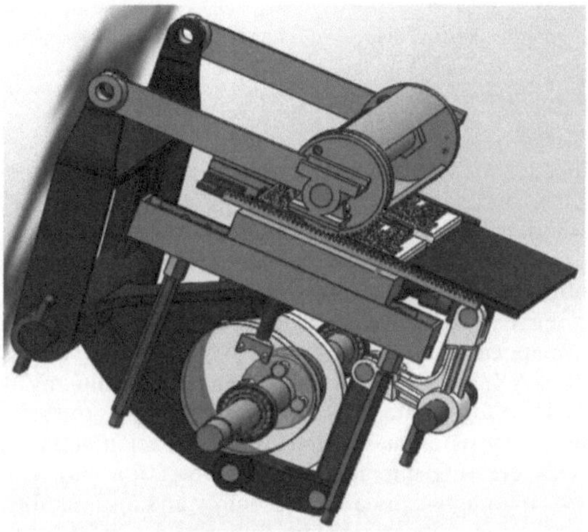

Figure 7. The printing unit model.

Using engineering analysis software, the agency model kinetic parameters are set, and then the dynamics of institutions are calculated.

3.2 *3D modelling of mechanism*

Three-dimensional digital models are built by surveying and mapping parts, and configuration can be achieved by using a series of different part types or by the secondary development of modelling software. In this modelling, the cam and the cam lever have some complex curved surfaces. The appearance features of the plate should be highlighted. Figure 6 shows the modelling of the part.

Parts are added to form an assembled model. The reasonableness of the added relationship has a direct impact on motion simulation. Agencies principle, adjustment, and motion simulation can be achieved by the secondary development of software[4]. The important things in the modelling agency are adding the correct relationship with the insertion of parts. The most common relationships are coaxial, face coincidence, tangency, the nearest point, and a combination of these. Figure 7 shows the result of an assembled model.

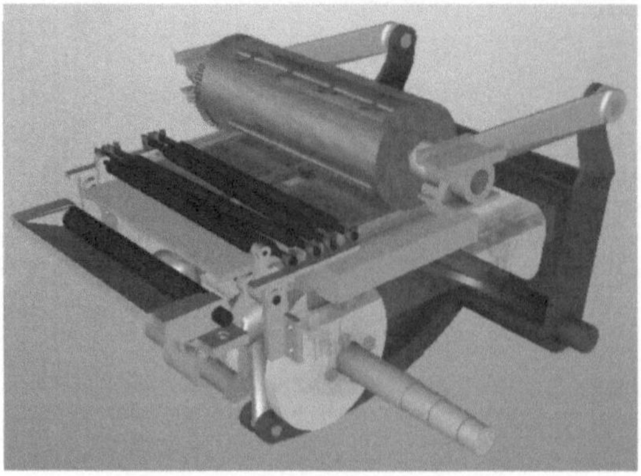

Figure 8. Virtual display of a flat-bed cylinder press.

3.3 *Virtual display*

Virtual reality technology is used to establish a virtual scene. The common software includes Virtools, EON, and so on. Virtools4.0 is used in this case. A Building Blocks (BB) flow chart is used to design the connection between them to form a flexible interactive effect[6].

Firstly, the three-dimensional model adds material and appearance through third-party software (such as 3DMAX, and so on). Secondly, we add lighting in Virtools editor, create the camera and set it, regroup the hierarchy parts, and script the initialization settings. After creating interaction, the observation path and viewing angle can be set. Assembly and disassembly, mechanism simulation, adjustment, and motion simulation can also be set up. Interaction function is a key process in establishing the model. For a better view, the final page file will also be linked to a web page file, such as shown in Figure 8.

4 SUMMARY

Today, we have entered the information age. Although it has lost its previous glorious position, machinery is still irreplaceable. Many new problems appear accompanied with intelligent control systems. Powerful, sophisticated equipment can be perfect only by improving certain aspects. Therefore, our thinking will be inspired with the previous compact design, and advanced display technology.

ACKNOWLEDGEMENT

Beijing Municipal Education Commission 18190114-015.

REFERENCES

[1] Elizabeth L.Eisenstein, The Printing press as an Agent of Change: Communications and Cultural Transformations in Early-Modem Europe, Cambridge, United Kingdom, 1979.

[2] S.H. Wang, X.Xu, Structure and Design of Presses, Printing Industry Press, Beijing, 1994.10.

[3] X.H. Wang, Y.S. Qi, Y.T. Li The Application of Virtual Design on Feeding Device of Precoated Laminating machine Advanced Materials Research, 2013, v765–767, p. 116–119.

[4] X.H. Wang, Y.S. Qi, Y.T. Li Virtual Design in Feeding Device of Precoated Laminating machine Information Technology and Computer Application Engineering 2013.11, p.185–188.

[5] Information on http://www.bpsnet.org.uk/history/18.pdf.

Future Communication Technology and Engineering – Chan (Ed.)
© 2015 Taylor & Francis Group, London, ISBN: 978-1-138-02777-0

Language model for assessing author similarity

J.Y. Chen, Z.C. Wang & J.L. Wang
Engineering Research Centre of the Ministry of Education on Enterprise Digitalization Technology, Tongji University, Shanghai, China

ABSTRACT: Currently, it is crucial for researchers to know if others have similar research objectives. Nevertheless, the identification of authors sharing the same motivations and interests may be complex, as the amount of research publications is growing rapidly. Furthermore, information about research papers is often fragmented and incomplete. The incomplete information, or metadata, in this paper refers to abstracts, keywords, journals, organizations, and so on. Thus, this paper analyses the metadata information about an author to find out similar authors. Author Similarity Model, a novel language model which is evolved from Author Topic Model, has been developed in this paper. For author similarity modeling, a four-dimensional vector has been set up to describe every author. Therefore an author's neighbours (a group of people who have the same direction of research) can be found out by calculating similarity between vectors.

1 INTRODUCTION

Due to the rapidly growing number of researchers, it can be complex to find similar authors in different journals or conferences. That is, it is not so easy to find the right people to work together for solving a scientific problem or forming a collaborative team. Among all the resources available to do such an assessment, an author's paper can be utilized to work out his/her interests. Different paper parameters have already been used to establish links between authors. For example, co-citation and bibliographic coupling are standard measurements in scientometrics for detecting author similarity (White & Griffith, 1981). Author recommendation (Heck et al., 2011) has been proposed by adopting Collaborative Filtering (CF).

The first step to discovering the interests of an author is to find his/her papers. However, the papers published by a single researcher are likely to be disseminated in multiple locations. It is far easier to get some incomplete, scattered information than to get a complete set. In this paper, metadata related to an author refers to abstracts, keywords, journals, and author's organizations. Nowadays, most researchers pay attention to full texts and do not make use of metadata. Furthermore, according to a few researches (Hurtado et al., 2013), metadata has improved the accuracy of results for paper similarity assessing. Therefore, metadata is an interesting choice and this paper takes metadata instead of comprehensive texts as sources of information.

Probabilistic language models have been used in many natural language processing applications. In order to find how similar two authors are, this paper also chooses to employ a novel set of semantic similarity methods relying on language modeling method Author Topic Model (ATM) (Rosen at al., 2010). A topic model has been put forward by David Blei, especially Latent Dirichlet Allocation (LDA) (Blei et al., 2003). After the introduction of LDA, many people have extended it by including other information. The Author Topic Model consists of authorship information. The main task in this paper is to verify how to use metadata to compare authors in the Vector Space Model (VSM) (Salton et al., 1975) and the Author Similarity Model (ASM). The VSM applied to an author's abstracts is a basic method. For Author Similarity modelling, this paper includes keywords, journals, and organizations to estimate the similarities between authors. Furthermore, the Author Topic Model is a part of the model and will be utilized to discover latent topics of authors and documents.

The structure of this paper is as follows. Section 2 describes related work and its background. In section 3, the Author Similarity Model is introduced. In Section 4 the results of experiments are analysed. Section 5 is a conclusion and proposal for future work.

2 RELATED WORK

The Author Topic Model simultaneously models the contents and the interests of authors. It extends LDA by using probabilistic topics-to-author modeling and allowing the mixture weights for different topics to be determined by the authors of the document. Conditioned on the set of authors and their distributions over topics, the process by which a document is generated can be summarized as follows: Assuming that we have T topics, we can parameterize the multinomial distribution over topics for each author by using matrix θ of size $T \times A$, with elements θ_{ta} to stand for the probability of assigning topic t to a word generated by author a. Thus $\sum_{t=1}^{T} \theta_{ta} = 1$ and θ_A stands for the ath column of the matrix. The multinomial distributions over words associated with each topic are parameterized by a matrix φ of size $W \times T$, with elements ϕ_{wt} to stand for the probability of generating word w from topic t. Thus $\sum_{w=1}^{W} \phi_{wt} = 1$ and ϕ_t stands for the tth column of the matrix. These multinomial distributions are assumed to be generated from symmetric Dirichlet priors with hyper-parameters α and β respectively.

A variety of algorithms have been applied to estimate the parameters of topic models, from basic expectation-maximization (Hoffmann, 1999), to approximate inference methods like variational EM (Blei et al., 2003), expectation propagation (Minka, 2001), and Gibbs sampling (Griffiths & Steyvers, 2004; Gersheman & Blei, 2012). The inference used throughout this paper is Gibbs sampling that is based on a Markov chain Monte Carlo algorithm. Our aim is to estimate the posterior distribution $P(\theta, \varphi | D^{train}, \alpha, \beta)$. Samples from this distribution can be useful in many applications. The inference is based on the observation that:

$$P(\theta,\phi \mid D^{train},\alpha,\beta) = \sum_{z,x} P(\theta,\phi \mid z,x,D^{train},\alpha,\beta) \square P(z,x \mid D^{train},\alpha,\beta) \tag{1}$$

We obtain an approximate posterior on θ and φ by using the Gibbs sampler to compute the sum over z and x.

The Author Topic Model has been used in many areas. For example, McCallum (McCallum et al., 2012) presents the Author-Recipient-Topic Model for social network analysis to predict people's roles. Okolica (Okolica et al., 2006) uses the Author Topic clustering algorithm, and tries to discern employees' interests from their daily emails and find out the insider threat. Steyvers (Steyvers et al., 2004) uses the Author Topic Model to analyse topic trends over time for finding the authors who are most likely to write on a given topic and finding the most unusual paper written by a given author.

3 AUTHOR SIMILARITY MODEL

In this paper, a novel approach to calculate the similarities has been put forward and it is the Author Similarity Model. Each author is a four-dimensional vector. As defined below:

$$A_i = (A_{i1}, A_{i2}, A_{i3}, A_{i4}) \tag{2}$$

Furthermore, each dimension is a vector that can be defined as follows:

A_{i1} is a vector which represents an author's distribution over words. The Author-Topic Model is applied to abstracts. The results are an author-topic distribution matrix and a topic-word distribution matrix. As is defined by Formula 3, it is easy to get an author's distributions over words:

$$A_{i1} = P(w_t \mid t) \times P(t \mid AD_i) \tag{3}$$

AD_i is the set of author i's papers, t is the topic, and k means the word is the kth word in the dictionary.

A_{i2} is a vector which represents an author's distribution over words. However, here we use the Organization-Topic Model instead of Author Topic Model, and pseudo-documents rather than abstracts. The Organization-Topic Model is derived from the Author-Topic Model, just replacing the author with organization. Here the pseudo-document for each organization combines with all the papers that come from the same organization. First, the organization-topic distribution matrix and the topic-word distribution matrix are obtained. Then by using Formula 4, a distribution of each organization in the dictionary word entries will be easy to calculate:

$$A_{i2} = P(w_k \mid t) \times P(t \mid AS_l) \tag{4}$$

AS_l refers to all the papers of the organization l, t is topic, and k means the kth word that can be found in the dictionary.

A_{i3} is a vector of keywords related to the author. A_{i3} combines with all the keywords that have appeared in author i's papers. For example, an author has written three papers. In the original paper, there are keywords k_1, k_2. In the second paper, there are keywords k_1, k_3. In the third paper, there are keywords k_2, k_3. So the author i's keywords vector is like this $(k_1, k_2, k_3, \ldots)$. It is more convenient to convert this vector into a numeric representation and then compute it. That is, if a keyword appears, the corresponding vector dimension is labelled as 1. If not, it is marked as 0.

A_{i4} is a vector of journals related to the author i. A_{i4} combines with all the journals where author i has published his/her papers. For example, the author has written two papers; the first paper published in journal j_1, and the second paper in journal j_2. The author's journal vector is like this $(j_1, j_2, \ldots)$. It is more convenient to convert this vector into a numeric representation and then compute it. That is, if the paper is published in the journal, the corresponding vector dimension should be labelled as 1. If not, it is marked as 0.

The next step is to define a formula to assess the similarity of two authors. μ_i is the weight of each factor. As each vector is defined differently, we have to calculate the similarities separately as follows:

$$\begin{aligned} sim(A_i, A_j) &= \mu_1 sim(A_{i1}, A_{j1}) + \mu_2 sim(A_{i2}, A_{j2}) \\ &+ \mu_3 sim(A_{i3}, A_{j3}) + \mu_4 sim(A_{i4}, A_{j4}) \end{aligned} \tag{5}$$

$$sim(A_{i1}, A_{j1}) = \frac{KLD(A_{i1} \mid \frac{A_{i1}+A_{j1}}{2}) + KLD(A_{j1} \mid \frac{A_{i1}+A_{j1}}{2})}{2} \tag{6}$$

$$sim(A_{i2}, A_{j2}) = \frac{KLD(A_{i2} \mid \frac{A_{i2}+A_{j2}}{2}) + KLD(A_{j2} \mid \frac{A_{i2}+A_{j2}}{2})}{2} \tag{7}$$

$$\mathrm{sim}(\overrightarrow{A_{i3}}, \overrightarrow{A_{j3}}) = \frac{\overrightarrow{A_{i3}} \bullet \overrightarrow{A_{j3}}}{\left\|\overrightarrow{A_{i3}}\right\| \left\|\overrightarrow{A_{j3}}\right\|} \tag{8}$$

$$\mathrm{sim}(\overrightarrow{A_{i4}}, \overrightarrow{A_{j4}}) = \frac{\overrightarrow{A_{i4}} \bullet \overrightarrow{A_{j4}}}{\left\|\overrightarrow{A_{i4}}\right\| \left\|\overrightarrow{A_{j4}}\right\|} \tag{9}$$

While Formulae 6 and 7 are trying to use symmetric KL distance to calculate the similarities, Formulae 8 and 9 are trying to use cosine similarity to calculate the similarities.

In the last step, the final results are obtained through Formula 5.

4 EXPERIMENTAL SETUP AND EVALUATION

4.1 *Experimental setup*

The VSM and Author Similarity Model have been applied to a corpus consisting of metadata from the VLDB Journal from 1992 to 2013, Machine Learning Journal from 1986 to 2013, and Data Mining and Knowledge Discovery Journal from 1997 to 2013. These files contain 1,552 abstracts, 2,199 authors, 1,918 organizations, and 4,309 keywords altogether. It corresponds to 1,552 papers from these three journals and for each paper we only consider the first two authors (only one author, if the paper has a single author). First, For the abstracts, stopwords have been removed. StanfordParser has been applied to work out the grammatical structure of sentences. Then a test collection using the VSM approach and Author Similarity Model approach has been implemented. For the Author Topic Model and Organization-Topic Model, topic numbers from 10 to 70 have been tested to get the best topic number and the final result is 50. For 2,199 authors, we calculate all the pairs and set the threshold to 0.8/1.

Five experts are invited to verify the truth of the results. For each approach, experts analyse the author pairs to calculate the accuracy. Formula 13 defined below is used to evaluate these pairs:

$$Ac = \frac{\Pr e(\mathrm{p})}{|p|} \tag{11}$$

$\Pr e(p)$ is the number of right pairs verified by experts and $|p|$ is the number of all the pairs resulting from the different approach.

4.2 *Results*

Table 1 shows the experimental results obtained from the Author Similarity Model and the VSM.

As we compare the first line and the last line, the Author Topic Model applied to abstracts (we say ATA for short in the following part) outperforms the traditional VSM. This suggests the Author Similarity Model achieves better performance by comparing pairs. As presented before, the Author Similarity Model is based on a four-dimensional vector. The four dimensions of this vector are involved with the Author Topic Model, the Organization Topic Model, keywords information and journal information. We use configuration described in Table 1 to control the weights of different factors. In Table 1, A_{i1} indicates Author Topic Model applied to abstracts, A_{i2} indicates Organization Topic model applied to abstracts, A_{i3} indicates keywords, and A_{i4} indicates journal information.

Table 1. The experimental results.

	Configuration				
	Ail	Ai2	Ai3	Ai4	Accuracy
1	0.9	0	0	0	0.638
2	0	0.9	0	0	0.431
3	0	0	0.9	0	0.547
4	0	0	0	0.9	0.213
5	0.45	0.45	0	0	0.625
6	0.45	0	0.45	0	0.697
7	0.45	0	0	0.45	0.614
8	0.4	0.4	0.05	0.05	0.636
9	0.4	0.3	0.1	0.1	0.715
10	0.5	0.2	0.15	0.05	0.758
11	0.5	0.15	0.2	0.05	0.780
12	Traditional VSM				0.534

The first four lines only consider one factor and it turns out that abstract, using the Author Topic Model, has the best result. When we only use keywords, the accuracy is higher than the line by only using organization information. However, 0.213 is obtained when only journal information is considered. We can conclude that journal information is too general.

In the rest of the lines, we consider two factors or four factors. When the abstract combines with other metadata, here it means A_{i2}, A_{i3}, A_{i4}, and it increases the accuracy compared to VSM as we observe line 12 and line 5–11 in Table 1. Therefore, it is not hard to draw a conclusion that the addition of metadata is useful and it should be taken into consideration.

It is not difficult to find out when using A_{i1} and A_{i3} together, that the performance is a little better than using them alone. The combination of the Author Topic Model and keywords can perform slightly better than the other two factors combined. The last four lines (here, we suppose the last line, that is VSM, is not included) give weight to every factor. We can observe that line 11 has the best result among the four, even among all the lines. To give a lion share to A_{i1}, second is A_{i3}, and A_{i2} is the third. The result stands out. By comparing the last two lines, it shows that the keywords have a bigger influence than A_{i2}. In a word, keywords improve the performance.

In Figure 2, there is an example of line 11. As depicted in Figure 2, the result demonstrates the

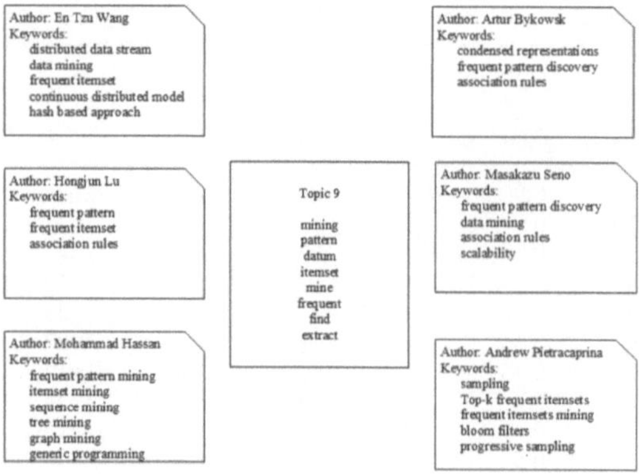

Figure 1. Authors related to Topic 9.

authors are linked to Topic 9. In the middle, the first eight words distributed in Topic 9 have been picked out. As we can see, this topic is about mining frequent patterns from a large datum set. The authors with the same topic are around Topic 9 and we pick out six authors, with their names and keywords from one of their papers. By comparing the keywords, almost 50% of the items are the same.

5 CONCLUSION AND FUTURE WORK

This paper proposes a different approach to compare authors and illustrate this with experiments. First it takes metadata into the model. Moreover, most of the time information is incomplete and scattered. While most of the research focuses on full texts, metadata should be made full use of; it seems more relevant to exploit metadata. Clearly, the Author Topic Model is more applicable to abstracts and performs better than the traditional VSM. Furthermore, after other metadata has been added to the language model, we can see there is a noticeable change.

However, some problems need to be solved. First of all, we do not take different people of the same name into consideration. In digital libraries such as Google Scholar, this type of name disambiguation is a significant and difficult problem. Second, it will be an interesting task to apply the Author similarity model to social networks, such as Twitter or Facebook.

REFERENCES

Blei D M, Ng A Y, Jordan M I 2003. Latent dirichlet allocation. *The Journal of machine Learning research*, 993–1022.

Gershman S J & Blei D M 2012. A tutorial on Bayesian non-parametric models. *Journal of Mathematical Psychology*, 2012, 56(1): 1–12.

Griffiths T L & Steyvers M 2004. Finding scientific topics. *Proceedings of the National academy of Sciences of the United States of America*, 101(Suppl 1): 5228–5235.

Heck T, Peters I, Stock W G 2011. Testing collaborative filtering against co-citation analysis and bibliographic coupling for academic author recommendation. *In: Proceedings of the 3rd ACM RecSys' 11 Workshop on Recommender Systems and the Social Web*, 16–23.

Hurtado MartíN G N, Schockaert S, Cornelis C, et al 2013. Using semi-structured data for assessing research paper similarity. *Information Sciences*, 221: 245–261.

McCallum A, Corrada-Emmanuel A, Wang X 2005. The author-recipient-topic model for topic and role discovery in social networks: Experiments with enron and academic email.

Minka T P 2001. Expectation propagation for approximate Bayesian inference. *In: Proceedings of the Seventeenth conference on Uncertainty in artificial intelligence. Morgan Kaufmann Publishers Inc*, 362–369.

Okolica J S, Peterson G L, Mills R F 2006. Using author topic to detect insider threats from email traffic. *Intelligence and Security Informatics. Springer Berlin Heidelberg*, 642–643.

Rosen-Zvi M, Griffiths T, Steyvers M, et al 2004. The author-topic model for authors and documents. *In: Proceedings of the 20th conference on Uncertainty in artificial intelligence. AUAI Press*, 487–494.

Rosen-Zvi M, Chemudugunta C, Griffiths T, et al 2010. Learning author-topic models from text corpora. *ACM Transactions on Information Systems (TOIS)*, 28(1).

Salton G, Wong A, Yang C S 1975. A Vector Space Model for Automatic Indexing. *Communications of the ACM*, 18(9): 613–620.

Steyvers M, Smyth P, Rosen-Zvi M, et al 2004. Probabilistic author-topic models for information discovery. *In: Proceedings of the tenth ACM SIGKDD international conference on Knowledge discovery and data mining. ACM*, 306–315.

White H D & Griffith B C 1981. Author cocitation: A literature measure of intellectual structure. Journal of the American Society for information Science, 32(3): 163–171.

Future Communication Technology and Engineering – Chan (Ed.)
© 2015 Taylor & Francis Group, London, ISBN: 978-1-138-02777-0

Emergency simulation for a drill system of a roller coaster in a virtual environment

X.Y. He, H.W. He & Z.Y. Hu
Guangdong Provincial Key Laboratory of Computer Integrated Manufacturing System, Guangdong University of Technology, Guangzhou, China

ABSTRACT: Roller coasters belong to specialist equipment. In order to improve the ability to deal with an accident, it is necessary to hold an emergency drill. A drill system combined with virtual reality technology can greatly improve the drill effect. We use Solid Works, 3ds Max, and Unity3D to develop the system. We realize scene roaming, action figure simulation, and rope motion simulation. The system presents a complete virtual emergency drill of a roller coaster, and the usage of important rescue tools.

Keywords: emergency drill, virtual reality, Unity3D.

1 INTRODUCTION

Roller coaster bring us a lot fun. As special equipment, it is necessary to hold emergency drills regularly. The traditional way is based on the physical device, and needs the roller coaster to stay out of action. Then a video is made of the drill process for subsequent study. This method costs too much, and the drill process cannot reappear. What is worse, is that later trainees cannot experience the real drill feeling by simply watching videos, and of course, this method cannot get a satisfactory effect.

A drill system combined with virtual reality technology can overcome the shortcomings of the traditional emergency drill. In the virtual emergency drill system, the models of the scene are completely in accordance with the real scene and the procedure has gone through the identification of relevant experts. The operator can interact with the virtual environment, and also repeat learning. Thereby, the system will greatly improve the effectiveness of training [2]. Meanwhile, the system can also be used for passengers' education to enhance passengers' awareness of an emergency.

The difficulty of developing the system is that the drill needs to be divided into a number of steps, and involves humans, rope, and other flexible body motion simulation. This paper will focus on these.

2 PROGRAM OF THE SYSTEM

2.1 *Emergency drill*

In the virtual emergency drill, we will simulate a roller coaster accident: passengers are trapped in the upgrade section; rescuers carry rescue tools, they use artificial rescuing modes, and bring the passengers to the ground. It mainly shows the relief norms, and the usage of some important rescue tools.

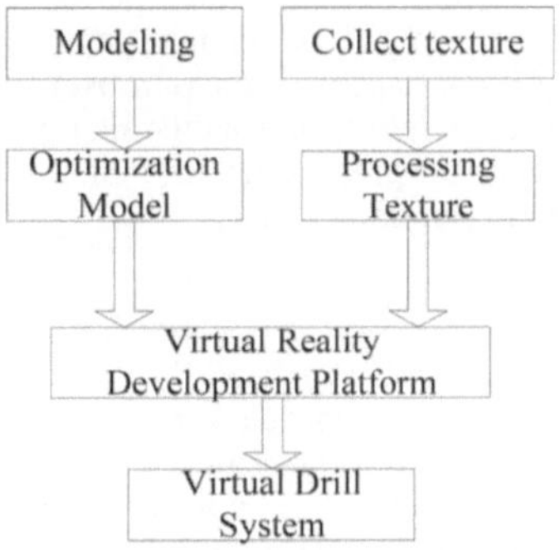

Figure 1. Development process of virtual roller coaster emergency drill.

2.2 *Difficulties of the development process*

The system developing process is shown in Figure 1. First, we complete the coaster mechanical models, the scene decorative models, and the human models in the modelling process. At the same time, we collect texture of the models from the real environment. Second, we import the models and textures to a virtual reality development platform. Third, we develop the system according to the real drill steps. Finally, we release the system to the appropriate using platform.

Throughout the development process, the difficulty is how to present the drill steps orderly in the system. Each step has a strict sequence in the real drill. Therefore, all steps should have triggered relations in the system. In addition, the virtual drill involves some simulation of humans, rope, and other flexible objects.

2.3 *Choosing a development platform*

For modelling and mapping work, we use Solid Works [3], 3ds Max, and other common development platforms. In virtual reality development, the drill is a multi-step, multi-scene activity, and relates to the simulation of a flexible body. Unity3D is suitable for multi-scene game development. Because of its cross-platform ability and excellent rendering effect, more and more game developers choose it as their development engine. Unity3D not only shows its powerful features in the field of entertainment, it also has its place in serious areas. Therefore, we choose Unity3D as our virtual reality development platform.

3 PROBLEM SOLUTION

3.1 *Ordered steps of the drill*

In reality, the drill steps are ordered. Therefore, in the system, we also need to ensure this characteristic. We use the script in Unity3D to control the order. The update function in JavaScript is performed in each frame of the scene, and we can take advantage of this function's feature to achieve the purpose. The following are the key codes to control the roller coaster start. When the operator presses the start button, the roller coaster should run immediately.

```
function Update()
{if(bPushButton) //whether can press the start button
  { Operator.animation.Play("PushButton"); //show the action of pressing start button
if(Operator.animation["PushButton"].normalized Time>0.99f) //whether the action is completed
  { bCarMove=true; //the roller coaster can be moving
      bPushButton=false; //the pressing action is completed
      }
  }
  if(bCarMove) //whether can the roller coaster moving
  { //call the code to drive the roller coaster moving}
}
```

In the above codes, since the function Update is executed in every frame, once the button is pressed, the value of variable "bCarMove" becomes true. Then the roller coaster will move. we can add an Update function in every code. You may add multiple scripts components in a scene, so that some controls may begin concurrently.

3.2 *Scene roaming and human-computer interaction*

In order to observe the objects in the scene, we need to add a camera. When the scene runs up, it can achieve the scene roaming by controlling the movement of the camera [4]–[5]. The key codes are as follows, where "move", "rotate", and "up" are added by input management.

```
//WASD keys control the camera move of front, back, left, right, QE keys to control the up and down movement
//"move" corresponds to WS keys, "rotate" corresponds to AD keys, "up" corresponds to DE,
MainCamera.transform.Translate(0,0,-6*Time.delta Time*Input.GetAxis("move"),Space.Self);
MainCamera.transform.Translate(-6*Time.delta Time*Input.GetAxis("rotate"),0,0,Space.Self);
MainCamera.transform.Translate(0,20*Time.delta Time*Input.GetAxis("up"),0,Space.Self);
//Mouse X, Y direction control the rotation of the camera
MainCamera.transform.Rotate(0,20*Time.delta Time*Input.GetAxis("Mouse X"),0,Space.Self);
MainCamera.transform.Rotate(-20*Time.delta Time*Input.GetAxis("Mouse Y"),0,0,Space.Self);
```

Once the scene is running, the operator can control the scene roaming through a keyboard and mouse, and observe the emergency drill from all directions. In addition, we preset a view navigation button, so the operator can quickly return to a normal learning view.

Except for the scene roaming, we need the operator to make decisions via the keyboard or mouse in some important drill processes, and only when the selection is correct, can the drill continue. The system will prompt if the selection is wrong. These actions enhance the operator's participation, and also play a role in the assessment of train. We can achieve human-computer interaction via GUI (Graphical User Interface) function in Unity3D such as GUI button, and GUI label for creating buttons and text labels. The operator interacts with the system by inputting or clicking.

3.3 *Simulation of rope and other flexible objects*

During the drill, we need to do the simulation of a safety belt. In Unity3D, the motion simulation of a rigid body is very simple; we only need to use the translation function to change the location of the object, or create some animations. In order to complete the rope simulation, we can use the chain principle, which is to divide the rope into many small rigid pieces. However, the motion of the rope is different from that of a chain. The trajectory of each segment of the chain is fixed; we just need to obtain the trajectory of each chain link. But for the rope, it is difficult to determine the trajectory. When you try to calculate its trajectory, the workload will be too large.

The reason why the rope presents special flexible movement is because of its soft material and gravity. Based on Unity3D physics engine, we can easily achieve rope simulation [6]; we add gravity and collision properties to each section of rope, and adjacent property to each two hinges. Then there will be a force between two rope sections. When we change the location of the rope section, the shape of the whole rope will change. Some of the key control codes are as follows:

```
rope[i]=GameObject.CreatePrimitive(PrimitiveType.
Cylinder); //A cylindrical as each shape rope
rope[i].transform.localScale=Vector3(Scale[0],
Scale[1],[2]); //Control scaling cylinder
  //define the position and the y-direction coordinate
as a variable
      rope[i].transform.position.x=x;
      rope[i].transform.position.y=dy;
      rope[i].transform.position.z=z;
//Adding fixed and hinge node
   wuti[i].AddComponent("Rigidbody");
   rg=wuti[i].GetComponent("Rigidbody");
   if(i==0)
       wuti[i].AddComponent("FixedJoint");
     if(i!=0)
       wuti[i].AddComponent("HingeJoint");
     //add adjacent property each two hinges
     hj=rope[i].GetComponent("HingeJoint");
     hj.connectedBody=wuti[i-1].rigidbody
```

By attaching the above codes to the object in the scene to make it run, the simulation results can be obtained as follows:

4 SYSTEM IMPLEMENTATION

To completing the virtual emergency drill system based on the above key techniques, the implementation is divided into the following steps:

(1) Modelling and collecting texture.

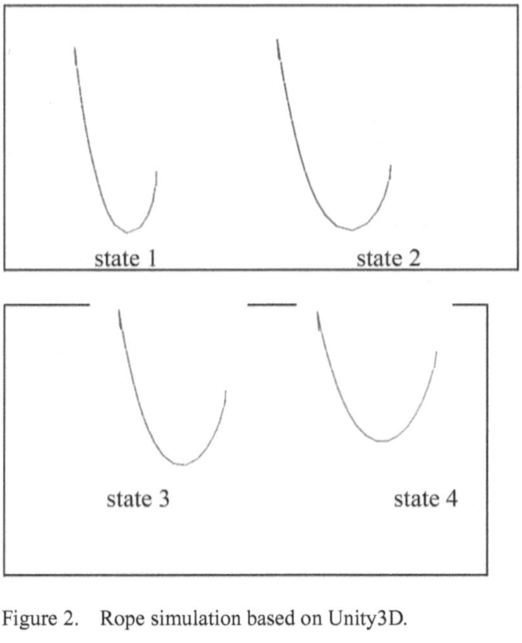

Figure 2. Rope simulation based on Unity3D.

Use SolidWorks to draw the mechanical part of the roller coaster. Before drawing, we should determine which parts or components need to move. For example, for parts or components needing to be rotated, the model's origin should be located at the centre of rotation, and when drawing the model, the important structure should be drawn carefully; while unimportant parts such as chamfer can be omitted.
Using 3ds Max [7]–[8] to establish the non-mechanical models will be more convenient. For instance, we can use 3ds Max to draw human models and decorative models, and use the pictures we collected to add the texture.

(2) Import Model
Since Unity3D can identify well the "fbx" format file, so, converting Solid Works and 3ds Max models into "fbx" format, and the model into "fbx" format can save animation and map information which we have done in 3ds Max. The imported file will be located in the Assets directory of the project window. Similarly, Unity3D

a. The drill scene.

b. Passengers trapped.

c. Prepare rescue tools.

Figure 3. Virtual emergency drill.

d. Prepare to rescue.

e. Rescuing.

f. Arrive on the ground.

Figure 3. Continued.

will create a file called “Asset” in the project folder. In order to facilitate the management, we establish the appropriate folders for different types of files.
(3) In a virtual reality development platform, divide the system into multiple scenes according to the realistic emergency drill
(4) In each scene, simulate the emergency drill in the virtual environment, and increase human-computer interaction.
(5) Link all scenes together orderly to complete the virtual drill.
(6) Hand the system to experts to complete the assessment, and modify the parts which do not meet the specification.
(7) Release the final system to the platforms.

Finally, we complete the system. The following pictures show the drill process.

5 CONCLUSION

This thesis introduces how we use virtual reality technology to develop the system of a virtual roller coaster emergency drill. We use Solid Works and 3ds Max as modelling tools, and Unity3D as the virtual reality development platform. By taking advantage of Unity3D, we achieve scene roaming and real-time interaction, as well as the simulation of rope and other flexible objects. Through this system, trainers can greatly improve the effectiveness of training, and the system development process is also applicable to other areas.

ACKNOWLEDGEMENTS

This work was financially supported by the National Natural Science Foundation of China (51275094), AQSIQ public projects (201210034), Research and Guide project in Guangdong Province (2012B091100131).

REFERENCES

[1] Xiaoqiang Hu. Introduction to Virtual Reality Technology [M]. Beijing: University of Posts and Telecommunications Press, 2005: 60–135.

[2] Fanya Weng, Xiangmei Zhou. Teaching game based on Experience learning [J]. Chinese education information 2007, 27(12): 26–27.

[3] Zheng Wei, Gong Zhao, Xiaoming Shong. Solid Works design and application guide [M]. Beijing: Tsinghua University Press, 2009.

[4] Unity Technologies. Unity 4. X from entry to the master [M]. Beijing: China Railway Publishing House, 2013.

[5] Aling Cheng, Zhaohui Hu, Xiangzhi Qi. 3D modeling and implementation of campus virtual reality scenes [J]. Chongqing Normal University: Natural Science, 2007, 24(4): 37–40.

[6] Yusong Xuan. Unity 3D game development [M]. Beijing: People Post Press, 2012.

[7] Mars Times. Autodesk 3DSMAX 2010 standard training materials [M]. Beijing: People Post Press, 2010.

[8] Hongling Zhao, Mingliang Chao. Three-dimensional virtual campus roaming system key technologies [J]. Henan Institute of Education: Natural Science, 2006, 15(4):27–29.

Future Communication Technology and Engineering – Chan (Ed.)
© 2015 Taylor & Francis Group, London, ISBN: 978-1-138-02777-0

Temperature state and residual deformations in a welding conditions study

A.B. Ivanchenko, L.V. Belyaev, A.V. Zhdanov & V.V. Morozov
Vladimir State University, named after Alexander and Nikolay Stoletovs, Vladimir, Russia

ABSTRACT: This article describes the numerical modeling technique of formation processes of the residual welding tension, based on the solution of a non-stationary task of heat conductivity in a three-dimensional statement with application of finite element method. The results of experimental studies of welded details in a thermal condition are given. Temperature measurement in the course of welding is carried out with application of chromel-copel thermocouples. The results of calculations are qualitatively and quantitatively coordinated with the results of the experiment. The maximum divergence of sizes of temperature in control points at a stage of heating does not exceed 8%. The probe of residual welding deformations is carried out through a method of boreholes using the strain gauge. The comparative analysis of calculated and experimental data of residual deformations has shown that the received maximum divergence does not exceed 30%. Results of the conducted research confirm a correctness of an offered technique of numerical modeling of a temperature condition of details in the course of welding, and processes of formation of residual deformations that allows consideration this factor, when the strength analysis of welded designs is carried out.

1 INTRODUCTION

The creation of effective welding technologies is one of the major problems in the modern aerospace industry.

Currently, in the production of complex welded body parts, complex science-intensive technologies of automated pulsed arc welding are introduced, based on inverter power sources with a processor control.

Despite the advanced welding technologies, in some cases it is necessary to increase the stability and manageability of the process. Most frequently, it is necessary to control this by thermal processes during the welding. Mainly this is related to the limitations on the temperature during the welding of bimetallic thin-walled tubes, mounting assembly parts with low thermostability, such as fluoropolymer, and consolidation of parts with strict requirements for the deviation form. Under an intensive forced heat sink from the welded parts, the level of temperature gradients has significantly increased, which leads to the appearance of high temperature compression stresses, and it exceeds the material yield point. As a result, in the material, a compression residual deformation is formed Upon cooling, the presence of compression residual deformation causes the apparition of tension residual stresses. In the component material's disposition towards brittle fracture (low plasticity characteristics and impact toughness, three-dimensional state of stress, work in the temperature interval of cold brittleness), the effect of residual stresses on the strength characteristics can be quite meaningful. Thereby, it becomes urgent to develop methods of forming a numerical modeling for the residual welding stresses.

2 MATERIAL AND METHODS

The proposed method consists of the following stages. The first stage is to solve the problem of non-stationary heat conduction. The decision is realized in a three-dimensional formulation by using the finite element method, and it is implemented in the modern CAE (Computer-aided Engineering) systems. The movement of the heat source in the welding and welded joint length during the modeling is considered at the expense of successive heat inputs along the length of the joint. Therefore, in the calculation model, the weld joint area is divided into length sections, which are comparable to the effective arc spot. The heat supply time for each section corresponds to the transit time through this distance by an electrode during the welding process. In the calculation, boundary conditions of type II are used, which are obtained based on existent models of heat distribution during welding [1]. The cooling joints using boundary conditions of type III are also considered, which correspond to heat sink conditions. During the resolution of the heat conduction, it is necessary to take into account the dependence of the thermal material properties on the temperature.

The calculation of stresses and deformation is carried out by using an elastic-plastic deformation of the model. Namely, the flow theory permits "trace" on the development of plastic deformation on the heat supply stages during the welding and cooling processes.

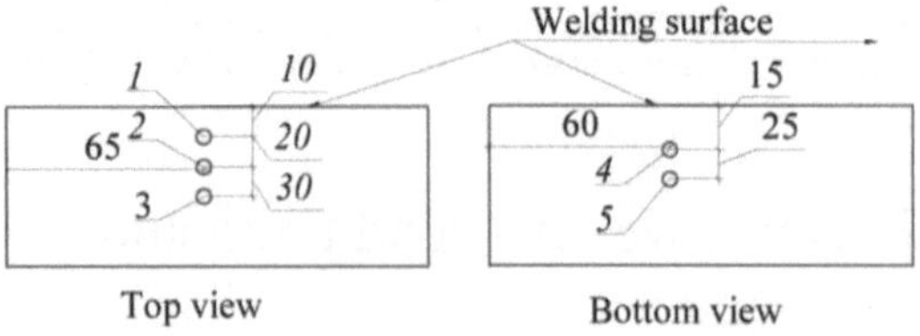

Figure 1. The location scheme of thermocouples.

To determine the effectiveness of the proposed method, it carries out numerical and experimental studies of the heat exchange processes and the formation of residual deformations during the welding of two steel plates 40 with dimensions of 100 × 200 × 8 mm. We have carried out the welding plates in a single pass. We have previously fixed the plates in vices, which allow us to simulate the effect on the welded plates of a constructions part, and they are not subject to the temperature effect. To solve the experimental problems, we have used the LabVIEW software package and the hardware of the National Instruments Company. This choice is made with the following LabVIEW values: high data collection accuracy from sensors, a full-fledged programming language, wide collection capabilities, data processing and analysis, instrument control, driver support for more than 2,000 devices, mobility, and ability to create databases of experimental data. Temperature measurement in the welding process is carried out by using chromel-copel thermocouples. The sealing-off depth of the thermocouples is 1.5–2 mm. The thermocouples are placed at a distance 10, 20, and 30 mm on top, and 15 and 25 mm on the bottom from the welded joint axis (1, 2, 3, 4, 5 points respectively in Figure 1).

3 RESULTS

Experimental and calculated temperature curves of the heating stage at the control points are shown in Figure 2. Both qualitatively and quantitatively, the calculation results mirror the experimental results. The maximum divergence of the temperature values at the control points on the heating step does not exceed 8%.

The maximal temperature values at the control points are obtained by calculations and they are experimentally close to each other. That maximal temperature values in a meaningful measurement determine the residual deformation magnitude. Thus, the comparative realized analysis of the thermal welded components' state confirms the correctness of the definition of the boundary conditions, and the acceptability of the obtained assumptions during the calculation.

The probe of residual welding deformation is carried out on a method of boreholes by using a strain gauge. [2]. During the strain measurement it uses the resistive strain sensor KF5P1-1-100 brand with a base of 1 mm and a bridge scheme connection. The disposition and quantity of measurement points of the residual deformation is stipulated by an expected nature of the field distribution of residual deformation. The aperture diameter accepted is 4 mm, and the adopted depth is 8 mm (Figure 3).

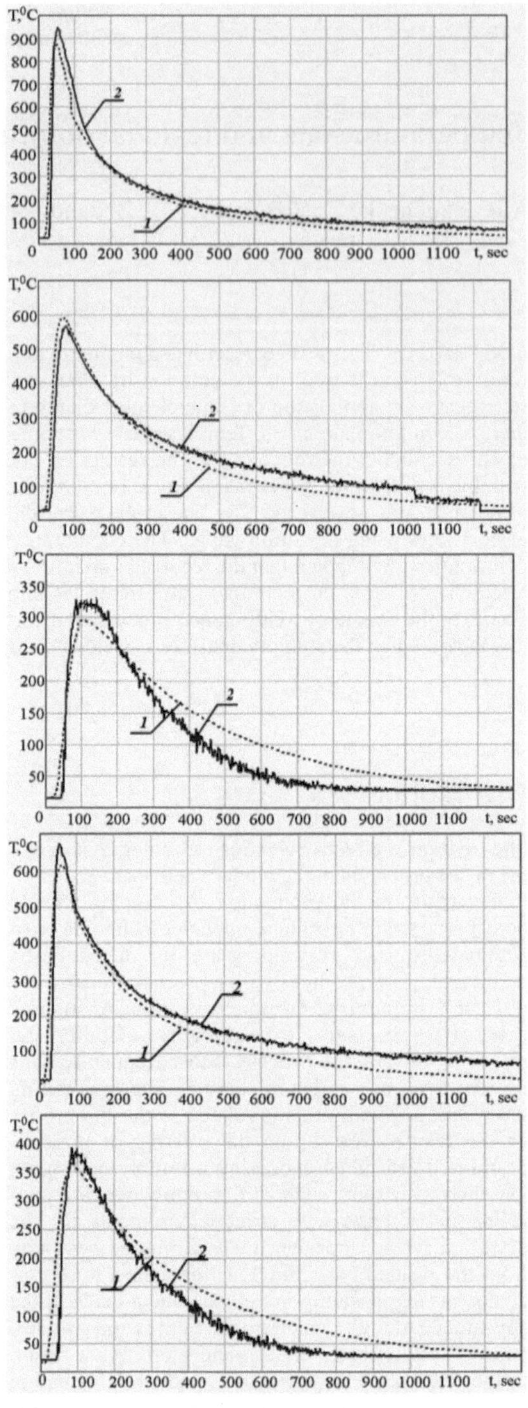

Figure 2. The change of temperature at control points during the welding process: 1 – calculated data; 2 – experimental data.

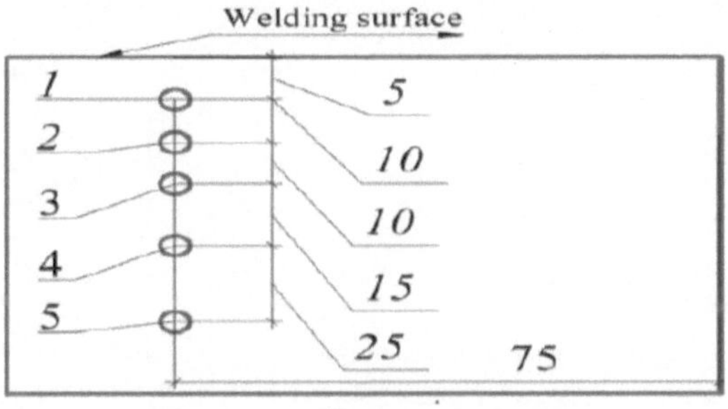

Figure 3. The location scheme of strain sensors.

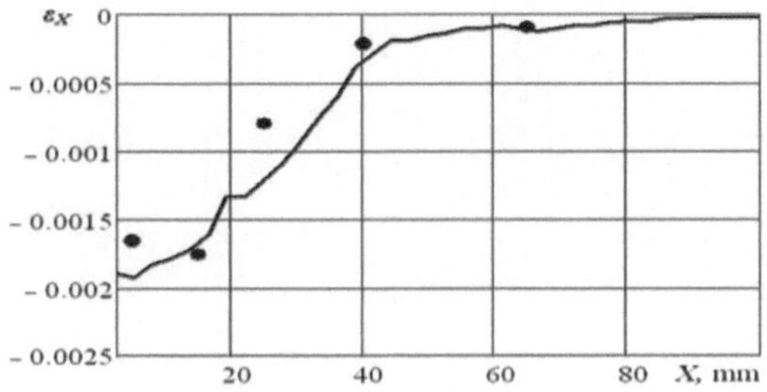

Figure 4. Residual deformation change on the X axis: ▬ - estimated data; • - experimental data.

Thus, research results confirm the correctness of the proposed method of numerical modeling of the components' temperature state and the formation residual deformation processes, which allows us to recognize this factor during the welded structures' stress analysis.

Figure 4 shows the calculated curve that reflects the changes of residual welding deformations (εH) perpendicularly to the weld axis, and the points present the experimental data.

The distribution analysis εH shows that the highest level of residual welding deformations can be observed in the transition area from the weld to the plate surface, and it reaches −0.2%. As the distance from the weld axis increases, the permanent stresses are reduced to zero values. A comparative analysis of the calculated and experimental values of deformations shows that the resulting maximal difference does not exceed 30%.

REFERENCES

Sudnik V. A., Erofeev V. A. Mathematical modelling of technological welding processes in mechanical engineering, Mechanical Engineering, Moscow, 1987 (in Russ.)

Birger I. A., Residual stress, Moscow State Scientific and Technical Publishing of mechanical engineering literature, 1963. (in Russ.)

Electrical engineering and information technologies

Future Communication Technology and Engineering – Chan (Ed.)
© 2015 Taylor & Francis Group, London, ISBN: 978-1-138-02777-0

Analysis and design of a micro-capacitor array detection circuit

W.D. Ge, D.Z. Xu & H.J. He
Anhui Key Laboratory of Advanced Numerical Control and Servo Technology, Anhui Wuhu, China
AHPU Institute of Technology Robotic Industry, Anhui Wuhu, China

ABSTRACT: During the measurement of a capacitive tactile sensor array, there always appear large and unknown parasitic capacitance and other capacitance conducted by the effect of the charge injection on analogue switches, which seriously limits the measuring accuracy of the tactile sensor. This article analyses the various performance indicators of MS3110P based on the experiments. A P controller used to match the charge injection effect conducted by the analogue switch has been investigated. The measurement mode of a micro-capacitor array detection circuit possessing the function of single capacitance array and the measurement mode of differential capacitance array, have been designed. Besides this, causes of parasitic capacitance have been discussed. The experimental results demonstrate that the designed circuit overcomes the impact of various parasitic capacitances on the sensor and enhances the measuring accuracy. It can meet the needs of the measurement requirements of a micro-capacitor array for MEMS sensors.

1 INTRODUCTION

Capacitive sensors have been widely used in various fields of industrial and agricultural production as they has the advantages of a simple structure, high resolution, reliable operation, quick dynamic response, non-contact measurement, and can work at high temperatures, with radiation and intense vibration, and other harsh conditions. For example, a capacitive sensor can be utilized to acquire the pressure signal in a tactile sensor system.

A wide variety of methods can be used for capacitance readout [1–3] such as charge sensitive amplifiers, charger distribution techniques, impedance measurements, RC-oscillators, or direct coupling to a field effect transistor gate. However, for these methods it is difficult to measure the micro-capacitor array. Besides this, the property of the use switches has not been analysed as well.

This paper investigates a P controller that can be used to match the charge injection effect conducted by the analogue switch [4]. Besides this, causes of parasitic capacitance have been discussed. The micro-capacitor array detection circuit is equipped with the function of single capacitance array measurement mode, and differential capacitance array measurement mode has been designed.

2 ANALYSIS AND DESIGN OF THE DETECTION CIRCUIT

2.1 *Basic principles of the measurement circuit*

The measurement circuit of a micro-capacitance array is based on MS3110P, and it consists of the main control chip, analogue switches, a virtual instrument, and the PC control interface. The system block diagram is shown in Figure 1.

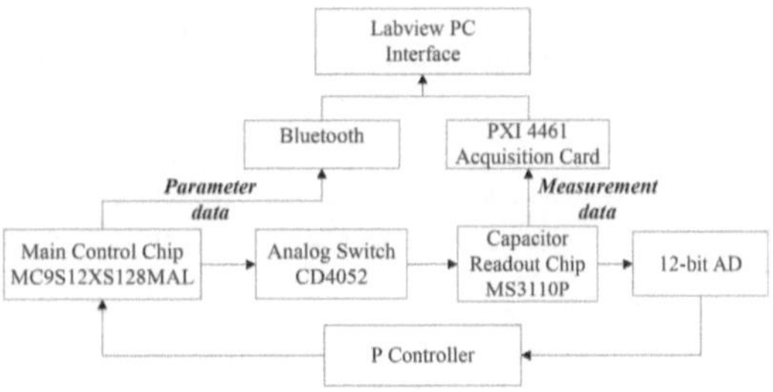

Figure 1. Block diagram of micro-capacitor array detection system.

MC9SXS128MAL, the main control chip produced by Freescale, has been used in the detection circuit board to control the analogue switches, MS3110P and Bluetooth. The detection result is read from the MS3110P and is sampled by 12-bit AD, which can make a feedback to the main control chip through the P controller to realize closed loop control. The parameter data for the entire system is transmitted to the PC via Bluetooth and the measurement data is transmitted to the PC through a PXI4461 acquisition card. The real product photo show of the measurement circuit is shown in Figure 2.

2.2 *Schematic and feature analysis of analogue switches*

The schematic for the connection of the analogue switches and MS3110P is illustrated in Figure 3, and there are two working modes in this circuit.

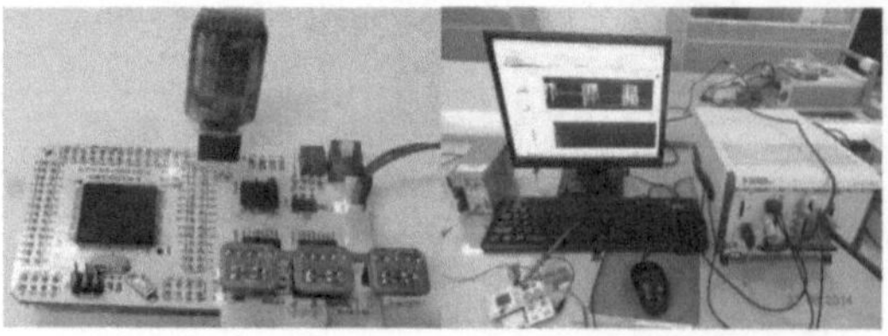

Figure 2. Micro-capacitor array detection circuit and data acquisition system.

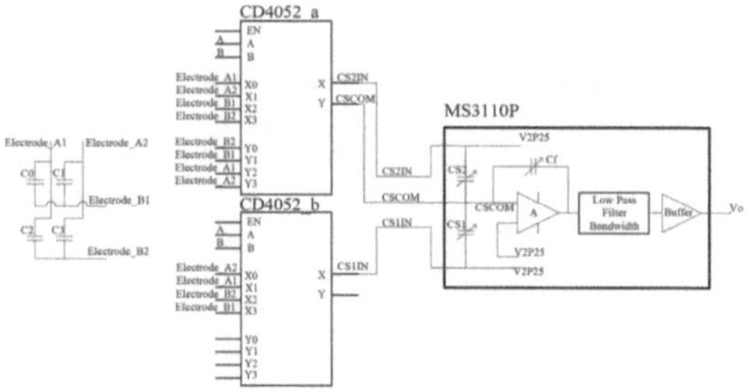

Figure 3. Circuit schematic diagram of logical switch and MS311P.

Table 1. Scan measurement mode of single capacitance array.

BA	00	01	10	11
Tested Capacitance	C2	C1	C0	C3

Table 2. Scan measurement mode of differential capacitance array.

BA	00	01	10	11
Tested Capacitance	C2-C3	C1-C0	C0-C2	C3-C1

2.2.1 *Scanning measurement of the single capacitor array mode*

To work in this mode, CD4052_a should be enabled and CD4052_b should be forbidden. Then the purpose of measuring the capacitor array can be achieved by controlling the analogue switch timing through MCU. The truth table is shown in Table 1.

2.2.2 *Scanning measurement of the differential capacitor array mode*

For working in this mode, both CD4052_a and CD4052_b should be enabled. Then the purpose of measuring the capacitor array can be achieved by controlling the two analogue switches timing simultaneously through MCU. The truth table is shown in Table 2.

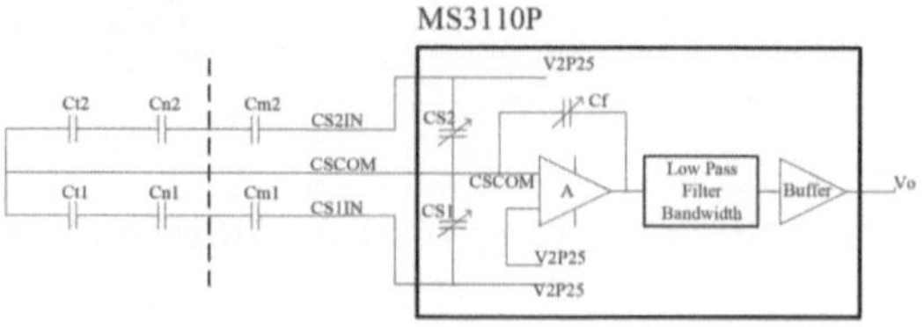

Figure 4. Equivalent measurement circuit of MS3110P.

2.3 *Performance analyses of the capacitance readout IC*

The performance of the capacitive tactile sensor is tested by measuring the capacitance change of the contact sensor from MS3110IC, which is used for measuring high resolution change in capacitance of MEMS (Micro Electro Mechanical Systems) devices. The complete internal circuitry in the form of a block diagram is shown in Figure 4, which includes a transimpedance amplifier with a programmable feedback capacitor CF, a low pass filter with programmable bandwidth, and an output buffer [8].

The output from MS3110IC is obtained in terms of voltage which is converted into change in capacitance by the following transfer function as shown in Equation 1:

$$Vo = \left[V_{2P25} \times 1.14 \times Gain \times \left(C_{S2T} - C_{S1T}\right)\right] / Cf + VREF \quad (1)$$

$$C_{S1T} = C_{S1IN} + C_{S1} = C_{T1} + C_{N1} + C_{M1} \quad (2)$$

$$C_{S2T} = C_{S2IN} + C_{S2} = C_{T2} + C_{N2} + C_{M2} \quad (3)$$

where V_{2P25} is reference voltage which is equal to 2.25 V, $VREF$ can be set to 0.5 V or 2.25 V. Gain can be set to 2 or 4. C_{S1T} is the total capacitance of one side including C_{T1}, C_{N1}, C_{M1}, which mean tested capacitance, connectors' parasitic capacitance, and the analogue switch's capacitance, respectively. Similarly, C_{S2T} is the total capacitance of the other side. Cf is the internal feedback capacitor.

In order to verify the performance of the MS3110 chip, the internal programmable capacitor array is changing by the step of 76fF. The three terminals of MS3110 for sensing capacitance change are connected to the output of the analogue switch, as shown in Figure 3.

The experiment result about the C-V curve is shown in Figure 5. The blue curve represents the original output signal. The shock wave means the implementation process of reloading the new parameter (CS2). The red star point represents the results of the filtered output signal, which can prove that the MS3110P has a great linearity.

In this experiment, Gain is set to 2, VREF is set to 0.5 V, and the feedback capacitor is set to 9.5pF. Therefore, the theoretical transfer function of MS3110P is indicated by Equation 4:

$$\begin{aligned} Vo &= \frac{2.25 \times 1.14 \times 2}{9.5} f + 0.5 \\ &= 0.54 f + 0.5 \end{aligned} \quad (4)$$

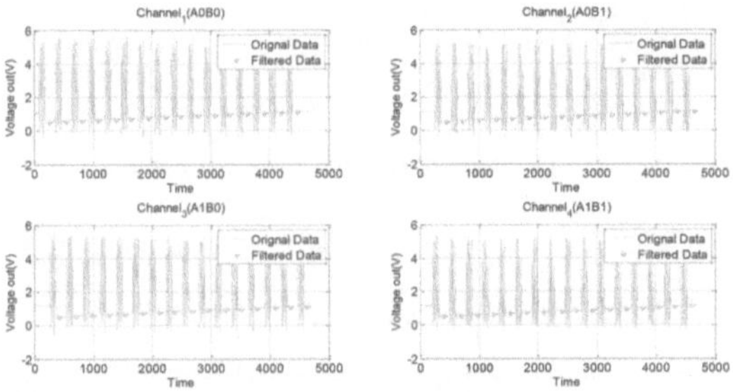

Figure 5. Linearity test of MS3110P.

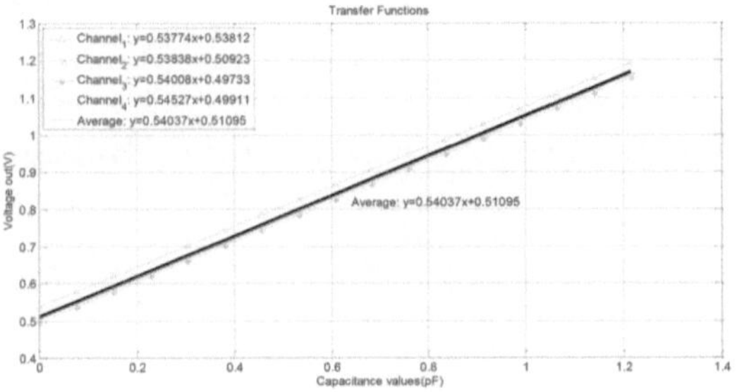

Figure 6. Input-output curve fitting of MS3110.

In the actual measurement, set CS1 = 0, and CS2 is increasing at the step of 76fF from 0 to the maximum. The input-output relationship is plotted in Figure 6. The x-axis in the figure represents the capacitance values, while the y-axis means the output value of the detection circuit. The average regression function is Vo = 0.54037f + 0.51095, which is basically consistent with the theoretical counterpart as shown in Equation 4. The average coefficient of linear correlation is 0.99978 [5] and the sensitivity of the system can be drawn as 0.54037 V/pF.

The test results demonstrate the detection system has a high accuracy. Since the integrated A/D switch in MC9SXS128MAL is 12-bit, when Cf is placed at 9.5pF, the resolution of the capacitance in the system is 0.45fF. Hence, it is obvious that the resolution of the A/D converter is an important factor for restricting the detection precision. Therefore, when the system accuracy needs to be improved, the A/D converter with a higher number of bits can be considered.

3 EXPERIMENTS AND RESULTS ANALYSIS

3.1 *Resolve the charge injection effect produced by the analogue switch*

There are two ways to solve the charge injection effect that is produced by the analogue switch. One is given by Bai Guohua, et al., who have proposed eliminating the charge injection effect by using a CPLD (Complex Programmable Logic Device) to design reasonable switch timing based on differential principle

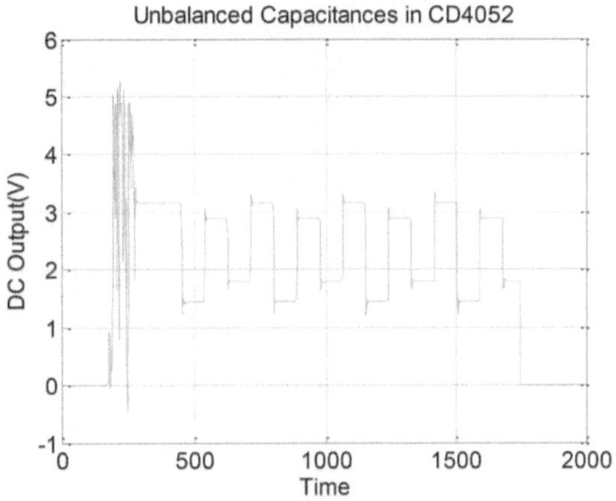

Figure 7. Unbalanced capacitances in CD4052.

[4]. The idea of this method can be clarified in Equations 5–7:

$$\Delta C=(C_{ml}+C_{T1})-(C_{m2}+C_{T2}) \tag{5}$$

$$\because C_{ml}=C_{m2} \tag{6}$$

$$\therefore \Delta C=C_{T1}-C_{T2} \tag{7}$$

However, even with the different channels in the same analogue switch, the charge injection effects are different, which is shown in Figure 7. This is because $C_{m1} \neq C_{m2}$. Therefore, this approach I is hard for reaching the desired purpose.

Another method is to match the internal adjustable capacitance of MS3110P and the charge injection effect produced by the analogue switch. In other words, the output of the four channels should be adjusted to the equal level. The idea of this method can be demonstrated in Equations 8–10:

$$\Delta C=(C_{ml}+C_{T1}+C_{S1IN})-(C_{m2}+C_{T2}+C_{S2IN}) \tag{8}$$

where C_{S1IN} and C_{S2IN} are the internal programmable capacitors. So it is easy to satisfy Equation 9 by the P controller, as shown in Figure 8.

$$C_{ml}+C_{S1IN}=C_{m2}+C_{S2IN} \tag{9}$$

$$\therefore \Delta C=C_{T1}-C_{T2} \tag{10}$$

3.1.1 *Theoretical analysis and experiment*

The signal is acquired by an A/D switch iterating repeatedly to the expected value through the P controller. As shown in Figure 9, the blue dotted line, green solid line, and the red solid line represent the adjustment process of the four channels, the adjusted output of the four channels, and the results after inserting the standard capacitance, respectively. The upper part of the figure is the adjusting and testing process for the measurement mode of a single capacitor

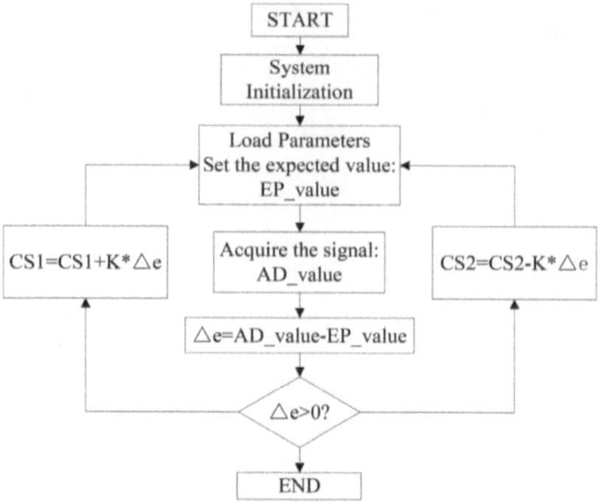

Figure 8. Flowchart of the P controller.

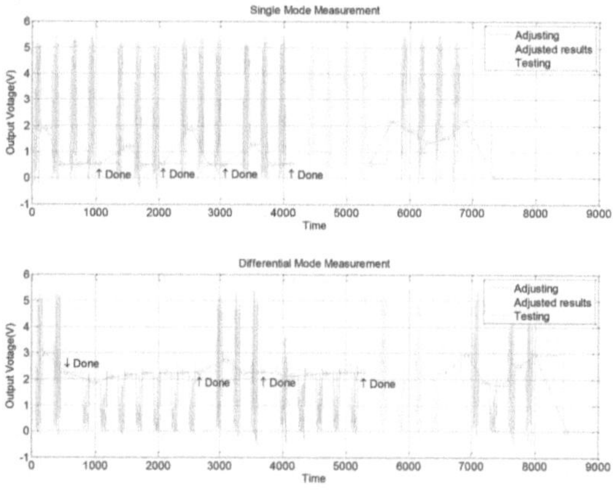

Figure 9. Adjustment and test process of the P controller.

array, while the expected value is set to 0.5 V. Similarly, the lower part of the figure is the adjusting and testing process for the measurement mode of a differential capacitor array while the expected value is set to 2.25 V.

3.1.2 *Analysis of experimental results*

Initially the capacitances of the analogue switch have been removed by setting internal programmable capacitors of MS3110 IC from the software interface. It is necessary to remove these capacitances which carry different sensing signals to MS3110IC as is shown in Figure 7, which may increase or decrease the total output capacitance. Succinctly, the designed P controller performs well as is shown in Figure 9.

However, the experimental results show 0.6pF more on average compared with the standard capacitance. The probable reason for this may be the parasitic capacitance of connectors, pads, and other parts of the circuit board.

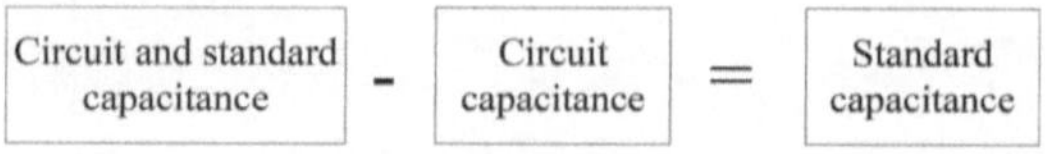

Figure 10. Measurement idea of the experiment.

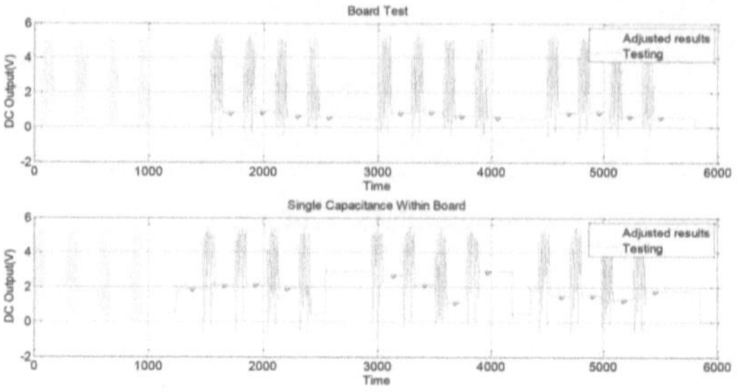

Figure 11. Circuit board capacitance and total capacitance.

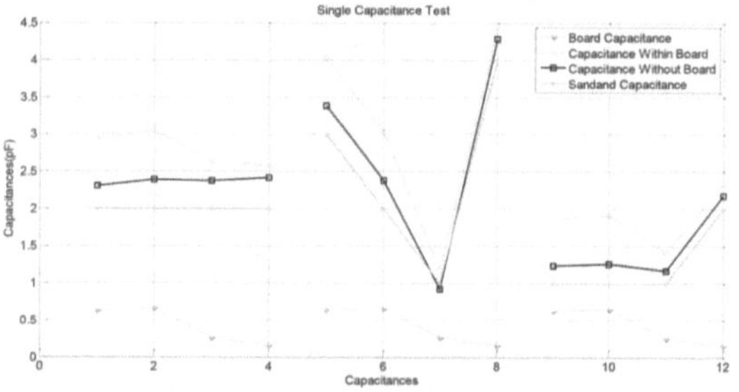

Figure 12. Comparison among the four kinds of capacitances.

3.2 *Explore the origin of the parasitic capacitance*

3.2.1 *Theoretical analysis and experiment*

The measurement ideas of this experiment are that the tested capacitance is equal to the total capacitance minus the circuit board capacitance, as is shown in Figure 10.

The results of the single capacitance measurement mode are shown in Figure 11, in which the solid green line and the red line represent the adjusted output of the four channels and the results after inserting the standard capacitance, respectively. The upper part of the figure is the measurement of the board capacitance, while the lower part is the measurement of the board and the standard capacitance in total.

Figure 12 shows the comparison of the experimental results of all kinds of capacitance that is mentioned above. The blue dash line, red line, black line, and green line represent the circuit board capacitance, total capacitance, tested capacitance, and the standard capacitance, respectively.

3.2.2 *Analysis of experimental results*

It is apparent that after eliminating the circuit board capacitance, the measurement result is much closer to standard value. However, the experimental results

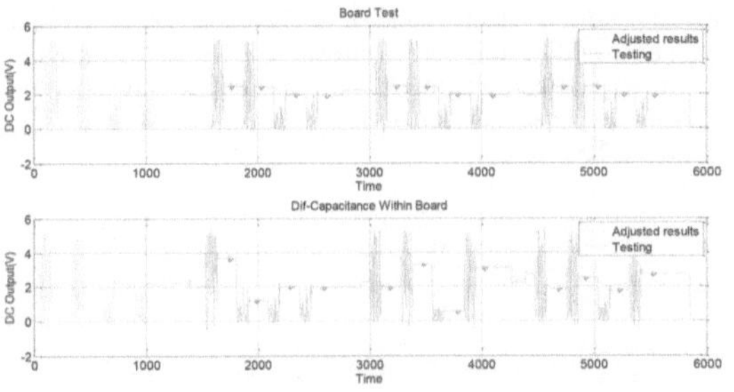

Figure 13. Circuit board capacitance and total capacitance.

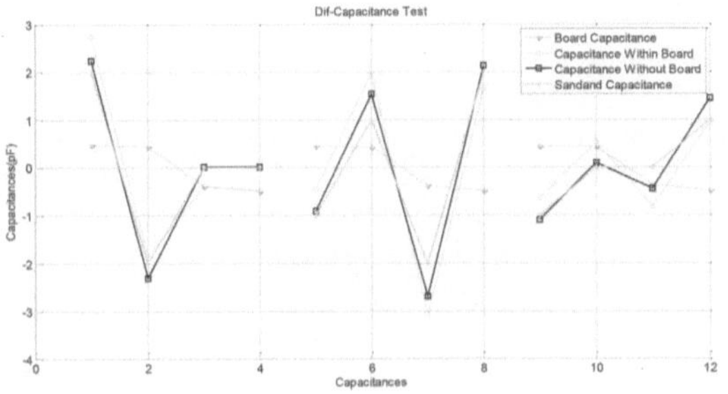

Figure 14. Comparison among the four kinds of capacitances.

Figure 15. Analysis of the blank circuit board.

show slightly bigger (0.2 pF on average) as compared to the standard capacitance. The probable reason for this is the other parasitic capacitance due to the pads and other parts of the sensor [5]. This increase in capacitance does not affect the performance of the differential capacitance measurement mode, as is shown in Figures 13–14.

Hence, we can safely draw the conclusion that the measurement accuracy of the differential mode is higher than the single mode. In addition, in the actual measurement of MEMS sensors, we always measure the variation of the capacitor, thus ensuring this error does not affect the measurement results.

The causes of the PCB parasitic capacitance are the pin headers which connect the PCB and the sensors. The sectional view of the connector is shown in Figure 15, and it has the contact area S as is shown in Equation 11:

$$S \approx 2.5*0.8*10 = 20mm^2 \tag{11}$$

The gap d between two adjacent electrodes is nearly 2 mm. Set the relative dielectric constant of the connectors to 6, so that the capacitance value of the parasitic capacitance can be obtained (approximately 0.531 pF) as shown in Equation 12:

$$C = \frac{\varepsilon_0 \varepsilon_r S}{d} = \frac{6*8.85*10^{-12}*20*10^{-6}}{2*10^{-3}} = 0.531pF \tag{12}$$

The parasitic capacitance value between the different pins in the circuit board is different, as shown in Figure 15. Adjacent pins are much bigger as they have a small distance between them. Also, the figure shows that the capacitance between pin 1 and pin 3 is slightly bigger than that between pin 2 and pin 4, due to the fact that their wire on the PCB is comparatively longer.

3.2.3 *Methods for suppressing the parasitic capacitance*

According to the experimental results, the following measures can be adopted to suppress the generation of parasitic capacitance:

On one hand, do not use the connector to connect the sensor and the circuit board if possible. In other words, it is best to integrate the circuit board directly to the sensor [6].

On the other hand, if method one is inevitable, the distance between the two pins should be increased so that the parasitic capacitance will be reduced.

4 CONCLUSION

1) Proved the capacitance readout chip MS3110P has a good linearity and high accuracy experimentally.
2) The designed P controller can correct the capacitance error caused by the analogue switch.
3) The generation of the parasitic capacitance of the circuit board and the way to suppress it has been discussed, which provides reference for the circuit design of capacitance sensors.
4) The experimental results show that the accuracy and the sensitivity of the detection circuit can meet the needs of the measurement requirements of a micro-capacitor array for MEMS sensors.

ACKNOWLEDGEMENT

The authors would like to thank the financial support from the National Natural Science Foundation of China (Grant No. 51175001) and Universities Natural Science Research Project of Anhui province (TSKJ2014B12).

REFERENCES

Rödjegård, H. and A. Lööf, A differential charge-transfer readout circuit for multiple output capacitive sensors. Sensors and Actuators A: Physical, 2005. 119(2): p. 309–315.

Qiu Guiping, Yu Xiaoyang, and Chen Deyun, Micro-capac itance measuring circuit. Heilongjiang Electric Power, 2006(05): p. 362–366.

Yu Nengsheng and Sun Shiping, Design of new capacitance measuring circuit for capacitance sensor. China Measurement & Test, 2005(05): p. 42–43.

Bai Guohua and Ma Tiehua, The Application of CPLD on Low Capacitance Measuring Circuit for Capacitance Sensor. Metrology & Measurement, 2005(02): p. 14–16.

Bazaz, S.A., F. Khan, and R.I. Shakoor, Design, simulation and testing of electrostatic SOI MUMPs based microgripper integrated with capacitive contact sensor. Sensors and Actuators A: Physical, 2011. 167(1): p. 44–53.

Future Communication Technology and Engineering – Chan (Ed.)
© 2015 Taylor & Francis Group, London, ISBN: 978-1-138-02777-0

Applied research on memristor's nonlinear characteristics in circuits

X.L. Cao, W. Cui & X.C. Ren
School of Physics and Electronics Information, Yan'an University, Yan'an, China

ABSTRACT: With the emergence of a memristor, a new opportunity has been brought for the further development of electronic information technology. In this study, the memristors are applied separately to a hysteretic comparator and arithmetic circuit based on self-contraction memristor hysteretic behaviour analysis. Furthermore, the performance of a memristor in the circuit is deduced by theoretical analysis. It shows that the memristor with non-linear characteristics can replace ordinary resistors to achieve the movement of a hysteretic comparator's characteristic curve and change a linear operation circuit into a non-linear operation circuit. It would provide an example for analysis of the new functions that the memristor produces in electronic circuits.

Keywords: memristors, hysteresis loop, nonlinear, applied circuit.

1 INTRODUCTION

Hewlett-Packard have scientists who have discussed a major breakthrough in the memristor technology and implementation in the United States on 8 April, 2010, which was published in *Nature* magazine[1]. Moreover, they had found that the memristor that could perform Boolean logic could be applied to data processing and storage, for example. Scientists predict that the memristor circuit could in future years replace silicon transistors to lead to new features in the field of electronics, which can replace the digital era with an analogy processing new era. The memristor technology promotes low-power computers, storage devices and the memristor generation and application of a new class of neuromorphic devices which can simulate learning, self-adaption, and spontaneous behaviour. This brings new opportunities for the further development of electronic information technology.

Memristors used for electronic information technology are based on these elements and have self-contraction hysteresis characteristics of the constitutive variable[2]. Memristor, as a new nanometre device, can be suitable for some novel function. Its exploration of a new application functionality has become an important issue.

2 HYSTERETIC BEHAVIOUR ANALYSIS FOR MEMRISTOR

2.1 *Analytical expression of analogy memory devices constitutive variables*

A characteristic of the memristor is the relationship between the definitions of charge and the magnetic flux. Later, it is described as the integral of voltage to time.

Generally, if $\dot{x}$ represents an N-dimensional vector, which is a description of the internal state of the system. $u(t)$ and $y(t)$ represent any relevant variables, and they are the input and output of systems. g represents the general response. A class of memory system is defined by the following[3]:

$$\begin{cases} y(t) = R(x,u,t)u(t) \\ \dot{x} = f(x,u,t) \end{cases} \tag{1}$$

where f is continuous N-dimensional vectors function. We suppose that the initial state value is $u(t = t_0)$, at a certain time t_0.

The memristor system is a special case of equation (1). By the current – voltage concrete reflection:

$$\begin{cases} V_M(t) = R(x,I,t)I(t) \\ \dot{x} = f(x,I,t) \end{cases} \tag{2}$$

where $V_M(t)$ and $I(t)$ represent the device's terminal voltage and the current flowing through the device. R is a coefficient (that is memristor). When $I(t) = \frac{dq(t)}{dt}$ is taken into account, the current-voltage relationship of memristor can be expressed as:

$$V_M(t) = R(q(t))I(t) \tag{3}$$

2.2 *Memristor model and represents the symbol*

The basic physical structure of the TiO_2 memristor in HP labs is shown in Figure 1. It is used to describe the relationship among electrical impedance,

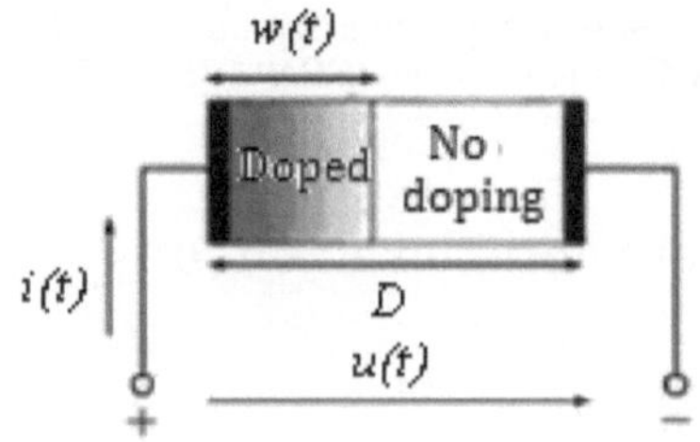

Figure 1. Memristor physical structure.

current and the ratio of doping, and to simplify the expression of memristor's basic characteristics mathematically.

$$R(q(t)) = R_0\sqrt{1-\frac{2\eta\cdot\Delta R}{Q_0R_0^2}\int_0^t u(\tau)d\tau} \tag{4}$$

η represents the polarity of the memristor. When "+1" is taken, the memristor is in the forward conduction, and when "−1" is taken, the memristor is in the negative conduction. In $Q_0 = D^2/\mu_D R_{ON}$, μ_D represents impurity ions' average mobility in a homogeneous field. D is the total length of the memristor. R_0 represents the memristor's value when $t=0$. In $\Delta R = R_{ON} - R_{off}$, R_{ON} is the ON resistance, and that is resistance value when $w(t)=D$, showing that elements are all doped. R_{off} is the OFF resistance, which is the resistance value when $w(t)=0$, That is a maximum resistance showing that elements are not doped.

Due to doped and undoped section, the resistance is equal to the components of the total resistance $R(t)$. The study shows a linear relationship[1] with the rate of change, $w(t)$ whilst current $i(t)$ goes through the memristor. The memristor mathematical model is as follows[6]:

$$\begin{aligned} u(t) &= \left[R_{ON}\frac{w(t)}{D} + R_{off}\left(1-\frac{w(t)}{D}\right)\right]i(t) \\ &= \left[R_{off} - (R_{off}-R_{ON})\frac{w(t)}{D}\right]i(t) \end{aligned} \tag{5}$$

$$w(t) = \mu_D\frac{R_{on}}{D}i(t) \tag{6}$$

where $w(t)$ denotes the width of the element's doped region. The resistivity of the doped section is smaller than the resistivity of the undoped section.

Due to the incorporation of different impurities, both positive and negative ions can be formed, and therefore there are two memristors: $\mu_D > 0$ is P-type memristor, whilst $\mu_D < 0$ is N-type memristor. According to the literature [3], the memristors are non-symmetrical components. It is appointed that the memristor is into a high impedance state when a positive voltage is applied to a terminal, and in that way this terminal uses black thick lines as expressed in Figure 2.

Figure 2. A symbol of memristor.

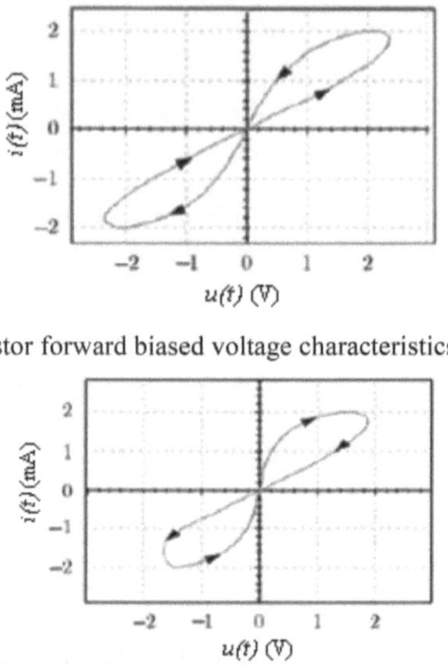

(a) Memristor forward biased voltage characteristics

(b) Memristor reverse biased voltage characteristics

Figure 3. Memristor voltage characteristics.

2.3 *Memristor nonlinear voltage characteristics*

A memristor has both positive and negative bias modes[4]. If doped section is widen when the current goes through the P-type memristor, this is called a forward bias. If doped section is narrower, this is called a reverse bias. Concerning the P-type memristor forward bias, if $w(t) \in (0, D)$, $u(t) > 0$, and $i(t) > 0$, a doped region expands right, $w(t)$ increases, and $R(t)$ diminishes. When $u(t) < 0$ and $i(t) < 0$, doped region shrinks left, $w(t)$ decreases and $R(t)$ increases. In case of $u(t) = 0$, in which way $i(t) = 0$, doped region $w(t)$ remains unchanged, and $R(t)$ remains unchanged as well. Resistance change of N-type memristor contrasts with P-type memristor.

The volt-ampere characteristics of P-type memristor in the case of forward and reverse biased voltage are shown in Figure 3 (a), and (b).

Therefore, in the memristor reverse bias, the resistance is continued increase to the maximum value. And hysteretic correlate with frequency. The hysteresis loop is suppressed with increased frequency of applied signal[5]. The hysteresis characteristic curve is shown in Figure 4.

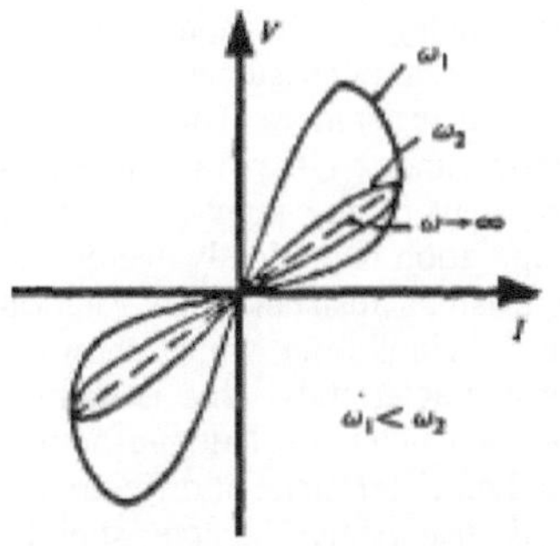

Figure 4. Characteristic curve of memristor in different frequencies.

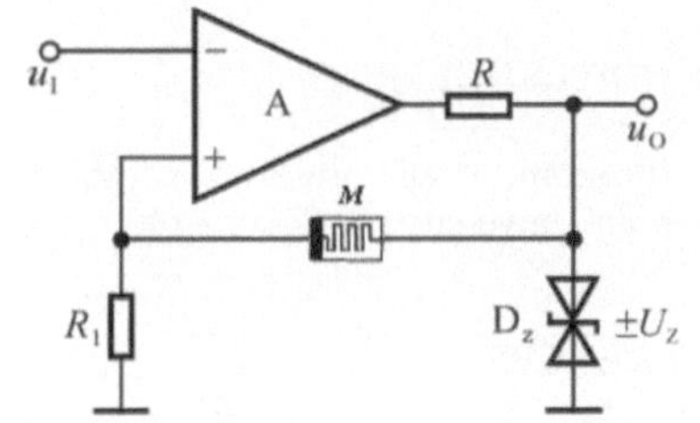

(a) Hysteretic comparator composed by memristor

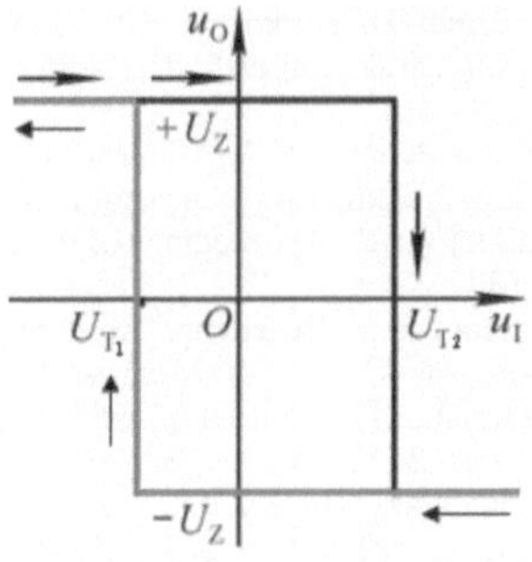

(b)Threshold asymmetric hysteresis characteristic curve

Figure 5. Hysteretic comparator and characteristics.

3 MEMRISTOR APPLICATIONS IN THE CIRCUIT

a) *Memristor use for changing threshold of hysteretic comparator*

In some electronic systems, non-symmetrical comparing window is often necessary. Memristor can be used for setting the threshold voltage to move the comparator hysteresis characteristic curve. The traditional approach is to use the non-inverting input terminal of comparator to connect the ground by a reference voltage (as R1 in Figure 5a) in order to reset the threshold voltage, in order to change the characteristic curve moves.

The memristor connection method is used in Figure 5(a). The reference voltage is omitted to set a threshold voltage reaching to move the characteristic curve. Besides, the bi-Zener voltage value determines the size of the hysteresis loop. Assuming that the input signal frequency is low enough, the regulator voltage value is sufficient to reach the resistance of the memristor R_{ON}, R_{off}. The two threshold voltages can be expressed as:

$$U_{T1} = -\frac{R_1}{R_1 + R_{off}} \cdot U_Z \tag{7}$$

$$U_{T2} = \frac{R_1}{R_1 + R_{ON}} \cdot U_Z \tag{8}$$

The characteristic curve is shown in Figure 5(b). In terms of the resistor circuit with the same configuration, the curve moves to the right. If polarity of memristor M in the Figure 5(a) is reversed, the characteristic curve moves to the left.

b) *The memristor can change linear arithmetic circuit into non-linear arithmetic circuit*

The partial resistors in phase summation circuit use memristors replacement, and the linear operation circuit would become a nonlinear operation circuit. Taking into account the symmetry of two input terminals, except access to the op amp's feedback section memristor, the non-inverting input terminal and ground are connected across the same parameters' memristor. Therefore, it is expressed as:

$$u_P = R_P\left(\frac{u_{I1}}{R_1} + \frac{u_{I2}}{R_2} + \frac{u_{I3}}{R_3}\right) \quad (R_P = R // M = R_1 // R_2 // R_3 // M) \tag{9}$$

$$u_O = R_f \cdot \left(\frac{u_{I1}}{R_1} + \frac{u_{I2}}{R_2} + \frac{u_{I3}}{R_3}\right) \tag{10}$$

The current through the memristor make resistance in two situations, that are on and off. From the formula (5), (6), and (10), it can be obtained:

$$u_O = \left[R_{off} - 2(R_{off} - R_{ON}) \cdot \frac{\mu_D R_{ON}}{D^2} \times \left(\frac{u_{I1}}{R_1} + \frac{u_{I2}}{R_2} + \frac{u_{I3}}{R_3}\right)\right] \times \left(\frac{u_{I1}}{R_1} + \frac{u_{I2}}{R_2} + \frac{u_{I3}}{R_3}\right) \tag{11}$$

The output is a nonlinear expression. it contains first term, squared term and cross-product terms about multi-input variables. The characteristic can act as a multiplier or mixer at a low frequency.

In case of single-input (That is, $u_{I2}, u_{I3} = 0$), the output will be a quadratic polynomial of a single variable. It is used to greatly reduce frequency component of the output signal when the circuit serves as frequency doubler.

A nonlinear output expression depends on a selected memristor. For example with HP's memristor. The data parameters that memristor fits are[1]: $R_{ON} = 100\,\Omega$, $R_{off} = 16\,\text{k}\Omega$, $\mu_D = 10^{-14}\,\text{m}^2/\text{V}\cdot\text{s}$, and $D = 10\,\text{nm}$.

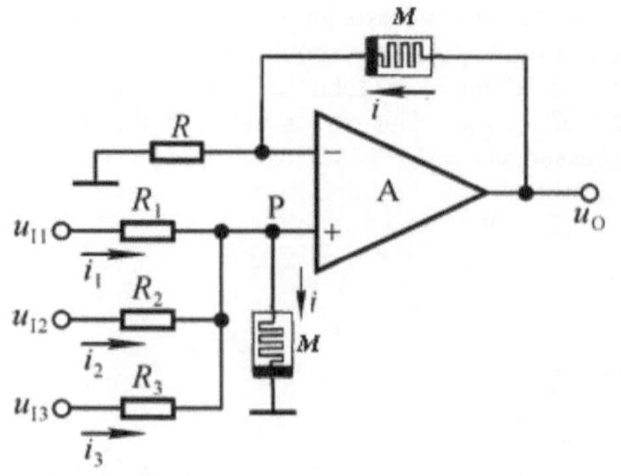

Figure 6. Memristor nonlinear operation circuit.

The output will be a quadratic polynomial, and the expression of a single variable is:

$$u_O = [16000 - 2\times 15900 \times 1000 \cdot u_I] \cdot u_I \tag{12}$$

In such a circuit that constitutes the memristor, quadratic term's coefficiency (that is amplitude) is much larger than first term's coefficiency. It can approximately replace an analogue multiplier for exponentiation.

4 SUMMARY AND EXTENSION

The study, developed by HP Labs, is based on the expansion of a memristor. The memristor's voltage characteristic is a hysteretic curve and its non-linear characteristic shows a quadratic function relationship. In hysteretic comparator applications, the positive and negative characteristics of a memristor would related with the past state. This will make two comparative thresholds asymmetrical, thus move the characteristic curve of the hysteretic comparator. With the improvement of the input-signal frequency, the two comparative valves tend to have positive and negative symmetry. In the nonlinear operation circuit applications, a memristor needs to work between the two limited resistances. This will make the output quadratic function have a limited range. With the improvement of the input-signal frequency, the nonlinear operation is gradually decreased.

The memristor's resistance has relations with the state of the previous time, which shows the hysteresis curve's characteristics. This is common for this type of devices which have analogue memory function. However, a memristor made of different materials and structures also has special characteristics. Some device can be realized cubic operation. Applications of circuit analysis and analytical method described herein are similar.

ACKNOWLEDGEMENTS

Fund Program: A key project of natural science fund of Yan'an University (YDZ2012-08)

REFERENCES

Strukov D B, Snider G S, Stewart D R, Williams R S. The missing memristor found [J]. Nature, 2008, 45(12): 80–83.

Biolek Z, Biolek D, Biolkova V. SPICE model of memristor with nonlinear dopant drift [J]. Radio Engineering, 2009, 18(2): 210–214.

Ventra M D, Pershin Y V, Chua L O. Putting memory into circuit elements: memristors, memcapacitors and meminductors [C]. Proceedings of the IEEE, 2009, 92(8): 1371–1372.

He Bao-xiang, Bao Bo-cheng. Research on the Equivalent Analysis Circuit of Memristors Network and Its Characteristics [J]. Journal of Electronics & Information Technology. 2012, 34(5): 1252–1256. (China)

Cai Kunpeng, Wang Rui, Zhou Ji. Research progress on memristor and its applications as the fourth passive component [J]. Electronic Components and Materials. 2010, 29(4): 78–82. (China)

Tour J M and H Tao. Electronics: the fourth element [J]. Nature, 2008, 453(5): 42–43.

Future Communication Technology and Engineering – Chan (Ed.)

A conformance testing device for an IEC 61850 configuration file based on an NI CRIO platform

J. Jiang, H.Q. Gou, Z.H. Liu & D. Liu
Chongqing Jiangjin Power Company, Chongqing, China

J. Bu & X.X. Ji
Nanjing University of Science & Technology, Nanjing, China

ABSTRACT: The IEC 61850 standard is the key for the devices from different manufacturers to interoperate; therefore carrying out the standard conformance testing on an IEC 61850 configuration file is important. According to the IEC 61850 standard, this article introduces the testing content of conformance testing for the configuration file, and it designs the conformance testing device for the configuration file on CRIO. It introduces the design process, structure, and function of the testing devices. The conformance testing tool of the configuration file is in compliance with the IEC61850 standard, which is a testing tool with engineering practicality.

Keywords: IEC 61850; configuration file; conformance testing; NI CRIO platform; testing device.

1 INTRODUCTION

The IEC 61850 standard is one of the core technologies of an intelligent substation system[1], and is the key for the devices from different manufacturers to interoperate[2]. Therefore, the standard of conformance testing has a very important significance. The configuration file is used to describe substation equipment object models[3], and its conformance testing is of great significance as an important test item of IEC 61850 testing[4]. But IEC 61850-10 only do the provisions in principle of conformance testing for the configuration file. It is very difficult to test in practical engineering for the less of specific test items for configuration file.

Conformance testing for a configuration file has an important significance for engineering applications. Literature[5] introduced the IEC 61850 configuration file, did necessary research on conformance testing as well as analysed concerns of several domestic widely used test tools, but it did not carry out any research on specific test items, or any specific implementation. In order to meet the specific engineering needs in the configuration file conformance testing, this article puts forward the conformance testing plan for an IEC 61850 configuration file on smaller portable test equipment based, on a CRIO platform.

2 THE FUNCTION OF A CONFORMANCE TESTING DEVICE FOR AN IEC 61850 CONFIGURATION FILE

The fundamental purpose of conformance testing for IEC 61850 is to enable people to do the conformance testing for tested devices[6]. As is described, the system file, the conformance testing for IEC61850 configuration file is important. In engineering, a variety of problems will appear in configuration files. According to the IEC 61850-10[7], the testing device is able to implement the conformance testing of a configuration file based on the CRIO platform. The conformance testing item of configuration file is shown in Table 1.

3 THE IMPLEMENTATION OF THE TESTING DEVICE FOR AN IEC 61850 CONFIGURATION FILE

3.1 *The design of the whole device*

According to the requirements of conformance testing for a configuration file, the testing device is designed based on the NI Compact RIO, and the software LabVIEW. The overall program design diagram of this device is shown in Figure 1. It includes the file acquisition of backplane FPGA, the analysis of the algorithms

Table 1. Test item of conformance testing for configuration file.

Test item	Specific implementation
Check communication parameters	Including the IP address, GOOSE MAC address, etc.
Checking the validity of reference of communication	Including the matching check the iedName, apName, GSE cbName, ldInst between communication and IED.
Checks whether duplicate defined data type template	Including existence of EnumType, LNType, DOType or DAType with different names but same content
Check if there is instantiated data type template	Check whether there is data type template not be instantiated in data module
Check the matching of examples and templates	Including the matching check between examples of intelligent electronic devices and data type referenced template
Check whether the data template matches standards	Including various types of data template such as DOType, Enum Type, DAType, Enum Type and other standard checks
Check the normative of logical device name	Including to detect whether each logical device name meets the requirements of IEC 61850.
Check the name of data set and the control block	Including to check whether the name of report control block and data set with uniform definition of IEC 61850.
Check whether the data set index is valid	Including the matching between the index of the data set and their reference data/data attributes,
Check the validity of the connection index of GOOSE	Including the matching between GOOSE connection index in the inputs of SCD file and the reference data/data attributes.

of CRIO controller integrated performance and the display module.

This test device tries to minimize the degree of coupling between the various modules, so that upgrading modification in the future can be easier. According to the structural characteristics of the system hardware, the software structure is divided into three: FPGA bottom layer, the controller layer, and the host computer PC layer. The design of the device is shown in Figure 2.

3.2 *FPGA acquisition module*

The function of FPGA acquisition module is to collect the required data, and its structure consists of a system configuration module, a data acquisition module, and a time calibration module.

Self-test: After the power starts on the board, the self-test program immediately starts and checks the function of the hardware.

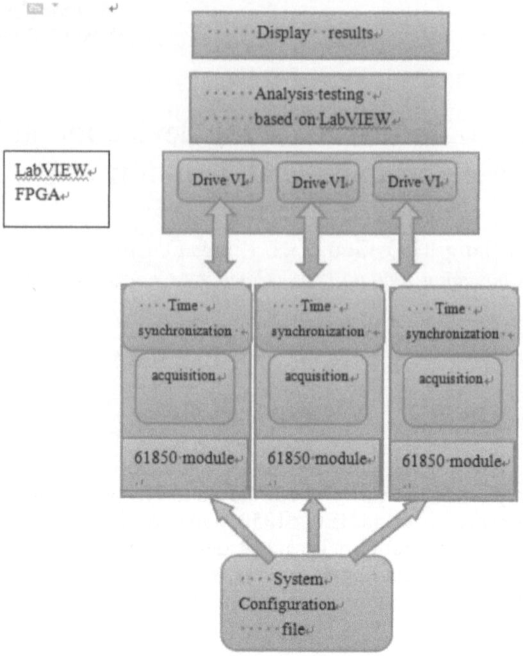

Figure 1. The overall structure based on NI CRIO platform schematic diagram.

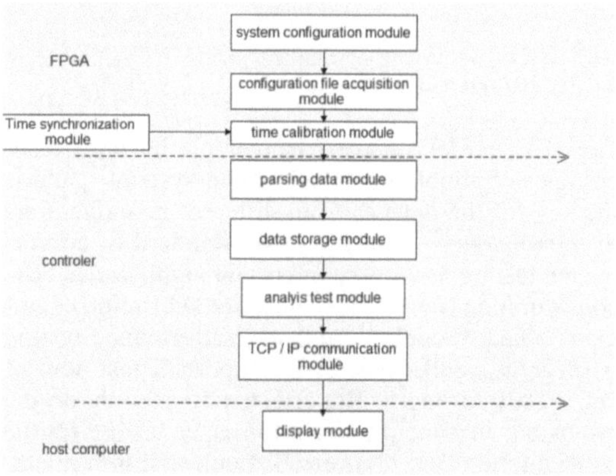

Figure 2. The structure of a function module schematic diagram.

The configuration module and the time calibration module: the FPGA starts the configuration module and waits for configuration commands. At the same time, the configuration module set the time of system clock to time calibration module.

Acquisition module: after configuration is completed, the collection module will start collecting data after receiving the acquisition order. During the process of data acquisition, the acquisition can stop if the end command is received. Program flowchart is shown in Figure 3.

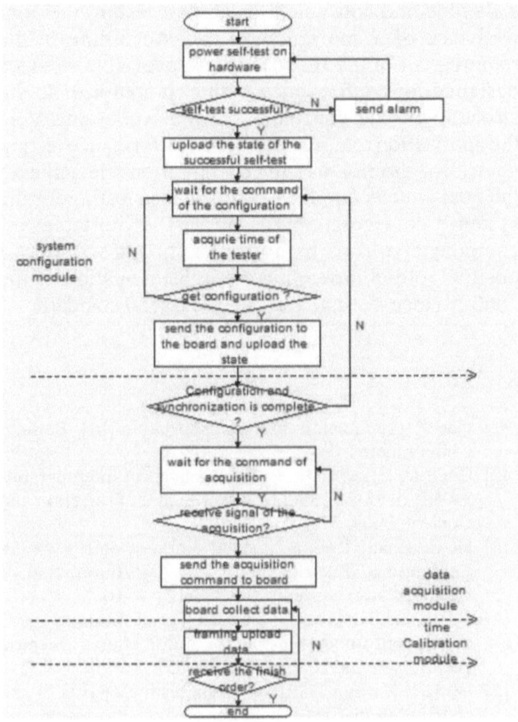

Figure 3. FPGA acquisition module flowchart.

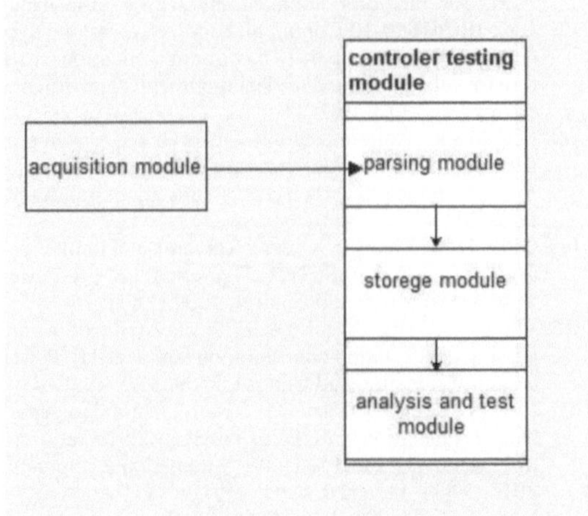

Figure 4. Controller test module software modules relations schematic diagram.

3.3 *Controller test module*

The main function of control test module of an NI CRIO platform is to import, parse, and analyse the configuration file. It includes: a data analysis module, a data storage module, and analysis and testing modules. The structural relationship is shown in Figure 4.

Self-test waiting modules: after the NI CRIO platform software starts it can upload the self-test status information and the framing synchronization state

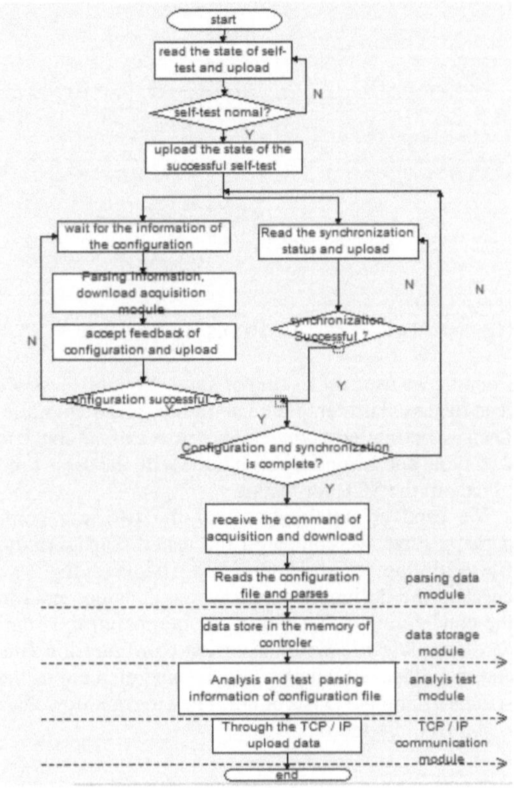

Figure 5. Analysis test module flowchart.

from the board to the host computer. After completing uploading the synchronization states and it will wait for the configuration messages.

Analysis Module: the analysis module works while the issued acquisition command is received. It will read, process, and parse the uploaded data from the board to make a good preparation for further analysis for the configuration file.

Data storage module: the information of the parsed file is stored in the memory of the controller, while it also stores the standard templates to include the basic data types as is specified in the literature [8]: 15 kinds of DA Type, 55 kinds of DO Type and 19 kinds of Enum Type.

Analysis of test modules: It carries out the useful data from the stored configuration file of the store module. The program flowchart is shown in Figure 5.

The analysis test module is the core part of the tester. The IEC61850 configuration file is based on Extensible Mark-up Language (XML1.0) of a Substation Configuration Description Language (SCL) as a description language[9]–[10]. The SCL file defines the three object models. The conformance testing needs to test each node[11]–[13].

The implementation is done by operating the parsed nodes and comparing them with the information of the node element of the storage module. It will issue a false alarm when the requirements are not met. As an

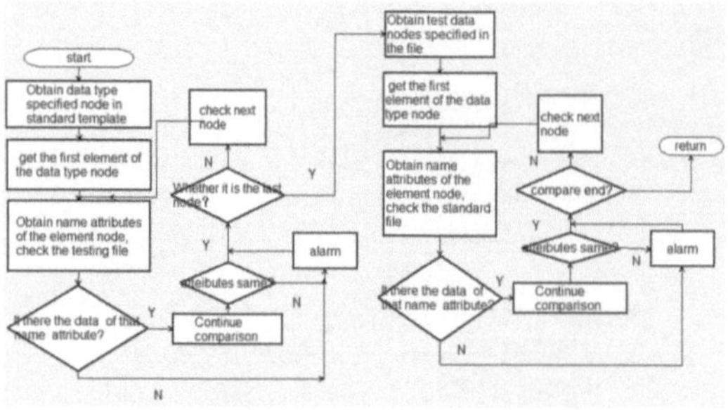

Figure 6. Data model analysis and test flowchart.

example we used the testing of data template. First, we read the standard template file in the storage module. Then, we extracted a collection of nodes of the specific data type, and compared each node with the same kind of data in the SCL for testing.

We used an approach called the two-way comparison. First, we compared the tested configuration file with the standard template file, and then we checked the defined element which does not exist in the configuration file. Then, we compared the standard template file with the tested configuration file, and then checked whether the defined element in the standard template exists or not. The program flowchart is shown in Figure 6.

3.4 *Display module*

A display interface of the test program is designed by a Lab VIEW. We recorded an error that appeared in the test, including an error type and index of the target in the configuration. After the program starts, the user can click the 'select a profile' button to select the configuration file. Meanwhile according to the test results of the analysis test module, the test results are displayed on the screen.

4 APPLICATION AND TEST RESULTS

This article implements a practical application of the conformance testing device. Moreover, it uses the device to test a configuration file of the items specified for need. It proves that the device can complete all the kinds of test items shown in Table 1, and that it is able to output the error type and show the wrong location correctly in the display module.

5 CONCLUSION

This article designs a test platform based on the Compact RIO of an NI company. The platform provides a flexible and convenient means for testing the conformance of a configuration file. According to the requirement of the IEC 61850 protocol, it is a device designed for conformance testing. It conforms to the structure of the configuration file while satisfying the application requirements. The conformance testing device for the IEC 61850 configuration file achieves the goal which completes conformance testing for the system's configuration file on smaller portable test equipment. It has a high practicality and will play a positive role in promoting the implementation of the conformance testing, and the IEC 61850 standard.

REFERENCES

[1] Gao X. Application of digital substation [M]. Beijing: China Electric Power Press, 2008.
[2] Chen J.C., Gao X.H., Pan, F. Analysis of interoperability test of IEC 61850[J]. Power System Protection and Control, 2009, 37(15): 121–123.
[3] He C., Zhang R. Experimental summary on IEC 61850 International Interoperability test [J]. Automation of Electric Power Systems, 2012, 36(2): 6–10.
[4] Cui H.K., Tang X.J., Liang Z.C. et al. Study on IEC 61850 conformance testing [J]. Automation of Electric Power Systems, 2006, 30(8): 80–83.
[5] Wang Z.Z., Sun D. Research on testing of IEC 61850 configuration files [J]. Power System Protection and Control, 2011, 39(6): 95–98.
[6] Pan Y.B., Shen D.Q., Yang C.S. Coherence test of IEC61850 based substation automation system [J]. Power Automation Equipment, 2009. 29(4): 143–146.
[7] IEC. IEC61850-10 Communication networks and systems in substations: part 10 basic communication structure for substations and feeder equipment; conformance testing [S]. [S.l.]; IEC.2005.
[8] IEC 61850. Communication networks and systems in substations-Part6: Configuration description language for communication in electrical substations related to IEDs[S].
[9] Qi Z., Du J., Zhang Z.X., et al. Research and implementation of IEC61850 SCL configuration tool [J]. Power System Protection and Control, 2009, 37(7): 76–81.
[10] Wu Y.C., Wang Z.P., Lv Y.S., et al. Application and parsing of substation configuration language [J]. Power System Protection and Control, 2009, 37(15): 38–41.
[11] Li Z.M., Zeng Y.J., Yuan D.F. General method to generate IED definition with IEC61850 SCD file for tele control system [J]. Electric Power Automation Equipment, 2010, 30(7): 117–120.
[12] Wang L.H., Ma J.H., Wang C.Q. et al. Research on application of substation configuration description language [J]. Power System Technology, 2006, 30(S): 93–96.
[13] Wang F.X., Fang C.E., Li W. Design of substation configuration tool based on IEC61850 standards [J]. Power System Protection and Control, 2010, 38(10): 106–109.

Future Communication Technology and Engineering – Chan (Ed.)
© 2015 Taylor & Francis Group, London, ISBN: 978-1-138-02777-0

A sustainable monitoring pulse wave ring detector

Y. Li, L. Yang, C.L. Lv, X.X. Zhao & L.X. Meng
College of Life Science and Bio-engineering, Beijing University of Technology, Beijing, China

ABSTRACT: Hypertension and cardiovascular pathology are the most important factors in the occurrence of cardiovascular diseases. A cardiovascular disease is closely linked to a pulse wave, so a simple and real-time portable pulse wave detector is necessary for users. In this paper, to collect a signal, a reflective infrared sensor is placed in a special ring. Through filtering and enlarging the pulse wave signal, it is transformed into a PC to be analysed and displayed. In this paper, according to the characteristics, we design a circuit, and then print and weld it. Finally, we debug circuit to make sure this detector is feasible. We found that the detector had a good dynamic response and a real-time waveform display. Therefore, it shows that this device has some advantages such as convenience to operate; has simple and intuitive results, long-time monitoring, and has real practical use in our daily lives. It can detect pulse wave fast and create conditions for home monitoring. Furthermore, the device shows that it has broad prospects for commercial use.

Keywords: pulse wave, portable device, reflective sensor.

1 INTRODUCTION

Chronic non-communicable disease is the main burden on global health, as 63% of deaths are caused by chronic diseases, and 48% of them are attributed to cardiovascular diseases. The morbidity and mortality rates of cardiovascular diseases in our country are still growing[1–2]. A pulse wave contains a wealth of physiological and pathological cardiovascular information, and the characteristics of a pulse wave forms are closely associated with cardiovascular functions[3]. Conformation about cardiovascular diseases in physiology and pathology which is extracted from a pulse wave is considered the basis of clinical detection and treatment, which has been brought to the attention of the Chinese and the fields of foreign medicine[4].

Although the human body manifests no obvious symptoms in the early stages of arteriosclerosis and hypertension, the parameters reflecting the cardiovascular, blood flow, blood pressure, and blood vessel elasticity, have begun to change. They first reflect the pulse wave change. Studies by Weber have shown that characteristic parameters of the pulse wave can be continuously and independently regarded as an important predictor of cardiovascular disease[5]. If the change of these parameters can be detected and analysed in time, we can diagnose cardiovascular diseases, such as arteriosclerosis and hypertension at an early stage, prevent, and treat them effectively, thus controlling the morbidity and mortality of cardiovascular diseases.[6]

Therefore, it is important to design a medical electronic instrument that can collect pulse wave signals for monitoring cardiovascular functions.

2 THE PRINCIPLE

2.1 *Device structure*

This device summarizes the application of a reflective infrared sensor as a basic function, and, in improving and optimizing this function, make it into a new type of portable pulse wave detector.

This device has the characteristics of working long hours and continuous monitoring without affecting people's daily operation. Therefore, this device is reduced portion, which can be omitted. Because we use PCs to read the values, there is no influence on the function of the physical location where the processing circuit is placed, so it can be isolated from the portable devices. The original data are measured and sent using a sensor, whilst a receptor receives data and transfers it to a PC. In conclusion, this device has the characteristics of having a small size and a high operability, and can realize continuous monitoring for a long time, record a series of pulse wave data for comparative analysis without affecting people's physiological activity and entertainment.

In this paper, that the design consists of a sustainable monitoring pulse wave detector ring made up of two

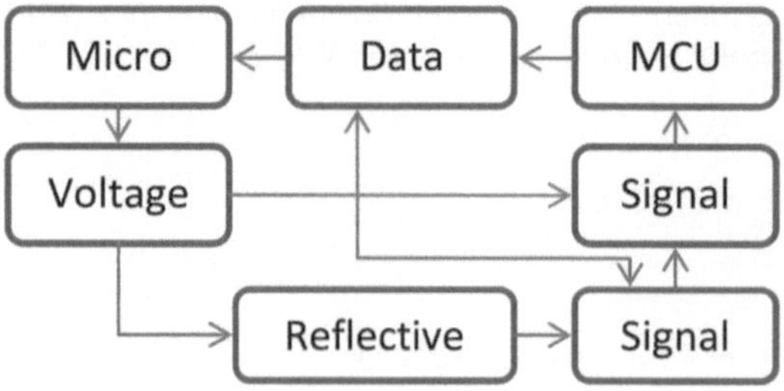

Figure 1. System structure diagram.

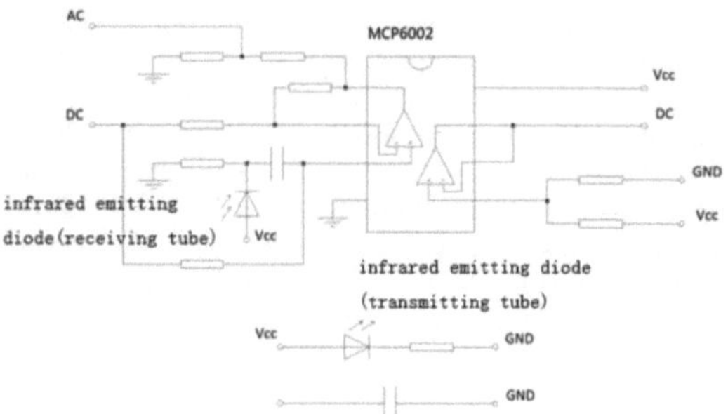

Figure 2. Circuit principle diagram.

parts: signal detection module and signal processing module. Signal detection module consists of a voltage conversion module, a pulse wave detection circuit, and a signal filter circuit. Its system structure diagram is shown in Figure 1.

2.2 *The main hardware circuit structure*

2.2.1 *Pulse wave detection module*

A pulse wave detection circuit mostly consists of a reflective infrared sensor which can convert visible light into electrical signals. An infrared emitting diode as a quick light source, and a photo triode of a 3DU5 series as a receiver tube are combined into an infrared launch-receive tube. Light is launched from a quick light source of infrared emitting diode, which goes through the skin and tissue. A portion of light is absorbed by the skin, muscles, blood, and the other part reflected back by the blood is converted into an electronic signal by photo triode. Moreover, in order to prevent an external light source interfering with the acquisition of a pulse wave, a filter, which is 1 cm in diameter is fitted outside of the sensor, in order to filter the external light. The circuit principle diagram is shown in Figure 2.

2.2.2 *Signal processing module*

A signal processing module mostly consists of a signal filter circuit, a signal amplification circuit, a MCU control module, a data conversion, and transmission module, and a micro USB. The collected signal was transmitted into filter circuit, and being amplified into MCU control module, and then transmit the signal into PC or memory card.

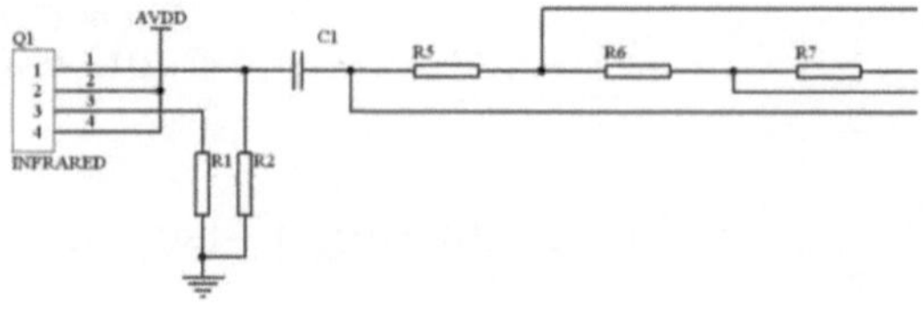

Figure 3. Pulse wave signal filter circuit.

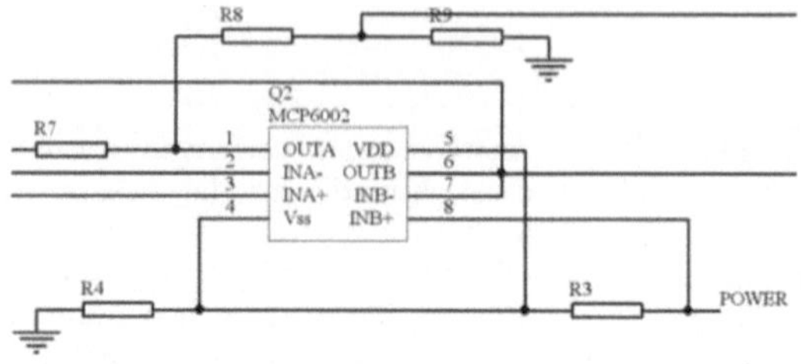

Figure 4. Pulse wave signal amplification circuit.

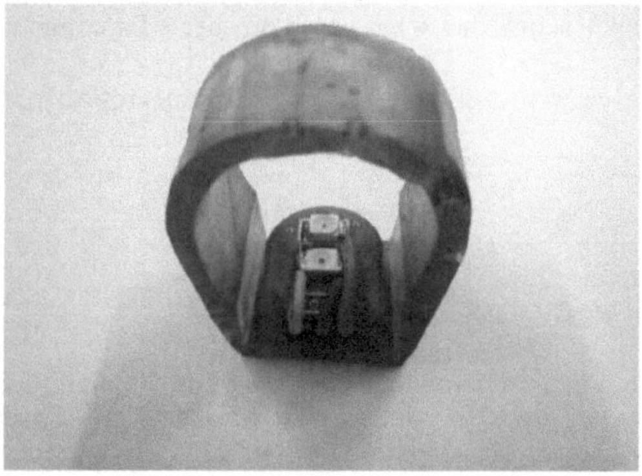

Figure 5. Final physical diagram.

2.2.3 *Signal filter circuit*

The interference picks up the pulse wave signal using photoelectric sensor, therefore we need to filter the signal collected. The frequency of a human pulse wave signal is low, and its spectrum distribution is between 0 and 20 Hz therefore using a low-pass filter can effectively eliminate the power line interference and other high-frequency noises. The input of the circuit of a pulse wave filter is connected with the output of a reflective infrared sensor, as described in this paper, which is used to filter the interference signal, and transfer it to the pulse wave signal amplification circuit. The diagram of the pulse wave signal filter is shown in Figure 3.

2.2.4 *Signal amplification circuit*

The signal collected by a sensor is weak, therefore we have designed a pulse wave signal amplification circuit module to amplify the signal after filtering it using a filter circuit. We have used a MCP6002 chip to amplify the signal and transmit it into an MCU control module. The circuit diagram of a pulse wave signal amplification is shown in Figure 4.

3 CONCLUSION

In this paper, we designed a sustainable monitoring ring by using a photo electric sensor to collect finger tip light-electricity pulse wave signals in an unconscious condition, and the collecting wave form can be displayed on PC.

This device has some advantages; for example, it is easy to operate, gives simple and intuitive results; has long-time monitoring ability, and is continuously useful in our daily lives. It can detect a pulse wave fast, and create conditions for home monitoring. In addition, it shows that it has a broad prospect for commercial use.

4 PROSPECT FORECAST

This device uses a PC power supply at present, it can be changed to an internal power supply in order not to interfere with the user's daily life. It can effect wireless transfer between a photo electric sensor and a processing module or between a processing module and a PC. In addition, the device can interact with a PC, and increase an alarm system for the PC feedback.

ACKNOWLEDGEMENTS

This work was supported by Beijing Municipal Commission of Education Foundation (PXM2013_014204_07_000069).

REFERENCES

[1] Schillaci, G., Verdecchia, P., Borgioni, C. et al. 1996, Association between persistent Pressure Overload and Ventricular Arrhythmias in Essential Hypertension [J]. Hypertension, 28(2): 284–289.

[2] Mitchell, G.F., Lacourciere, F., Ouellet, G.P., et al. 2003, Determinants of Elevated Pulse Pressure in Middle-aged and Older Subjects with Uncomplicated Systolic Hypertension: the Role of Proximal Aortic Diameter and the Aortic Pressure-flow Relationship [J]. Circulation, 108(13): 1592–1598.

[3] Westerhof, N., Bos, W.J., Stergiopulos, N., et al. 2007, Arterial pressure transfer characteristics: effects of travel [J]. American Journal of Physiol Heart Circulatory Physiol, 292: H800–H807.

[4] Wesseling, K.H. 1995, A century of noninvasive arterial-pressure measurement-from marey to penaz and finapres [J]. Homeostasis in health and disease, 36(2–3): 50–65.

[5] Weber, T., O'Rourke, M.F., Lassnig, E., et al. 2010, Pulse-waveform characteristics predict cardiovascular events and mortality in patients undergoing coronary angiography [J]. Journey of Hypertension. 28(4): 797–805.

[6] Tsuneo, T., Kenji, K., et al. 2004, Pulse wave velocity as an indicator of arteriosclerosis in hemodialysis patients [J]. Atherosclerosis, 176: 406–409.

Future Communication Technology and Engineering – Chan (Ed.)
© 2015 Taylor & Francis Group, London, ISBN: 978-1-138-02777-0

An angle of arrival estimation using the projection method

Solomon Wu, Daniel Kerr, Hung Lu, Nick Wilkins, Zekeriya Aliyazicioglu & H.K. Hwang
Electrical and Computer Engineering, California State Polytechnic University-Pomona, CA, USA

ABSTRACT: Angle of Arrival (AOA) estimation techniques such as Multiple Signal Classification (MUSIC), Estimate Signal Parameters via Rotational Invariant Technique (ESPRIT), Polynomial Root Intersection for Multiple Dimensional Estimation (PRIME), and Maximum Likelihood Estimation (MLE) require large number computations. They require either an eigen analysis or likelihood function computation. This paper proposes using the Projection Method (PM) to estimate the signal's AOA. The computation load of the PM is much lower than the above mentioned algorithms. Results of the PM are presented from a simulation study. The antennae used in this simulation study are Multiple Input Multiple Output (MIMO) and Uniformly-Spaced Linear Array (ULA) antennae. The performance is investigated under varied parameters, including varied array size, Pseudo Random (PN) sequence length, a number of snapshots, and Signal to Noise Ratio (SNR). The results of the projection method are compared to the MUSIC and MLE methods.

1 INTRODUCTION

The projection of any arbitrary vector $\mathbf{s}$ onto the space spanned by vector $\mathbf{x}_1, \mathbf{x}_2, \ldots, \mathbf{x}_K$ can be obtained by premultiplying the vector $\mathbf{s}$ by the matrix $\mathbf{P}$. The matrix $\mathbf{P}$ is the projection matrix and it is defined by the following equation [1]:

$$\mathbf{P} = \mathbf{A}\,(\mathbf{A}^H\mathbf{A})^{-1}\mathbf{A}^H \tag{1}$$

where matrix $\mathbf{A} = [\mathbf{x}_1, \mathbf{x}_2, \ldots, \mathbf{x}_K]$.

Premultiplying the vector $\mathbf{s}$ by matrix $\mathbf{P}_\perp = \mathbf{I} - \mathbf{P}$ yields the projection of vector $\mathbf{s}$ onto the space orthogonal to the space spanned by vectors $\mathbf{x}_1, \mathbf{x}_2, \ldots, \mathbf{x}_K$.

This projection has many engineering applications in adaptive filtering, a linear prediction and an antenna pattern design [1, 2]. This paper presents the use of the PM in order to identify a signal's AOA.

2 MIMO ARRAY ANTENNA

The use of array antennae to estimate a signal's AOA using advanced signal processing algorithms such as MUSIC, ESPRIT, PRIME and MLE usually requires extensive numerical computation [3–7]. The idea of PM is very similar to MUSIC algorithm. If there is only one target, the projection matrix $\mathbf{P}$ is constructed by the normalized outer product of the received vector $\mathbf{v}$. Premultiplying any vector by matrix $\mathbf{P}$ results in its projection along the received vector $\mathbf{v}$. Another projection matrix that projects any arbitrary vector $\mathbf{s}$ onto the space orthogonal to space spanned by $\mathbf{v}$ is $\mathbf{P}_\perp = \mathbf{I} - \mathbf{P}$. $\mathbf{P}_\perp$ is the orthogonal projection matrix. Premultiplying any vector by this matrix yields its projection onto the space orthogonal to the received vector $\mathbf{v}$.

The angular resolution of antennae is limited by the antenna Main Lobe Beam Width (MLBW), where the MLBW is proportional to the signal wavelength and inversely proportional to the antenna aperture size. A good angular resolution requires an antenna with a large aperture size. Mobile systems have physical size limitations. Thus, it is difficult to achieve a fine angular resolution. A MIMO antenna has the capability to create a virtual array with a larger aperture size. Thus, it is highly desirable to have a MIMO antenna to improve the signal's angular resolution.

Various processing algorithms estimate the AOA using Single-Input Multiple-Output (SIMO) array antennae and MIMO have been investigated by the research team at California State Polytechnic University, Pomona [5, 8–11, 14].

MIMO [12–14] antennae systems have shown promising improvements in angular resolution without increasing their physical size. With a MIMO antenna system, synthesized virtual arrays can effectively increase the aperture size. This paper presents the results of using a MIMO antenna and the PM to estimate a signal's AOA. The performance is investigated under varied parameters, including varied array sizes, the PN sequence length, a number of snapshots, and the SNR. The results of the PM and orthogonal projection method (OPM) are compared to MUSIC and MLE. Computation load of PM, and OPM are considerably lower than MUSIC and MLE. Although this simulation study is based on ULA, it can be easily extend to a two-dimensional array.

A ULA antenna consists of M uniformly-spaced antennae elements. The inter-element spacing is d. The 5 element ULA is shown in Figure 1.

The MIMO antenna has multiple transmitters which send orthogonal signals and multiple receivers which

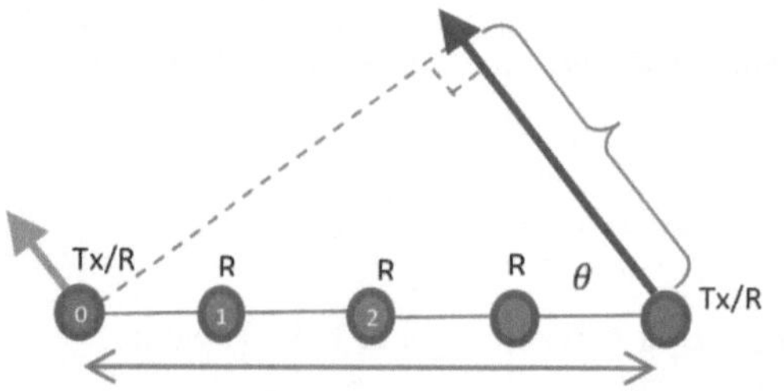

Figure 1. MIMO transmitter/receiver.

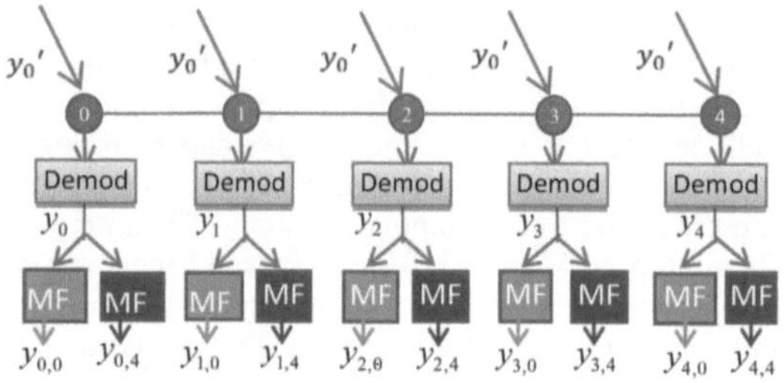

Figure 2. MIMO receiver.

collect the returned waveforms. The returned waveform is the transmitted waveform reflected from the target plus white noise. In this simulation study, a MIMO antenna is assumed to be a ULA with 5 elements. The 0th and 4th element transmit narrowband orthogonal signals, and all the elements serve as receivers.

Assume the transmitted signals from 0th and 4th transmitters are $s_0(n)$ and $s_4(n)$, where $s_0(n) = p_0(n) e^{j2\pi fn}$ and $s_4(n) = p_4(n)e^{j2\pi fn}$, where $p_0(n)$, $p_4(n)$ are orthogonal PN sequences and f is the carrier frequency. The signal received from the target at angle θ by the mth array element is

$$x_m(n) = \left[p_0(n) + p_4(n)e^{-j4\beta}\right]e^{-jm\beta}e^{j2\pi fn}$$
$$\text{for } m = 0, 1, \ldots, 4 \tag{2}$$

where $\beta = \frac{2\pi d}{\lambda}\cos\theta$ is the phase factor due to a relative propagation delay of the adjacent element.

The receiver block diagram is shown in Figure 2. After each antenna receives a reflected waveform, the baseband waveform is recovered by demodulators and matched filters that correspond to the PN sequences $p_0(n)$ and $p_4(n)$. The matched filter output $y_{i,j}(n)$ is the output of the ith element due to jth transmitted signal.

Using the received waveform of the 0th receiving antenna $y_0'(n)$ as the reference, the received waveform of the mth antenna $y_m'(n)$ is

$$y_m'(n) = \left[p_0(n)e^{-jm\beta} + p_4(n)e^{-j(m+4)\beta}\right]e^{j2\pi fn} + w_m'(n)$$
$$m = 0, 1, \ldots, 4 \tag{3}$$

where $w_m'(n)$ is the additive white Gaussian noise. The demodulated waveform of the mth antenna $y_m(n)$ is

$$y_m(n) = \left[p_0(n)e^{-jm\beta} + p_4(n)e^{-j(m+4)\beta}\right] + w_m(n)$$
$$m = 0, 1, \ldots, 4 \tag{4}$$

The output waveform at the first matched filter of each demodulator output $y_{m,0}(n)$ is

$$y_{m,0}(n) = \sum_{k=0}^{N-1} p_0^2(k)e^{-jm\beta} + \sum_{k=0}^{N-1} p_0(k)p_4(k)e^{-j(m+4)\beta} + \sum_{k=0}^{N-1} p_0(k)w_m(k) \tag{5}$$

Since $p_0(n)$ and $p_4(n)$ are orthogonal sequences, the second term of Equation (5) theoretically is zero. Equation (5) can be rewritten as:

$$y_{m,0}(n) = Ne^{-jm\beta} + w_{m,0} \tag{6}$$

where

$$w_{m,0} = \sum_{k=0}^{N-1} p_0(k)w_m(k)$$

Similarly,

$$y_{m,4}(n) = Ne^{-j(m+4)\beta} + w_{m,4} \tag{7}$$

where

$$w_{m,4} = \sum_{k=0}^{N-1} p_4(k)w_m(k)$$

Equations (6 and 7) show the average signal power is $N^2\sigma_s^2$ where σ_s^2 is the average signal power before the matched filter. The average noise power P_w is:

$$P_W = E\left[\sum_{l=0}^{N-1}\sum_{k=0}^{N-1} p_4(k)p_4(l)w_n(k)w_n^*(l)\right] = \sum_{l=0}^{N-1}\sum_{k=0}^{N-1} p_4(k)p_4(l)E\left[w_n(k)w_n^*(l)\right] = N\sigma_W^2 \tag{8}$$

The SNR at the output of the matched filter is improved by a factor of N, where N is the length of the PN sequence.

Define the virtual array received baseband waveform $v_m(n)$ as:

$$v_m(n) = \begin{cases} y_{m,0}(n) & m = 0,1,2,3 \\ \dfrac{y_{4,0}(n) + y_{0,4}(n)}{2} & m = 4 \\ y_{m-4,4}(n) & m = 5,6,7,8 \end{cases} \tag{9}$$

Received waveforms $v_m(n)$ for $m = 0, 1, \ldots, 8$ are the equivalent to a ULA antenna with 9 elements.

Thus, the length of ULA effectively increased from $4d$ to $8d$. Estimating the signal's AOA by processing the waveform $v_m(n)$ should result in a smaller estimation variance.

The research team at Cal Poly Pomona has investigated various methods to estimate the signal's AOA [6–10] using the array antenna. This paper presents the AOA estimation using the MIMO antenna system and a PM.

Define a scan vector

$$\mathbf{s} = [1, e^{-j2\pi(d\cos\theta)/\lambda}, e^{-j2\pi(2d\cos\theta)/\lambda}, \ldots, e^{-j2\pi(8d\cos\theta)/\lambda}]^T$$

scans over all possible angle θ. If this scan vector matches the signal's direction vector, its projection $|\mathbf{Ps}|$ should reach the peak value. Similarly, its orthogonal projection $|\mathbf{P}_\perp \mathbf{s}|$ should reach the minimum value. The procedures of PM and OPM are:

1. Normalize the received data vector by multiplying the received data vector with the conjugate phase of the first array element. The normalized received data vector at sample time k is

$$\mathbf{v}(k) = e^{-j\beta_0(k)} [v_0(k), v_1(k), \ldots, v_8(k)]^T \quad (10)$$

where $\beta_o(k)$ is the phase of the first array element.

2. Compute the average received data vector by the following equation

$$\mathbf{v} = \frac{1}{K}\sum_{k=0}^{K-1} \mathbf{v}(k). \quad (11)$$

3. Projection matrix $\mathbf{P}$ is defined as $\mathbf{P} = \mathbf{v}(\mathbf{v}^H\mathbf{v})^{-1}\mathbf{v}^H$ and orthogonal projection matrix is defined as

$$\mathbf{P}_\perp = \mathbf{I} - \mathbf{P}. \quad (12)$$

4. Compute the spectrum of PM by the following equation. $S_{PM}(\theta) = ||\mathbf{Ps}(\theta)||^2$
5. Compute the spectrum of OPM according to

$$\mathbf{S}_{OPM}(\theta) = \frac{1}{||\mathbf{P}_\perp \mathbf{s}(\theta)||^2} \quad (13)$$

6. Peaks of $S_{PM}(\theta)$ and $S_{OPM}(\theta)$ correspond to signals' AOA.

3 COMPUTER SIMULATION

In this simulation study, assume the signal's AOA is 100°, the SNR at the input of each antenna is −10 dB, and the length of PN sequence is 15. The received data vector is derived by averaging the input data vector over 15 snapshots. Figure 3 shows the spectrum of MLE, PM and the OPM for a five-element ULA antenna with inter-element space equal to half a wavelength.

This result indicates that the peak of the spectrum of three estimation methods is roughly at the signal's AOA, $\theta = 100°$. The orthogonal projection method has the narrowest peak.

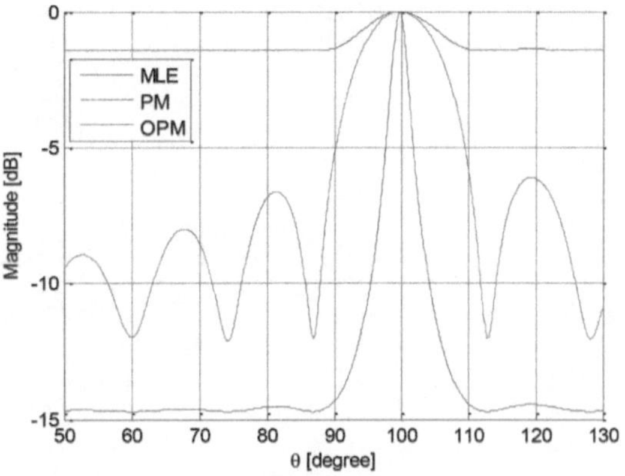

Figure 3. Spectrum of MLE, PM, and OPM.

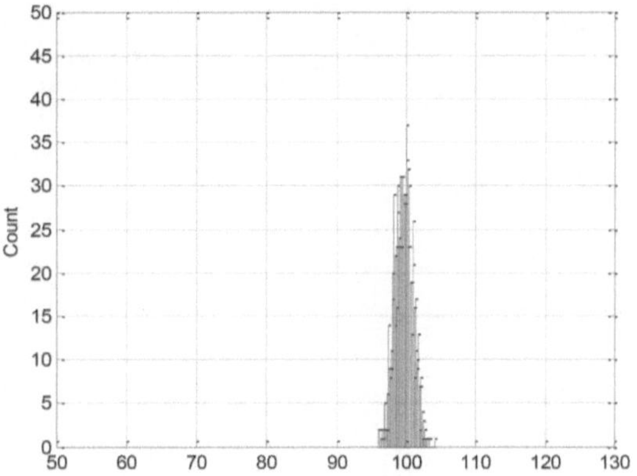

Figure 4. Histogram of PM.

The histogram of AOA estimation using the PM is shown in Figure 4. This result is based on 500 independent simulations. The SNR is −20 dB, length of PN sequence is 15 and vector $\mathbf{v}$ is computed by averaging over 15 snapshots. The histogram of OPM is very similar to this figure.

One of the performance measurements is the estimation variance. Figure 5 shows the AOA estimation variance of MLE, PM, and OPM for a different SNR. The SNR is the raw SNR at the input of each receiving array element. After averaging the vector $\mathbf{v}$ over 15 snapshots and processing the receiving match filter with length 15, the processed SNR is 23.5 dB better than the raw SNR. This Figure shows that for SNRs higher than −17 dB, all the processing algorithms are very close to the theoretical Cramer Rao bound (CRB). The performances of the PM, and OPM are actually better than the MLE for a low SNR region.

The processed SNR can be improved by using a longer PN sequence. Figure 6 shows the AOA estimation variance of OPM using the PN sequence length of 15 and 31. The performance improvement of using length 31 PN sequence is approximately 3 dB better than PN sequence of length 15.

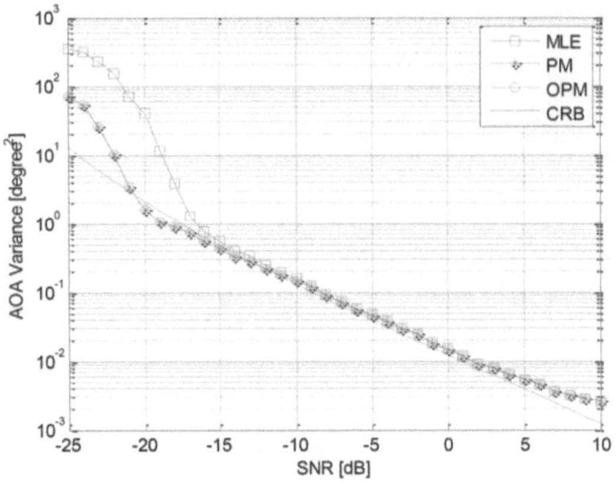

Figure 5. Estimation variance.

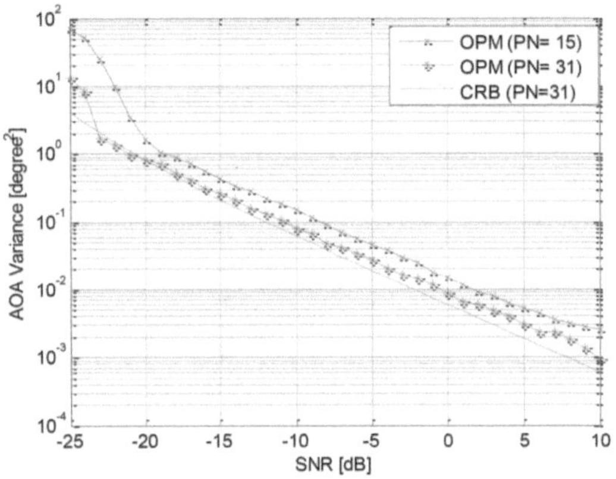

Figure 6. Estimation variance with PN length of 15 and 31.

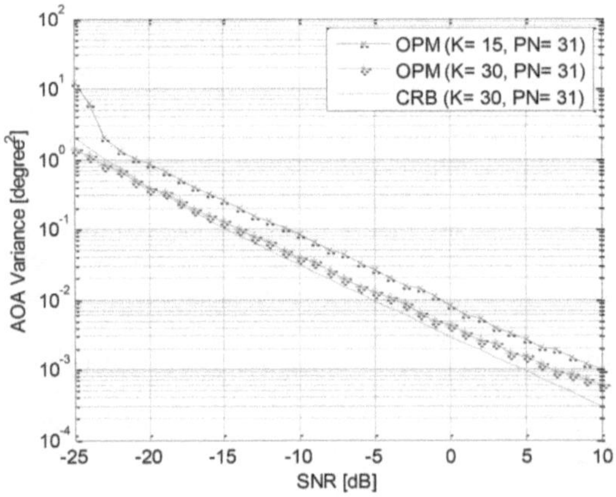

Figure 7. Estimation variance for 15 and 30 snapshots.

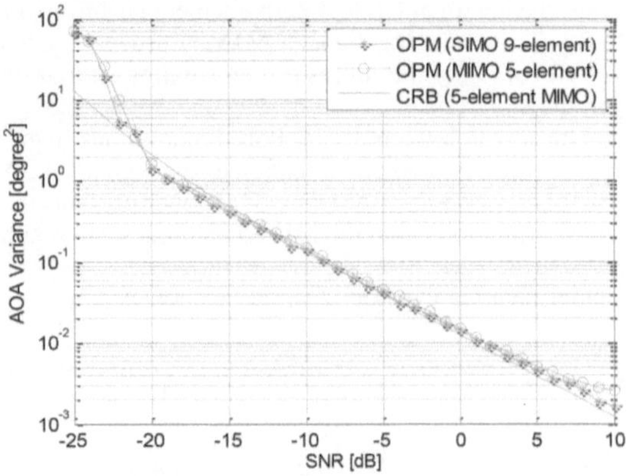

Figure 8. Performance comparison of 5 element MIMO and 9 element SIMO Array Antenna.

Table 1. Simulation computation time.

Algorithm	MUSIC	MLE	PM	OPM
Average Time	0.570	1.675	0.226	0.320

Increasing the snapshots improves the accuracy of the AOA estimation. Figure 7 shows the AOA estimation variance of OPM using snapshots equals 15 and 30. The performance improvement of using 30 snapshots is approximately 3 dB better than 15 snapshots. However, more snapshots require a longer time to collect data.

The virtual array of 5-element MIMO antennae have the same size as 9-element SIMO antennae, therefore they should have similar AOA performance. Figure 8 shows the AOA estimation variance for a 5-element MIMO antenna and a 9-element SIMO antenna. The received data vector **v** is based on averaging over 15 snapshots, and the PN sequence length is 15. The two curves are relatively identical. This result shows the enhanced resolution of a MIMO antenna without having to extend its physical size.

The PM and OPM require neither eigen-analysis nor the likelihood of function computation. Thus they are very attractive in real time application. Table 1 is the summary of computation time in our simulation study.

Table 1 shows there is a large factor of computation saving in AOA estimation using the PM or the OPM.

4 CONCLUSION

Several important conclusions from this simulation study are listed below:

- MIMO antennae significantly improve the AOA resolution while maintaining the same physical antenna aperture. Thus, MIMO may be highly desirable for use in mobile systems.
- Estimated AOA is obtained by the peak of the PM or OPM spectrum.
- Estimated angle error decreases exponentially with SNR.

- Signal processing gain can be improved by using a longer PN sequence. However, a much wider transmission bandwidth is required for a longer PN sequence.
- SNR can be enhanced by averaging the received data vector over multiple snapshots and performing match filtering to the demodulated waveform.
- The computation time of PM and OPM is only a fraction of the time requirement of MUSIC and MLE. There is no performance difference compared with MUSIC or MLE. At a low SNR environment, PM and OPM even provide less estimation variance.

REFERENCES

[1] Simon Haykin, Adaptive Filter Theory 5th Ed" Pearson, 2014.

[2] Bhattcharyya, A. K., "Projection Matrix Method for Shaped Beam Synthesis in Phased Arrays and Reflectors," IEEE Transactions on Antennas and Propagation, March 2007.

[3] Amin G. Jaffer, "Maximum Likelihood Direction Finding of Stochastic Sources: A Separable Solution," IEEE ICASSP Conference, 1998.

[4] R.O. Schmidt, "Multiple Emitter Location and Signal Parameter Estimation," Proceedings of RADC Spectrum Estimation Workshop, Rome, NY, 1979.

[5] H.K. Hwang, et al., "Direction of Arrival Estimation using Innovative Signal Processing," International Signal Processing Conference, 2006.

[6] Hatke, Gary, and Keith Forsythe, "A Class of Polynomial Rooting Algorithms for Joint Azimuth/Elevation Estimation Using Multidimensional Arrays," 28th Asilomar Conference on Signals, Systems and Computers, p. 694–699, 1994.

[7] R. Roy, T. Kailath, "ESPRIT-estimation of signal parameters via rotational invariance techniques," IEEE Transactions on Acoustics, Speech and Signal Processing, 1989.

[8] M. Grice, J. Rodenkirch, A. Yakovlev, H.K. Hwang, Z. Aliyazicioglu, Anne Lee, "Direction of Arrival Estimation using Advanced Signal Processing," IEEE RAST Conference, 2007.

[9] H.K. Hwang, Z. Aliyazicioglu, M. Grice, A. Yakovlev, P. Lu, "Direction of Arrival Estimation using Polynomial Root Intersection for Multidimensional Estimation (PRIME)," IMECS Conference, 2008.

[10] J. Bermudez, R.C. Chin, P. Davoodian, A.T. Yin Lok, Z. Aliyazicioglu, H.K. Hwang, "Simulation Study on DOA Estimation using ESPRIT Algorithm," WCECS Conference 2009.

[11] A.T. Yin Lok, Z. Aliyazicioglu, H.K. Hwang, "Performance Analysis of Multiple Signal Detection using ESPRIT Algorithm," IEEE International Signal Processing Conference, 2010.

[12] Jian Li, Petre Stoica, MIMO Radar Signal Processing, Wiley, 2009.

[13] Chun-Yang Chen, Signal Processing Algorithms for MIMO Radar Signal Processing Algorithms for MIMO Radar, Cal Tech Thesis, 2009.

[14] Hung Lu, Nick Wilkins, Solomon Wu, Daniel Kerr, Zekeryia Aliyazicioglu, H.K. Hwang, "AOA Estimation using Maximum Likelihood Method," International Conference on Signal Processing, Communication and Networking, Brazil, February, 2014.

Future Communication Technology and Engineering – Chan (Ed.)
© 2015 Taylor & Francis Group, London, ISBN: 978-1-138-02777-0

Assessing and evaluating Arabic morphological analyzers and generators

T. Elghazaly & A.M. Maabid
Department of Computer and Information Sciences, Institute of Statistical Studies and Research, Cairo University, Egypt

ABSTRACT: Morphology analysis is an essential part in many applications of natural language processing, NLP. Applications of NLP include Machine Translation, MT., Information Retrieval, IR., etc. While natural language processing applications are based on the Morphology part, it has to meet some criteria in order to satisfy the required functionality. In order to measure the accuracy of Arabic morphological system, it has to be evaluated regarding the input words and their output according to assessment criteria to study and analyze its strength and weakness, trying to find an Arabic morphological analyzer free from all mistakes. Many stemming algorithms and systems have been developed to solve the problem of Arabic morphology; researchers have proposed their own gold standard, testing methodology and accuracy measurements to test and compute the accuracy of their algorithms. Therefore, we cannot make comparisons between these algorithms. In this paper, we have developed precise assessment criteria for Arabic morphological analyzers, which is adapted to measure Arabic Morphological Analyzers with some features intended for integration with lager applications in natural language processing.

1 INTRODUCTION

Morphology in linguistics concerns with the study of the structure of words (Kiraz 2004). In other words, morphology is simply a term for that branch of linguistics concerned with the forms words take in their different uses and constructions (Beesley 1998a).

Arabic is one of the languages having the characteristics that from one root the derivational and inflectional systems are able to produce a large number of words (lexical forms) having specific patterns and semantics for each of them (Buckwalter 2002). The root is a semantic abstraction consisting of two, three, or (less commonly) four consonants from which words are derived through the superimposition of templatic patterns (Watson 2007). Unfortunately if understanding is considered, undiacritized words may make problems of meaning, where many words can have more than one meaning when they appear in undiacritized text; these different meanings rise problems of ambiguity (Mohammed 2006).

In Arabic, like other Semitic languages, word surface forms may include affixes, concatenated to inflected stems. In nouns, prefixes include conjunctions ("و" "and", ف "and, so"), prepositions ("ب" "by, with", "ك" "like, such as", "ل" "for, to") and a determiner, and suffixes include possessive pronouns. Verbal affixes include conjunction prefixes and negation, and suffixes include object pronouns. Either object or possessive pronouns can be captured by an indicator function for its presence or absence, as well as by the features that indicate their person, number and gender (Darwish 2002). A large number of surface inflected forms can be generated by the combination of these features, making the morphological generation of these languages a non-trivial task (Soudi, Cavalli-Sforza et al. 2001).

Natural languages processing and analysis improved substantially in recent years due to applying data intensive computational techniques (Soudi, Bosch et al. 2007). However, state of the art approaches are essentially language specific stemmer (Morphology), considering every surface word in the language (Shaalan and Rafea 1993). A shortcoming of this word-based analysis of the Arabic language is that it is sensitive to lack of data and information about Arabic words and it morphemes. This is an issue of importance as aligned corpora are expensive resources, which are not abundantly available for many language analysis levels. This is particularly problematic for morphologically rich languages, where word stems are realized in many different surface forms, which exacerbates the hindering in language analysis of higher level.

Morphological analysis can be performed by applying language specific rules. These may include a full-scale morphological analysis, or, when such resources are not available, simple heuristic rules, such as regarding the last few characters of a word as its morphological suffix. In this work, we will adapt some major assessment criteria for measuring advantage or drawback of any Arabic morphological system (Roark and Sproat 2007).

2 BACKGROUND AND PREVIOUS WORK

We believe that this is the first proposed work to sum up assessment criteria for Arabic morphological analyzers and Generators. Several researches talked about building powerful stemmers for the Arabic language with accuracies normally exceeding 90% but none of these stemmers offer the source code and/or the datasets used. It is therefore difficult to verify such claims or make a comparison between different stemmers without having the full description of the proposed method or the source code for the implementation of the algorithm (Al-Kabi, Al-Radaideh et al. 2011). In this section we review some efforts in this direction.

Mohammed N. Al-Kabi and Qasem A. Al-Radaideh (Al-Kabi, Al-Radaideh et al. 2011) proposed analysis of the accuracy and strength of four stemmers for the Arabic language using one metric for accuracy and four other metrics for strength as following:

- The first metric called empirical evaluation (EE), which represents a percentage of the correct roots produced by the stemmer under consideration.
- The mean number of words per conflation class (MWC) depends on the number of words processed.
- Index compression factor (ICF) represents the extent to which a collection of unique words is reduced (compressed) by stemming.
- Word change factor (WCF) represents the proportion of the words in a sample that have been changed in any way by the stemming process.
- The mean number of characters removed in forming stems (Average CR): Usually strong stemmers remove more characters from words to form stems.

Azze Al-din Al-Mazroui, et al. (Mazrui, Meziane et al. 2010) proposed a specification of morphological analysis system in the Arabic language. In this study the researcher outlined the general characteristic that has to be considered during processing and building Arabic morphological system in terms of input, analysis and output. The study doesn't provide any criteria or automation for comparing different systems.

Dassouki (Desouki 2009) proposed a tabulate items as mechanism for assessing morphological analyzer in terms of development of the system speed, input, output, integrating with other applications and capabilities of analyzing new and non-Arabic words. The study doesn't provide any criteria for those selected terms.

William B. Frakes and Christopher J. Fox (Frakes and Fox 2003) evaluated the strength and similarity among, four affix removal stemming algorithms. Strength and similarity were evaluated in different ways, including new metrics based on the Hamming distance measure. Data was collected on stemmer outputs for a list of 49,656 English words derived from the UNIX spelling dictionary and the Moby corpus. The study doesn't provide any criteria for those selected measurements and it is specific to English stemmers.

3 PROPOSED ASSESSMENT CRITERIA OF ARABIC MORPHOLOGICAL SYSTEMS

Assessing and evaluating Arabic morphological systems depends on the *input* words and resulted *output* (Mazrui, Meziane et al. 2010) according to a predefined criteria to measure and analyze given system in order to study its weakness and strength, trying to find an Arabic morphological analyzer free from all mistakes. Then we will apply these criteria on some of the existing available systems; these criticisms will not detract from its value and effectiveness.

3.1 *Input*

A very fundamental problem with software testing is that testing under all combinations of inputs and preconditions (initial state) is infeasible, even with a simple product. The input can be considered as bulk of text passed to the system in form of word or phrase fully or partially diacritized.

3.1.1 *The possibility of analyzing the modern standard texts*

Most western scholars distinguish two standard varieties of the Arabic language: the Classical Arabic (CA) of the Qur'an and early Islamic (7th to 9th centuries) literature, and Modern Standard Arabic (MSA), the standard language in use today (Mushira Eid 1993). The modern standard language is based on the Classical language. Most Arabs consider the two varieties to be two registers of one language, although the two registers can be described in Arabic as (MSA) and (CA) (Elgibali, Versteegh et al. 2009).

3.1.2 *The possibility of analyzing the common error words*

Common typing errors "common error words" are those words mistyped but are traditionally considered correct; typically a feminine ending character "ـة" written without dots "ـه", the dotless "ى" instead of "ي" and the letter "ا" without hamza instead of "أ"; for e example word "احمد" can be read and understood correctly as "أحمد" while the first one is linguistically mistyped (Eid, Cantarino et al. 1994).

3.1.3 *The possibility of analyzing new words (Neologisms)*

Neologisms are often created by combining existing words or by giving words new and unique suffixes or prefixes. Portmanteaux "حقائب السفر" are combined words those are sometimes used commonly. Neologisms also can be created through abbreviation or acronym, by intentionally rhyming with existing words or simply through playing with sounds.

Neologisms can become popular through memetics, by means of mass media, the Internet, and words of mouth, including academic discourse in many fields renowned for their use of distinctive jargon, and often become accepted parts of the language. Other times, however, they disappeared from common use just as readily as they appeared. Whether a neologism continues as part of the language depends on many factors, probably the most important of which is acceptance by the public. It is unusual, however, for a word to enter into common use if it does not resemble another word or words in an identifiable way.

When a word or phrase is no longer "new", it is no longer a neologism. It may take decades for Neologisms to become "old", however. Opinions differ on exactly how old a word must be to cease being considered as a neologism.

Neologisms analysis in morphological system measures the capability of processing the new Arabic words which can be added later to morphological systems' predefined knowledge base.

3.1.4 *Processing of Arabized and transliterated words*

Transliteration is a subset of hermeneutics. It is a form of translation, and is the practice of converting a text from one script into another. From an information-theoretical point of view, systematic transliteration is a mapping from one system of writing into another, word by word, or ideally letter by letter. Transliteration attempts to use a one-to-one correspondence and be exact, so that an informed reader should be able to reconstruct the original spelling of unknown transliterated words. Ideally, reverse transliteration is possible.

Transliteration is opposed to transcription, which specifically maps the sounds of one language to the best matching script of another language. Still, most systems of transliteration map the letters of the source script to letters pronounced similarly in the goal script, for some specific pairs of source and goal language. If the relations between letters and sounds are similar in both languages, a transliteration may be (almost) the same as a transcription. In practice, there are also some mixed transliteration/transcription systems that transliterate a part of the original script and transcribe the rest (Desouki 2009).

In Arabic transliteration is writing non-Arabic words by Arabic alphabet characters as 'فاكس' "Fax" in English and "انترنت" "Internet" In English.

3.1.5 *Processing of non-tripartite verbs*

Arabic verbs, as the verbs in other Semitic languages, are more complicated than those in most other languages. A verb in Arabic is based on a set of three or four consonants called a root (trilateral or quadrilateral according to the number of consonants).The root carries the basic meaning of the verb, e.g. "كتب" k-t-b "write", "قرأ" q-r-' "read", and "أكل" '-k-l "eat". Changes of the vowels between the consonants, along with prefixes or suffixes, specify grammatical functions such as person, gender, number, tense, mood, and voice.

Arabic words are divided into three types: noun, verb, and particle. Nouns and verbs are derived from a closed set of around 10,000 roots. The roots are commonly three or four letters and are rarely five letters. Arabic nouns and verbs are derived from roots by applying templates to the roots to generate stems and then introducing prefixes and suffixes (Darwish 2002).

Assessing and evaluating Arabic considering the system capability of analyze quadrilateral and quinqueliteral verbs like "طمأن" "Reassure" and all possible cases of their forms of transitivity and weakness (Mazrui, Meziane et al. 2010).

3.2 *Output*

Morphology output is all possible combination of affixes that produced a valid Arabic word, roots and patterns.

3.2.1 *Covering analysis of all input words*

- The system should cover all cases of analysis.
- Determining word types (pattern, root, stem and attached affixes) (Mazrui, Meziane et al. 2010).
- Analyzing the words in all domains of the language (Geographic, Historical, Religion, and Math).
- Considering syntactic case of input word (within phrase)

3.2.2 *Meet all possible cases for analysis*

The system has to assume that the input word is a verb, name and character so it has to determine the followings:

- Verb: it has to cover non-tripartite, quadrilateral, quinqueliteral with their forms of transitivity, augmentation, hollow...etc. (Watson 2007).
- Name: it has to cover names, infinitives, adjectives and adverbs.
- Particle: it has to cover prepositions, conjunctions, vowel, and vocative particles.

3.2.3 *Express grammatical function of the affixes*

Affixes are those characters attached to the stem. (prefix, suffix and infix) each has its own grammatical alternation of the stem attached.

3.2.4 *Ambiguity and overlapping of syntactic cases*

Many words in Arabic are homographic (Mohammed 2006): they have the same orthographic form, though the pronunciation is different. There are many recurrent factors that contributed to this problem. Among these factors are:

- Orthographic alternation operations (such as deletion and assimilation) frequently produce inflected forms those can be divided into two or more different lemmas.
- Some lemmas are different only in that one of them has a doubled sound which is not explicit in writing. Arabic Form I and Form II are different only in the aspect that Form II has the middle sound doubled.
- Many inflectional operations underlie a slight change in pronunciation without any explicit orthographical effect due to the lack of short vowels (diacritics).

- Some prefixes and suffixes can be homographic with each other. The prefix can indicate 3rd person feminine or 2nd person masculine.
- Prefixes and suffixes can accidentally produce a form that is homographic with another full form word. This is termed as "coincidental identity".
- Similarly, clitics can accidentally produce a form that is homographic with another full word.
- There are also the usual homographs of uninflected words with/without the same pronunciation, which have different meanings and usually have different POS's.

That means determining the lack of morphological knowledge of the word analyst; in case of partially diacritized or non-diacritized words, the ambiguity problem may appear, so the better way is to determine all possible cases of the input word; as an example the work "رب" may be either "رَبٌّ" (God) or "رُبَّ" (maybe).

3.2.5 *Identifying the root of the word and determining all possible roots for the analyzed word*

Right root identification of the input word, and with all generated words, the system has to be capable to determine their roots and patterns.

3.2.6 *Grammatical errors and misspellings in the context of the expression of results of the analysis*

The output representation of the system has to be error free in terms of expression and representation of output result.

3.2.7 *Cover all possible cases of syntactic word analyst*

The system also should represent and explain the analysis result of each analyzed word and the generated words.

3.2.8 *Consistency between analyzed word and its patterns*

The system should produce correct and consistent patterns for the analyzed and generated words.

3.2.9 *The result has to be coming from Arabic dictionary*

While processing the word the system should combine the Arabic morphological rules with its knowledge base in order to reflect a better analysis and generation which measures the trust of morphological analysis result.

3.3 *System architecture and design*

3.3.1 *Percentage of non-reliance on predefined knowledgebase of affixes, roots and patterns*

An affix is a morpheme that is attached to a word stem and then a new word forms. Affixes may be derivational, like English -ness and pre-, or inflectional, like English plural -s and past tense -ed. They are bound morphemes by definition; prefixes and suffixes may be separable affixes. Affixation is, thus, the linguistic process which speakers use to form different words by adding morphemes (affixation) in the beginning (prefixation), in the middle (infixation) or in the end (suffix) of words.

As in other Semitic languages, Arabic has a complex and unusual morphology (i.e. method of constructing words from a basic root). Arabic has a nonconcatenative "root-and-pattern" morphology. A root consists of a set of bare consonants (usually three), which are fitted into a discontinuous pattern in order to form words. For example, the word for "I wrote" is constructed by combining the root k-t-b "write" with the pattern -a-a-tu "I X'd" to form katabtu "I wrote". Other verbs meaning will typically have the same pattern but with different consonants, "قرأت" "I read", "اكلت" "I ate", "ذهبت" "I went", although other patterns are possible (e.g. "شربت" "I drank", "قلت" "I said", "تكلمت" "I spoke", where the sub pattern used to signal the past tense may change but the suffix -tu is always used).

3.3.2 *Percentage of non-reliance on common words (Stop List)*

In Natural Language Processing (NLP), stop words are words those are filtered out prior to, or after, processing of natural language data. Any group of words can be chosen as the stop words for a given purpose. Common words (stop word) are the words that are frequently used in Arabic text with the same meaning such as day names, month names, numbers names, adverbs, etc.

3.3.3 *Processing speed*

In software engineering, in general performance testing is testing performed to determine how a system performs in terms of responsiveness and stability under a particular workload condition. It can also serve to measure, investigate, validate and verify other quality attributes of the system, such as scalability, reliability and resource usage.

Performance testing is a subset of performance engineering, an emerging computer science practice which strives to build performance into the implementation, design and architecture of a system.

The processing speed can be measured by how many words processed per second.

3.3.4 *Ease of use and integration with larger applications*

In engineering, system integration is bringing the component subsystems together into one system and ensuring that the subsystems function together as one system. In information technology, systems integration is the process of linking different computing systems and software applications physically or functionally, in order to act as one coordinated whole.

- How much the system is capable for use and what are the prerequisites for the system to run.
- The ability to integrate the system within larger applications.

- The ability of modifying some of the system behavior of output or even input procedures and functions. (Customization).
- The ability to add inputs to the system knowledge base.

3.3.5 *Availability and documentation*

Software documentation or source code documentation is written text which accompanies computer software. It either explains how it operates or how to use it, and it may mean different things to people in different roles.

In terms of Arabic morphological system, it measures the availability of the system and its algorithms for newcomers and researchers considering the cost of commercial systems.

3.3.6 *User interface (English – Arabic)*

The user interface, in the industrial design field of human–machine interaction, is the space where interaction between humans and machines occurs. The goal of interaction between a human and a machine at the user interface is effective operation and control of the machine, and feedback from the machine which aids the operators in making operational decisions. Examples of this broad concept of user interfaces include the interactive aspects of computer operating systems, hand tools, heavy machinery operator controls, and process controls. The design considerations are applicable when creating user interfaces are related to or involve such disciplines as ergonomics and psychology.

A user interface is the system by which people (users) interact with a machine. The user interface includes hardware (physical) and software (logical) components. User interfaces exist for various systems, and provide a means of:

- Input, which allows the users to manipulate a system;
- Output, which allows the system to indicate the effects of the users' manipulation.

Generally, the goal of human-machine interaction engineering is to produce a user interface which makes it easy, efficient, and enjoyable to operate a machine in the way which produces the desired result. This generally means that the operator needs to provide minimal input to achieve the desired output, and also that the machine minimizes undesired outputs to the human.

There are two major factors in judging morphological system interface as follows:

- The interface language of system itself;
- The language used to represent the output of the system in case of analysis or generation.

3.3.7 *Encoding and word representation*

Identify the character encoding used in the system itself for processing and representing the data. As Arabic letters need to be represented in Unicode set; some systems need to transliterate the input as a preparation for processing step and then revert the transliterated results into Arabic to match user input and user interface.

3.4 *The procedure of applying the proposed assessment criteria*

Assessments are carried out by executing some of the available Arabic morphological analyzers on a randomly selected Arabic political news article, an Arabic Sport News article "from Al-Ahram newsletter" and the Chapter number 36 of the Holy Qur'an "سورة يس Surah Yassin" with total of 11000 distinct words. We then manually extracted the roots from the test documents' words to compare results from different analyzers, thus creating our baseline test set. Roots extracted were then checked manually in an Arabic dictionary. Voting weights are assigned to each assessment item (assigned Score) in order to accurately make comparisons between those algorithms. Each assessment item has to be applied and calculated as per the result of applying the analysis to the sample input words. Table 1 shows that assessment items where the voting mark of each individual item is assigned 100 points. Here is the step by step procedure of executing the assessment criteria:

- Manually extracting the roots from the test documents' words.
- Assigning voting mark for each assessment item.
- Manually checking the extracted roots against Arabic dictionary.
- Applying each assessment item separately on each of Arabic Morphological Analyzer.
- For the output results, checking them manually against Arabic dictionary.

Finally, the assessment factors can be separately applied on each of Arabic Morphological Analyzer where all factors can be assigned score with a maximum value of 100 marks. Each assessment factor will be applied and calculated as per Analyzer result of applying the analysis of the sample document words.

4 EXPERIMENTS AND RESULTS

Experiments are done by executing some of existing and available Arabic morphological systems on a randomly selected contemporary Arabic political news article, Arabic Sport News article "from Al-Ahram newsletter" and the first 15 verses of chapter number 36 of the Holy Qur'an "Souraht Yassin". Each test document contains domain specific words and represents contemporary and standard Arabic. The test documents contain 540 distinct tokens. We manually extracted the roots from the test documents' words to compare results for each stemming algorithm. Roots extracted have been checked against Arabic dictionary.

The analysis also shows that function words such as "في" "fi", "من" "min", "بين" "bian" are most frequent words in any Arabic text. In the other hand, nonfunctional words with high frequency such as "الإفريقية"

Table 1. Assigned scores of the assessment factors.

Cat.	No.	Assessment Criteria	Score %
Input	1	The possibility of analyzing the standard and modern texts	100
	2	The possibility of analyzing the common error words	100
	3	The possibility of analyzing new words	100
	4	Processing of Arabized and transliterated words	100
	5	Processing of non-tripartite verbs.	100
Output	6	Covering analysis of all input words	100
	7	Meet all possible cases for analysis	100
	8	Express grammatical function of the affixes	100
	9	Ambiguity and Overlapping of syntactic cases	100
	10	Identifying the root of the word and determining all possible roots	100
	11	Grammatical errors and misspellings in the context of the results of the analysis	100
	12	Cover all possible cases of syntactic word analyst	100
	13	Consistency between analyzed word and its patterns	100
	14	The result has to be coming from Arabic dictionary	100
System Architecture and design	15	Percentage of non-reliance on predefined knowledgebase of affixes	100
	16	Percentage of non-reliance on common words	100
	17	Processing Speed	100
	18	Ease of use and integration with larger applications	100
	19	Availability, documentation and customization	100
	20	User Interface (English – Arabic)	100
	21	Encoding and word representation	100
Sum			2200

"al-afiriqiah", "القمة" "al-Qemah" and other words out of 30 most frequent tokens as shown in table 1 gives a general idea about the main topic of the article.

Simple tokenization is applied for the text of the gold standard documents and can be used to test any algorithm smoothly and correctly.

4.1 *Conclusion and future research*

The proposed assessment criteria are adapted to measure Arabic Morphological Analyzers with some features intended for integration with lager applications in natural language processing. Many other criteria can be added to the proposed items and may vary in weight and phase of testing; similar to the source code related metrics used for measuring the system as a product.

Table 2. Assessment result.

	Morphology System			
Factor	*Al-Khalil*	*Sarf*	*AMA*	*Khoja*
1	75	NA	80	50
2	85	NA	90	20
3	30	NA	20	0
4	10	NA	5	0
5	90	NA	85	80
6	75	NA	80	70
7	87	NA	85	0
8	92	NA	80	0
9	90	NA	35	30
10	85	NA	95	30
11	85	NA	98	90
12	45	NA	40	0
13	80	NA	95	0
14	86	NA	97	80
15	0	0	0	0
16	0	0	0	0
17	35	0	0	30
18	60	60	30	60
19	70	85	0	70
20	50	50	50	50
21	50	50	10	10
Total	1280	245	1075	670

The stemming algorithms involved in the experiments agreed and generated analysis for simple roots that do not require detailed analysis. So, more detailed analysis and enhancements are recommended as future work.

Most stemming algorithms are designed for information retrieval systems where accuracy of the stemmers is not important issue (Sawalha and Atwell 2008). On the other hand, accuracy is vital for natural language processing. The accuracy rates show that the best algorithm failed to achieve accuracy rate of more than 65%. It proves that more research is required.

REFERENCES

Al-Kabi, M. N., Q. A. Al-Radaideh, et al. (2011). "Benchmarking and assessing the performance of Arabic stemmers." Journal of Information Science 37 (111).

Beesley, K. R. (1998a). Arabic Morphological Analysis on the Internet. 6th International Conference and Exhibition on Multi-lingual Computing, Cambridge.

Buckwalter, T. (2002) Buckwalter Arabic Morphological Analyzer Version 1.0. Linguistic Data Consortium.

Darwish, K. (2002). Building a Shallow Morphological Analyzer in One Day. 40th Annual Meeting of the Association for Computational Linguistics (ACL-02), Philadelphia, PA, USA.

Desouki, M. S. (2009). Mechanism for assessing morphological analyzer (In Arabic). Meeting of experts in computational morphological analyzers for the Arabic language. Damascus, The Arab League Educational, Cultural and Scientific Organisation (ALECSO) – King Abdulaziz City for Science and Technology.

Eid, M., V. Cantarino, et al. (1994). Perspectives on Arabic Linguistics VI: Papers from the Sixth Annual Symposium on Arabic Linguistics, Volume 4, John Benjamins Publishing Company.

Elgibali, A., K. Versteegh, et al. (2009). Encyclopedia of Arabic Language and Linguistics, Brill Academic Pub.

Frakes, W. B. and C. J. Fox (2003). Strength and Similarity of Affix Removal Stemming Algorithms. Proceedings of the Annual Conference on Research and Development in Information Retrieval, ACM SIGIR Forum

Kiraz, G. A. (2004). Computational Nonlinear Morphology; *With Emphasis on SemiticLanguages*. The Edinburgh Building, Cambridge CB2 2RU, UK, The press syndicate of the University of Cambridge, the Pitt building, Trumpington street, Cambridge, United Kingdom.

Mazrui, A., A. Meziane, et al. (2010). Morphological analysis system specifications. Meeting of experts in computational morphological analyzers for the Arabic language. Damascus.

Mohammed, A. A. (2006). An Ambiguity-Controlled Morphological Analyzer for Modern Standard Arabic Modelling Finite State Networks, School of Informatics, The University of Manchester.

Mushira Eid, C. H. (1993). Perspectives on Arabic Linguistics V: Papers from the Fifth Annual Symposium on Arabic Linguistics, Volume 5, John Benjamins Publishing Company.

Roark, B. and R. Sproat (2007). Computational Approaches to Morphology and Syntax. United States, Oxford University Press, New York.

Sawalha, M. and E. Atwell (2008). Comparative Evaluation of Arabic Language Morphological Analysers and Stemmers. COLING 2008 22nd International Conference on Comptational Linguistics, Manchester, Coling 2008 Organizing Committee.

Shaalan, K. F. and A. A. Rafea (1993). "Lexical Analysis of Inflected Arabic Words using Exhaustive Search of an Augmented Transition Network." Software Practice and Experience 23 (6).

Soudi, A., A. v. d. Bosch, et al. (2007). Arabic Computational Morphology; Knowledge-based and Empirical Methods. P.O. Box 17, 3300 AA Dordrecht, The Netherlands, Springer.

Soudi, A., V. Cavalli-Sforza, et al. (2001). A Computational Lexeme-Based Treatment of Arabic Morphology. Arabic Natural Language Processing Workshop, Conference of the Association for Computational Linguistics (ACL 2001), Toulouse, France.

Watson, J. C. E. (2007). The Phonology and Morphology of Arabic. New York, United States, Oxford University Press.

Future Communication Technology and Engineering – Chan (Ed.)
© 2015 Taylor & Francis Group, London, ISBN: 978-1-138-02777-0

A comparative analysis of free and open source tools for electronic evaluations in sociology

M. Joukl & L. Hosák
Faculty of Arts, University of Hradec Králové, Czech Republic

ABSTRACT: Currently, the most often carried out evaluation in schools is based on paper questionnaires. Acquisition of these data and their further processing is laborious and time-consuming, even if the process can be partially automated. It was also necessary that respondents were physically present during the implementation of the evaluations. Therefore the obtained data can, in addition, lose anonymity. Electronic evaluation can solve many problems and eliminate some problems in collecting and processing survey data. This paper presents selected options for open source and free tools which can be used for electronic evaluation surveys in schools. It points to the possible real use at easy acquisition and fast data evaluation. This paper is based on the project of specific research of the University of Hradec Králové called "Experimental verification of study evaluation in correspondence with current trends and technologies in social research".

Keywords: LimeSurvey, Google Forms, evaluation, data processing, open source software, free software

1 INTRODUCTION

Current commonly used methods of conducting evaluations in schools may not always be appropriate. Evaluations are usually conducted on paper. Students get standardized paper questionnaires from a teacher and then are asked to manually rate a lesson that they have just had and the teacher's work. Ideally, those sheets are then sealed in an envelope and submitted to a third party who evaluates the information questionnaires. This work is time-consuming, and it is not always possible to guarantee absolute anonymity, though it is necessary. Educators who create evaluations themselves can get valuable information, but their ability to deal with them further, are limited (Gošová, 2011). Another problem we may encounter is the low return of evaluation forms. Not all students are always attending the classes and thus the questionnaire investigation may not be objective. In addition, teachers often forget that they have to distribute those and pick them up again. Finally, the big disadvantage of paper evaluation questionnaires is that it is not always possible to read (poor handwriting, incorrect or ambiguous completion). Their final processing involves an evaluation which depends on other people. Students can rarely learn what effects their ratings have on teachers upon completing the course. The publication of the results can be evaluated in the case of paper evaluations but it is a complicated process. Whilst it is possible for paper questionnaires to automatically extract the entire solution, it is costly and time-consuming (Myška, Celá & Rybenská, 2014). In doing so, the solution may be simple, as paper evaluation forms should be put aside, and electronic forms should be used instead. In this case it is necessary to consider several factors. The used instrument should be affordable, user friendly, and also safe to use.

The text aims to compare the possibilities of using open source software Lime Survey and a freely available tool, Google Forms, as a possible replacement of paper-based evaluation forms for use in an electronic survey. Students at the University of Hradec Králové, especially sociology students, were approached as survey objects. The presented results and findings are based on the research, which was carried out at the Faculty of Arts, University of Hradec Králové. There were fifty respondents interviewed, and the first preliminary research was conducted in the form of an interview investigation. Questions were focused on the possible use of students' own smart technology devices when implementing pan electronic evaluation. Further research tested the possibility of using appropriate technologies in electronic evaluations, as is partly described in this article. The completed research was a qualitative research focused on the issue of electronic evaluations, and their strengths/weaknesses from the perspective of students.

2 THE ELECTRONIC EVALUATION

2.1 *The strengths and weaknesses of the electronic evaluation*

Electronic data collecting and processing is faster and more convenient than filling out paper forms. One

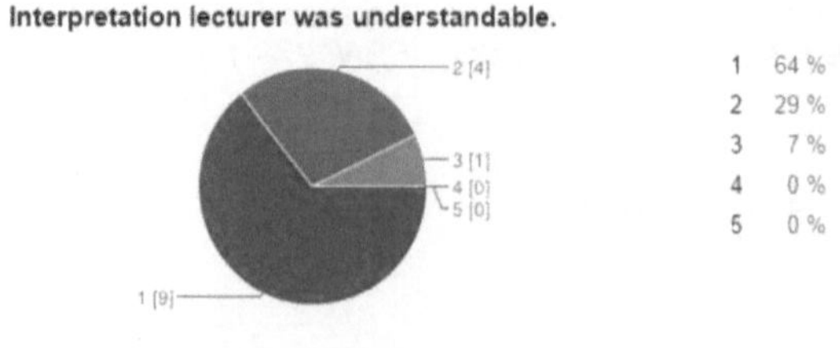

Figure 1. Example of automatically generated graph of the evaluation questionnaire being made by Google Forms.

of the biggest advantages is its undeniable speed and anonymity. Each student, at the start of the assessment, for example, can get an electronic questionnaire through the email link, which he/she will be able to immediately fill in. The questionnaire can be very elaborate or take identical form that is commonly used in schools or similar institutions. The questionnaire is completely anonymous and it is impossible to see who filled it in, unless the people taking part chose to voluntarily show their names. When evaluating the questionnaires, it may give a teacher immediate access not only to the actual data, but in many cases, to the graphs which show the responses (Figure 1). Many students can also appreciate that it is possible to fill in an evaluation questionnaire not only on the computer. Lime Survey, and Google Forms can support display on mobile devices (tablet, and smartphone), but can also use different operating systems. Compatibility testing including evaluation tools is made both for mobile devices and for tablets. We selected individual devices according to various diagonals, so the operating system, which uses the facilities. We tested the following software smart technologies and used the web browser Google Chrome or Safari.

- Samsung Galaxy Note (4,8″, operating system Android 4.1.2),
- iPhone 4S (3,5″, operating system iOS 5.0),
- Lenovo IdeaTab (7″, operating system Android 4.1.2),
- Acer Iconia Tab W3 (8″, operating system Windows 8),
- Google Nexus 7 (7″, operating system Android 4.4.3).

Of those smart technologies mentioned, investigated open source and free software work without any problems. Another advantage of using electronic evaluations is the fact that individual users can access electronic forms anywhere, if they have internet access. It is for the educator to send a link to enable the student to fill out the form, via e-mail. Finally, the student can still fill the form at home.

However, the use of electronic evaluation has some weaknesses: for example, if the technology fails, it is not possible to create an electronic evaluation; also if the student does not have internet access or the electric current fails when tested in the class, the evaluation can't be effected. Similarly, it is necessary to rely on the fact that he must chose the appropriate software. In the best cases, with open source or free licenses, it

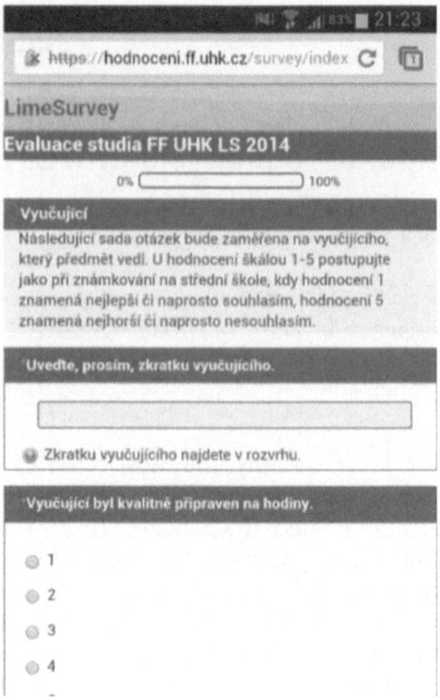

Figure 2. A questionnaire in the program LimeSurvey. The questionnaire is displayed on the smartphone Samsung Galaxy Note. The display is easy to read and operate.

ensures that it will be free of any charge. It also allow the student to make any adjustment, at least partially, as necessary. The use of evaluation instruments should be easy for users, since their deployment by the school must not be forgetting the training in such a system. It takes time, and it is also necessary to motivate individual users, who want to learn how the system works, to use it.

2.2 *LimeSurvey*

The first of the test systems that can be used to create evaluation and assessment questionnaires is open source LimeSurvey. It is a program that can create and process questionnaires (Figure 2). Those can be created in any number, restrictions are almost nonexistent and hundreds of respondents can be working simultaneously.

The questionnaires can then be converted to formats such as PDF, CSV or MS Excel. In addition, there are also modules for integration with the CMS systems like Drupal or WordPress, so it is possible not only to create questionnaires and the process, but also to distribute or export them to the target group. The data (statistics) can then export the file into the SPSS commands, allowing them to continue working on other statistical software (Šín, 2013).

When creating questions, it is necessary to categorize them into different groups, which provides a better overview. For each question also determines its code, the text itself, and possibly help to the question. Mandatory or optional questions can also be set. LimeSurvey offers many advanced features that offers

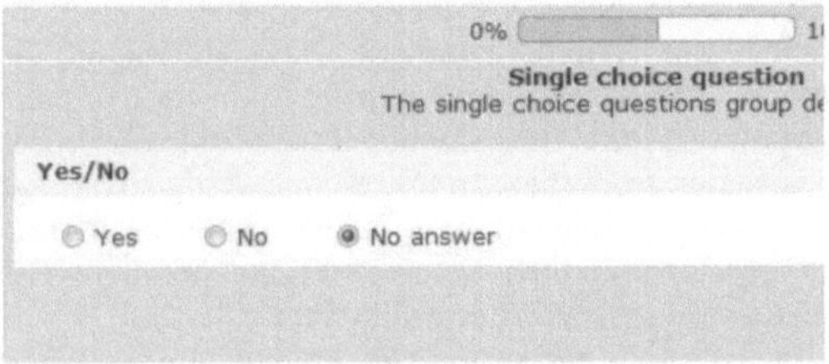

Figure 3. Example of questions with yes/no answer in LimeSurvey.

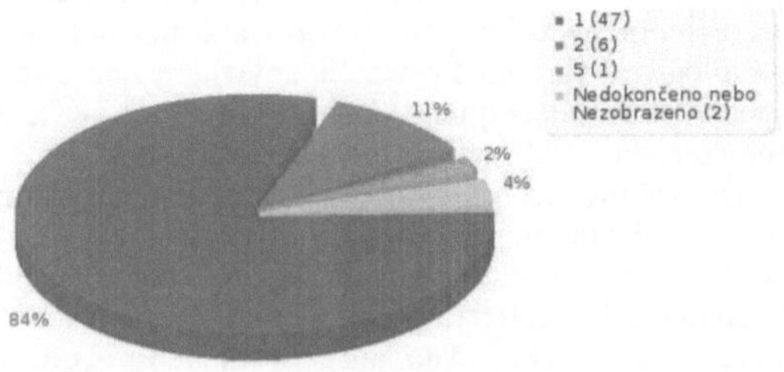

Figure 4. Graph as the output display data in LimeSurvey. Design electronic evaluation in a group of sociology students.

options such as hide issues in statistics or printed versions, the timeout setting, and checks the correct filling boxes and the like. A question in LimeSurvey can be directed according to the type of response. For every question there is also a preview option available, so the questionnaire creator can easily see how does the question change under different settings Options Deposits may be yes/no questions (see Figure 3), with multiple possible answers, with the possibility of comments, and school evaluations (1-5). It is possible to choose time and date, upload a file or choose a question from the drop-down list.

In the case of using an electronic questionnaire, the computer takes care of data processing. The resulting display in the program LimeSurvey is relatively simple to use. We can see the statistics and graphs (Figure 4) and their dispositions, also converting the data into another format, as mentioned above.

In the case of LimeSurvey data processing, we obtain the necessary information in the form of an HTML file, PDF or Excel, as well as also graphs. That way all outputs (statistical data, graphs) can easily be processed even further. All outputs (statistical processing, graphing) but we can do themselves. One of the biggest advantages of this questionnaire system is the fact that this software is a licensed open source, so it can be further customized according to the needs of an organization (LimeSurvey, 2014).

2.3 *Google Forms*

When we were looking for alternatives, it was not only those that could be used in the case of replacing paper with electronic evaluation questionnaires at the University of Hradec Králové, but I liked to recommend

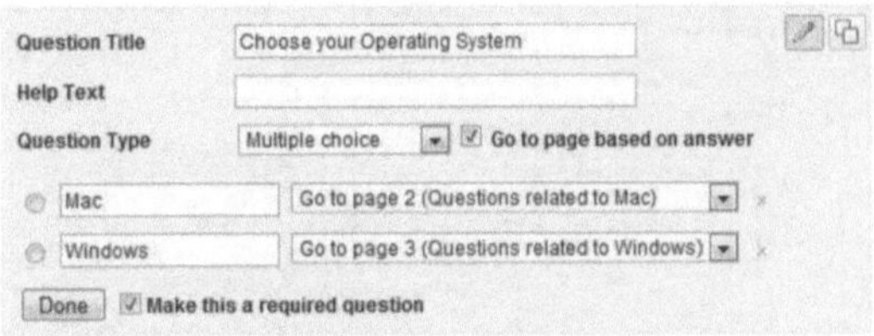

Figure 5. An example of creating questions through Google Forms (Agarwal, 2014).

students of sociology simple tools for creating questionnaires, which would be compatible with mobile devices, and we also tested Google Forms. At first glance, Google Forms has a poorer impression. This is a part of free Google Docs office suite, and it is a tool for creating questionnaires, surveys, and surveys' feedback. Forms can be sent directly through email (in a message body), located on the site or as an application which consists of a separate page. Answers are entered in the spreadsheets, which facilitate further work with the data. Google Forms include and graphically evaluate the given questions. It is also possible to export data as Excel, PDF or CSV, and others. Processing input from the form to an arbitrarily automation, using Apps Script is also possible. That is a popular application which is confirmed by the fact that its collection of customer feedback is also used by companies such as IBM and Vodafone (APP Satori, 2014).

Moreover, it is possible to use spreadsheets, because Google Forms are part of Google Docs. These tables contain processor functions and graphs, having the facility to publish a table and the table mainly establishes the forms. Thus, the resulting process is easier. Creating forms is more user-friendly at first glance than the aforementioned variant LimeSurvey. Especially for students who need to study, processing a simple questionnaire will work better with simple Google Forms. There is no installation, and everything is available immediately online. Creating forms requires several steps. It is necessary to fill in a web form name and description, after creating each question. We can choose from nine different offers, including, for example, time and date. By default, we can choose multiple options, check boxes, or select from a list or text.

From the figure above (Figure 5), I can get help on the issues. As with LimeSurvey, we can create any number of forms, which are completely free (free license). Google Forms are viewable and readable on mobile devices (Figure 6). Thus, it is possible anytime and anywhere, not only to fill in the form, but still to create one. Individual responses are the same as for automatically collected LimeSurvey forms automatically collected. Thus, it is easier to analyse the resulting data such as graphs or complex functions. Google Forms also support branching logic (that can be set according to an answer to the question, where the respondent is directed to the next, logical follow-up question). One benefit is the ability to change the background and engage in the form. When creating

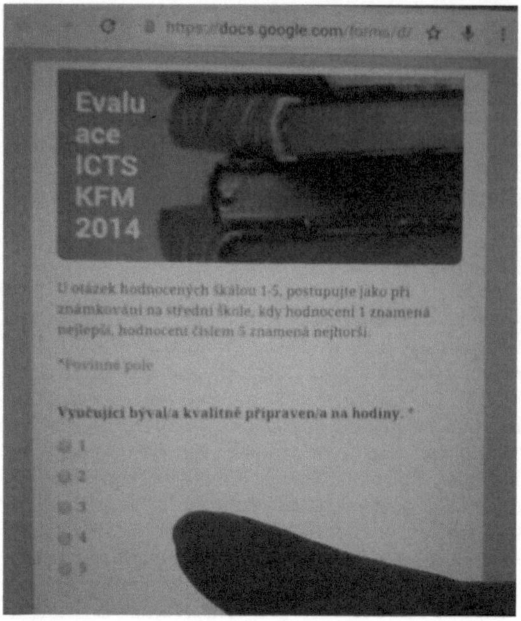

Figure 6. Questionnaire in Google Forms displayed on the tablet. It allows convenient use of the evaluation questionnaire with mobile devices.

the background, the user can choose from a variety of templates, or use their own photos (Google, 2014). Data evaluation is performed automatically, not only as a table, but also as a graph. It is possible to preselect the desired chart. All results are then stored in the cloud on Google Docs' user (Agarwal, 2014).

As an evaluation tool, Google Forms are a great choice. Building a questionnaire and sending it to the class or classes is very fast. By using Google Forms we can do tests or just find out the majority opinion on any issue. It can be distributed, as already mentioned above, to an unlimited number of respondents. Unlike LimeSurvey there is no server overload, because Google Forms, Google and the entire process of creating and extending is through the use of their servers.

3 CONCLUSION

Paper evaluation questionnaires certainly have their advantages. Respondents and lecturers are not dependent on the internet and other facilities and can fill them in with just a pencil on the printed questionnaire. In addition, further processing is time-demanding. In contrast, an electronic type evaluation questionnaire is indeed dependent on other devices (which usually are not available, because even in the absence of computers they can fill in the questionnaire on student mobile devices). When there is no internet connection or there is a power failure, the respondents can fill in electronic forms later, from their mobile device. A big advantage of an electronic evaluation is that it can be carried out anywhere: in the quietness of home or wherever a network infrastructure is available.

In this study we tested two tools which would be appropriate to use in an electronic evaluation. The first, LimeSurvey is a professional questionnaire tool that excels in setting up the questionnaires, the possibility of professional evaluation and treatment. It can clearly be recommended both for schools and not only for sociologically oriented research, anywhere where you need to collect data quickly and efficiently in the field and connect it with one of the mobile devices.

The second test tool, Google Forms, also proved to be a tool for the implementation of electronic evaluations or surveys. It could be definitely recommended to students for collecting data. Building a questionnaire, and collecting data and its evaluation is quick and easy. The whole form and its setting is user friendly and there is no need to install anything. The data can be obtained anytime owing to the cloud (Google Docs) and internet access. It also offers many templates and allows its users to create highly customized forms to make special form or take the respondent.

We can recommend both of the tested tools to be used in university environment and collecting data. For the evaluation questionnaires, we are more inclined to use LimeSurvey. For use in teaching and student data collection, we can make full use of both tools.

REFERENCES

Agarwal, A. 2014. Why Forms in Google Docs are Perfect for Creating Online Surveys. Digital Inspiration. Online. http://www.labnol.org/software/google-docs-forms-for-surveys/10056/

Create a Free Form. 2014. Google. Online. http://www.google.com/forms/about/

Google Formuláře (Google Forms). APP Satori 2014. Online. http://www.appsatori.eu/slovnik-pojmu/google-formulare?language=cz

Gošová, V. Evaluace. 2011. *Metodický portál: RVP*. Online. http://wiki.rvp.cz/Knihovna/1.Pedagogicky_lexikon/E/Evaluace

LimeSurvey. 2014. Online. https://www.limesurvey.org/en/

Myška, K., B. Celá & K. Rybenská. 2014. Using open source software and mobile devices for collecting research data in terrain. 2014 International Conference on Future Communication Technology and Engineering, Shenzhen China. In printing.

Šín, M. 2013. LimeSurvey – dotazníky snadno a rychle. LinuxExpress. Online http://www.linuxexpres.cz/software/limesurvey-dotazniky-snadno-a-rychle

Future Communication Technology and Engineering – Chan (Ed.)
© 2015 Taylor & Francis Group, London, ISBN: 978-1-138-02777-0

An experimental study on using electronic forms for evaluation in sociology

M. Joukl, L. Kudová & L. Hosák
Philosophical Faculty, University of Hradec Králové, Czech Republic

ABSTRACT: At many schools a traditional form of evaluating education prevails, namely by means of paper question forms being filled in by students in terms of their education. If a school uses an electronic evaluation form, there are electronic question forms intended for being filling in by means of a common computer. There is a disadvantage of such evaluation which consists in a fact that a student has to fill in the question form at school or in a computer classroom or at home or elsewhere with access to the computer. Main arguments against the electronic evaluation consist in its dependence on technical equipment and lesser return compared to the paper-form. This article will introduce a method which is trying to reveal main disadvantages of the electronic evaluation by means of using smart devices.

Keywords: Evaluation, data processing, forms, smart devices, sociology

1 INTRODUCTION

An anonymous student appraisal of education, i.e. the evaluation, is best performed at the end of term shall be suitable to perform always after finishing a given term. It represents a certain feedback for a teacher about his/her course and in an ideal case, even a student would obtain knowledge about the results of his/her subjects or of the teacher who has been evaluated by a class (Svobodová, 2012). At many institutions, the evaluation has been performed by means of standardized evaluation paper question forms. However, this form has been proven substantially obsolete. In addition, it has many flaws. One of them consists in the necessity of devoting longer time to filling evaluation question forms for subsequent processing. In most cases an extra person is needed, eventually a special software. An evaluation of the particular question form takes a lot of time. There are many hours of work awaiting an evaluator, if he/she has dozens or even hundreds of evaluation question forms to process. In addition, it is not possible to guarantee that all the question forms have been really anonymous and that all of them will be delivered. There is not ever an ideal situation that all the class comes to fill in the evaluation question forms. Consequently, there can be a substantially distorted result because of the low number of participants.

Our previous research focusing on performing the electronic evaluation, we have found that it can be performed quite easily. Tools, which we determined and their function tested, seem to be suitable for performing the electronic evaluation. There is especially the LimeSurvey, by means of which advanced versions of question forms can be created for professional data evaluations (LimeSurvey, 2014). There are also Google Forms representing a less sophisticated tool, although there are a very suitable alternatives for electronic evaluation (Google, 2014). Especially for students and fields of studies such as sociology which perform quantitative analyses for which it is necessary to effectively collect data by means of some simple tools which are free of charge (Joukl & Hosák, 2014).

However, we carried out the research further. We already know that there are tools suitable for creating the electronic evaluations. Nevertheless, we started to deal with a problem, namely whether there would be any real interest in using the electronic evaluations instead of the paper ones. This research had been performed in terms of a specific research in the field of sociology study at the University of Hradec Králové.

2 RESEARCH

2.1 *Methodology*

As already mentioned, the research had been performed in terms of the specific research at the University of Hradec Králové. Students of the university had been questioned, majority of them were studying at the Faculty of Arts in the Sociology study program. field of sociology. A quantitative part of this research was also performed by means of a non-standardized question form survey. The question forms were filled in by students of various fields of study at the university. A qualitative part was performed by means of a colloquy with selected sociology students. Data having been collected, we carried out an experiment in

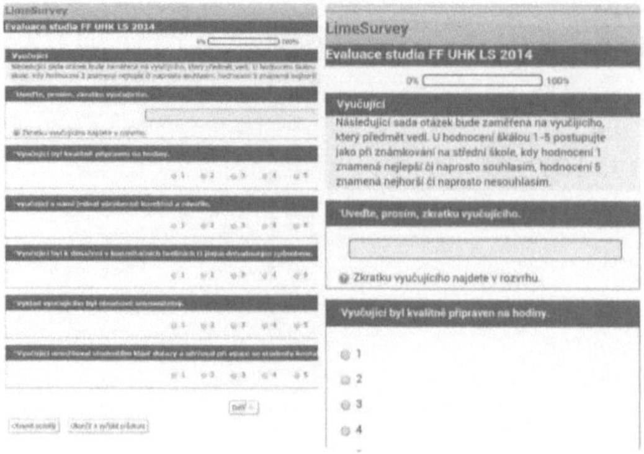

Figure 1. Demonstration of the electronic evaluating question form. Left on the computer, right on the smart device (smart-phone).

Table 1. Necessary time for filling in the particular forms of the question forms.

Type of form	Time to fill in
Electronic version for smartphone	65–120 s
Electronic version for tablet	65–110 s
Electronic version for PC	52–70 s
Standardized paper form	59–68 s

which we let the students fill in a standardized paper evaluation question form first, and then its electronic version. The electronic version has been filled in by the students first by means of using a computer and then by means of using a tablet and a smart- phone (Figure 1).

Filling in the paper-evaluating-question form was the quickest (about 59–68 sec.). Nevertheless it is necessary to say that the filling in time for electronic equipment was different for particular users. However, there are still users, who are not able to operate smart devices very well. For this reason, what sometimes happened was that filling in an electronic version of the question form took a lot longer than filling in other versions. Similarly, the filling in time for paper question forms was different amongst students who saw the evaluation question form for the first time, and those who read the evaluation question form really carefully.

2.2 *Experimental study*

The practical evaluation had been carried out in a computer class. The particular respondents could use even a smart device and a paper form of the standardized question form. We had recorded the experiment on a video camera and then analysed the users' behaviour by means of the video recording. Especially the filling in time for particular question forms (Table 1).

In addition, we had followed responses of the respondents during their work on particular device

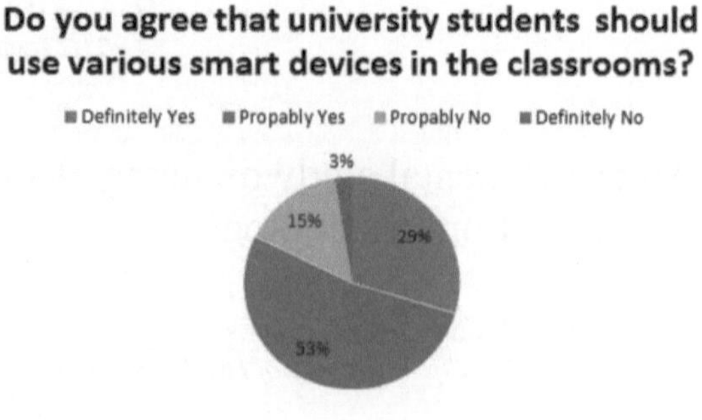

Figure 2. Do you agree that students should use diverse smart devices for their education?

types. There were thirty five respondents available for the qualitative research. We even had a colloquy with some of them later.

As mentioned here already, the question form survey preceded the experiment. According to the question form survey, the students would welcome an option of the electronic evaluation supposed that they could fill in the electronic forms on the computer, and during a particular subject class (78% of the respondents wanted to fill in the evaluation question form mostly during a lecture in a given subject). Most of the students had been content conciliated with the paper evaluation form (39%), and 34% of respondents would definitely want the electronic form of the evaluation question form. Both of the versions would be welcomed by 27% of respondents (i.e. the paper version, and the electronic one).

2.3 *Main results of the quantitative study*

Smart devices or at least a laptop are possessed by nearly every university student nowadays (89% respondents have their own laptop, 2% a tablet and 40% a smart phone), nevertheless knowing how to operate the equipment is substantially different. That has been evident even in our experiment. Some students had a little problem with a basic operation in the electronic question form whilst they used a smart phone. Especially, if they should find and start a given link to the web containing the evaluation question form. The students mostly stated in the question form that they agree with a fact that university students should use smart devices not only for their personal needs, but also for their education as well (Figure 2).

A problem, which may occur, during filling in the electronic evaluation form, is the case that a student cannot operate a particular smart device well enough. However, this could be solved if the students would learn to work with those technologies and use them not only for their entertainment, but also for their education, and eventually for work duties in their future lives. The students (47%) had agreed in the question form with a statement that at any study program there should be a specific subject focused on

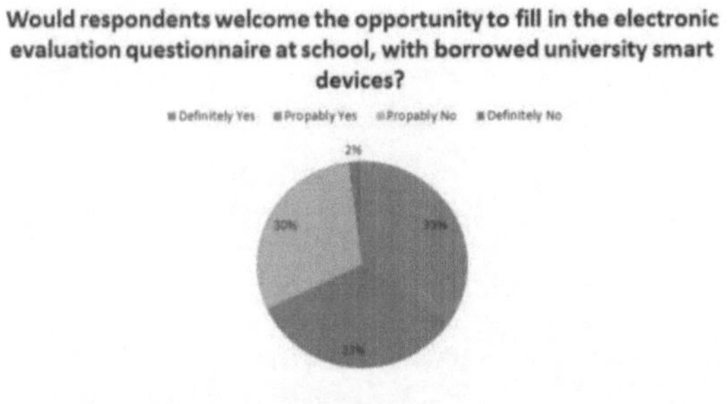

Figure 3. The graph showing an interest in borrowing a smart device from the university.

implementing smart devices, For example, the sociology students agreed that they should learn to use advanced technologies connected to the field of their study, i.e. a statistical software and ways of data measurements, and also learn how to create. electronic question-form surveys or use them for collecting or evaluation data). A question, whether the respondents (students) would welcome the option of filling in the electronic evaluation form by means of using their own mobile device, they responded mostly positively ((24% agreed definitely, 36% rather agreed and 36% rather disagreed. Others either disagreed definitely or they were not able to imagine what it means to own a mobile device). We assume that students are accustomed to using their own smart devices. Therefore, if they could use them in education for filling in the evaluation question forms, it could contribute to a higher economic return of those question-forms. On the other hand, a lot of students agreed with an option to borrow a school smart device (35%); all 30% have not been interest in this device. Others stated that they are not interested in borrowing this equipment at all (Figure 3).

The respondents' interest in taking part in testing systems of electronic evaluations was comparatively high, at least amongst students of the Faculty of Arts. In reference to the whole university, 16% of respondents were definitely in favour of taking part in testing. 43% were moderately interested rather for it, 34% did not probably want to take part in it and 7% definitely did not want to take part in it. The respondents also answered our questions relating to filling in the electronic question forms by means of using a smart device. The respondents could circle more answers in the question form. The most given answers were the following (see below and Figure 4):

- flexible approach to question forms (52%);
- option of filling in and importing the question-forms according to their own needs (48%);
- simple processing of large collections of (41%);
- ease of obtained and presented feedback (30%);
- guarantee of anonymity (27%);
- option of using own smart devices for filling in forms (23%).

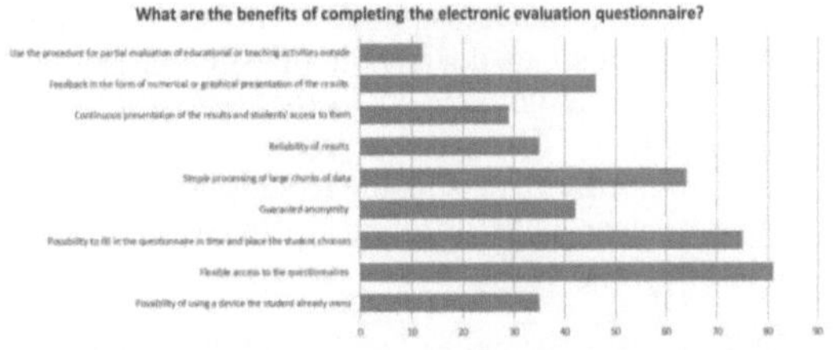

Figure 4. The most given answers relating to the advantage of filling in the electronic question form by means of using a smart device.

Figure 5. The talks with the students had been performed immediately after the experiment. The talks were recorded as well.

We supposed that the electronic forms would be popular for filling in the evaluation question forms. We failed in the confirmation of this assumption. According to our research, paper evaluation forms (81%) are still very popular with the students; the electronic one is preferred by 8% and a certain compromise, i.e. the possibility to select between the electronic form or the paper one of evaluation question forms, is supported by 11% of the university students.

2.4 *Main results of the quantitative study*

All the experiment was recorded by means of a video camera enabling the researchers to see some key situations again and to evaluate them. Having performed the experiment, we interviewed the selected students (Figure 5).

We had given diverse questions to the respondents which had been focused on their opinion in reference to the electronic equipment. Their answers were different. We give a selection of some questions and the most interesting answers below:

What do you think about the fact that the university students should learn to use diverse smart devices?

- "Technology development in general is comparatively quick and it is important to keep up with the times" (a female student);
- "If we learn to use smart devices well, then we shall communicate, and work more effectively" (a female student);

– "I think it is unnecessary to combine education and work on smart devices in a particular field of study which is not related to that, e.g. sociology (a student);
– "It would be more reasonable to support work with computers or with different software than with smart devices" (a student).

What are the advantages of filling in the electronic evaluation question form by means of using smart devices?

– "It is quicker, the time will be saved" (a female student);
– "It is more comfortable and less expensive compared with the paper form" (a female student);
– "It is easier to process, and we can find out about the results on the subjects we have been evaluating." (a student);
– "The question-form can be filled simply at home, in any time convenient for us and we don't have to go to school because of that" (a female student);
– "I think that the electronic form is more anonymous" (a female student).

Do you agree mean that evaluations support the quality of your study?

– "Yes, a teacher can change his/her approach to education if there is a problem, and the teacher knows about it" (a student);
– "No, we do not even get feedback. We do not know what evaluation score a teacher had received according to our appraisal. I think it is pointless (a female student);
– "I cannot see any improvements during other classes with the same teachers in other subjects" (a female student);
– There is a feedback missing in reference to the results and the information about subsequent processing of the question forms, and about a correction process that follows (a student).

What do you believe should specifically be contained in an evaluation question form?

– "An evaluation of comprehensibility of the classes" (a student);
– "An evaluation of the student satisfaction in the subjects they are studying" (a female student);
– "Space for expressing own opinions about the classes besides concrete questions" (a student);
– "An evaluation of the subject's difficulty" (a student).

Do you have any remarks or any appraisal in reference to this topic?

– I am afraid of progressive electronic technologies and of rising dependence on newer technology forms (a female student);
– I expect that if the electronic question forms are used, then the return will be also lower and the students will not fill in this question form unless under a direct supervision of a teacher. I would recommend a fine for those who do not return the question forms" (a student);
– I would recommend a parallel data collection during an evaluation, i.e. to hand out paper question forms, and to provide disinterested students with an option to evaluate by means of the electronic question-form (a student);
– I think each of the students should have an option to select by himself/herself whether they use an electronic evaluation form or a paper one (a student).

Figure 6. A female student filling in an electronic version of the question form by means of using a tablet.

3 CONCLUSION

The paper evaluation form is still very popular and will be certainly used a lot in the future. In spite of the time consumption connected to processing the results, it bears a lot of advantages which prevail at university premises. Nowadays technologies of the smart device type are always more advanced and offer possibilities. By using them we can increase the electronic evaluation to a higher level and because of that we can significantly enhance their popularity among teachers and students which can lead to higher return of the evaluation question forms. However, it is necessary that the electronic question forms themselves shall be highly applicable, not only to computers, but to smart devices as well. The particular results of our experiment have proved that, if correctly optimized question forms along with suitable technologies are used, then the electronic evaluation form is comparable with the paper one in terms of easy applications and it can be connected with a lot of advantages and eliminates or simplifies the processing of the result processing. However it shall be necessary to prepare not only a lot of equipment for this revolution, but also for new users who will at least at the beginning be very suspicious of this new way of electronic evaluation.

REFERENCES

Create a Free Form. 2014. Google. Online. http://www.google.com/forms/about/

Joukl, M., Hosák, L. 2014. Comparative analysis of free and open source tools for electronic evaluation in sociology. FCTE2014. In print.

LimeSurvey. 2014. Online. https://www.limesurvey.org/en/

Svobodová, M. 2012. Electronic students' evaluations are starting. FFakt: Journal of the FF UK students. Online. http://ffakt.ukmedia.cz/startuji-elektronicke-studentske-evaluace

Future Communication Technology and Engineering – Chan (Ed.)
© 2015 Taylor & Francis Group, London, ISBN: 978-1-138-02777-0

The design and application of a high power factor voltage regulator in polysilicon production

L.P. Gao, J. Wang, T. Li & S.G. Yin
Beijing Nari Smartchip Microelectronics Company Limited, Beijing, China

ABSTRACT: This paper introduces the basic principles and the practical applications of a reduction voltage regulator—the major equipment of the polysilicon reducing furnace electrical system, and proposes a method of designing high power factor reduction voltage regulators. It shows the harmonic content and the power factor in high power factor voltage regulators. It also proves that this voltage regulator can reduce harmonic content and improve the power factor. Finally, the article introduces a classification methods for multi-tapped high power factor voltage regulators. Through this paper, we can clearly see the advantages of high power factor voltage regulators.

1 INTRODUCTION

With the development of power electronics, power electronic equipment is playing a more important part in the power system, in the industry, its traffic and in the family, and there are more and more reactive powers and harmonics produced. Therefore, China stipulates that the average electrical power factor should be higher than 0.9, and has issued People's Republic of China national standards GB/T 14549-1993 Electrical Energy Quality—Public Electrical Network Harmonic (GB) which limits the harmonics' current values of the electrical network.

The major equipment of the polysilicon reduction furnace electrical system is a reduction voltage regulator, whose role is actually to heat the load resistance, and to maintain a constant surface temperature of silicon sticks (generally 1080°C). The load of the voltage regulator is pure resistance, which consists of polysilicon sticks in series and it changes along with the temperature and diameter of the silicon sticks. This paper researched and analysed the reduction voltage, and proposed a new reduction voltage regulator—High Power Factor Voltage Regulator (HPFVR), which could satisfy the power factor and the harmonic request stipulated by the GB directly, and does not need to attach the compensation equipment.

2 BASIC PRINCIPLES

The reduction voltage regulator first outputs power to heat silicon sticks, then the silicon sticks transmit the power to the gas of the reduction furnace and the coolant of the furnace wall by the ways of radiation, conduction, and convection. With the growth of the silicon sticks' diameter, the flow of gas increases and the heat is dispersed in response to the gas being increased. As a result, the voltage regulator outputs more and more power. Therefore, the reduction voltage regulator must satisfy the power supply requested in craft which changes along with the silicon sticks' diameter Φ. The output of voltage V, electric current I and the power P would change accordingly, as is shown in Figure 1.

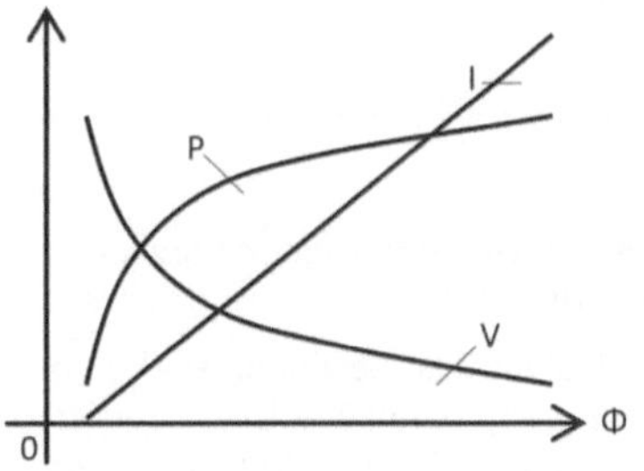

Figure 1. Craft curve of polysilicon reduction furnace.

Because the silicon sticks in the direct-current electric conduction area is round, the speed of its centre giving off heat is slower than its surface; the temperature of the silicon stick centre is higher than that of its surface, and the mechanical strength of a silicon stick would deteriorate and even have a core melt phenomenon. Therefore it generally uses an AC voltage regulator, because the polysilicon skin depth is approximated at 26 mm to the 50 Hz AC current. When the silicon stick diameter is bigger, the electric conduction area becomes ring's area, which can reduce the silicon stick centre temperature, and solve the problem of melted core.

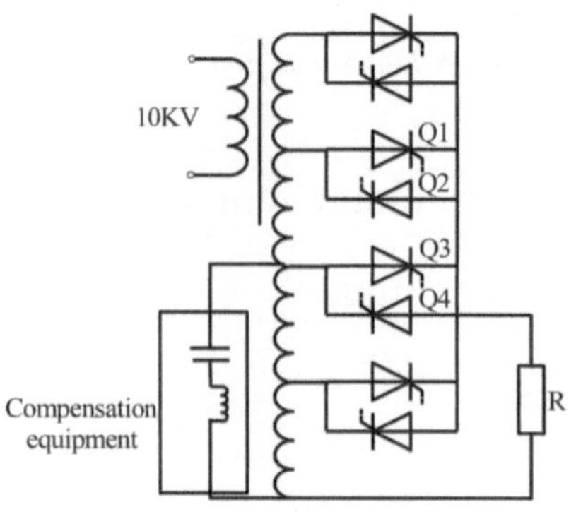

Figure 2. Circuit diagram of the multi-tapped voltage regulator.

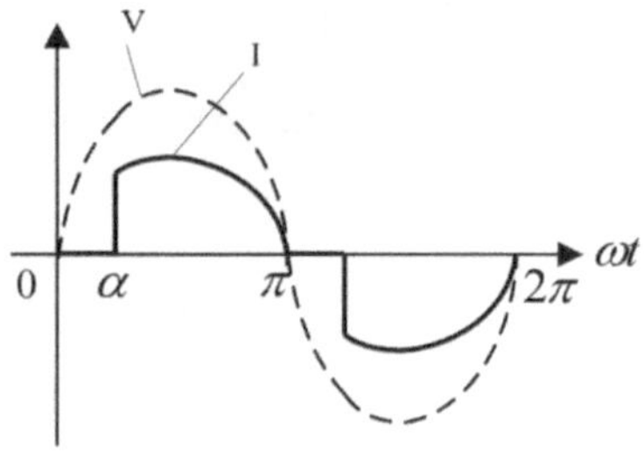

Figure 3. Current wave of the general regulator.

2.1 *General reduction voltage regulator principle analysis*

The craft decides that the silicon stick's voltage regulation range is approximately 10:1. In order to improve the power factor and reduce the harmonics, it usually adopts the vice-side multi-tapped transformer.

The circuit diagram of the general vice-side multi-tapped transformer, named multi-tapped voltage regulator, is shown in Figure 2. There is only one SCR working at one time. The current waveform of the primary-side transformer is shown in Figure 3, and the current waveform is composed of a zero-current line and part of a sine wave. Moreover, the electric current lags in the voltage in this voltage regulator's electric circuit, so the power factor is low, and the electric current distorts seriously in order to produce higher harmonics [1]. Therefore, it needs installation compensation equipment to improve the power factor and to reduce harmonics on the vice-side transformer in order to make the power factor higher than 0.9 and the harmonics into the standard electrical grid to meet the GB/14549-93 standards.

2.2 *High power factor reduction voltage regulator principle analysis*

In order to improve the power factor and the distorted electric current, this paper designs a new reduction voltage regulator—a high power factor voltage regulator (HPFVR). The neighbouring two SCR of the

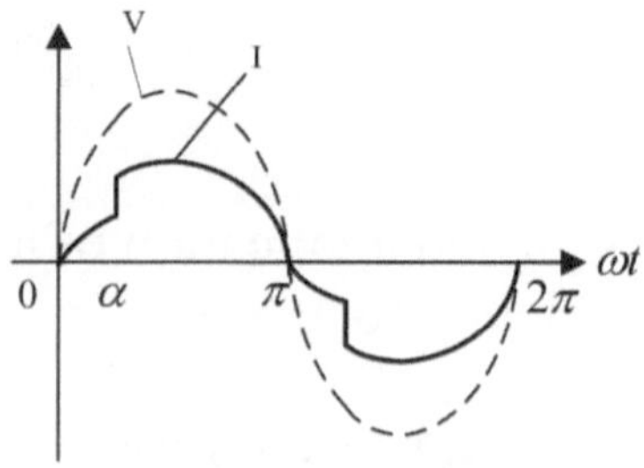

Figure 4. Current wave of the HPFVR.

HPFVR switch-on in turning each cycle, and the current waveform is shown in Figure 4. However, in this circuit, it presents the probability of a short-current on the vice-side transformer when the SCR is in different two taps switch-on at the same time. Therefore, the key technology is to guarantee that other SCR gate pulses are blocked when one SCR is still working

The main circuit of HPFVR had the same circuit as Figure 2, but there were two SCRs in the workplace at each cycle, for example The Q3 conducted and produced the small subtractive sine current wave first, then the Q1 conducted and produced the big subtractive sine current wave, then the Q4 conducted and produced the small negative sine current wave, and finally the Q2 conducted and produced the big negative sine current wave.

Notice: in the project, whether the power factor or harmonic content of the electrical system attains to the GB, it should be guaranteed that the harmonics in the standard electrical grid achieve GB standards, and the power factor in China is higher than 0.9. Therefore, it should transform the electric current of the transformer vice-side to the primary-side, in order to calculate and analyse the harmonic content and the power factor.

The ratio of the voltage effective values of the transformer vice-side's two taps is named as proportionality factor in this paper, assuming ρ, that is

$$\rho = \frac{E_2}{E_1}$$

Supposition: the first tap's voltage in the transformer primary-side E1 = E, then E2 = ρE, thus the electric current expression in the transformer primary-side is as follows:

$$i = \begin{cases} \dfrac{\sqrt{2}\rho^2 E}{R}\sin\omega t, k\pi < \omega t \le k\pi + \alpha \\ \dfrac{\sqrt{2}E}{R}\sin\omega t, k\pi + \alpha < \omega t \le (k+1)\pi \end{cases} \quad (k = 0,1,2,...) \tag{1}$$

Considering the half-wave symmetry current, but not DC component, the electric current expression in the transformer primary-side is the following form:

$$i = \sum_{n=1,3,5...}^{\infty} (a_n \cos n\omega t + b_n \sin n\omega t) \tag{2}$$

The Fourier coefficients, which calculate with Fourier decomposition, are as follows:

$$a_1 = \frac{\sqrt{2}E}{2\pi R}(\cos 2\alpha - 1)(1-\rho^2) \tag{3}$$

$$b_1 = \frac{\sqrt{2}E}{2\pi R}[\sin 2\alpha + 2(\pi-\alpha) + \rho^2(2\alpha - \sin 2\alpha)] \tag{4}$$

$$a_n = \frac{\sqrt{2}E}{\pi R}(1-\rho^2)\{\frac{1}{n+1}[\cos(n+1)\alpha - 1] + \frac{1}{n-1}[1-\cos(n-1)\alpha]\} \tag{5}$$

$$b_n = \frac{\sqrt{2}E}{\pi R}(1-\rho^2)[\frac{1}{n+1}\sin(n+1)\alpha - \frac{1}{n-1}\sin(n-1)\alpha] \tag{6}$$

Furthermore, it could calculate the fundamental wave's current effective value in the transformer primary-side, the initial phase angle, harmonics electric current effective value, as well as the total electric current effective value respectively:

$$I_1 = \frac{E}{2\pi R}\sqrt{\begin{matrix}[(\cos 2\alpha - 1) + \rho^2(1-\cos 2\alpha)]^2 + \\ [\sin 2\alpha + 2(\pi-\alpha) + \rho^2(2\alpha - \sin 2\alpha)]^2\end{matrix}} \tag{7}$$

$$\theta_1 = arctg\frac{a_1}{b_1} = arctg[\frac{(\cos 2\alpha - 1) + \rho^2(1-\cos 2\alpha)}{\sin 2\alpha + 2(\pi-\alpha) + \rho^2(2\alpha - \sin 2\alpha)}] \tag{8}$$

$$I_n = \frac{1}{\sqrt{2}}\sqrt{a_n^2 + b_n^2} \tag{9}$$

$$I = \frac{E}{R}\sqrt{\frac{1}{2\pi}[2(\pi-\alpha) + \sin 2\alpha + \rho^4(2\alpha - \sin 2\alpha)]} \tag{10}$$

Thus, the harmonic content and the power factor are as follows:

$$I_n / I = \frac{\frac{1}{\sqrt{2}}\sqrt{a_n^2 + b_n^2}}{E\sqrt{\frac{1}{2\pi}[2(\pi-\alpha) + \sin 2\alpha + \rho^4(2\alpha - \sin 2\alpha)]}} \tag{11}$$

$$\cos\varphi_1 = \cos\theta_1$$
$$= \cos arctg[\frac{(\cos 2\alpha - 1)(1-\rho^2)}{\sin 2\alpha + 2(\pi-\alpha) + \rho^2(2\alpha - \sin 2\alpha)}] \tag{12}$$

2.3 *Three-phase voltage regulator principle analysis*

In the actual project, a three-phase voltage regulator is usually constituted by three single phase voltage regulators. If the three-phase three-wire voltage regulation circuit is completely symmetrical at the public-junction of the electric grid, the three-time-harmonics would not flow, but the three-time-harmonics could not be fully offset in the general project because the three-phases are not symmetrically completed. The three-time-harmonics of polysilicon reduction furnace voltage regulator are offset only by 70%. Therefore, we should consider the offset factor δ ($\delta = 1$-offset factor) when we calculate the three-time-harmonics content. The electrical current has no double-time harmonics because of the half-symmetrical wave. In addition, there are only odd-time harmonics in the three-phase three-wire circuit.

Table 1. The allowed harmonic current of the public junction.

Standard voltage (kV)	Short-circuit capacity (MVA)	Time and permissible value of harmonic (A) 5	7	11	13	17	19	23	25
0.38	10	62	44	28	24	18	16	14	12
6	100	34	24	16	13	10	9.0	7.4	6.8
10	100	20	15	9.3	7.9	6.0	5.4	4.5	4.1
35	250	12	8.8	5.6	4.7	3.6	3.2	2.7	2.5
66	500	13	9.3	5.9	5.0	3.8	3.4	2.8	2.6
110	750	9.6	6.8	4.3	3.7	2.8	2.5	2.1	1.9

Assuming the voltage effective value E in a primary-side transformer and the load resistance R are constant (for instance 1), the harmonics' electrical current could be obtained with (9), then the maximum harmonics' current I5, I7, I11, ... would be founded by programming, and the ratio of them are as follows (reserving 2 bit RMS):

I7/I5 = 0.77; I11/I5 = 0.46; I13/I5 = 0.38; I11/I7 = 0.60; I13/I7 = 0.49; I17/I7 = 0.37; ......

The maximum GB harmonic current values are listed in Table 1, and the ratio of them are as follows:

I7/I5 < 0.77; I11/I5 < 0.467; I13/I5 >= 0.38; I11/I7 > 0.60; I13/I7 > 0.49; I17/I7 > 0.37; ......

From the analysis above, it could be clear that as long as the 3,5,7-time harmonics satisfies GB standards, the other high time harmonics of the HPFVR will satisfy the standards, for example. The ratio of 13-time harmonic and 5-time harmonic required by GB standards are equal or greater than 0.38 which is the ratio of the HPFVR, therefore if the 5-time harmonic content satisfies GB standards, the 13-time harmonic could satisfy the GB standards as well. Similarly, if the 7-time harmonic could satisfy GB standards, the other higher time harmonics could satisfy GB standards as well.

The expressions of the 3, 5, and 7-time harmonics content and the power factor in the primary-side transformer of the HPFVR are as follows:

$$I_3/I = (1-\rho^2)\sqrt{\frac{[\frac{1}{4}(\cos 4\alpha - 1)]^2 - [\frac{1}{2}(\cos 2\alpha - 1)]^2}{\frac{\pi}{2}[2(\pi-\alpha) + \sin 2\alpha + \rho^4(2\alpha - \sin 2\alpha)]}} \tag{13}$$

$$I_5/I = (1-\rho^2)\sqrt{\frac{[\frac{1}{6}(\cos 6\alpha - 1)]^2 - [\frac{1}{4}(\cos 4\alpha - 1)]^2}{\frac{\pi}{2}[2(\pi-\alpha) + \sin 2\alpha + \rho^4(2\alpha - \sin 2\alpha)]}} \tag{14}$$

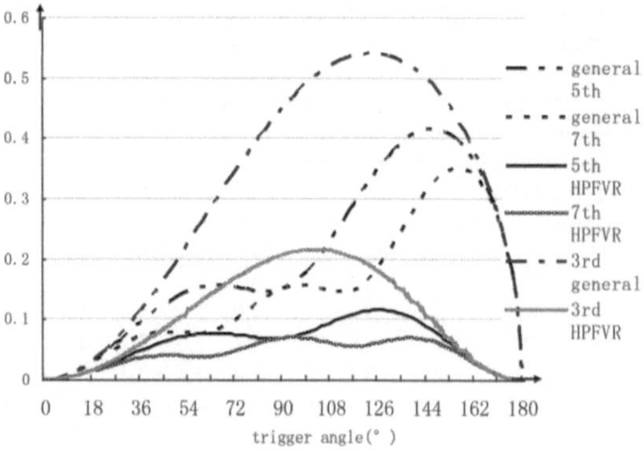

Figure 5. Harmonics content both of HPFVR and general voltage regulator.

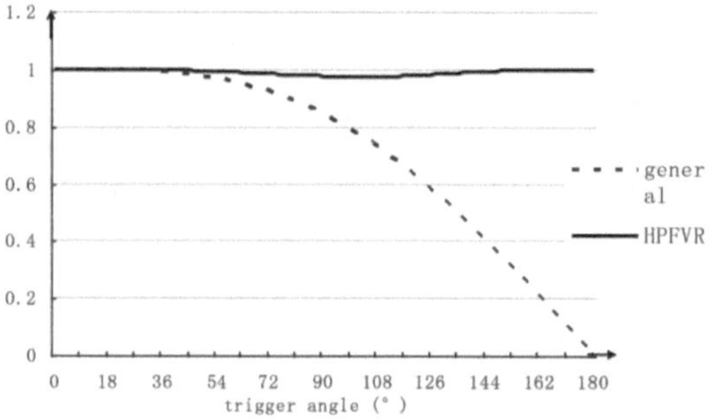

Figure 6. Power factor both of HPFVR and general voltage regulator.

$$I_7/I=(1-\rho^2)\sqrt{\frac{[\frac{1}{8}(\cos8\alpha-1)]^2-[\frac{1}{6}(\cos6\alpha-1)]^2}{\frac{\pi}{2}[2(\pi-\alpha)+\sin2\alpha+\rho^4(2\alpha-\sin2\alpha)]}} \quad (15)$$

$$\cos\varphi=\cos arctg[\frac{(\cos2\alpha-1)(1-\rho^2)}{\sin2\alpha+2(\pi-\alpha)+\rho^2(2\alpha-\sin2\alpha)}] \quad (16)$$

The curves about 3, 5, and 7-time harmonics contents and the power factor of the HPFVR changing with the trigger angle are drown from (13)–(16) which are shown in Figure 5 and Figure 6, and the proportional factor ρ is 0.7. Both of the figures are compared with the general voltage regulator's curves.

From (13)–(16), it is easy to obtain the conclusion: the proportional factor ρ would become bigger along with the reduction of the harmonics' content and the improvement of the power factor, which would lead the voltage regulation range to become smaller. Therefore, it is necessary to adopt a multi-tapped voltage regulator for a wide range of voltage regulation system, which would have more obvious difficulties.

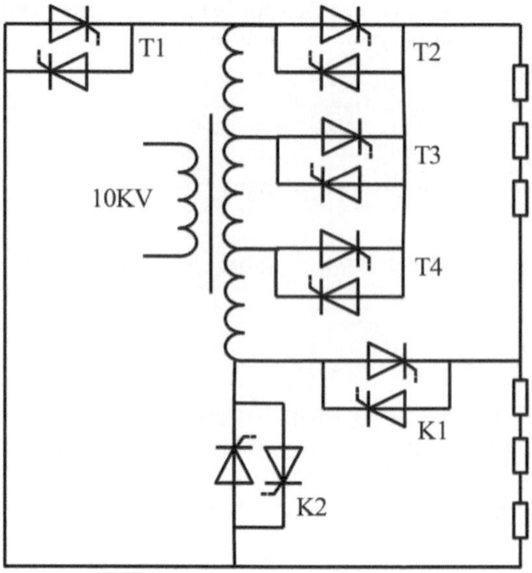

Figure 7. The 'parallel before series' voltage regulation system.

3 MULTI-TAPPED VOLTAGE REGULATOR'S CLASSIFYING METHOD

The process of finding the appropriate voltage values of the taps in a vice-side transformer is called classification of the multi-tapped voltage regulator. The purpose of the classification is to reduce the number of the transformer's taps to the fewest.

If the maximum voltage value in the craft is too high, which is determined by the length and the number of the silicon sticks in series, it is necessary to take the "parallel before series" power supply plans to resolve the voltage resistance problem of the CSR which is shown in Figure 7, for example. In the 18-pair-stick polysilicon reduction furnace electrical system, each single phase of the transformer supplies power to 6 pairs of silicon sticks. In the 'parallel state', K1 switches-on, then K2 should be switched-off first, and the voltage regulator T2 supplies power to the 3 pairs of silicon sticks above, and T1 supplies power to the next three ones. Along with the dropping of the load resistance which is caused by the silicon stick's bigger diameter and smaller resistance, the voltage becomes lower, then it enters the 'series state', while K2 switches-on; K1 switches-off, and T2, T3, T4 supplies power to the 6-pair-stick one-by-one. The reduction voltage regulator is an auto-adapted regulator, and its regulation parameters change along with the load resistance automatically.

From the section 2.3, it would be known that the most important standard to classify the voltage regulator is whether to guarantee that the instantaneous value of the 3, 5, 7-time harmonics current is not greater than the value stipulated by the GB standards and the power factor is not less than 0.9. The classification flow chart

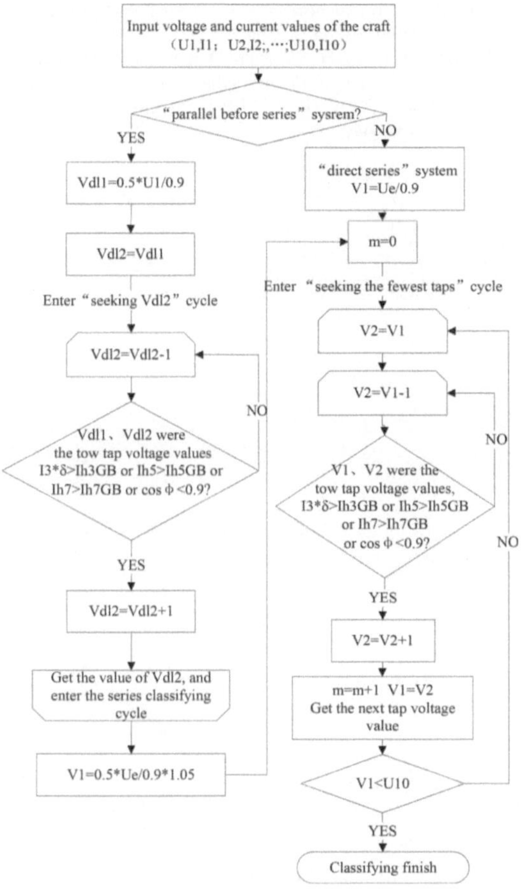

Figure 8. The classifying flow chart of the multi-tapped voltage regulation.

is shown in Figure 8, and the programming thinking is as follows:

(1) Get parameters about the voltage and current in the craft (U1, I1; U2, I2; ……; U10, I10).
(2) Assume Ih3GB = x, Ih5GB = y, Ih7GB = z (x, y, z are the maximum allowable values of the 3, 5, 7-time harmonic current calculated by GB/T 14549-1993).
(3) make sure that the power supply plan will be adopted to regulate voltage: 'parallel before series' enters 4, and "direct series" enters 8.
(4) Seek Vdl2: assumption Vdl1, and Vdl2 are the two tap voltage values of the voltage regulator in "parallel state", Vdl1 = 0.5*U1/0.9 (considering the voltage fluctuation rate is 10%).
(5) Vdl2 = Vdl1.
(6) Vdl2 = Vdl2 − 1 (the calculation accuracy was 1 V, which could be adjusted according to the practical situation.).
(7) If I3*δ > Ih3GB or Ih5 > Ih5GB or Ih7 > Ih7GB or cos $\phi < 0.9$, then get the second tap voltage value of the voltage regulator Vdl2 = Vdl2 + 1, and turns to 8; if not, it turns to 6.
(8) Classify voltage regulation system in "series state": if it is "parallel before series" system, then V1 = 0.5*U1/0.9*1.05; if not V1 = U1/0.9*1.05 (considering the voltage fluctuation rate is 10%, and the safe coefficient is 5 %.); Vmin = U10; m = 0 (the tap number in series began with 0.).
(9) Classify calculation: V2 = V1.
(10) V2 = V2 − 1 (the calculation accuracy is 1 V, which could be adjusted according to the practical situation.), if V2 <= Vmin, then it turns to 12.
(11) If I3*δ > Ih3GB or Ih5 > Ih5GB or Ih7 > Ih7GB or cos $\phi < 0.9$, V1 = V2, and m = m + 1, and get a new tap voltage value of the voltage regulation in series dangwei (m) = V1, then it turns to 9, otherwise it is turned to 10.
(12) Classification finish: in the 'parallel before series' system, the tap voltage values in parallel are Vdl1 and Vdl2, and the tap voltage values in series are from dangwei(1) to dangwei(m); in the "direct series" system, the tap voltage values are from dangwei(1) to dangwei(m).

4 EPILOGUE

Through the principle and application analysis about HPFVR in the polysilicon reduction furnace electrical system, it comes to the conclusions as follows:

(1) The HPFVR will improve the power factor, and reduce the harmonic content; it even could satisfy the national standards without any compensation equipment if the proportion factor ρ is appropriate.
(2) In the three-phase HPFVR system with pure resistive load, as long as 3,5,7 time harmonics satisfies the national standards, the other high time harmonics will satisfy the national standards as well.
(3) Using the HPFVR in a wide voltage range system, it usually needs to adopt the multi-tapped voltage regulator.

REFERENCES

[1] Wang Z.A, Yang J. & Liu J.J. Harmonic Control and Reactive Power Compensation. Beijing: China Machine Press, 2004. 2: 100–104.
[2] Li M., Wang Z.A., & Zhuo F. Ultraharmonics and Reactive Power Detect Based on the Theory of Instantaneous Reactive Power. Power Electronics, 1992(2).
[3] Read J C. The calculation of rectifier and converter performance characteristics. Journal IEF, 1945, 92, pt. II: 495–590.
[4] Hauth R L, Moran R J. The performance of thyristor controlled static var systems in HVDC application. IEEE Tutorial Course Text 78Eh0135-4-PWR, 1978, 56–64.
[5] Wu J.C., Sun S.Q., & Song W.N. Power System Harmonic. Water Conservancy and Power Press., 1988.
[6] Sakui M, Fujita H. Harmonic analysis of a capacitor-filtered three-phase diode-bridge rectifier with DC filter. IEEE Trans Ind Eletron. 1992, 39(1): 80–81.

[7] Xu F., Sheng Y. Modern Control Engineering. Scientific Publisher, 1978.6 Cha. 2, 13–34.
[8] Chen M T, Chu H Y, Huang C L et al. Power-component definitions and measurements for a harmonic-pollutied power circuit. IEEE Proc, pt. B, 1991, 138(4): 299–306.
[9] Lv R.Y. Ultraharmonics of Power System. China Power Press, 1999.5. Cha. 1, Cha. 4.
[10] Sakui M, Fujita H, Shioya M.A method for calculating harmonic currents of three-phase bridge uncontrolled rectifier with DC filter. IEEE Trans Ind Eletron. 1989, 36(3): 434–440.

Future Communication Technology and Engineering – Chan (Ed.)
© 2015 Taylor & Francis Group, London, ISBN: 978-1-138-02777-0

A study of the performance of an amplify-and-forward relay system with an interference at the relay and at the destination

M.R. He
Northwest University for Nationalities, Lanzhou, Gansu, China

ABSTRACT: This paper investigates the outage performance of an Amplify and Forward (AF) cooperative diversity system with co-channel interference at the relay and the destination. The closed form expression for the outage performance is derived from the analysis. The simulation results show that the outage performance will become worse with the effect of multiple interference at the relay and destination points.

Keywords: cooperative diversity, relay, interference, outage probability.

1 INTRODUCTION

In recent years, cooperative diversity communication as a promising solution to high–data coverage and the reliability required for high transmission in wireless networks is investigated in-depth[1]–[3]. The research work shows that the performance of wireless system can be improved tremendously, but most of the work has considered the ideal situation where there is no interference. However, the wireless system will be subjected to some interference in practice. At present the interference has been considered in some works. In [4] and [5] studies the outage performance for a two-way relay channel by using an Amplify-and-Forward (AF) strategy with an interference at the relay. In [6] the outage performance of AF and DF is studied with an interference limited at the destination. In [7] and [8], the outage probability and bit error rate is studied with interference at the destination.

The main purpose of this paper, is to study the outage probability of an AF cooperative diversity system with limited interference at all relays and destinations. Then it derives an exact expression of outage probability for AF cooperative diversity communication system with multiple interferes. The result of the simulation can present and verify the influence of multiple interferences on an AF cooperative diversity communication system.

2 SYSTEM MODEL

We consider a communication system, where the source S communicates with a destination D through two links, a direct link and an indirect (relay) link. It is assumed that relay R and destination D are interference-limited, and the communication is divided into two orthogonal time intervals. In the first time slot, S sends a symbol s_0 to R and D, in the second time slot, R sends an amplified symbol to D.

At the first time, the relay received signals with N_R number of interferes, each with an average power $P_{R,i}$, and additive white Gaussian noise (AWGN), which can be expressed as:

$$y_r = \sqrt{P_0} h_{sr} s_0 + \sum_{i=1}^{N_R} \sqrt{P_{R,i}} f_i s_{R,i} + n_r \tag{1}$$

The destination received signals with N_I number of interferes, each with an average power $P_{I,k}$, and the additive white Gaussian noise (AWGN), which can be expressed as:

$$y_{sd} = \sqrt{P_0} h_{sd} s_0 + \sum_{i=1}^{N_R} \sqrt{P_{I,k}} f_k s_{I,k} + n_0 \tag{2}$$

where P_0 is the transmit power of S. The information symbol s_0, the ith interference symbol $s_{R,i}$ and the kth interference symbol $s_{I,k}$ are zero-mean and power-normalized. n_r, n_0 are the AWGN at R and D which satisfy $E[|n_r|^2] = \sigma_r^2$, $E[|n_0|^2] = \sigma_0^2$. h_{sr} is the channel from S to R, h_i $i = 1, 2, \ldots, N_R$ is the channel from interference to R, h_{sd} is the channel from S to D, f_k is the channel from interference to D, and they are subjected to an independent zero-mean complex-Gaussian-distribution. $E[|h_{sr}|^2] = \Omega_{sr}$, $E[|h_i|^2] = \Omega_i$, $E[|f_k|^2] = \Omega_k$, $E[|h_{sd}|^2] = \Omega_{sd}$, The expectation operator is $E[\cdot]$.

At the second time, R amplifies and forwards received signal y_r to D with a fixed gain G. Then, the received signal at D is given by

$$y_d = h_{rd} G y_r + \sum_{j=1}^{N_D} \sqrt{P_{D,j}} g_j s_{D,j} + n_d \tag{3}$$

where N_D is the number of interference at D, and $P_{D,j}$ is the transmit power of jth interference at D. The jth interference symbol $s_{D,j}$ is zero-mean and power-normalized, n_d is the AWGN at D satisfying $E[|n_d|^2] = \sigma_d^2$. g_j $j = 1, 2, \ldots, N_D$ is the channel from interference to D, h_{rd} is the channel from R to D, and they are subjected to the independent zero-mean complex Gaussian distribution. $E[|h_{rd}|^2]=\Omega_{rd}$, $E[|g_j|^2]=\Omega_j$. G can be expressed as:

$$G=\sqrt{\frac{P_r}{P_0\Omega_{sr}+\sum_{i=1}^{N_R}P_{R,i}\Omega_i+\sigma_r^2}} \tag{4}$$

where P_r is the transmitted power of R. As a result, the SINR at D from the signal received through the direct link for the first time can be expressed by

$$\gamma_s=\frac{P_0|h_{sd}|^2}{\sum_{k=1}^{N_I}P_{I,k}|g_k|^2+\sigma_0^2} \tag{5}$$

The SINR at D from the signal received through the indirect link for the second time can be expressed by

$$\gamma_b=\frac{P_0G^2|h_{rd}|^2|h_{sr}|^2}{G^2|h_{rd}|^2\left(\sum_{i=1}^{N_R}P_{R,i}|h_i|^2+\sigma_r^2\right)+\sum_{j=1}^{N_D}P_{D,j}|g_j|^2+\sigma_d^2} \tag{6}$$

In (4), (5) and (6), they are assumed that D and R are interference-limited (n_r, n_d, n_0 can be neglected), γ_s can be derived as:

$$\gamma_s=\frac{\gamma_0}{\gamma_I+1} \tag{7}$$

γ_b can be derived as:

$$\gamma_b=\frac{\gamma_1\gamma_2}{\gamma_2\gamma_R+M\gamma_D} \tag{8}$$

where $\gamma_0=\frac{P_0|h_{sd}|^2}{\sigma_0^2}$, $\gamma_I=\sum_{k=1}^{N_I}P_{I,k}\frac{|g_k|^2}{\sigma_0^2}$, $\gamma_1=\frac{P_0}{\sigma_r^2}|h_{sr}|^2$, $\gamma_2=\frac{P_r}{\sigma_d^2}|h_{rd}|^2$, $\gamma_R=\sum_{i=1}^{N_R}P_{R,i}\frac{|h_i|^2}{\sigma_r^2}$, $\gamma_D=\sum_{j=1}^{N_D}P_{D,j}\frac{|g_j|^2}{\sigma_d^2}$, and $M=P_0\frac{\Omega_{sr}}{\sigma_r^2}+\sum_{i=1}^{N_R}P_{R,i}\frac{\Omega_i}{\sigma_r^2}+1$.

h_{sd}, h_{rd}, h_{sr} are zero-mean complex Gaussian variables, so the Probability Density Function (PDF) of the corresponding signal-to-noise ratio (SNR) γ_i, $i=0,1,2$, is given as:

$$f_{\gamma_i}(\gamma)=\frac{1}{\bar{\gamma}_i}\exp\left(-\frac{\gamma}{\bar{\gamma}_i}\right) \tag{9}$$

3 OUTAGE PERFORMANCE ANALYSIS

Outage probability which can measure wireless system performance is defined as the probability of SINR as D falls below a given threshold value γ_{th}, specifically in this two link relay system, as an outage event occurs if both of the two links are in outage, so it can be expressed as:

$$P_{out}=\Pr(\gamma_s<\gamma_{th},\gamma_b<\gamma_{th})=F_{\gamma_s}(\gamma_{th})F_{\gamma_b}(\gamma_{th}) \tag{10}$$

In order to calculate (10), the PDF of γ_I, γ_R, γ_D are needed respectively. According to paper [7], the PDF of γ_I is shown as:

$$f_{\gamma_I}(\gamma)=\sum_{i=1}^{\rho(A)}\sum_{j=1}^{\tau_i(A)}\chi_{i,j}(A)\frac{P_{I,i}^{-j}}{(j-1)!}\gamma^{j-1}\exp\left(-\frac{\gamma}{P_{I,i}}\right) \tag{11}$$

where $A=diag\left(\frac{P_{I,1}}{\sigma_0^2}\Omega_1,\frac{P_{I,2}}{\sigma_0^2}\Omega_2,\ldots,\frac{P_{I,N_R}}{\sigma_0^2}\Omega_{N_I}\right)$, $\rho(A)$ denotes the number of distinct diagonal elements of A. $P_{I,1}>P_{I,2}>\ldots>P_{I,[\rho(A)]}$ are the distinct diagonal elements in decreasing order, $\tau_i(A)$ is the multiplicity of $P_{I,i}$, and $\chi_{i,j}(A)$ is the (i,j)th characteristic coefficient of A.

The PDF of γ_R is shown as:

$$f_{\gamma_R}(\gamma)=\sum_{i=1}^{\rho(B)}\sum_{j=1}^{\tau_i(B)}\chi_{i,j}(B)\frac{P_{R,i}^{-j}}{(j-1)!}\gamma^{j-1}\exp\left(-\frac{\gamma}{P_{R,i}}\right) \tag{12}$$

where $B=diag\left(\frac{P_{R,1}}{\sigma_r^2}\Omega_1,\frac{P_{R,2}}{\sigma_r^2}\Omega_2,\ldots,\frac{P_{R,N_R}}{\sigma_r^2}\Omega_{N_R}\right)$, $\rho(B)$, $P_{R,i}$, $\chi_{i,j}(B)$ are defined the same as (11).

The PDF of γ_D is given by

$$f_{\gamma_D}(\gamma)=\sum_{i=1}^{\rho(C)}\sum_{j=1}^{\tau_i(C)}\chi_{i,j}(C)\frac{P_{D,i}^{-j}}{(j-1)!}\gamma^{j-1}\exp\left(-\frac{\gamma}{P_{D,i}}\right) \tag{13}$$

where $C=diag\left(\frac{P_{D,1}}{\sigma_d^2}\Omega_1,\frac{P_{D,2}}{\sigma_d^2}\Omega_2,\ldots,\frac{P_{D,N_D}}{\sigma_d^2}\Omega_{N_D}\right)$, $\rho(C)$, $P_{D,i}$, $\chi_{i,j}(C)$ are defined the same as (11).

With the help of (9), (11) and the integral formula [9, Eq. (3.351.3)], $F_{\gamma_s}(\gamma_{th})$ can be formulated as:

$$\begin{aligned}F_{\gamma_s}(\gamma_{th})&=\Pr\left(\frac{\gamma_0}{1+\gamma_I}<\gamma_{th}\right)\\&=1-\int_0^\infty\exp\left(-\frac{(1+\gamma_I)}{\bar{\gamma}_0}\gamma_{th}\right)f_{\gamma_I}(\gamma_I)d\gamma_I\\&=1-\exp\left(-\frac{\gamma_{th}}{\bar{\gamma}_0}\right)\times\sum_{i=1}^{\rho(A)}\sum_{j=1}^{\tau_i(A)}\chi_{i,j}(A)\left(\frac{\gamma_{th}}{\bar{\gamma}_0}P_{I,i}+1\right)^{-j}\end{aligned} \tag{14}$$

With the help of (9) and some manipulation, $F_{\gamma_b}(\gamma_{th})$ can be formulated as:

$$\begin{aligned}F_{\gamma_b}(\gamma_{th})&=\Pr(\gamma_b<\gamma_{th})=\Pr\left(\frac{\gamma_1\gamma_2}{\gamma_2\gamma_R+M\gamma_D}<\gamma_{th}\right)\\&=1-\int_0^\infty\exp\left(-\frac{\gamma_{th}}{\bar{\gamma}_1}\gamma_R\right)f_{\gamma_R}(\gamma_R)d\gamma_R\\&\quad\times\int_0^\infty\int_0^\infty\exp\left(-M\gamma_{th}\frac{\gamma_D}{\bar{\gamma}_1\gamma_2}\right)f_{\gamma_2}(\gamma_2)f_{\gamma_D}(\gamma_D)d\gamma_2d\gamma_D\end{aligned} \tag{15}$$

$$=1-U_1\times U_2$$

With the help of (9), (12) and the integral formula [9, Eq. (3.351.3)], U_1 can be formulated as

$$\begin{aligned}U_1&=\int_0^\infty\exp\left(-\frac{\gamma_{th}}{\bar{\gamma}_1}\gamma_R\right)f_{\gamma_R}(\gamma_R)d\gamma_R\\&=\sum_{i=1}^{\rho(B)}\sum_{j=1}^{\tau_i(B)}\frac{\chi_{i,j}(B)P_{R,i}^{-j}}{(j-1)!}\int_0^\infty\frac{\gamma_R^{j-1}}{\exp\left[\gamma_R\left(\frac{\gamma_{th}}{\bar{\gamma}_1}+\frac{1}{P_{R,i}}\right)\right]}d\gamma_R\\&=\sum_{i=1}^{\rho(B)}\sum_{j=1}^{\tau_i(B)}\chi_{i,j}(B)\left(\frac{\gamma_{th}}{\bar{\gamma}_1}P_{R,i}+1\right)^{-j}\end{aligned} \tag{16}$$

With the help of (9), and (13) and the integral formula [9, Eq. (3.471.9)], U_2 can be formulated as:

$$U_2 = \int_0^\infty \int_0^\infty \exp\left(-M\gamma_{th}\frac{\gamma_D}{\bar{\gamma}_1\gamma_2}\right) f_{\gamma_2}(\gamma_2) f_{\gamma_D}(\gamma_D) d\gamma_2 d\gamma_D$$
$$= \frac{1}{\bar{\gamma}_2}\int_0^\infty f_{\gamma_D}(\gamma_D)\int_0^\infty \exp\left(-M\gamma_{th}\frac{\gamma_D}{\bar{\gamma}_1\gamma_2} - \frac{\gamma_2}{\bar{\gamma}_2}\right) d\gamma_2 \quad (17)$$
$$= \frac{2}{\bar{\gamma}_2}\int_0^\infty f_{\gamma_D}(\gamma_D)\sqrt{M\gamma_{th}\frac{\bar{\gamma}_2\gamma_D}{\bar{\gamma}_1}} K_1\left(2\sqrt{M\gamma_{th}\frac{\gamma_D}{\bar{\gamma}_1\bar{\gamma}_2}}\right) d\gamma_D$$

where $K_n(x)$ is the nth-order modified Bessel function of the second kind [9, Eq .(3.471.9)]. Substituting (13) for (17), and after some algebraic manipulations, (17) can be calculated with the help of the integral formula [9, Eq. (6.631.3)], then it can be obtained that:

$$U_2 = \sum_{i=1}^{\rho(C)}\sum_{j=1}^{\tau_i(C)} \chi_{i,j}(C)(j!)\exp\left(\frac{M\gamma_{th}P_{D,i}}{4\bar{\gamma}_1\bar{\gamma}_2}\right) W_{-j,\frac{1}{2}}\left(\frac{M\gamma_{th}P_{D,i}}{2\bar{\gamma}_1\bar{\gamma}_2}\right) \quad (18)$$

where $W_{\lambda,\mu}(\chi)$ is the Whittaker function defined in [9 Eq. (9.222.1)], substitutes (16) and (17) for (15), it can be obtained that:

$$F_{\gamma_b}(\gamma_{th}) = 1 - \sum_{n=1}^{\rho(B)}\sum_{m=1}^{\tau_n(B)} \chi_{n,m}(B)\left(\frac{\gamma_{th}}{\bar{\gamma}_1}P_{R,n}+1\right)^{-m} \times \sum_{i=1}^{\rho(C)}\sum_{j=1}^{\tau_i(C)} \chi_{i,j}(C)(j!)\exp\left(\frac{M\gamma_{th}P_{D,i}}{4\bar{\gamma}_1\bar{\gamma}_2}\right) W_{-j,\frac{1}{2}}\left(\frac{M\gamma_{th}P_{D,i}}{2\bar{\gamma}_1\bar{\gamma}_2}\right) \quad (19)$$

Then, substituting (14) and (19) for (10), the outage probability of the AF cooperative diversity system with interferes can be given as (20).

In this paper, the relay and destination are subjected to the equal-power interferes, so according to the paper [6]: $PI = P_{I,k}\frac{\Omega_k}{\sigma_0^2}, \forall k = 1,2,\ldots,N_I$, $P_R = P_{R,j}\frac{\Omega_j}{\sigma_r^2}, \forall j = 1,2,\ldots,N_R, P_D = P_{D,i}\frac{\Omega_i}{\sigma_d^2}, \forall i = 1,2,\ldots,N_D$ and $\rho(A) = \rho(B) = \rho(C) = 1$, $\tau_k(A) = N_I, \tau_j(B) = N_R, \tau_i(C) = N_D$, $\chi_{1,N_I}(A) = 1$, $\chi_{1,k}(A) = 0, k = 1,2,\ldots,N_I - 1$ and, the same as: $\chi_{1,j}(B)$, $\chi_{1,i}(C)$, $\chi_{1,N_R}(B)$, $\chi_{1,N_D}(C)$.

Therefore, (20) can be reduced to (21):

4 SIMULATION RESULTS

In this section, we will present the simulation results to show the effect of interferes for the AF cooperative diversity system. For simulation, it is assumed that $\bar{\gamma}_0 = \bar{\gamma}_1 = \bar{\gamma}_2$, $\gamma_{th} = 5$ dB, $E[\gamma_R] = E[\gamma_D]$, and the signal-to-interference ratio (SIR) in the direct link is defined as $\beta_1 = E[\gamma_0]/E[\gamma_I]$, the SIR in indirect link which is defined as $\beta_2 = \frac{E[\gamma_1]}{E[\gamma_R]+E[\gamma_D]}$. Figure 1 shows the outage performance as a function of the average $SNR(\bar{\gamma}_0)$ for different SIRs and a number of interferes. It shows that the outage performance appears to be worse when the number of interferes increase at different SIRs. It also finds that the effect becomes smaller with the number of interferes increasing with high SIRs, and in the high SNR range the outage performance appears a floor effect for the impact of interference.

5 CONCLUSIONS

In this paper, the outage performance of an AF cooperative diversity system with fixed gain is investigated, and in this system there are multiple equal-power interference at direct link and indirect link, where the closed

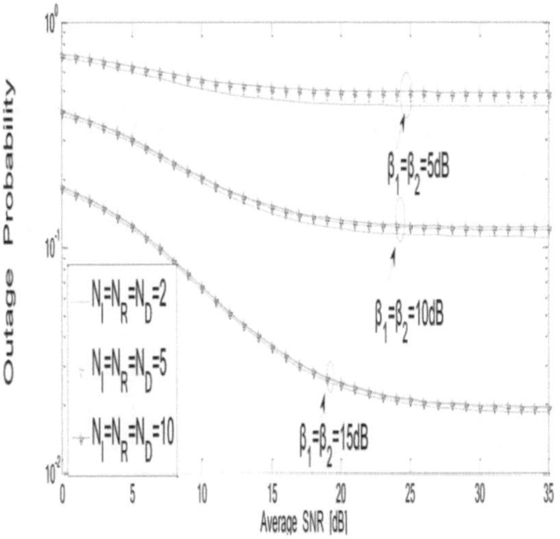

Figure 1. Outage probability for different SIR and number of interferes.

$$P_{out} = \left[1-\exp\left(-\frac{\gamma_{th}}{\bar{\gamma}_0}\right)\times\sum_{k=1}^{\rho(A)}\sum_{l=1}^{\tau_k(A)}\chi_{k,l}(A)\left(\frac{\gamma_{th}}{\bar{\gamma}_0}P_{I,k}+1\right)^{-l}\right] \times\left[1-\sum_{n=1}^{\rho(B)}\sum_{m=1}^{\tau_n(B)}\chi_{n,m}(B)\left(\frac{\gamma_{th}}{\bar{\gamma}_1}P_{R,n}+1\right)^{-m}\times\sum_{i=1}^{\rho(C)}\sum_{j=1}^{\tau_i(C)}\chi_{i,j}(C)(j!)\exp\left(\frac{M\gamma_{th}P_{D,i}}{4\bar{\gamma}_1\bar{\gamma}_2}\right)W_{-j,\frac{1}{2}}\left(\frac{M\gamma_{th}P_{D,i}}{2\bar{\gamma}_1\bar{\gamma}_2}\right)\right] \quad (20)$$

$$P_{out} = \left[1-\exp\left(-\frac{\gamma_{th}}{\bar{\gamma}_0}\right)\times\left(\frac{\gamma_{th}}{\bar{\gamma}_0}P_I+1\right)^{-N_I}\right]\times\left[1-\left(\frac{\gamma_{th}}{\bar{\gamma}_1}P_R+1\right)^{-N_R}\times(N_D!)\exp\left(\frac{M\gamma_{th}P_D}{4\bar{\gamma}_1\bar{\gamma}_2}\right)W_{-N_D,\frac{1}{2}}\left(\frac{M\gamma_{th}P_D}{2\bar{\gamma}_1\bar{\gamma}_2}\right)\right] \quad (21)$$

expression of outage performance is derived. The result shows the outage performance will deteriorate when the system is subjected to multiple interferences effects.

ACKNOWLEDGMENT

This work is supported by a project of the Fundamental Research Funds for the Central Universities (No. 31920140066).

REFERENCES

[1] Suchitra, S. and Kyung Hi, C. 2010. Closed-Form Solution of Outage Capacity for Cooperative DF and AF Relay Network. In Wireless Pers Commun.54: 651–665.

[2] Qijia, L., Xiaoli, M.G. and Tong Z. 2011. A General Diversity Gain Function and Its Application in Amplify-and-Forward Cooperative Networks. In IEEE Transactions on signal processing 59 (2).

[3] Selvaraj, M. D. and Ranjan, K. M. 2008. Performance Analysis of Two-User Cooperation Diversity. IEEE: 619–623.

[4] Xuesong, L., Shi, J., Wenjin, W., Xiqi, G. and Kai-Kit, W. 2012. Outage Probability of Amplify-and-Forward Two-Way Relay Interference-Limited Systems. In IEEE Trans on vehicular technology 61(7): 3038–3049.

[5] Daniel, B. C., Haiyang, D., Michel, D. Y. and Jianhua Ge. 2012. Two-Way Relaying in Interference-Limited AF Cooperative Networks Over Nakagami-m Fading. In IEEE Trans. 61(8): 3766–3771.

[6] C. Zhong, S. Jin, and K.-K. Wong, "Dual-hop systems with noisy relay and interference-limited destination," IEEE Trans. Commun., vol. 58, pp. 764–768, Mar. 2010.

[7] Suraweera, H. A., Garg, H. K. and Nallanathan, A. 2010. Performance analysis of two hop amplify-and-forward systems with interference at the relay. In IEEE Commun. Letters 64:692–694.

[8] Suraweera, H. A., Michalopoulos, D. S. and Yuen, C. 2012. Performance analysis of fixed gain relay systems with a single interfere in Nakagami-m fading channels. IEEE Trans. Veh. Technol, vol. 61: 1457–1463.

[9] Gradshteyn, I. S. and Ryzhik, I. M. 2007. Table of Integrals, Series, and Products, 7th ed. Burlington, MA, USA: Academic Press.

Future Communication Technology and Engineering – Chan (Ed.)
© 2015 Taylor & Francis Group, London, ISBN: 978-1-138-02777-0

An implementation of robustness testing for a TTCN-3 test system

S.M. Liu
College of Science and Information, Qingdao Agriculture University, Qingdao, China

X.M. Liu & X.F. Shao
Information Department, Beijing City University, Beijing, China

Y.P. Liu
School of Computer Science and Engineering, BUAA, Beijing, China

ABSTRACT: In software testing, robustness refers to the ability to ensure that the software system executes correctly under normal or abnormal conditions. Robustness testing needs a large number of illegal and effective inputs which is one of the keys to the researches. The data variance is based on mutation testing which is a type of white box testing through destructing the system under a test for robustness testing. The variance data is a type of data which can determine the robustness by applying some interference to the test input, in order to create abnormal data which is then used by the system under test. As TTCN-3(Testing and Test Control Notation) is used for black box testing, it can get the data rule description from the system under test. The data variance is also used for TTCN-3 testing. Therefore, the automatic generation of robust illegal inputs is realized on the basis of the data variance in this paper.

Keywords: TTCN-3, mutation testing, data variance, system under test, robustness testing, test case.

1 INTRODUCTION

Traditional software testing methods are mostly used to validate whether the software's behaviour accords with its demands, namely function testing or not. The abnormal test is often not paid enough attention. With the software used in the important fields of politics, military and finance, the robustness of software has been paid more and more attention. Robustness describes the degree of correctly running its functions of a system or a component in high intensity input environment or in an invalid data input. As the input data from different systems are different, the abnormal input of robustness is different too. Therefore it leads to the low reliability of the robustness test case. At the same time, robustness testing needs the test case which must cover all possible abnormalities or attack modes. The number of test cases is usually very high, which makes the development of robustness test cases more difficult.

Testing the distributed systems and protocols by using the illegal input technology of robustness is always a important research topic. Early studies focused on the illegal input technology of robustness based on hardware. Then, the illegal input technology of robustness based on software (SWIFI—Software Wireless Fidelity) was introduced in order to reduce costs and the difficulty of development. For SWIFI, the implementation technology and the method which is used to describe the faults of the illegal input of robustness have been the two main problems which draw more people's attention during the time of studying the robustness. A common method of implementing the illegal input of robustness is to insert an additional illegal input layer of robustness, which is responsible for injecting various communication faults between the IUT(Implementation Under Test) and its underlying protocol. There are two options of inserting the software: the EFA (Electronic Failure Analysis) and the virtual wire select the specific location in the OS(Operating System) kernel to insert it or the Orchestra directly inserts it under the IUT. Regardless of the way, the illegal input layer of robustness is usually developed as a kernel component OS By defining specific test scripts, the faults can be described as the specific test scripts which can control the test and the illegal input of robustness. The languages used by the test scripts include TCL (Tool Command Language, such as Orchestra), C (such as EFA), and self-defined languages (such as FIAT and Virtual Wire). The disadvantage of test script is that its usability is poor. First, the testers need to learn the syntax rules. At the same time, manual scripting also causes all types of errors easily.

Based on the above analysis, a fault injection is commonly implemented by inserting the illegal input layer of robustness into the current research. However, it is closely related with the operation platform and OS

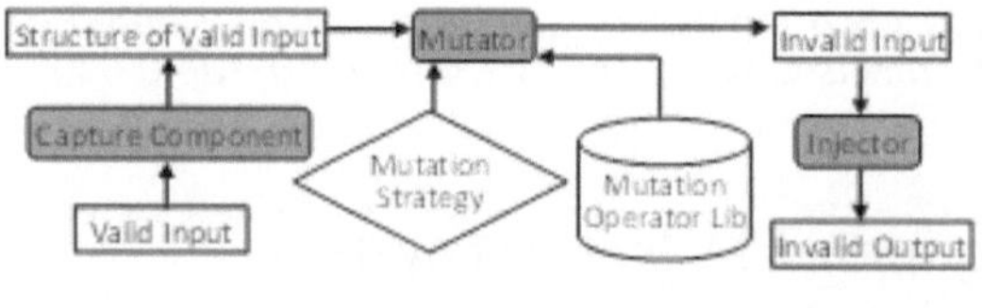

Figure 1. The module framework to generate the illegal input.

which leads to poor portability. Therefore, a new technique to implement the illegal input of robustness is presented in this paper, which can effectively separate the independent parts from the system and improve its portability. Simultaneously, in order to solve the usability of test scripts, this paper presents a method which can be used to describe faults based on a model. The fault activity is defined by the way of visualization in the test, which can automatically generate test scripts from the model and avoid writing the test scripts manually.

2 GENERATE THE ILLEGAL INPUT OF ROBUSTNESS TESTING

Traditional software testing methods focus on verifying whether the software behaves under test according to the demands which is called the function test or not. The abnormal condition tests do not draw more people's attention. Different software has different input data which makes a different illegal input of robustness, and leads to the low reusability of test cases. At the same time, robustness testing needs the test cases, which must cover all possible abnormalities or attack modes. The number of test cases is usually very high, which makes test case development more difficult. The most typical method of robustness testing is to generate a large number of illegal inputs to test the SUT (System Under Test) tolerance to abnormal data.

2.1 *The module framework to generate the illegal input*

In TTCN-3, the abstract layer is very high and its test set is a type of self-specification. Therefore in this paper, mutation operators are defined as being based on TTCN-3 data types. The legal data unit defined in the set of TTCN-3 test cases is made as input, which is mutated by selecting the mutation operator. By using some mutation strategies, a large number of illegal inputs are produced. Therefore, the method to generate the illegal input of robustness testing is also based on a specification. The module framework to generate the illegal input is shown as Figure 1.

The captured component is used to obtain the grammatical structure of a legal input. A mutation component is used to generate an illegal input by the mutation operator, according to the grammatical structure of legal data. The injector sends the illegal input to the SUT. In addition, the mutation operator library contains all the mutation operators which have already been realized. The mutation strategy describes the rules to select the mutation operators. Both of them will serve as the input of a mutation component.

When generating the illegal input, it is very important to capture the legal input and analyse its semantic structure, which is related to selecting a mutation operator, and implementing the mutation. There are two methods to capture the legal input. They are static analysis, and dynamic capture. The static analysis means to analyse the abstract test set of TTCN-3 by using the parser and then extract a semantics tree. The dynamic capture is used to capture the calling of an encode function by inserting codes between TE and CD. The parameter value of the function is the message instance with a semantic structure, as shown in Figure 2.

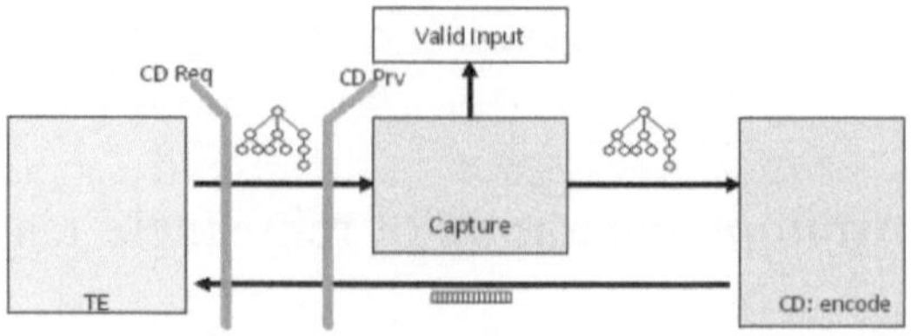

Figure 2. Capture the legal input dynamically.

Table 1. The differences between the mutation testing and the data variation.

Types of Mutation	The Objects of Mutation Operator	The Output of Mutation	Function
Mutation Testing (Program Variation)	The source code of the SUT	The variants of program	Detects the adequacy of test
Data Variation	The input data of SUT	The abnormal input data	Detects the robustness of SUT

2.2 *The method used to select the mutation operator*

The concept of a mutation operator derives from mutation testing which is a type of testing method oriented towards detecting defects. The data variation method proposed in this paper is different from the mutation testing because the data variation focuses on the data sent to the SUT, whilst the mutation testing is a type of program variation which focuses on the source code of the system under test. The differences are shown in Table 1. Because the objects of the mutation operator are different, the factors concerned are different when designing the mutation operators.

The data variation focuses on motivating the SUT. Therefore, it is regarded as an operation with sensitive data types. A message instance is made up of the structure and value which can be mutated, that is, breaking the data structure and updating the data value, or combining both of them. On the other hand, what type of mutation is applied to the data structure and the data value? It is also concerned with the factors in

Table 2. The mutation operator corresponding to the data type.

Mutation Operator	Data Type	Remarks
Boundary Value	integer	be possible to lead to data overflow
++/−−	integer	be possible to lead to data overflow
Insert the special characters repeatedly	char string	lead to the string buffer overflow
Insert the format characters repeatedly	char string	break the data string structure
Insert the special characters	char string	break the data string structure
Null string	char string	be possible to lead to the abnormal null pointe
Lack elements	record, set, record of, set of	be possible to lead to the abnormal null pointe
duplicate elements	record of, set of	lead o the abnormal data type
Elements disorder	record if	break the original order of data domain

designing the mutation operation. The mutation operators designed in this paper and the TTCN-3 data type are shown in Table 2. To be explained, because the focus of this paper is the method of generating illegal inputs based on data variations and the design of the mutation operator is not very comprehensive.

As different data types are corresponding to the different mutation operators, and a mutation operation may adapt to a variety of data types, the mutation operation is determined according to the data type and then chosen before executing the data variation. Therefore, in this paper the operation selection rule is described by use of a variation strategy. The focus of this paper is not how to select the mutation operator. For ease of implementation, the mutation strategy used in this paper is a random selection. First, determine the set of mutation operators according to the data type, and then select an operation from the set randomly.

Algorithm: Select the Mutation Operator

```
INPUT:   Operators := {operatori | operatori is one of mutation operators in the mutation
                       operator library, operator, ∈Operators}
         InputType := {field, | field, is the element of valid input}
         Map := {<Operator, Type>}
OUTPUT: Operator := {operator | operator ∈Operators}
ALGORITHM BEGIN
    Set<Operator> operators;
    Count := 0;
    FOR EACH entry<Operator, Type> IN Map
        Set<Type> types = entry.getTypes (InputType);
        IF InputType in types
            operators.add (entry.getOperator ());
            Count++;
        END IF
    END FOR
    Index := random(Count);
    Operator := operators.getOperator(Index);
    RETURN Operator;
ALGORITHM END
```

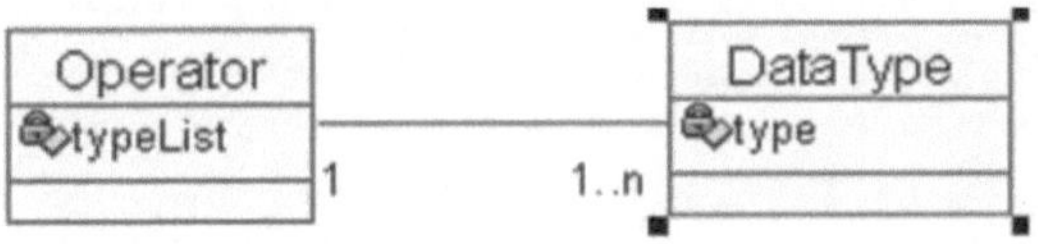

Figure 3. The relationship model between the mutation operator and the data types.

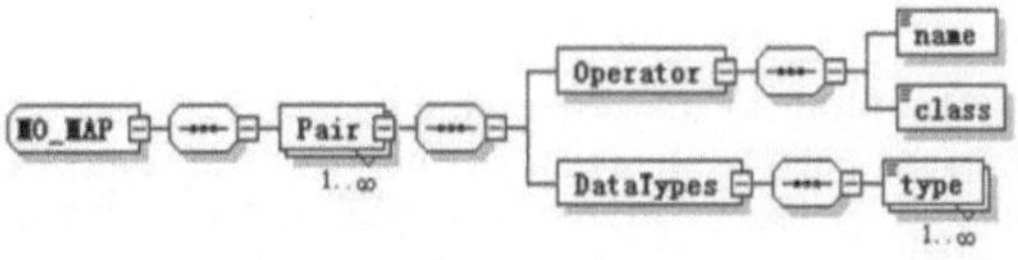

Figure 4. The schema of the relation between the mutation operator and data types.

Figure 5. The execution process of robustness testing.

In order to play the role of mutation strategy, in this paper, the corresponding relationship model is built between the mutation operator and the data type, which is shown in Figure 3.

The corresponding relation between them is based on the consistency of data type properties which is described in XML, as shown in Figure 4.

3 EXPERIMENTS

The object under test is, in this case, the TCP subsystem. The subsystem receives messages sent by the client, and then the processed messages are sent to the client. The data type of test data defined in the function test set of the TCP subsystem is shown as follows:

type record tcpPacket {

charstring packetType length(2),

charstring packetLength length(2),

charstring packetData

During the test, the test data are mutated by use of the abnormal input generation module with robustness, and the execution process is shown as Figure 5.

In order to clearly observe the implementation process of the data variation, the testing information collection module records the interaction information between the testing system and the SUT, as shown in Figure 6.

			… := "03", packetData := "TCP" }
2	OUT	1229187789187	{ packetType := "01", packetLength := "03", packetData := "TCP" }
3	IN	1229187789203	{ packetType := "01", packetLength := "03", packetData := "TCP" }

Figure 6. The interaction information log of robustness testing.

According to the test execution process, and the message interaction records, the first test data are applied by the structured data type mutation operators. Lack of its pack type and pack length leads to the first data case execution failing. But then the test data is not mutated and the test is executed successfully, and the correct response will be obtained. It can be seen that the TCP subsystem under test has a certain tolerance for the abnormal input whilst lacking data domain value.

4 SUMMARY

The abnormal input generation module and the test information collection module with robustness greatly enhance the applicability of TTCN-3 testing technology. In this paper there are better innovations in the implementation of the technologies of the test adapter framework, the mapping relations between the test adapter entity and the SUT, and the methods generating the abnormal input based on data variations.

The method of generating the abnormal input based on TTCN-3 is one of the main researches in this paper, which can effectively improve the efficiency of developing TTCN-3 test system, and reduce the test cost. During the implementation, many related technologies were effectively used, such as reverse engineering, data modelling, and code generating, all of which achieved good results.

ACKNOWLEDGEMENTS

This paper is contributed by National Key Laboratory of Software Development Environment Fund (SKLSDE-2010ZX-15) and High Technology Research and Development Program of China 863 Fund (2009AA01Z145) and Beijing City Special Finance Fund.

REFERENCES

Arlat J, Aguera M, Crouzet Y, et al. Experimental evaluation of the fault tolerance of an atomic multicast system [J]. IEEE Trans Reliability, 1990, 39(4): 455–467.

Dawson S, Jahanian F, Mitton T. ORCHESTRA: a fault injection environment for distributed systems[C]//26th Int'l Symposium on Fault- Tolerant Computing (FTCS), Jun 1996.

De P, Neogi A, Chiueh T-C. VirtualWire: a fault injection and analysis tool for network protocols[C]//Proc IEEE 23rd International Conference on Distributed Computing Systems, Providence, Rhode Island, USA, 2013.

Du Wenliang, Mathur A P. Vulnerability Testing of Software System Using Fault Injection. Coast TR 98-02, 1998.

Echtle K, Leu M. The EFA fault injector for fault- tolerant distributed system testing [C]//Proc Workshop on Fault-Tolerant Parallel and Distributed Systems, Amherst, USA, 2012.

FOKUS. Unified modeling language: testing profile, version 2.0. OMG Adopted Specification. Deutschland, 2004–04.

Gábor Ziegler, György Réthy. Performance testing with TTCN-3 [R], TTCN-3 User Conference, 2006

Gamma E, Kent Beck. Contributing to eclipse [M]. [S.l.]: Adision Wesley, 2013.

Rosenberg J, Schulzrine H, Camarillo G, et al. SIP: Session intiation protocol, RFC3261 [S], 2012.

Segall Z, Vrsalovic D, Siewiorek D P, et al. FIAT-Fault injection based automated testing environment [C]/ /Proc FTCS-18, Tokyo, Japan, 1988: 102–107.

Future Communication Technology and Engineering – Chan (Ed.)

The design and implementation of a hierarchical agriculture information resource management system

M.M. Ju, H.H. Zhang, C.Y. Zhai, P.P. Xin & C. Gao
College of Mechanical and Electronic Engineering, Northwest A&F University, Yangling, Shaanxi, China

ABSTRACT: With the development of agricultural information applications in recent decades, many agricultural information resources have emerged. Due to the lack of a uniform standard, those resources are characterized by heterogeneity, mass, and distribution. How to manage them efficiently has become an important research topic. This paper presents a Hadoop-based hierarchical Agriculture Information Resource Management System (AIRMS), through which heterogeneous information management can be realized. The architecture of the system includes four layers: a resource layer, a physical resource management layer, a logical resource management layer, and a resource reveal layer. The resource layer is made up of heterogeneous resources such as documents, database, service and a knowledge base, etc. The physical resource management layer consists of a Hadoop Distributed File System (HDFS), an Enterprise Service Business (ESB), and a device interaction management system. The logical resource management layer provides interfaces by calling the physical resource management to achieve metadata management, information sort management, resources description information management, and safety management. The resource reveal layer describes the information display forms, and access interfaces. An index-based (Extensible Markup Language) XML parser is used to interact among the modules. The AIRMS has been applied in Beijing, Shandong, Zhejiang and the other four provinces of China, and 16 types of first-class, and 126 types of second-class resources have been integrated into this system.

1 INTRODUCTION

Currently, agricultural information resources are widely distributed in different regions' personal computers, servers, and data storage devices. The result is a low integration and a poor sharing of resources, and it has such characteristics as diversity, and isomerization resources. How to effectively integrate and manage the mass heterogeneous resources needs to be settled. However, heterogeneous and effective organization, and the management of agricultural information resources are the basis of resource integration, and sharing.

Currently, there has been a lot of research into information resources management. Pan Yuchun designed a Web GIS-based system for crop quality monitoring, and planting optimization. The system was supported by a component, (Geographic Information System) GIS, and the grid GIS and spatial database technology, and it realized the data source of multi-source data analysis, model, and the flexible management of knowledge [1]. The technology of distribution storage in cloud computing is mostly researched data storage, organization and management, and it provided the upper application of a safe, reliable, scalable, and efficient data storage service [2]. Cui Jie proposed a distributed computing technology to manage and store massive amounts of data methods in order to construct massive scientific data storage platform solutions, using Linux cluster technology to design a Hadoop-based mass data storage platform [3].

Traditional large-scale data processing mostly uses distributed high performance computing and grid computing technology, which needs expensive resources. In order to achieve large-scale data for effective segmentation, and a reasonable allocation of computing tasks, it requires tedious programming. This paper puts forward a design of a heterogeneous agricultural information resource management system for effective management of agricultural information resources in Beijing, Shandong, Hunan, Zhejiang, Hebei, etc. which realizes the effective management of agricultural information resources in heterogeneous fields through HDFS [4] and ESB [5].

2 SYSTEM ARCHITECTURE

In order to manage and integrate resources effectively, this paper presents a heterogeneous agricultural information management system architecture, which mostly describes functions by their stratifications. The system's architecture includes four layers: a resource layer, a physical resource management layer, a logical resource management layer, and a resource reveal layer (Figure 1).

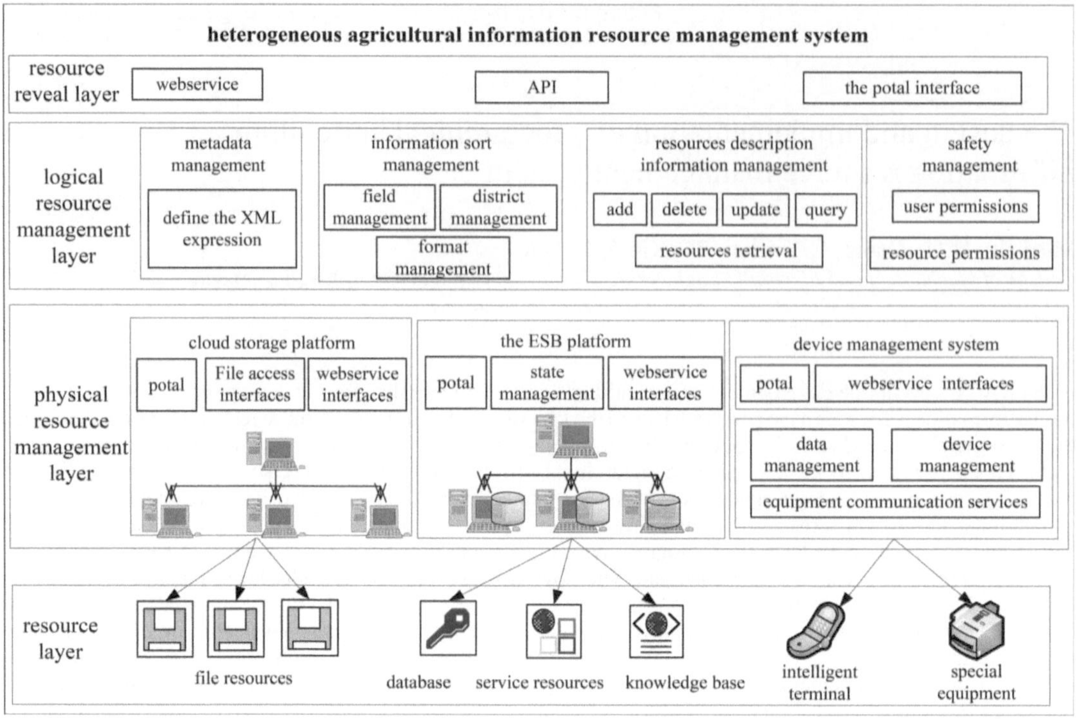

Figure 1. Heterogeneous agricultural information management system architecture.

The heterogeneous resource pool is a provider of real resources, including file resources, database resources, service resources, knowledge resources, and equipment resources, etc. All of these resources have the characteristics of heterogeneousness and diversification.

The physical resource management layer consists of an HDFS, an ESB platform, and a device interaction management system. It provides a universal description and a universal access interface to the resources. It will serve as a common behaviour of abstract resources presented to the logical resource management layer.

The logical resource management layer provides interfaces by calling the physical resource management to achieve metadata management, information sort management, resources description information management, and safety management.

The resource reveal layer describes the information display forms and the access interfaces. Developers can use the Portal, Web Service, and API functions, provided by the resource reveal layer as follows: resource classification management, add, delete, modify, and view.

3 RESOURCES PHYSICAL ORGANIZATION

3.1 *Cloud storage system*

This platform uses HDFS to realize the integration of document resources. A distributed file system is mostly used to store a massive growth of agricultural information data, such as equipment, intelligent search, and log data.

HDFS uses the master/slave architecture, and the cluster consists of a namenode and some datanodes. The namenode performs file system namespace operations, such as opening, closing, and renaming a file or a directory [6]. The datanodes under the namenode scheduler perform data blocks to create, copy, delete, etc. This module mainly introduces basic file operations, such as reading, creating, writing, deleting, and querying. This system uses Axis 2 [7] components to release the file operations function into a web service interface.

3.2 *ESB architecture*

The ESB includes an ESB container, a service catalogue, an adapter framework, a news intermediary, and a channel. This module will switch heterogeneous resources, such as web applications, a database, and knowledge, into the corresponding format using the corresponding adapter as follows: the HTTP protocol, the database adapter, and the knowledge adapter. Then, it will reuse message mediation for heterogeneous data integration, mostly including data structures, types, and form transformations.

For the integration of heterogeneous data, we switch all kinds of heterogeneous data types into the XML schema, and then use JMS [8] of the asynchronous communication mechanism to realize the XML message transmission.

3.3 *Device management system*

According to specific requirements there are the following five types of terminal equipment, such as: portable agricultural information service terminals, terminal data acquisition and processing, an agricultural information mobile search intelligent terminal, the agricultural enterprise specialized information terminal, and a precise automatic control terminal. This module realizes all the terminal equipment registration, data interaction, and the data management capabilities unified management. The equipment interaction and the management system accomplished an interworking of all the data from the equipment and servers, provided the information push function to the particular device, and a unified interface to the cloud platform to obtain equipment data.

4 RESOURCE MANAGEMENT

The logic resource management consists of four modules such as: metadata management, classification information management, resource description information management, and security management (Figure 2).

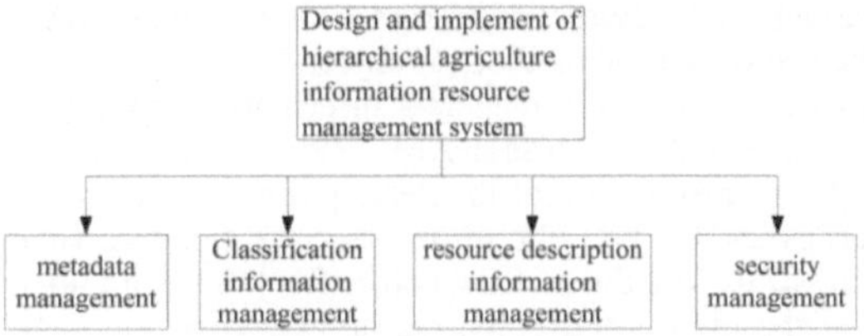

Figure 2. Logic resource management module.

4.1 *Metadata management*

Metadata management is the basis of resource registration, of the classification of information management, of resource description information management and of security management. According to the description of the metadata standard, metadata management is for cloud storage data of the relational database and a distributed file type achieves the effective management of metadata.

Metadata in agriculture mostly involves the following three types of information: First the basic information on resources: district, field, resource type, and user permissions. Second, a metadata functional description which should involve the descendant, date of upload, creation time, etc, when uploading files. Third, a specific metadata description, namely resource name, resource keyword description information resource description and whether this should be for the public or not. The basic information of resources and specific metadata description belongs to each metadata and contains, metadata functional description metadata for each specific field.

At present, the hash algorithm or its improved algorithm is one of the commonly used techniques in the implementation of a cloud storage system. However, a cloud storage system based on a consistent hashing algorithm often shows the key/value pair, which itself does not have hierarchy tree directory structure. Therefore, it does not apply to traditional file systems' applications. Because of the above technical shortcomings, this paper builds a hierarchical directory on the basis of consistency hash data distribution [9] by using the solutions of the metadata management cloud storage directory [10]. It has large-scale deployment, as well as good scalability and flexibility.

The Dublin Core (DC) metadata is one of the most influential metadata formats in many metadata standards all over the world [11], because the DC is simple, scalable, repeatable, and easy to understand and use. It can solve the problems of the information resources management well, and it can be widely used in various fields.

4.2 *Classification of information management*

Classification information management mainly uses three classification methods as follows: The first is the classification by resource format. The resources were divided into data, database, text resources, multimedia resources, spatial data, knowledge, services, etc. The second is the classification by geography which is mainly according to the province and the city level administrative region coding table: Beijing 110000, Tianjin 120000, and Hebei 130000. The third is classifications by field, based on the agricultural resources classification method which was put forward by Xu Yong [12]. The 16 types of first-class and 126 types of second-class agricultural resources can be divided into database statistics, market information, agricultural policy regulations, agricultural resources, agricultural standard procedures, agricultural production technologies, agricultural quality products security, agricultural production processing enterprises, farm machinery enterprises and the identification of products; as well as farmers' professional cooperation organizations, agricultural experts, agricultural video education, farmers' brokers and the large planting, agricultural projects, leisure agricultural, agricultural parks, practical technologies, agricultural question-and-answers, and technology results.

4.3 *Resources description information management*

The module realized the functions including add, delete, modify, view, and retrieve resources. Taking into account the diversity of agricultural resources, this module classified and managed the agricultural resources by region and format. This module used the keyword retrieval method to achieve fast retrieval. This paper extracted the metadata information from each resource and saved it as an XML file which covered the critical information of the description resource properties and content, the resources corresponding field classification, the metadata classification ID number, and the address of the physical storage. Then, the XML documents were parsed by the parser, in order to get the meta-information. After the meta-information was extracted, the metadata information was split into different words. In addition, the keywords were added to the master server to retrieve the metadata information.

4.4 *Safety management*

On the basis of the existing permissions system analysis, this paper presents a J2EE-based authorization model [13] based on role-based access control (RBAC) [14]. In this scenario, we proposed two methods of authorization: direct authorization and indirect authorization. Researchers have proved that the scheme has good flexibility and manoeuvrability.

The system managed the agricultural information resources by four basic roles: such as system administrators, provincial regional administrators, ordinary users, and anonymous users. Meanwhile, the system administrators customized user roles and assigned roles to users. When the user enters the system, we must first determine whether the user name and password are correct. For security reasons, we encrypt user passwords through the MD5 algorithm; if the user name and password are correct, then we will determine whether the user belongs to a particular role. If the role is not empty, then query the table on the role of authority. If the role is empty, then query the user permissions' table. Finally, different users entering the same system will present different interfaces because of different permissions.

5 SYSTEM IMPLEMENTATION

Based on the above research of a hierarchical heterogeneous agricultural resources management framework, a heterogeneous agricultural information management system was designed using a J2EE platform. The system has effectively achieved the management of heterogeneous agricultural information systems such as databases, files, knowledge, equipment and other resources, by HDFS, and ESB. For example, if a user enters the keyword Zhejiang Academy of Agricultural Sciences, the system will return the result by retrieving the information required.

6 CONCLUSION

Aiming at the characteristics of agricultural information resources which are massive, distributed, and heterogeneous, this paper gives full consideration to the differences in management techniques and the access of information resources in different regions of the world, and it designs a hierarchical agriculture information resource management system. It utilizes a cloud computing platform and enterprise service bus, in order to manage and store resources, and at the same time supports resource registration with a high level of reliability, good expansibility and maintainability. In addition, the system achieves heterogeneous agricultural information resources statistical analysis classification and management, and the safe and rapid retrieval and utilization of resources. However, the system is currently implemented in only a few provinces. The resource service ability in the future researches should be further improved.

ACKNOWLEDGEMENTS

All the work above is supported by the Science funds for youth F010406): a multi-dimensional service quality model in the heterogeneous resources integration and key technology research. I would like to express my gratitude to those who have helped me during the writing of this paper. I gratefully acknowledge the help of teachers Haihui Zhang and Changyuan Zhai. I do appreciate their patience, encouragement, and professional instructions during this research. Also, I thank Kun Yao, who kindly gave me a hand in software design. Cheng Gao and Pingping Xin.

REFERENCES

[1] Pan Yuchun, Wang Jihua, Zhao Chunjiang. Web GIS-based system for crop quality monitoring and planting optimization [J]. Transactions of the CSAE, 2004, 20(6): 120–123.

[2] Wang Yijie, Sun Weidong, Zhou Song, et al. cloud computing environment, the key technologies of the [J]. Software distribution storage Sinica, 2012, 23 (4): 962–986.

[3] Cui Jie, Li Taoshen, Lan hongxing. Design and Development of the Mass Data Storage Platform Based on Hadoop [J]. Journal of Computer Research and Development, 2012, 49(z1): 12–18.

[4] Tao Yong cai, Zhang Ningning, ShiLei, et al. Researching on Dynamic Management of Data Replicas of Cloud Computing in Heterogeneous Environments [J]. Journal of Chinese Computer System, 2013, 34(7): 1477–1492.

[5] Ceng Wenying, Zhao Yuelong, Qi Deyu. ESB principle, architecture, implementation and application [J]. Computer Engineering and Application, 2008, 44(25): 225–228.

[6] Land Jiaheng. Hadoop actual [M]. Second version ed.: Mechanical Industry Press.

[7] Axis 2 [M].

[8] Li Ningning, Design and Research of a new kind of MOM-JMOM which bases 0n the Java Message Service technology.

[9] Zhao Zhifeng, Zhang Wei. A cloud storage directory metadata management system [J]. Radio and television technology, 2013, 40 (2): 66–69.

[10] Lewin. Consistent hashing and random trees: Algorithms for caching in distributed networks [M]. Massachusetts Institute of Technology, 1970.
[11] The Dublin Core Metadata Initiative [M].
[12] Xu Yong, An Xiangsheng, Wang Zhiqiang. The agricultural resource ontology framework of agricultural resource classification of agricultural network information based on the design of [J], 2009, (10): 8–12, 27.
[13] Fan Xiaokang, He Lianyue, Wang Xiaochuan, et al. A Role Manage Method Based on RBAC [J]. Journal of Computer Research and Development, 2012: 211–215.
[14] Wang Chunquan and implementation [J]. Journal of Shandong Light Industries College of user authority management based on J2EE (Natural Science Edition), 2010, 24 (3): 60–63.

Geoscience and environmental engineering

Future Communication Technology and Engineering – Chan (Ed.)
© 2015 Taylor & Francis Group, London, ISBN: 978-1-138-02777-0

The ozone depletion theory of global warming

P.L. Ward
U.S. Geological Survey, Teton Tectonics, Jackson, Wyoming, USA

ABSTRACT: Mean global surface temperatures have remained essentially constant since 1998 while carbon-dioxide concentrations continue to increase. Record high temperatures and drought were common in North America during 2012–2013, while record rains flooded England. Here we show that ozone depletion caused by anthropogenic chlorofluorocarbons and small, effusive volcanic eruptions explains these and other climate anomalies clearly. The highest-energy ultraviolet-B radiation from the sun is normally absorbed by the ozone layer in the lower stratosphere. When ozone is depleted, as observed since 1970, more ultraviolet-B energy is measured reaching Earth. Greenhouse-gas theory underestimates the thermal effects of ozone depletion because it assumes electromagnetic radiation propagates as waves in space. Radiation transfers thermal energy as frequency, not wavelength. The thermal energy in ultraviolet-B radiation is 48 times the thermal energy in infrared radiation absorbed by greenhouse gases. There simply is not enough thermal energy absorbed by greenhouse gases to cause observed global warming.

1 INTRODUCTION

Mean global surface temperatures changed little from 1945 to 1970, rose sharply until 1998, and have remained nearly constant since then (bars, Fig. 1) (NOAA, 2014b). Meanwhile atmospheric concentrations of carbon dioxide increased monotonically and are now increasing at the highest rates ever observed (NOAA, 2014a). This divergence in trends suggests that increasing concentrations of greenhouse gas may not be the primary cause of global warming.

By 1965, production of chlorofluorocarbon gases (CFCs), used for refrigerants, propellants, and solvents, began increasing rapidly, leading to an increase in total tropospheric anthropogenic chlorine (gray line) (Solomon, 1999). In 1974, scientists discovered that CFCs, when exposed to ultraviolet sunlight in very cold environments in the lower stratosphere, release chlorine atoms that destroy ozone (Molina and Rowland, 1974). It takes CFCs three to five years to be moved up into the lower stratosphere (WMO, 1999). Ozone concentrations in the lower stratosphere began decreasing by 1970 (black line, y-axis reversed) (Staehelin et al., 1998, WOUDC, 2014). Discovery of the ozone hole over Antarctica in 1984 added urgency to negotiate the Montreal Protocol on Substances that Deplete the Ozone Layer by 1989. Under this protocol, emissions of CFCs began to decline by 1993, and increases in ozone depletion stopped after 1996.

Both the onset of rapid warming by 1970 and the cessation since 1998 can be explained by ozone de-pletion caused by anthropogenic CFCs (Fig. 2b). CFCs are very stable, persisting in the atmosphere for as long as a century (Solomon, 1999). Therefore ozone remains depleted, explaining the continuing increase in ocean heat content (Levitus et al., 2012).

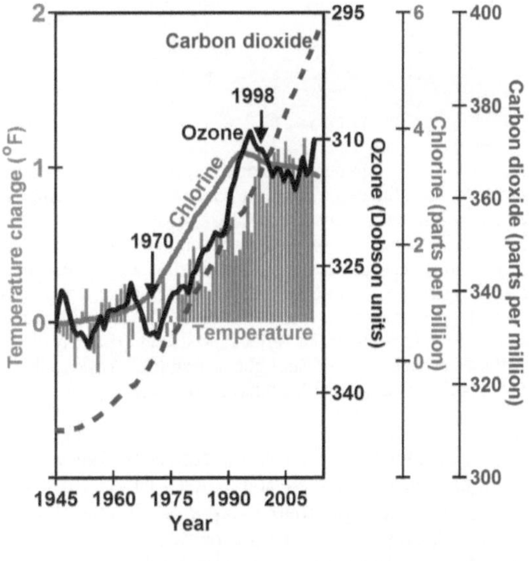

Figure 1. The trend of carbon dioxide diverges from trends in temperature and ozone depletion. Trends in temperature (bars), anthropogenic tropospheric chlorine (gray line) and ozone (black line) over the past 70 years are distinctly different from trends in concentrations of greenhouse gases such as carbon dioxide (dashed line). Increases of chlorine concentrations in the troposphere tend to precede by 3 to 5 years increases in the stratosphere that deplete ozone (Solomon, 1999).

Mean global surface temperatures result from a delicate balance between net thermal energy lost from

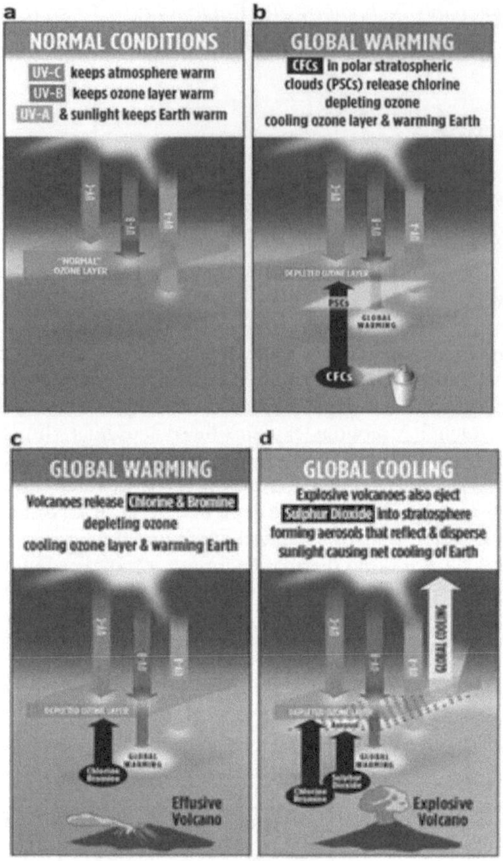

Figure 2. The effects of ozone depletion and volcanic aerosols on climate. When the ozone layer is depleted either by human-manufactured CFCs (**b**) or by chlorine and bromine emitted by volcanic eruptions (**c**), more ultraviolet-B sunlight than normal reaches the ocean, warming Earth. When large, explosive volcanic eruptions eject megatons of sulphur dioxide into the lower stratosphere forming sulphuric-acid aerosols that reflect and scatter primarily ultraviolet-B components of sunlight, thermal energy reaching Earth is decreased, causing global cooling (**d**).

Earth's surface via radiation and convection versus net thermal energy received at Earth's surface from the sun. Greenhouse-gas theory explains climate change by modulating thermal energy lost while ozone depletion theory explains climate change by modulating solar energy received (Fig. 2). There is much more thermal energy in play radiated by the sun than radiated by Earth.

Radiation transfers thermal energy from warmer matter, through air or space, to cooler matter. Thermal energy (heat) in matter is proportional to frequency as discussed below. In 1900, Planck (Planck, 1900) had to postulate that thermal energy (**E**) in radiation is equal to frequency ($\boldsymbol{\nu}$, the Greek letter nu) times the Planck constant (**h**), the number of joules (or electron volts) per cycle per second, in order to write an equation describing observed spectral radiance. $\mathbf{E} = \mathbf{h}\boldsymbol{\nu}$ is used in most photochemical equations to signify the thermal energy required to cause a photochemical reaction typically driven by ultraviolet radiation. When the frequency (energy) is high enough, molecules are "shaken apart", or destroyed.

$\mathbf{E} = \mathbf{h}\boldsymbol{\nu}$ explains why infrared radiation does not have enough energy to be detected by our eyes, but visible light does. $\mathbf{E} = \mathbf{h}\boldsymbol{\nu}$ explains why visible light has enough energy to cause photosynthesis but infrared radiation does not. $\mathbf{E} = \mathbf{h}\boldsymbol{\nu}$ explains why too much ultraviolet light would destroy any rods and cones sensitive to those frequencies in our eyes. As frequency (energy) increases to levels that damage DNA, cumulative length of exposure (dosage) becomes important. This is why extremely small doses of X-rays have enough energy to penetrate our bodies without harm, yet higher doses with the same energy can be focused to destroy cancer cells. Similarly for very high-frequency nuclear radiation, exposure above some minimum dosage causes illness and death. $\mathbf{E} = \mathbf{h}\boldsymbol{\nu}$ is also the fundamental quantum of energy in modern physics.

$\mathbf{E} = \mathbf{h}\boldsymbol{\nu}$ states that a narrow sliver of ultraviolet-B radiation received on Earth when ozone is depleted has frequency (energy) 48 times larger than the thermal energy of a much broader band of infrared radiation absorbed most strongly by carbon dioxide, because thermal energy is a function of frequency, not wavelength or bandwidth as currently assumed based on Maxwell's wave equations (Maxwell, 1865). Waves inhabit matter, not space or some luminiferous aether. Infrared radiation emitted from a pot of lukewarm water feels warm while ultraviolet-B radiation feels hotter than an oxy-acetylene torch. A torch warms Earth faster and to a higher temperature than a pot of lukewarm water. A little ultraviolet-B radiation burns our skin – something an ocean full of lukewarm water could never do.

Ozone is also depleted by chlorine and bromine gases emitted by explosive, effusive, and even quietly degassing volcanoes (von Glasow, 2010). Large explosive eruptions also eject megatons of sulphur dioxide into the lower stratosphere, forming an aerosol that reflects and scatters primarily the ultraviolet components of sunlight, causing net global cooling of $\sim$0.5°C for three years (Fig. 2d). All other volcanic eruptions not explosive enough to form aerosols, deplete ozone, causing net global warming for up to one decade (Fig. 2c).

2 THERMAL ENERGY IS FREQUENCY

Heat (temperature) in matter results from very high-frequency oscillations of all chemical bonds that hold matter together. At room temperature, these oscillations are on the order of 30 terahertz (30 million cycles per second). These very high frequencies reflect the very small atomic dimensions of the oscillators. Each bond oscillates around a minimum in potential energy between electrostatic forces that attract oppositely charged atomic and molecular components when they are near and electrostatic forces that repel similarly charged atomic and molecular components when they are too near.

Each chemical bond oscillates at some frequency over some amplitude determined by the strength of the bond and how much the bond expands and contracts as a function of temperature. Heating bonds increases both frequencies and amplitudes – increasing volume, decreasing density. Because all molecules in matter touch contiguous molecules, these frequencies and amplitudes can flow mechanically via conduction, becoming homogenised over some portion of matter, reaching thermal equilibrium where heat no longer flows. Heat flows spontaneously from hot to cold, from higher frequency and amplitude to lower frequency and amplitude. The rate of flow is proportional to the differences.

In 1900, Max Planck (Planck, 1900) formulated empirically an equation (Planck's law) describing the observed spectral radiance, the amount of radiation (the brightness) measured at each frequency as a function of the absolute temperature of matter. His equation assumes matter is a black body, a perfect absorber and emitter of radiation, whose surface is at thermal equilibrium. While these curves represent spectral radiance just above the surface, they must also represent the frequency and amplitude of oscillations on the surface. The motion of charge is well observed to generate an electric field that induces a magnetic field, that induces an electric field, ad infinitum, forming electromagnetic radiation. This is how radio signals are transmitted.

The solid curves in Fig. 3 show spectral radiance emitted from bodies at various temperatures. Note (1) temperature of matter, as we perceive it, is a function of a broad spectrum of oscillations, (2) the width of this spectrum increases with increasing temperature, (3) spectral radiance (amplitude) at each frequency is determined by the temperature of the emitting body and increases with increasing temperature, (4) the frequency of the peak spectral radiance increases rapidly with temperature, (5) the broad spectrum of radiation from the surface of a body at one temperature does not overlap, at any frequency, the broad spectrum of radiation from the surface of another body at any other temperature, and (6) the broad spectrums do overlap when the bodies are at different distances so that their amplitudes (spectral radiance) are reduced by one over the square of the distance travelled.

Note that a body can only be warmed by radiation from a warmer body that contains higher frequencies and higher amplitudes. The dashed black line shows Wien's displacement law, the frequency (ν) of peak spectral radiance as a function of a body's temperature (**T**) where $\nu = 5.88 \times 10^{10}\,\mathbf{T}$. Radiation consisting of a narrow band of frequencies close to this value or larger could warm a body to temperature **T** if absorbed in sufficient quantity. Radiation with peak spectral radiance less than ν can only warm a body to a lower

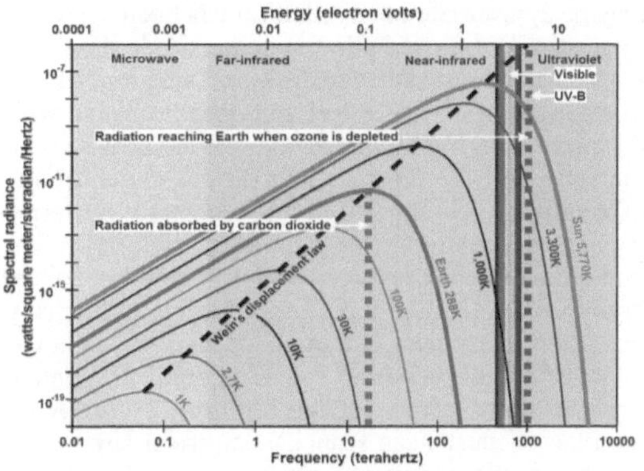

Figure 3. Planck's law for spectral radiance, the amount of radiation at each frequency. These curves show the spectral radiance for bodies of matter at different absolute temperatures. The dashed black line shows the peak spectral radiance for each temperature and can be thought of as the temperature of the radiation. A body can only be warmed by radiation with higher peak spectral radiance. The dotted grey line in the middle of the plot shows the peak spectral radiance for Earth (288K) absorbed by greenhouse gases such as carbon dioxide. The dotted grey line to the right shows the peak spectral radiance for ultraviolet-B radiation (16,500K) that reaches Earth when ozone is depleted. Spectral radiance should not include units of watts as explained in the text.

temperature. This means that it is physically impossible for Earth to be warmed by its own radiation as assumed, for example, in energy budget calculations for greenhouse gases (Trenberth et al., 2009). Ultraviolet-B radiation that reaches Earth when ozone is depleted has the potential to warm Earth to 16,500K, but luckily only very small amounts are available and only during daytime.

Radiation just above the emitting surface consists of frequencies of oscillation with the amplitude at each frequency equal to the amplitude of oscillation on the surface of the emitting body. While conduction homogenises frequency and amplitude within matter to form a Planck distribution, in air or space, frequency and amplitude are completely independent. Frequency (colour) does not change even over galactic distances, except for Doppler effects, while amplitude decreases as the inverse square of the distance travelled because amplitude becomes spread out over a surface area that increases with the square of the distance. For radiation, amplitude is the brightness of the light that we see, the radiance, and it is set by the amplitude of oscillations on the surface of the emitting body.

In the 19th century, spectral radiance was measured by using a prism to separate frequencies spatially in order to measure the amount of radiation in each narrow frequency band using a thermocouple, a thermopile, or a temperature sensitive resistor. These measurements did not measure radiation directly – they measured the change in temperature of matter caused by radiation. This is why spectral radiance was thought of in terms of watts per square meter. This may be appropriate once radiation has been absorbed by matter. But for radiation in space or air shown by Planck's law, frequencies on the x-axis are proportional to energies (watts) (**E** = **h**ν) and do not change with distance while amplitudes on the y-axis decrease with the square of the distance and do not contain units of energy or power (watts). Unfortunately all terms we have describing the amount (amplitude) of radiation include units of power (watts): spectral radiance, brightness, intensity, flux. Even luminous intensity measured in candelas is thought of as watts. Spectral radiance on the y-axis in Fig. 3 should not contain units of watts, which can be removed by replacing the Planck constant in Planck's empirical law with a constant without units of Watts.

3 OZONE ABSORBS ULTRAVIOLET-B RADIATION

Ozone makes up only about 0.6 parts per million of the atmosphere by volume, but it is continuously created and destroyed in an endless photochemical Chapman cycle (Chapman, 1930), absorbing ultraviolet radiation to maintain an "ozone layer" between 15 and 35 km above Earth's surface. The stratosphere is heated daily primarily by continual dissociation of molecular oxygen (O_2) and ozone (O_3) where the molecular/atomic pieces fly apart at high velocity, increasing the average translational kinetic energy of all gas molecules, which, according to the ideal gas law, increases the temperature of air. The optical thickness of this layer varies regionally by the minute, time of day, season, and latitude (WOUDC, 2014). Ozone tends to accumulate at the poles between mid-winter and early-spring when increasing sunlight forms ozone faster than it can be destroyed (Fig. 4) (Fioletov, 2008). More ozone during winter heats the lower stratosphere, causing very cold temperatures on Earth. Ozone depletion since 1970 is most noticeable during these times with total column ozone decreasing ~50% each winter in the Antarctic ozone hole and less regularly in the Arctic but reaching 45% in 2011 (Manney et al., 2011). Less ozone cools the lower stratosphere and warms minimum surface temperatures. The greatest warming observed on Earth since 1970 was along the Antarctic Peninsula (Hughes et al., 2007, Mulvaney et al., 2012), central West Antarctica (Bromwich et al., 2013), and in southern oceans (Waugh et al., 2013) in late winter, early spring, all within the Antarctic ozone hole. The second greatest warming was in the Arctic (Jeffries and Richter-Menge, 2012), well-known as Arctic amplification.

Between 1978 and 2008, ozone depletion increased ultraviolet-B radiation received on Earth (Herman, 2010) 23% at 50°S, 9% at 35°S, 7% at 39°N, 9% at 50°N, decreasing temperatures in the lower stratosphere ~1°C (Hadley Centre, 2013, Remote Sensing Systems, 2012) (Fig. 5, grey line, upper right) and increasing Earth's mean surface temperatures ~0.6°C (NOAA, 2014b) (Fig. 1, bars).

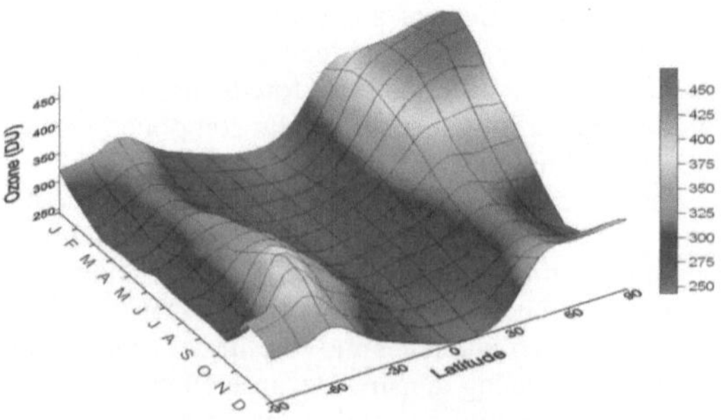

Figure 4. Ozone accumulates during the winter in polar regions. Zonal monthly mean total column ozone in Dobson Units shown as a function of latitude and month estimated from ground-based data for the period 1964 to 1980 (Fioletov, 2008). Excess ozone means more solar ultraviolet-B radiation is absorbed in the lower stratosphere, causing temperatures to become particularly cold near the ground. Ozone depletion is greatest during these late winters to early spring peaks.

Substantial daily changes in ozone, especially north of 40°N, appear closely related to changes in atmospheric and surface temperatures, the depths and locations of surface pressure highs and lows, and position of jet streams (Reed, 1950, Fioletov, 2008) suggesting a link between long-term climate and short-term weather.

4 OZONE DEPLETION AND VOLCANOES

Chlorine and bromine gases emitted by volcanoes deplete ozone (von Glasow, 2010) with one atom destroying ~100,000 molecules of ozone because of the catalytic nature of the Chapman cycle (Molina and Rowland, 1974). Total column ozone, the amount of ozone in a vertical column above a point on Earth, varies rapidly (WOUDC, 2014), but annual averages at each station tend to be similar. It is these averages that relate most directly to annual average surface temperatures.

The longest continuous measurements of total column ozone have been made since 1927 at Arosa, Switzerland (black line, Fig. 5) (Staehelin et al., 1998, WOUDC, 2014). The dashed grey line with black dot data markers shows, for 1964–2009, the annual mean area-weighted total ozone deviation from 1964–1980 means in northern mid-latitudes (30–60°N) scaled from −8% at the bottom of the figure to 10% at the top (Douglass et al., 2011). Years of increasing or decreasing ozone are nearly identical at Arosa and for this mean with small differences in amplitude. Thus the Arosa data provide a reasonable approximation for annual mean total column ozone throughout northern mid-latitudes since 1927.

Ozone at Arosa averaged 331 DU (Dobson units) until 1974, fell 9.4% to 300 DU by 1993 and began rising again until 2011. This long-term decrease has been reliably associated with an increase in anthropogenic

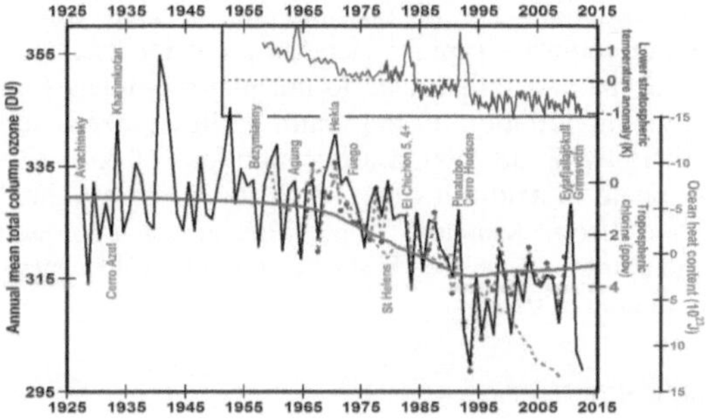

Figure 5. Ozone concentration decreases in the year following most volcanic eruptions. Annual mean total column ozone (black line) (Staehelin et al., 1998, WOUDC, 2014) decreases ~6% during the year following volcanic eruptions, cooling the lower stratosphere (line upper right) (Hadley Centre, 2013, Remote Sensing Systems, 2012) and warming the ocean (dashed line, y-axis inverted) (Levitus et al., 2012). Names of the erupting volcanoes are shown in grey. The solid, relatively straight line shows annual mean tropospheric anthropogenic chlorine with the y-axis inverted (Solomon, 1999) and has been scaled to have a slope similar to the slope for ozone between 1970 and 1991. This figure is described in detail at ozonedepletiontheory.info/depletion.html.

tropospheric chlorine (solid relatively straight line, y-axis inverted) through chlorine catalysed destruction of ozone (Solomon, 1999) (Fig. 2b). Under the Montreal Protocol, production of CFCs was phased out, halting the increase in tropospheric chlorine by 1993.

Note the substantial drop in total column ozone the year following most volcanic eruptions (labelled in grey). Ozone depletion is accompanied by decreasing temperatures in the lower stratosphere (line upper right) (Hadley Centre, 2013, Remote Sensing Systems, 2012) especially following major explosive eruptions of Agung (1963), El Chichón (1982), and Pinatubo (1991) (Randel, 2010, Thompson and Solomon, 2009).

In 1991, the major eruption of Pinatubo depleted ozone 6% but exploded 17 megatons of sulphur dioxide into the lower stratosphere, forming a sulphuric-acid aerosol 15–25 km above Earth (Self et al., 1996), spreading globally within months. The aerosol particles grew large enough to Rayleigh scatter the highest frequencies of sunlight (primarily ultraviolet-B), cooling Earth ~0.5°C for three years (Fig. 2d). The ozone layer was most depleted during late winter, causing warming over continents (Robock, 2002), but the globally distributed aerosols caused net cooling (Fig. 2d). Such cooling has been observed after all major explosive volcanic eruptions throughout human history. The largest, Tambora in 1815, preceded the last cold phase of the Little Ice Age around 1850. When such eruptions happen frequently, their cumulative effects decrease ocean temperatures substantially (Gleckler et al., 2006, Gregory et al., 2006), ultimately ushering in an ice age (Ward, 2009).

The much smaller effusive eruptions of Eyjafjallajökull (2010) and Grímsvötn (2011) also depleted ozone ~6% each but did not eject sulphur dioxide into the stratosphere (Fig. 2c). Record warm temperatures and drought followed in North America during 2012–2013 related in time and location to 14% ozone depletion measured over Toronto Canada (WOUDC, 2014): 3% due to CFCs and ~5.5% due to each of these small eruptions. Fig. 6 shows ozone concentrations over Toronto by month and year and compares winter-spring ozone depletion with minimum surface temperatures (Environment Canada, 2014). Ocean temperatures east of New England reached the highest temperatures ever recorded (NOAA, 2012) likely causing Great Britain's record flooding in 2012–2014. The previous major "Dust Bowl" drought in North America, which peaked in 1934–1936, was also preceded by a highly unusual sequence of seven small volcanic eruptions around the Pacific Ocean from 1931 through 1933.

The greatest warming at the end of the last ice age was contemporaneous with major, basaltic, effusive eruptions in Iceland (Licciardi et al., 2007) between 11,750 to 9,350 years ago (Ward, 2009).

Every 22 million years, on average (Courtillot and Renne, 2003), massive eruptions of basaltic lava covered areas as large as Australia or the contiguous United States. Some of these events erupted up to four million cubic kilometers of basaltic lava over hundreds of thousands of years, contemporaneous with lethally hot temperatures, acidic oceans, major ozone depletion, increased rates of mutation, and major mass extinctions (Reichow et al., 2009, Sun et al., 2012).

Climate, throughout Earth's history, appears controlled by the frequency of major explosive volcanic eruptions causing ozone depletion but forming aerosols that caused net cooling (Fig. 2d) and the duration of effusive, basaltic volcanic eruptions not explosive enough to form aerosols but causing prolonged ozone depletion and associated warming (Fig. 2c). The relative proportion of these two types of volcanism is determined by plate tectonics processes that control the relative motion of tectonic plates. For example, 56.1 million years ago, global temperatures rose 6°C in less than 20,000 years associated with voluminous basaltic eruptions during the initial opening of the Greenland-Norwegian Sea (Storey et al., 2007).

Ozone depletion caused by volcanoes and CFCs appears to explain qualitatively most observed global warming. It will take considerable data gathering and modelling to determine quantitatively whether ozone depletion explains all global warming.

5 WHAT ACTIONS SHOULD WE TAKE?

We humans caused the world to warm ~0.6°C from 1970 to 1998. Heat is accumulating in the ocean (Levitus et al., 2012), resetting Earth's thermostat. The only natural way to lower the thermostat is through increased numbers of major explosive volcanic eruptions.

We have halted the increase in human-caused global warming by reducing emissions of CFCs via the

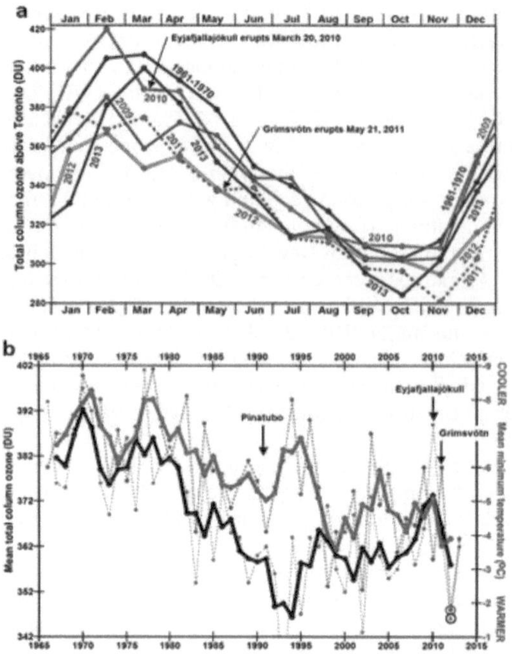

Figure 6. Ozone concentrations above Toronto, Canada, reached a minimum in 2012 when minimum temperatures reached a maximum. a, Total column ozone (WOUDC, 2014) per month in November, 2011 (dotted line), was 12% below the average for Novembers in 1961 through 1970 (upper line) and has remained unusually low throughout 2012. b, When mean total column ozone measured during the months of December. through April (lower lines) decreased, mean minimum temperature (Environment Canada, 2014) for the same months typically warmed (upper lines, y-axis inverted) except following the eruption of Pinatubo in 1991. The dashed lines show annual means; the solid lines are smoothed using a 3-point centered running mean.

Montreal Protocol. We need to strengthen the prohibition against manufacturing gases that deplete ozone and to seek ways to remove these gases from the atmosphere.

We also need to understand better exactly how volcanoes deplete ozone and what our options will be if the rates and types of global volcanism suddenly change.

Reducing emissions of carbon dioxide is not likely to reduce global warming but may slow ocean acidification.

Rapid increases in pollutants such as black carbon formed by incomplete combustion, sulphur dioxide from burning fossil fuels, and ground-level ozone formed as a result of nitrogen oxides, volatile organic compounds, and other pollutants occurred 30 years prior to major warming (ozonedepletiontheory.info/trends.html) suggesting that they do not have much effect on global warming other than absorbing some ultraviolet-B radiation. But these pollutants do have major deleterious effects on public health and many cause acid rains. Most developed countries have reduced pollution to relatively healthy levels, but improvements are still possible and of value. Developing countries rapidly increasing energy use, such as China and India, need to put major emphasis on reducing pollution for the health of their citizens and others living downwind. Concentrations of corrosive sulphate in acid-rains that reach the western United States from Asia and from ships at sea approach and sometimes exceed US maximum allowable levels (Lin et al., 2014).

ACKNOWLEDGMENTS

Thanks to Michael MacCracken and Peter Molnar for critical reviews of many drafts over many years, challenging many of my findings. Thanks to Huiming Bao, James Bjorken, Terrance Gerlach, Peter Giles, Zach Hall, William Happer, George Helz, David Laing, Bertram Raynes, and Adrienne Ward for critical comments. Thanks to the many contributors to Wikipedia for making knowledge so accessible.

REFERENCES

Bromwich, D. H., Nicolas, J. P., Monaghan, A. J., Lazzara, M. A., Keller, L. M., Weidner, G. A. & Wilson, A. B. 2013. Central West Antarctica among the most rapidly warming regions on Earth. *Nature Geoscience,* 6, 139–145.

Chapman, S. 1930. A theory of upper-atmospheric ozone. *Memoirs of the Royal Meteorological Society,* 3, 103–125.

Courtillot, V. E. & Renne, P. R. 2003. On the ages of flood basalt events. *Comptes Rendus Geoscience,* 335, 113–140.

Douglass, A., Fioletov, V., Godin-Beekmann, S., Müller, R., Stolarski, R., Webb, A., Arola, A., Burkholder, J., Burrows, J. & Chipperfield, M. 2011. Chapter 2: Stratospheric ozone and surface ultraviolet radiation. *In:* ENNIS, C. A. (ed.) *Scientific Assessment of Ozone Depletion: 2010.* World Meteorological Organization Global Ozone Research and Monitoring Project – Report No. 52.

Environment Canada. 2014. *National climate data and information archive,* www.climate.weatheroffice.gc.ca/climateData/dailydata_e.html?StationID=5097.

Fioletov, V. 2008. Ozone climatology, trends, and substances that control ozone. *Atmosphere-Ocean,* 46, 39–67.

Gleckler, P. J., Achutarao, K., Gregory, J. M., Santer, B. D., Taylor, K. E. & Wigley, T. M. L. 2006. Krakatoa lives: The effect of volcanic eruptions on ocean heat content and thermal expansion. *Geophysical Research Letters,* 33, L17702.

Gregory, J. M., Lowe, J. A. & Tett, S. F. B. 2006. Simulated global-mean sea level changes over the last half-millennium. *Journal of Climate,* 19, 4576–4591.

Hadley Centre. 2013. *HadAT: globally gridded radiosonde temperature anomalies from 1958 to present*]. www.metoffice.gov.uk/hadobs/hadat/images.html.

Herman, J. R. 2010. Global increase in UV irradiance during the past 30 years (1979–2008) estimated from satellite data. *J. Geophys. Res,* 115, D04203.

Hughes, G. L., Rao, S. S. & Rao, T. S. 2007. Statistical analysis and time-series models for minimum/maximum temperatures in the Antarctic Peninsula. *Proceedings of the Royal Society A: Mathematical, Physical and Engineering Science,* 463, 241–259.

Jeffries, M. O. & Richter-Menge, J. A. 2012. State of Climate in 2011, Chapter 5: The Arctic. *Bulletin of the American Meteorological Society,* 93, S127–S147, www.ncdc.noaa.gov/bams-state-of-the-climate/2011.php.

Levitus, S., Antonov, J., Boyer, T., Baranova, O., Garcia, H., Locarnini, R., Mishonov, A., Reagan, J., Seidov, D. & Yarosh, E. 2012. World ocean heat content and thermosteric sea level change (0–2000 m), 1955–2010. *Geophysical Research Letters,* 39, L10603.

Licciardi, J. M., Kurz, M. D. & Curtice, J. M. 2007. Glacial and volcanic history of Icelandic table mountains from cosmogenic ^{3}He exposure ages. *Quaternary Science Reviews,* 26, 1529–1546.

Lin, J., Pan, D., Davis, S. J., Zhang, Q., He, K., Wang, C., Streets, D. G., Wuebbles, D. J. & Guan, D. 2014. China's international trade and air pollution in the United States. *Proceedings of the National Academy of Sciences,* 111.

Manney, G. L., Santee, M. L., Rex, M., Livesey, N. J., Pitts, M. C., Veefkind, P., Nash, E. R., Wohltmann, I., Lehmann, R. & Froidevaux, L. 2011. Unprecedented Arctic ozone loss in 2011. *Nature,* 478, 469–475.

Maxwell, J. C. 1865. A dynamical theory of the electromagnetic field. *Philosophical Transactions of the Royal Society of London,* 155, 459–512.

Molina, M. J. & Rowland, F. S. 1974. Stratospheric sink for chlorofluoromethanes: Chlorine catalysed destruction of ozone. *Nature,* 249, 810–814.

Mulvaney, R., Abram, N. J., Hindmarsh, R. C. A., Arrowsmith, C., Fleet, L., Triest, J., Sime, L. C., Alemany, O. & Foord, S. 2012. Recent Antarctic Peninsula warming relative to Holocene climate and ice-shelf history. *Nature,* 489, 141–144.

NOAA. 2012. *Sea surface temperatures reach record highs on northeast continental shelf,* www.nefsc.noaa.gov/press_release/2012/SciSpot/SS1209/.

NOAA. 2014a. *Annual mean CO_2 at Mauna Loa* ftp. cmdl.noaa. gov/ccg/co2/trends/co2_annmean_mlo.txt

NOAA. 2014b. *The global surface temperature is rising,* www.ncdc.noaa.gov/indicators/.

Planck, M. 1900. Entropie und temperatur strahlender wärme. *Annalen der Physik,* 306, 719–737.

Randel, W. J. 2010. Variability and trends in stratospheric temperature and water vapor, in The Stratosphere: Dynamics, Transport and Chemistry. *Geophysical Monograph,* 190, 123–135.

Reed, R. J. 1950. The role of vertical motion in ozone-weather relationships. *Journal of Meteorology,* 7, 263–267.

Reichow, M. K., Pringle, M., Al'mukhamedov, A., Allen, M., Andreichev, V., Buslov, M., Davies, C., Fedoseev, G., Fitton, J. & Inger, S. 2009. The timing and extent of the eruption of the Siberian Traps large igneous province: Implications for the end-Permian environmental crisis. *Earth and Planetary Science Letters,* 277, 9–20.

Remote Sensing Systems. 2012. *Description of MSU and AMSU data products,* www.remss.com/msu/msu_data_description.html.

Robock, A. 2002. Pinatubo eruption: The climatic aftermath. *Science,* 295, 1242–1244.

Self, S., Zhao, J.-X., Holasek, R. E., Torres, R. C. & King, A. J. 1996. The atmospheric impact of the 1991 Mount Pinatubo eruption. *In:* Newhall, C. G. & Punongbayan, R. S. (eds.) *Fire and Mud: Eruptions and lahars of Mount Pinatubo, Philippines.* Philippine Institute of Volcanology and Seismology and University of Washington Press.

Solomon, S. 1999. Stratospheric ozone depletion: A review of concepts and history. *Reviews of Geophysics,* 37, 275–316.

Staehelin, J., Renaud, A., Bader, J., Mcpeters, R., Viatte, P., Hoegger, B., Bugnion, V., Giroud, M. & Schill, H. 1998. Total ozone series at Arosa (Switzerland): Homogenization and data comparison. *Journal of Geophysical Research,* 103, 5827–5841.

Storey, M., Duncan, R. & Tegner, C. 2007. Timing and duration of volcanism in the North Atlantic Igneous Province: Implications for geodynamics and links to the Iceland hotspot. *Chemical Geology,* 241, 264–281.

Sun, Y., Joachimski, M. M., Wignall, P. B., Yan, C., Chen, Y., Jiang, H., Wang, L. & Lai, X. 2012. Lethally hot temperatures during the early triassic greenhouse. *Science,* 338, 366–370.

Thompson, D. W. J. & Solomon, S. 2009. Understanding recent stratospheric climate change. *Journal of Climate,* 22, 1934–1943.

Trenberth, K. E., Fasullo, J. T. & Kiehl, J. 2009. Earth's global energy budget. *Bulletin of the American Meteorological Society,* 90, 311–323.

Von Glasow, R. 2010. Atmospheric chemistry in volcanic plumes. *Proceedings of the National Academy of Sciences of the United States of America,* 107, 6594–6599.

Ward, P. L. 2009. Sulfur dioxide initiates global climate change in four ways. *Thin Solid Films,* 517, 3188–3203.

Waugh, D. W., Primeau, F., Devries, T. & Holzer, M. 2013. Recent changes in the ventilation of the southern oceans. *science,* 339, 568–570.

WMO 1999. Scientific Assessment of Ozone Depletion: 1998. World Meteorological Organization, Global Ozone Research and Monitoring Project—Report No. 44.

WOUDC. 2014. *World Ozone and Ultraviolet Radiation Data Center,* Available: www.woudc.org/data_e.html.

Future Communication Technology and Engineering – Chan (Ed.)
© 2015 Taylor & Francis Group, London, ISBN: 978-1-138-02777-0

Ore-controlling factors of the bauxite deposits in eastern part of Guizhou province

Y. Kong
Kunming University of Science and Technology, Kunming, Yunnan, China

ABSTRACT: The Kaili bauxite deposit is of the sedimentary type which was formed on the eroded surface of carbonate. This area underwent many times of movements of the earth's crust during the geological history. Based on ore-forming background and bauxite deposit characteristics in Kaili, this article analyses the main ore-controlling factors of the formation process. It is suggested that the process can be roughly divided into three phases: the deposition and preservation phase, the material forming phase, and the secondary enrichment phase. The primary ore-controlling factors are weathering, paleoclimate, and hydrogeology for the first phase, and marine transgressions and karst for the second phase, respectively.

1 INTRODUCTION

As a great province of bauxite, Guizhou holds nearly 400 million bauxite resource, which amounts to 16% of the total reserves of our country, ranks secondary in China (Wu Guohui 2006, 2008). Kaili region, has been preliminary proved a middle-large size deposit and many middle and small deposits, which is the key bauxite region found in Guizhou recently, Researching the ore-controlling factors of these deposits has theoretical and practical meaning in exploring larger scale bauxite ore resources.

2 GEOLOGICAL SETTING OF ORE-FORMING PROCESSES

Kaili area is located in the east of Guizhou province, Its exposed stratum and lithology include the Loushanguan Group dolostones of Cambrian; limestone, dolomite of Lower Ordovician; quartz sandstone and clay rock of Silurian Weng Formation; dolomite of upper Devonian Gaopo; limestone of Middle Permian Qixia and Maokou Formation (Figure 1).

Structure of study area presents North-East direction. The faults present North-North-East direction, near South-North direction and North-East direction; folds include Dapaomu syncline, Yudong syncline and Kulijing syncline, which respectively controls the distribution of ore-bearing rock series.

During geological times, this area went through several complex movements of the earth's crust. Although experienced repeated crust movement from Cambrian to Devonian, sedimentation and transgression still played the main role; Early Carboniferous began to rise as the land, accepted weathering and peneplain; when the late Hercynian movement of the Permian Liangshan period,this area occured settlement and transgression; Late Triassic crust raise to land and experienced longterm of weathering and denudation till now.

3 GEOLOGIC FEATURES OF DEPOSIT

3.1 *Ore-bearing rock series and features*

The ore-bearing rock series of the Kaili area's bauxite is middle Permian Liangshan Formation (P2l). The bauxite is stored in sedimentary rocks stratum over paleokarst surface at the bottom of Liangshan Formation, which lower strata is dolomite of upper Devonian high field Formation while overlaying strata is limestone of Permain Qixia Formation.

Permian Middle Liangshan Formation can be divided into three strata: lower iron layer, centre bauxite layer, and upper coal-bearing (carbonaceous) layer (Chart 2), amongst which the thickness of lower iron layer and centre bauxite layer varies greatly. The lower iron layer (P2l1) is made of purple (maroon) clay shale and alumina clay shale with nodular siderite strata, between 0 m–16 m thick; The centralbauxite layer (P2l2) is made of bauxite, bauxite shale with clasolite, pisolitic and oolitic bauxite, semi-earthy bauxite, earthy bauxite, and massive bauxite, which is between 0 m–40 m thick; The upper coal-bearing (carbonaceous) layer (P2l3) is made of brown block fault coal, carbonaceous shale, quartz sandstone, some fragments of bauxite, and rocks, soil here is porous with poor sticky.

3.2 *Ore body output features*

The distribution of ore body is mainly effected by the syncline structure, the ore-bearing rock series

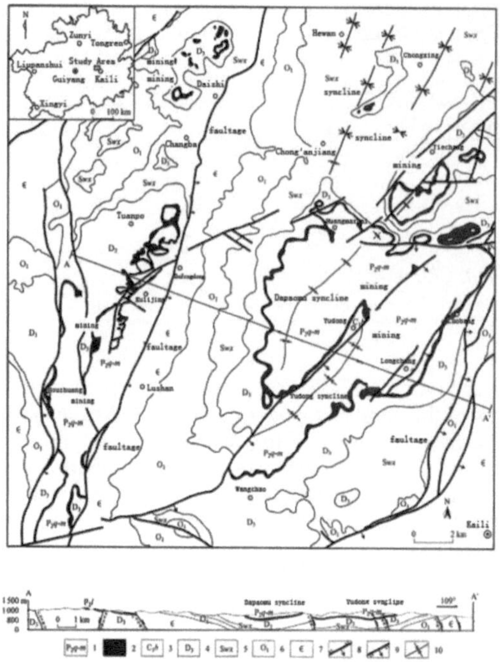

1-limestone of Middle Permian Qixia and Maokou Formation; 2-dolomite of Upper Devonian; 3-quartz sandstone and clay rock of Silurian Wengxiang Formation; 4-Syncline limestone and dolomite of Lower Ordovician;5-dolomite of Cambrian; 6-reverse fault; 7-normal fault; 8-fold; 9-ore-bearing rock series; 10-geologic boundary

Figure 1. Geological map of the Kaili area.

presented as semi-ellipse, and disconformable contact to the underlayer. Morphology of ore body is simple with banding, strip and irregular outcrops and stratiform, stratiform-like, lenticular, lentoid and funnel-form in section. Occurrence and strata occurrence is consistent, the angle of 10°–15°. The length of single ore body extends from tens to over 5,000 meters, wide from tens to over 900 meters and thickness from 0.90 to 17.80 meters, generally 2 to 3 meters. Small amount large thickness ore blocks are commonly distributed in carbonate dolines and karrens of metallogenic section.

3.3 *Ore features*

3.3.1 *Ore types and constitution*

There are two different industrial types of bauxite ore: diaspore low sulphur and low iron aluminium ore, diaspore low sulphur and high iron aluminium ore. Diaspore low sulphur and low iron aluminium ore is the main ore type in Kaili area, generally is pisolitic, oolitic structure, earthy structure, clastic structure, and massive structure. Based on its Al_2O_3 content and A/S value, diaspore low sulphur and low iron aluminium ore is grade 3. Diaspore low sulphur and high iron aluminium ore mainly distribute in No. 1, No. 4 and No. 5 ore body of Yudong ore zone, accounts for 7% of overall ore; High iron aluminium ore of Kulijing ore zone take up 22% of the total. It is generally bean

Rock Member	Thickness (m)	Column Diagram Of Lithoface	Lithology Description
P_2q	80~120		Pale grey medium bed to thick layer limestone
P_2l^3	1.2~12.7		Coal(carbon) bearing stratum; Ash black macker、clay rock with poor quality coal stratum or coal streak, bottom is pale、grayish yellow medium bed small or medium grain quartz sandstone;
P_2l^2	0.8~37.6		Bauxite bearing stratum; Pale、light grey clastic structure、pisolitic, oolitic structure or massive structure bauxite with light grey、grayish yellow medium bed massive structure bauxite rock and bauxite shale;
P_2l^1	0~16		Iron bearing stratum; Amaranth irony clay shale or bauxite shale; tawny、brownish red concretion siderite; light grey、light red clay shale;
D_3gp	130		light grey、off-white to yellow medium bed dolomite;

Figure 2. Column diagram of bauxite-bearing rock series in the Kaili area.

massive ore, the content of Al_2O_3 in the ore generally 43%–60%, SiO_2 content in general is 8%–18%, the content of Fe_2O_3 in 16%–27%. Its A/S value is between 2.32 to 6.60 with an average at 4.3. This type of bauxite is usually of poor quality.

4 ORE-CONTROLLING FACTORS

The Kaili area bauxite is made of a diasporic sedimentary deposit on a carbonatite erosion surface (Liu Ping, 2001; Wang Cuizhi, 2005; Zhai Dongxing, 2002). Formation of bauxite is an extremely complex process, in which structure, weathering, climate, fluid, geomorphology, and biology all play important roles. According to the research on bauxite deposits in Kaili area, this process can be divided into three phases: the material forming phase, the deposition and preservation phase, and the secondary enrichment phase. Each phase has its unique ore-controlling factor of which tectonism is the most important, because crustal movement has a critical role in the entire forming process of bauxite deposits.

Before the early carboniferous epoch, crust movement of Kaili area was mainly on sedimentation, which sedimentated vast terrigenous fragmentary material and carbonate material, especially in Devonian epoch, general trend of sedimentary evolution in this area had changed from coastal debris terrigenous clast sediment to carbonate sediment. During the late Devonian Period, it had been replaced by carbonatite rocks, which provided the basis for the later formation of aluminous material as a product of weathering. Since the early Carboniferous epoch, as a result of uplifted from marine environment to archicontinent by effect of stable epeirogenetic movement, this area was exposed at weathering and denudation during the entire Carboniferous epoch and early Permian epoch. Long-term sedimentation and weathering formed vast aluminous material; in the Middle Permian epoch, under the formation of massive transgression, coast, tidal flat, swamp-lagoon environment, aluminous material deposited and overlying rocks sedimentated over bauxite layer; During the Late Triassic, earth crust of Kaili area uplifted again, after many times of tectonic movements, bauxite layer gradually rising which created conditions for later transformation and optimization (Chart 1).

4.1 *Aluminium source rock analyses*

The formation mechanism of bauxite deposit is fairly complex. The mechanism of the formation of the bauxite deposit is quite complex. Bauxite deposit is a multi stage, multi factor atmospheric conditions formed residual, accumulation or migration accumulation bauxite deposit (Liu Changling, 2005). According to the research on bauxite metallogenic geological background, ore-bearing rock series profile and rate of titanium, we come to a conclusion that material source of bauxite deposit is mainly of basal carbonate rocks weathering residual material with little silicate rocks. After short-distance transferred, aluminous material sediment in karst depression near-by. The above conclusion is based on the following evidence:

(1) According to the research on the Guizhou modern carbonate rocks' weathering crust (Zhu Lijun, 2004; Li Jingyang, 2004), the Al2O3 content in the surface soil of the dolomite red weathering crust is 20.99% to 34.56%. In Carboniferous, latitude of Guizhou was closer to equator than nowadays, the climate environment was easier for enrichment of carbonate rock laterization and aluminous. According to Liu Ping's research in 2001, most Guizhou bauxite is dolomite basement with little shale,

In the favorable climatic conditions, and and calcium laterization, carbonate rock can also form bauxitemetallogenic parent – laterite weathering crust

(2) The output of bauxite deposits in Kaili area is strictly controlled by the distribution of Devonian system. Sedimentary stratum around Kaili area formed before Permian were all carbonate rocks except shale of lower Cambrian system (Guizhou province Geology & Mineral Bureau, 1997). After Ziyun movement occurred from last stage of late Devonian to initial stage of early Carboniferous, the whole area raised steadily, and in the following 60Ma years till middle Permian, it had been continually weathered and denuded. there was no aluminosilicate archicontinent. Then Even in middle Permian, that area was still just a denuded zone covered by carbonate rocks of Devonian.

4.2 *Ore-controlling factors in material formation stage*

Material formation stage occurred in Carboniferous period and early Permian epoch. Kaili area underwent uplifting in Carboniferous Period and transgression in the Middle Permian Liangshan Formation. After the period of sixty million years, of sedimentation hiatus and weathering denudation, this area presented favourable conditions for formation of bauxite material in both space and time.

Table 1. Evolution of crustal movement in the Kaili area.

Geologic age	Crustal moveme	Basic characters	Forming stage of bauxite
Now | Late Triassic	uplift	Overlying rock was weathring and denudation, bauxite deposits were enriched.	Secondary enrichment phase
Middle Triassic | Middle Permian Liangshan	sedimentary	aluminous material deposited and covered by sedimentary rock layers.	Deposition and preservation phase
Early Permian Early | Carboniferous	Uplift	In weathering and denudation stage:Long-term sedimentation hiatus and weathering formed vast aluminous material.	Material forming phase
Devonian	sedimentary	from coast terrestrial sources clast sediment to carbonate sediment, sedimentated vast terrige fragmentary material and carbonate material	
Silurian Carmbrain	Mainly sedimentary	Uplift crustal movement in Ordovician to early Silurian made stratum missing and disconformity relationship	

Table 2. Chemical compositions of stratigraphic column of the bauxite deposits in the Kaili area.

Stratum Code	lithology	Chemical compositions					A/S	Titanium rate
		Al_2O_3	SiO_2	Fe_2O_3	TiO_2	Ignition		
P_2l^3	Carbonaceous layer							
	Earthy bauxite	67.05	9.03	4.46	3	15.22	7.42	22.35
P_2l^2	Pisolitic, oolitic bauxite	57.74	14.73	5.55	1.98	16.35	3.92	29.16
	Bauxite rock	32.6	41.17	4.07	1.58	14.29	0.79	20.63
P_2l^1	Iron caly rock or high iron Bauxite rock	25.67	31.89	18.7	1.16	17.54	0.8	22.13
	Weathering remained	36.39	38.36	3.93	1.36	16.54	0.95	26.76
D_3gp	Weathering dolomite	5.78	7.61	1.17	0.25		0.76	23.12
	Dolomite	0.65	2.26	0.28	0.03		0.25	21.67

4.2.1 *Weathering*

The paleocrust of weathering aluminous material, which constituted the bauxite deposit, was formed by the effect of weathering. Early crustal movement of Kaili area sedimented massive terrigenous detrital and carbonate material that provided favourable conditions for later weathering action. After Carboniferous, earth crust of this area uplifted, till middle Permian Liangshan period before transgression, the sedimentated carbonate rocks suffered 60 Ma years weathering and denudation. During long time physical and chemical weathering, elements such as Na, Ca, Mg and Si in carbonate rocks of surface all run off. Liu Baojun (1980) considered in final stage of weathering (allite- lateritization weathering stage), aluminium silicate minerals were completely decomposed and all removable elements were taken away. The remained iron and aluminum oxide and partial SiO_2 formed lateritic weathering crust rich in aluminous, silicic and irony, in which aluminous enrichment to provide material sources of bauxite deposits formation.

4.2.2 *Paleoclimate*

Paleoclimate is indispensable for bauxite deposits formation. Modern bauxite deposits are mostly distributed in tropical, and rainy areas which are influenced by tropical, and rainy climate. Under high temperature and rainy condition, chemical weathering process is accelerated and abundant. Atmospheric precipitation continually takes weathering compound away to enrich aluminous form aluminous material. An ancient magnetic survey (Wang Junda, 1998; Liao Shifan, 1999) showed Guizhou was located in a tropical zone (south latitude 8 degree to 14 degree) in Carboniferous, during that period, hot, humid and long sunshine duration in Kaili area offered some conditions for further weathering and pedogenesis of carbonate rocks. The paleocrust provided massive aluminous material for bauxite deposits formation.

4.2.3 *Hydrogeology*

Research indicated that modern lateritic bauxite deposits mainly developed in tropical rainy areas and the ore body mainly developed in the higher massif, which has plenty of atmospheric precipitation and good discharge conditions. In Carboniferous, Kaili area was the tropical rain forest areas near the equator, slightly higher then thesea level. Development of karst depression, high precipitation and advanced drainage system benefited laterization process, and caused metallogenic materials gathered. According to the research, bauxite ore of Kaili area was mainly occurred in karst depression.

4.3 *Ore-controlling factors in sedimentary and preserved stage*

The sedimentary and preserved stages mostly occurred between the Middle Permian Epoch (P2) and the Middle Triassic Epoch (P3). In Middle Permian Epoch Liangshan Formation, aluminous material sedimentated, and the following sedimentation has preserved the bauxite deposits until today.

4.3.1 *Transgression*

In Middle Permian Epoch, Kaili area was characterized by transgression, and its function on bauxite deposits were mainly in two aspects. On the one hand, aluminous material formed by weathering under atmospheric conditions on the other hand after the bauxite material deposition, transgression continued the late Triassic Epoch, transgression stopped and overlying sedimentary rocks stratum prevented ore-bearing rock series from suffering erosion, denudation and transition.

4.3.2 *Karst*

Kaili area bauxite deposits were formed by ancient weathering surface erosion of carbonate rocks of Devonian system. On this surface, karst landform development was mainly included karst depression funnel, karst depression and karst cave. There are obvious relations with the thickness, grade and depth of karst depression. As primarily enrichment and mineralization places of bauxite deposits karst landforms has important significance for the preservation of bauxite deposit.

4.4 *Secondary enrichment stage of ore-controlling factors*

The Secondary Enrichment Stage is mainly from the late Triassic epoch till now. Through the two stages aluminous material had sedimentated to rocks but not bauxite deposits because of low grade, which needed further optimizing. In the later period, qualities of aluminous ores were better by secondary enrichment action. The main internal factors of secondary enrichment is the different Chemical propertie between in Al_2O_3 and SiO_2; and the external factor is the water fluidity.

5 CONCLUSION

(1) Kaili area bauxite is formed by a diasporic sedimentary deposit on carbonatite surface erosion. It is an extremely complex process, in which structure, weathering, climate, fluid, geomorphology, and biology all play important roles. According to earlier research, this paper has divided this progress into three stages: the formation stage, the sedimentary and preserved stage, and the secondary enrichment stage.

(2) Amongst all these factors, crustal movements have the most important role in the formation of bauxite deposits, which is the precondition for the above stages. During the formation stage, the main ore-controlling factors are weathering, paleoclimate, the hydrogeological process, ancient landform, and biology. During the sedimentary and preserved stage, the main ore-controlling factors are transgression and karst. During the secondary enrichment stage, the main ore-controlling factors are the hydrogeological process and breakage.

REFERENCES

J.L. Dong. 2004. Geological feature of bauxite in Guizhou province and thoughts of exploration & development [J]. Mineral resources and geology, 18(6): 555–558. (in Chinese)

J.Y. Li & L.T. Zhu. 2004. On modern weathering crust and palaeo-weathering crust of Carbonate rock [J]. Carsologica Sinica, 23(1): 56–62. (in Chinese)

Y. Li et al. 1996. Several problems in the research of metallogenic theories of bauxite [J]. Mineral Resources and Geology, 10(1): 22–26. (in Chinese)

Liao Shi-fan. 1999. Some advances in study of sedimentary deposits [J]. Guizhou Geology, 16(2): 122–129. (in Chinese)

B.J. Liu. 1980. Sedimentary Petrology [M]. Beijing: Geological Publishing House: 8–97. (in Chinese)

C.L. Liu. 2005. Recent progress in study on kaolinite clay and bauxite [J]. Acta Sedimentologica Sinica, 23 (3): 467–474. (in Chinese)

Future Communication Technology and Engineering – Chan (Ed.)
ISBN: 978-1-138-02777-0

Re-Os geochronology of Panjia polymetallic deposit, Zhejiang province, China

B.J. Jia
Zhejiang Institute of Geology and Mineral Resources, Hangzhou, Zhejiang, China
China University of Geosciences, Wuhan, Hubei, China

Z.G. Chen
Tianjin North China Geological Exploration Bureau, Hedong district, Tianjin, China
China University of Geosciences, Beijing, China

J.F. Qian, C. Jin, L.Y. Zhou & X.Y. Zhu
Zhejiang Institute of Geology and Mineral Resources, Hangzhou, Zhejiang, China

ABSTRACT: The Panjia large-middle scale Polymetallic deposit is newly discovered in Zhejiang Province in China. In this paper, five molybdenite samples were collected from Datongkeng ore section of Panjia Ming Area for Re-Os compositional analyses and Re-Os isotopic dating. We got an isochron age of 142.88 ± 0.78 Ma (2σ) with an initial ^{187}Os of 0.02 ± 0.48 (MSWD $= 1.3$). The age reveals that Panjia Deposit formed in a period of the tectonic transition of large-scale lithosphere, and the period is also an outbreak period of mineralization in southern China.

1 GEOLOGICAL SETTING

The Datongkeng ore section of Panjia Ming Area is a newly-discovered, large-middle scale W-Cu-Mo polymetallic in the northwest of Zhejiang Province in China. Tectonically it is located at the combinative region of Yangtze and Cathaysia blocks to the north of Jiangxhan-Shaoxing Fracture zone.

The deposit is located in the Datongkeng-Panjia syncline in northwestern limb of Jiemeng-Qinchuan NE-SW direction synclinorium. The strata include Upper Sinian Lantian Formation in the core and Lower Sinian Nantuo-Xiuning Formation in the limbs of the syncline. The Datongkeng granodioritic porphyry intruded into the core of the syncline along the NE-SW direction thrust fault. The ore bodies mainly occur in the intrusive rock and Lantian Formation. The W-Mo mineralizations mainly occur in the intrusive rock, and W-Cu mineralizations mainly occur in the Lantian Formation. The molybdenite commonly appears in the intrusive body, the contact zone and surrounding formation presenting as fine veins and star-like structures.

2 RE-OS ISOTOPE ANALYSIS

2.1 *Results and discussion*

Five molybdenite samples used for Re-Os isotope dating were collected from the drilling cores of no. 1206 in Datongkeng ore section of Panjia Ming Area. Re-Os isotope dating was performed by the instrument of Thermo Scientific X Series2 ICP-MS at State Key Laboratory of Isotope Geochemistry, Guangzhou Institute of Geochemistry, Chinese Academy of Sciences. The analytical results are listed in Table 1.

Five molybdenite samples used for Re-Os isotope dating were collected from the drilling cores of no. 1206 in Datongkeng W-Cu-Mo polymetallic ore section of Panjia Ming Area. The isochron age calculated by Isoplot software (Ludwig, 2003) is 142.88 ± 0.78 Ma (2σ) with an initial ^{187}Os of 0.02 ± 0.48 (MSWD=1.3) (Figure 1), the weighted mean values of five Re-Os model ages is 142.9 ± 0.78 Ma (MSWD $= 1.03$) (Figure 2). The two ages are very consistent, whichindicates the isochron age is reliable. It is considered that the mineralization was mainly occurred in a transition period of the Jurassic and cretaceous in Datongkeng W-Cu-Mo polymetallic ore section of Panjia Ming Area.

The Mesozoic are considered to be an outbreak period of mineralization in southern China (Hua et al., 1999; Mao et al., 2004). There formed a large quantity of wolfram, stannum, copper, molybdenum, gold, lead and zinc polymetallic deposits related to Yanshan period magmatic movement. The Panjia deposit was formed in the outbreak period and had a direct relationship with the intrusion of Yanshanian Datongkeng granodioritic porphyry.

Early Yanshanian (J_2–J_3) and Late Yanshanian (K_1–K_2) are two important periods showing subduction

Table 1. Molybdenite Re-Os data of the Panjia Cu-W polymetallic deposit, Zhejiang Province.

		Re/μg g^{-1}		^{187}Re/μg g^{-1}		^{187}Os/ng g^{-1}		Age/Ma	
Sample	Weight/g	Content	2σ	Content	2σ	Content	2σ	ModelT	2σ
Ma-4	0.06005	63.37	0.35	39.83	0.22	94.84	0.49	142.7	1.1
Ma-6	0.0184	160.05	0.71	100.6	0.44	239.91	1.93	143	1.3
Ma-10	0.0285	182.48	0.81	114.69	0.51	272.44	1.04	142.4	0.8
Ma-3-2	0.0268	37.97	0.13	23.86	0.08	56.86	0.21	142.9	0.7
Ma-15	0.0162	101.85	0.39	64.01	0.25	153.35	0.79	143.6	0.9

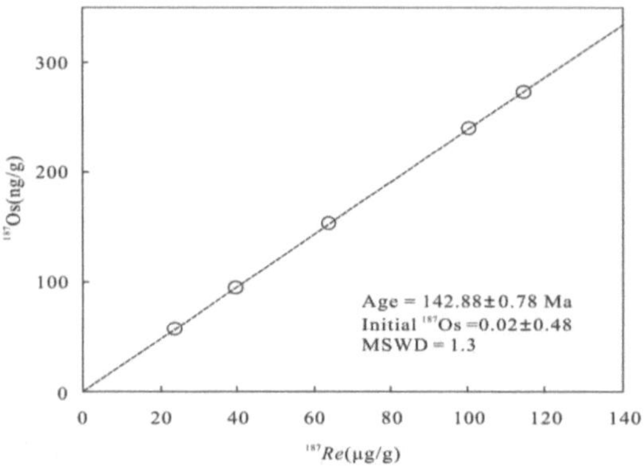

Figure 1. Molybdenite Re-Os isochron of the Datongkeng W-Cu-Mo polymetallic ore section.

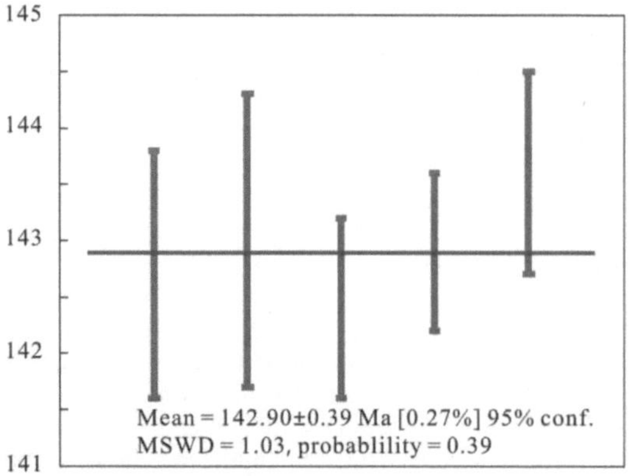

Figure 2. Molybdenite Re-Os model ages of the Datongkeng W-Cu-Mo polymetallic ore section.

and orogenic process of the paleo-Pacific plate and Eurasian continent in southeastern China. Li et al. (1998) suggested that the tectonic transition period of southern China happened between the Late Jurassic and Early Cretaceous, about 145 Ma ago. The tectonic transition and extension events of large-scale lithosphere in Southern China led to large scale granitic magmatism and mineralization during Mesozoic (Ren et al., 1997; Hua et al., 1999; Mao et al., 1999, 2000, 2004). Datongkeng granodioritic porphyry just intruded during the tectonic transition period. The complicated tectonic setting caused the intrusive rock to have very complicated characteristics which were different from the intrusive rocks of early Yanshanian and Late Yanshanian period in Northwestern Zhejiang Province and formed a unique large-middle scale W-Cu-Mo polymetallic deposit in Panjia, Chun'an County.

2.2 *Re content and the material source of the deposit*

Mao et al. (1999) have compared the rhenium contents in molybdenite from many deposits around the world. They concluded that rhenium contents in molybdenite decrease from a mantle sourced ($n \times 10^{-4}$), to mantle-crust mixing sourced ($n \times 10^{-5}$), and then to crust sourced ($n \times 10^{-6}$) origin.

The rhenium contents of molybdenite in the Datongkeng W-Cu-Mo polymetallic ore section are generally high (ranging from 37.97 to 182.48 ppm). Consequently, we can suggest that the ore material probably originated from the mantle to mantle-crust mixing source.

3 CONCLUSIONS

Based on the available data, the following conclusions are discussed herein.

The metallogenic age of Panjia polymetallic deposit is 142.88 ± 0.78 Ma (2σ) with an initial 187Os of 0.02 ± 0.48 (MSWD = 1.3).

Panjia polymetallic deposit form in a tectonic setting of changing from compression to extension.

The source of ore-forming materials comes from mixture of the crust and mantle, and is dominated by those from the mantle.

ACKNOWLEDGEMENTS

The authors would like to thank Sun Changyi and Feng Tao from the first geological unit of Zhejiang province in china for their assistance in the field work. This study was financially supported by the Ministry of Land and Resources of the People's Republic of China (Grant No. 201211013).

REFERENCES

Hua Renmin, Mao Jingwen, 1999. A preliminary discussion on the Mesozoic metallogenic explosion in East China. Mineral Deposits, 18(4): 300–307 (in Chinese with English abstract).

Li Wenda, Mao Jianren, Zhu Yunhe, Xie Huaguang, 1998. The mesozoic igneous rocks and deposits of southeastern China. Beijing: Seismological Press, 1–156 (in Chinese).

Li Xianhua, 2000. Cretaceous magmatism and lithospheric extension in Southeast China. Journal of Asian Earth Sciences, 18(3): 293–305.

Ludwig K., 2003. Isoplot/Ex 3.0. A geochronogical toolkit for Microsoft Excel. Berkeley Geochronology Center, Berkeley, CA, USA, Special Publication No. 4.

Mao Jingwen and Wang Zhiliang, 2000. A preliminary study on time limits and geodynamic setting of large-scale metallogeny in east China. Mineral Deposits, 19, 289–296 (in Chinese with English abstract).

Mao Jingwen, Huarenmin and Li Xiaobo, 1999. A preliminary study of large-scale metallogenesis and large clusters of mineral deposits [J]. Mineral Deposits, 18(4): 291–299 (in Chinese with English abstract).

Mao Jingwen, Xie Guiqing, Li Xiaofeng, Zhang Changqing, Mei Yanxiong, 2004. Mesozoic large scale mineralization and multiple lithospheric extension in South China. Earth Science Frontier, 11(1): 45–55 (in Chinese with English abstract).

Mao Jingwen, Zhang Zhaochong, Zhang Zuoheng and Du Andao, 1999. Re-Os isotopic dating of molybdenites in the Xiao liu gou W(Mo) deposit in the northern Qilian Mountains and its geological significance. Geochimica Cosmochimica Acta, 63(11–12): 1815–1818.

Ren Jishun, Niu Baogui, He Zhengjun, Xie Guanglian and Liu Zhigang, 1997. Tectonic frame and geodynamic evolution of eastern China. Geological Research, 29–30: 43–55 (in Chinese with English abstract).

Future Communication Technology and Engineering – Chan (Ed.)
© 2015 Taylor & Francis Group, London, ISBN: 978-1-138-02777-0

Rock fracture closing moderated by pressure solution

B. Mahmoudzadeh, L. Liu, L. Moreno & I. Neretnieks
Department of Chemical Engineering and Technology, Royal Institute of Technology, Stockholm, Sweden

ABSTRACT: Fracture apertures may decrease or increase by different mechanical and chemical mechanisms when the fractures are subject to stress. A model is presented to describe fracture closure/opening that accommodates pressure dissolution at contacting asperities as well as free-face dissolution/precipitation at free faces of the fracture and of the rock matrix. The derived analytical model accounts for the fact that dissolved minerals carried by flowing water along the fracture can not only diffuse into and out of the adjacent rock matrix but also at first diffuse into the stagnant water zone existing in part of the fracture plane and then from there into and out of the rock matrix adjacent to it. The analytical solution is used to study fracture closure/opening rate in a pseudo steady state, PSS, procedure. This simple model allows us to gain some insights into which processes and mechanisms have the larger impact on the fracture aperture under different circumstances.

1 INTRODUCTION

Rock fractures can close or open due to mechanically and chemically mediated changes. Mechanically mediated changes occur faster compared to chemically mediated changes, which are time-dependent and take longer time to reach a steady state (Yasuhara & Elsworth 2008). Aperture closure/opening over time in stressed natural fractures has been experimentally observed (Polak et al. 2003, Yasuhara et al. 2006b), however the main causes have still been under discussions. Evolution of fracture permeability has also been studied theoretically and modelled by accounting for pressure dissolution and precipitation (Yasuhara et al. 2003, Revil 1999, 2001, Revil et al. 2006, Raj 1982, Yasuhara & Elsworth 2006, Elsworth & Yasuhara 2006, 2010, Yasuhara et al. 2006a). In the involved mechanisms in modeling, the mineral basically tends to dissolve where the higher stress is applied and precipitate where the stress is lower.

In fractured rocks, the two surfaces of a fracture are in contact with each other in some asperities but are open in other locations. At the contact asperities, the local stress is higher than the average stress on the whole fracture since stress which is concentrated in a small part of entire fracture surface. At points where the local stress is much larger than the grain strength, mechanical crushing may occur. This will produce small mineral fractions within the fracture. When fractured rocks contain water, the minerals may dissolve if the water is not saturated. At contacting asperities with high concentrated stress, the solubility of minerals is higher than at fracture voids. The minerals, then, tend to dissolve there and precipitate where the stress is lower. This so-called pressure dissolution process may lead to closure of the fracture after some time. In contrast, free-face dissolution at fracture voids with under-saturation fluid may result in local fracture widening (Liu et al. 2006).

Another phenomenon that contributes to changes in solute concentration, which in turn affects dissolution and precipitation, is dissolution of small mineral fragments which have larger solubility than large minerals (Kahlweit 1975). The small dissolving fragments increase the concentration in the fracture. As a result, the fragments decrease the driving force for pressure dissolution. This effect has only been studied recently by Neretnieks (2014) even though the formation of these small fragments has been modeled (Yasuhara & Elsworth 2008).

Any advective flow through the fracture that can carry in or away solute and even colloidal mineral fragments would increasingly complicate the situation. Yasuhara et al. (2004) have showed that the water flow rate through a fracture can affect the fracture closure rate.

Matrix diffusion has been found to be an important transport process in fractured crystalline rocks (Neretnieks 1980, Neuman 2005). It has been used in modeling solute transport in fractured crystalline rocks (Tang et al. 1981, Park et al. 2001) and it is a very important process that retards contaminant transport in fractured rocks. Recently Neretnieks (2014) has showed that matrix diffusion effect is also very important for chemically mediated closing of fractures in crystalline rocks. Until recently, Neretnieks (2006) and Mahmoudzadeh et al. (2013) have taken the influence of the stagnant water zone in fracture plane into account. They have found that solute diffusion into the stagnant water zone adjacent to the flowing channel can significantly contribute to the retardation of solute. In this paper, it shows that stagnant water zones in the fracture plane can also affect fracture closure in crystalline rocks. This effect, which has not been addressed yet, will be considered together with matrix diffusion effect in a water flowing fracture.

2 CONCEPTUAL MODEL

To highlight the most important phenomena affecting on fracture evolution, we try to develop a reasonable simplifying model by combining all stressed areas and all free-face areas. This includes the mineral grains in the flowing part of the fracture as well as those in the porous rock matrix and in stagnant water zones in the fracture. Figure 1 illustrates the conceptualized model wherein the flow is assumed to take place between two parallel surfaces separated by $2b_f$, forming a rectangular channel of width $2W_f$, with constant mean velocity, u. The stagnant water zone next to the flowing fracture is similarly conceptualized as another rectangular of $2b_s$ aperture and $2W_s$ width. It is assumed that stressed and unstressed crystals are lying in the stagnant water zone adjacent to the flowing channel. The rock matrix is porous and the mineral grains there can also dissolve or act as precipitation surfaces depending on local over- or under-saturation. The model accounts not only for free-face dissolution in the flowing channel but for both pressure dissolution and free-face dissolution in stagnant water zone. The model also accounts for the fact that material dissolved by either pressure or free-face dissolution can directly diffuse from the flowing or stagnant water into and out of the adjacent rock matrix, in addition to advection through the flowing fracture and diffusion into the stagnant water zone.

In order to devise a model that can be solved analytically, a first step is to consider the case when the size of stressed surface and free-face surface is constant in

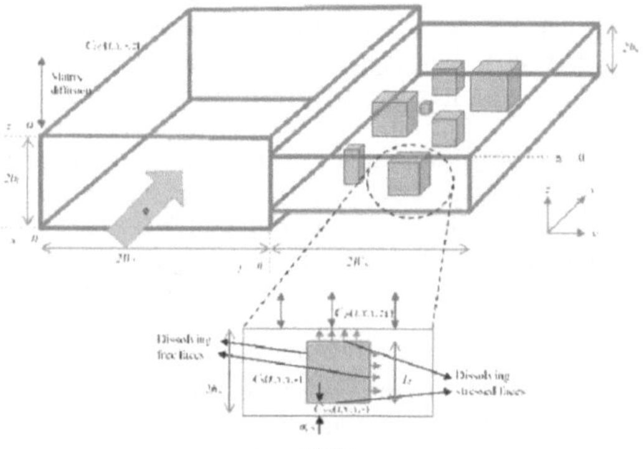

Figure 1. Flow in a channel in a fracture where crystals are stressed in the stagnant water zone. The dissolved crystals by pressure dissolution, precipitate on the unstressed surfaces and also diffuse into the adjacent porous rock matrix.

time. The strength of the analytical approach is that the rates of different competing processes can be summarised in a limited number of parameter groups. The analytical model is used in a pseudo steady state, PSS, procedure to study the rate of fracture closure/opening due to stress and the conditions that may lead to the growing of the flowing channel aperture.

3 MATHEMATICAL MODEL

We start from a case with constant aperture in flowing channel and stagnant water zone. According to the coupled 1-D approach, the system shown in Figure 1 can be identified as four subsystems; flowing channel, stagnant water zone, the rock matrix adjacent to the flowing channel and rock matrix adjacent to the stagnant water zone, for which equations of continuity can be formulated individually. Transport equations will be then coupled at boundaries between subsystems.

3.1 *Transport in the flowing channel*

The transport equation for the aqueous concentration in the flowing channel, C_f, can be written as

$$\frac{\partial C_f}{\partial t} = -u\frac{\partial C_f}{\partial x} + \frac{k_{diss}}{b_f}\left(C_{eq,F} - C_f\right) + \frac{b_s}{b_f}\frac{D_w}{W_f}\left.\frac{\partial C_s}{\partial y}\right|_{y=0} + \frac{D_{ef}}{b_f}\left.\frac{\partial C_{pf}}{\partial z}\right|_{z=0} \tag{1}$$

with

$$C_{eq,F} = C_{eq,1}\exp\left(V_m\frac{\sigma_F - \sigma_1}{RT}\right) \tag{2}$$

where $C_{eq,1}$ and $C_{eq,F}$ are the equilibrium concentrations of the mineral at the reference stress σ_1 1 bar and at the stress σ_F the pressure applied at free faces, respectively.

The initial condition is

$$C_f(t=0,x) = C_{eq,F} \tag{3}$$

and the boundary condition is

$$C_f(t,x=0) = C_{f,in} \tag{4}$$

In equation (1), x, y and z are coordinates along the flowing fracture, into the stagnant water zone and into the rock matrix adjacent to the flowing channel, respectively. $k_{diss} = k_{+1}/C_{eq,1}$ is the same constant in all four regions. k_{+1} is the dissolution/precipitation rate constant of the mineral at the reference stress 0.1 MPa. A complete list of symbols and units is given in notation section.

3.2 *Transport in the stagnant zone*

Assuming that diffusion in the direction parallel to the flow is negligible, the transport equation for aqueous concentration in the stagnant water zone, C_s, is given by

$$\frac{\partial C_s}{\partial t} = D_w \frac{\partial^2 C_s}{\partial y^2} + (1-\alpha)\frac{k_{diss}}{b_s}\left(C_{eq,F} - C_s\right) + \alpha \frac{k_{diss}}{b_s}\left(C_{eq,P} - C_s\right) + \frac{D_{es}}{b_s}\frac{\partial C_{ps}}{\partial z_1}\bigg|_{z_1=0} \quad (5)$$

with

$$C_{eq,P} = C_{eq,1}\exp\left(V_m \frac{\sigma'_P - \sigma_1}{RT}\right) \quad (6)$$

where $C_{eq,P}$ is the equilibrium concentrations of the mineral at the effective stress σ'_P applied on the stressed faces of the crystals (Terzaghi, 1923). The ratio α is a fraction of wetted surface in stagnant water zone representing all lumped stressed surfaces of crystals.

The initial condition is

$$C_s(t=0, y) = C_{eq,F} \quad (7)$$

and the boundary conditions are

$$C_s(t, y=0) = C_f \quad (8)$$

and

$$\frac{\partial C_s}{\partial y}\bigg|_{y=2W_S} = 0 \quad (9)$$

The transport processes in the flowing channel and in the stagnant water zone are coupled through equation (8) describing the continuity of the aqueous concentration of dissolved material.

3.3 *Transport in the rock matrix adjacent to the flowing channel*

The 1-D transport equation for the concentration in the pore water in the rock matrix adjacent to the flowing channel is

$$\varepsilon_{pf}\frac{\partial C_{pf}}{\partial t} = D_{pf}\varepsilon_{pf}\frac{\partial^2 C_{pf}}{\partial z^2} + k_{diss}a_f\left(C_{eq,F} - C_{pf}\right) \quad (10)$$

where a_f is specific surface of dissolving crystals in the rock matrix adjacent to the flowing fracture.

The initial condition is

$$C_{pf}(t=0, z) = C_{eq,F} \quad (11)$$

and the boundary conditions are

$$C_{pf}(t, z=0) = C_f \quad (12)$$

and

$$\frac{\partial C_{pf}}{\partial z}\bigg|_{z=\delta_f} = 0 \quad (13)$$

3.4 *Transport in the rock matrix adjacent to the stagnant water zone*

Similar to equation (10), transport equation for concentration in the pore water in the rock matrix adjacent to the stagnant water is

$$\varepsilon_{ps}\frac{\partial C_{ps}}{\partial t} = D_{ps}\varepsilon_{ps}\frac{\partial^2 C_{ps}}{\partial z_1^2} + k_{diss}a_s\left(C_{eq,F} - C_{ps}\right) \quad (14)$$

The initial condition is

$$C_{ps}(t=0, z_1) = C_{eq,F} \quad (15)$$

and the boundary conditions are

$$C_{ps}(t, z_1=0) = C_s \quad (16)$$

and

$$\frac{\partial C_{ps}}{\partial z_1}\bigg|_{z_1=\delta_s} = 0 \quad (17)$$

4 LAPLACE-TRANSFORMED SOLUTION

The governing equations can be solved by applying Laplace transformation approach (Watson 1981). This approach removes the time variable leaving a system of ordinary differential equations, solution of which yields aqueous concentration as a function of space variable. One shall begin with the Laplace transformed equations in the rock matrices and continue with equations in the stagnant water zone and the flowing channel. In this section, however, we only present analytical solutions in the Laplace domain for the flowing channel and the stagnant water zone. The Laplace-transformed concentration in the stagnant water zone can be written as,

$$\overline{C_s} = \left(\overline{C_f} - G\right)\left\{\frac{\cosh(2\sqrt{\Omega_s} - \sqrt{\Omega_s}\,y/W_s)}{\cosh(2\sqrt{\Omega_s})}\right\} + G \quad (18)$$

with

$$G = \frac{\dfrac{1}{s\tau_{Pdiss,s}}\left(C_{eq,P} - C_{eq,F}\right)}{s + \dfrac{1}{\tau_{Pdiss,s}} + \dfrac{1}{\tau_{Fdiss,s}} + \dfrac{D_{es}}{b_s}\tanh(\sqrt{A_s}\delta_s)\sqrt{A_s}} \quad (19)$$

and

$$\Omega_s = \tau_s\left(s + \frac{1}{\tau_{Pdiss,s}} + \frac{1}{\tau_{Fdiss,s}}\right) + \mathrm{F}_s D_{es}\tanh(\sqrt{A_s}\delta_s)\sqrt{A_s} \quad (20)$$

with

$$A_s = \frac{1}{D_{ps}}\left(s + k_{diss}\frac{a_s}{\varepsilon_{ps}}\right) \quad (21)$$

where we have introduced two characteristic parameters $\tau_{Pdiss,s} = b_s/(\alpha k_{diss})$ and $\tau_{Fdiss,s} = b_s/((1-\alpha)k_{diss})$ for characteristic pressure dissolution time in stagnant water zone and for characteristic free-face dissolution time in stagnant water zone, respectively. Other characteristic parameters presented in Equations 19–21 and in the following equations are similar to those introduced by Mahmoudzadeh et al. (2013).

The Laplace-transformed concentration in the flowing channel can be written as,

$$\overline{C}_f(x,s) = C_{eq,F}/s + \mathrm{H} + \left\{\overline{C}_{f,in} - C_{eq,F}/s - \mathrm{H}\right\} \\ \exp(-\Omega_f)\exp(-\mathrm{N}\sqrt{\Omega_s}\tanh(2\sqrt{\Omega_s})) \quad (22)$$

with

$$\mathrm{H} = \frac{\mathrm{N}\sqrt{\Omega_s}\tanh(2\sqrt{\Omega_s})}{\Omega_f + \mathrm{N}\sqrt{\Omega_s}\tanh(2\sqrt{\Omega_s})}G \quad (23)$$

and

$$\Omega_f = \tau_f\left(s + \frac{1}{\tau_{Fdiss,f}}\right) + \mathrm{F}_f D_{ef}\tanh(\sqrt{A_f}\delta_f)\sqrt{A_f} \quad (24)$$

where $\tau_{Fdiss,f} = b_f/k_{diss}$ is characteristic free-face dissolution time in the flowing channel and A_f is defined in the same way as A_s except for changing subscripts from s to f.

5 FRACTURE CLOSURE RATE

In our model we assume that crystals are cubs of size L_z and L_y and have two stressed sides parallel to the fracture surfaces. The closure rate of stagnant zone will be determined by the size of the stressed shrinking crystal assuming that wall is inert where the crystal is in contact. The fracture aperture, which determines the hydraulic conductivity, will close with the rate that the stressed crystals shrink but will on the other hand grow with the rate that the free-face dissolution determines. The free-face dissolution in the stagnant zone will not influence the closure of the fracture in the flowing channel.

Fracture closure rate in stagnant water zone can then be obtained by shrinking rate of crystal as,

$$\frac{db_s}{dt} = \frac{1}{2}\frac{dL_z}{dt} = \frac{1}{2}\left[-2V_m k_{diss}\left(C_{eq,P} - C_s\right)\right] \quad (25)$$

and in the flowing channel the aperture evolution can be determined by

$$\frac{db_f}{dt} = \frac{1}{2}\left[\frac{dL_z}{dt} + 2V_m k_{diss}\left(C_{eq,F} - C_f\right)\right] \quad (26)$$

Table 1. Data used in examples.

Notation	Value	Notation	Value
b_f (m)	1e–4	D_{pf} & D_{ps} (m^2/s)	5.2e–10
b_s (m)	1e–4	ε (%)	30
x (m)	50	T (K)	353
W_f (m)	0.1	k_{+1} (mol/m^2/s)	4.78e–12
W_s (m)	0.5	$C_{eq,1}$ (mol/m^3)	5.57e–4
u (m/yr)	5	σ' (MPa)	8.1
δ_f (m)	1	α (–)	0.11
δ_s (m)	1	$M_{crystal}$ (g/mol)	60.01
D_w (m^2/s)	1e–9	$\rho_{crystal}$ (g/m^3)	2.65e6

There may be free-face dissolution or precipitation depending on the driving force direction, and the aperture as a whole may close if the stressed crystals in the stagnant zone dissolve faster than the free-face crystals do.

6 SIMULATION AND DISCUSSION

In this section a series of simulations are presented and discussed to explore the effect of different processes and mechanisms on the concentration and fracture aperture. The data used in the examples are tabulated in Table 1. The geometrical and physical properties of the flowing channel, the stagnant water zone and the rock matrix are representative of those from field observations (Crawford & Moreno 2006, Lindgren and Lindström 1999) whereas the effective diffusivity is calculated according to Archie's law, $D_w\varepsilon^{0.6}$. The solubility related data are taken from Rimstidt & Barnes 1980, and the other data have been chosen to resemble the condition of water-saturated sandstone rock matrix with 30% porosity hosting a nuclear waste repository in about 500 m depth. The grains in the matrix have the same size of stressed crystal which is initially the same as aperture in stagnant zone.

6.1 *Pseudo steady state approach*

We start from a simplified case where there is no flowing water i.e. the system consists only of stagnant water zone and its adjacent rock matrix. This system resembles the model studied by Neretnieks (2014) where the oversaturated solute from the stressed crystal diffuses out to the surrounding water from which the solute re-precipitates on the unstressed surfaces of the same crystal or diffuses into the porous rock matrix to precipitate on other crystals there. His numerical results show that the sink due to matrix diffusion is important and affect crystal dissolution and fracture aperture evolution. We account here for combined effect of pressure dissolution, free-face dissolution/precipitation and matrix diffusion on the dissolution rate of stressed crystals. By using parameter values tabulated in Table 1, our pseudo steady state approximation is able to reproduce the shrinking rate

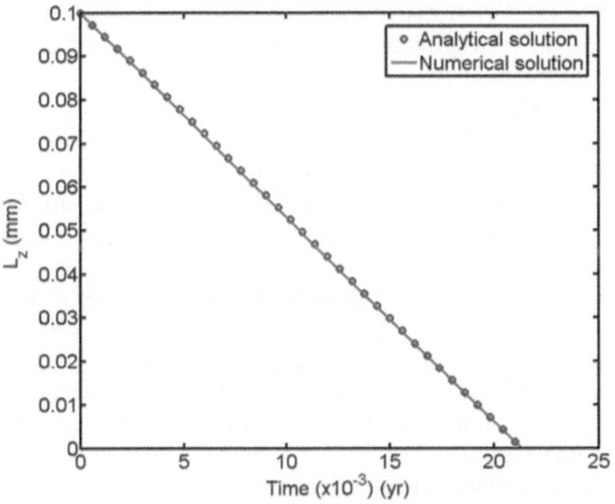

Figure 2. Comparing shrinking rate of crystal obtained by pseudo steady state analytical solution and by numerical solution given by Neretnieks, 2014.

of crystal obtained numerically by Neretnieks (2014) with an acceptable agreement. The results illustrated in Figure 2 indicate that our model can be used with confidence. The power of the analytical approach is that it gives us some insights into which processes or mechanisms have larger impact on fracture closure rate since the rate of different competing processes can be summarized in a number of characteristic parameters.

6.2 *Effect of three mechanisms on concentration evolution*

To gain some insights into how the three main mechanisms, i.e. pressure dissolution, free-face dissolution/precipitation and matrix diffusion, affect the rate of concentration increase/decrease in the fracture, we consider the complete system containing both flowing fracture and stagnant water zone and their adjacent rock matrices where the initial concentration is equilibrium concentration at free faces. The mineral concentration in stagnant water is illustrated in Figure 3, for four different conditions. First when all three processes are considered and then when the effect of either matrix diffusion, free-face dissolution or pressure dissolution is emphasized. As can be seen from the figure, the rate of concentration increase is dominated mostly by matrix diffusion which fails to account for this process has increased the concentration in water considerably. This is obviously for the role of matrix in up-taking dissolved minerals from the stagnant water. The figure shows that the concentration is also highly affected by pressure dissolution but to some extent by free-face dissolution. It indicates that pressure dissolution has a large effect on the concentration compared to free-face dissolution. The reason might be that the initial concentration is the same as concentration at free faces and this makes the driving force for free-face dissolution to be so small that it hardly affect mineral concentration in the water. The same results are drawn

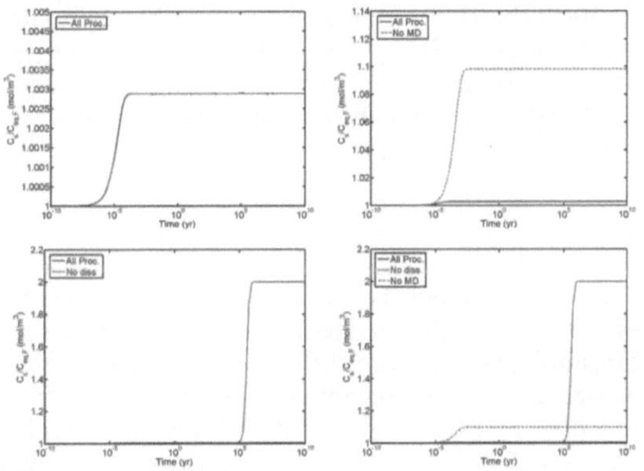

Figure 3. Concentration in stagnant water at $y = 1.0$ m and $C_{f,in} = 2C_{eq,F}$ with $C_{eq,F} = 5.8\text{e}{-}04\ \text{mol/m}^3$; when all processes are included or either matrix diffusion or dissolution is not accounted for.

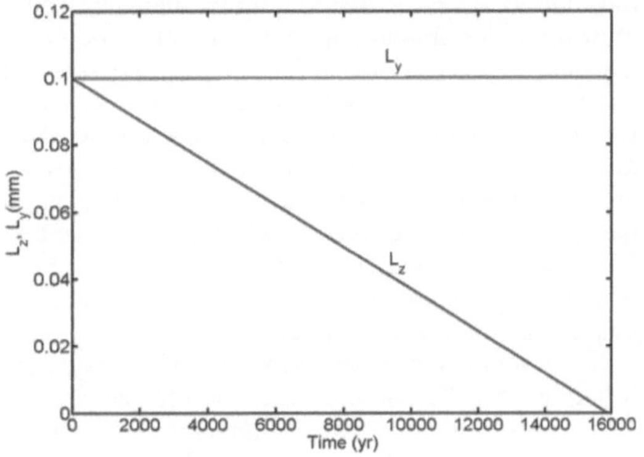

Figure 4. Change of crystal dimensions L_z and L_y with time. All processes are active.

for effect of these three mechanisms on output mineral concentration from the flowing channel.

6.3 *Shrinking rate of crystal*

The system studied in the previous section is used to get the shrinking rate of crystals in a pseudo steady state approach. We start from a short time step in which the concentration is considered constant and calculate the shrinking rate of crystal. The new crystal size and also new stress would be starting point for a new calculation. The procedure is repeated for a large number of time steps.

Figure 4 shows how the crystal sizes L_z and L_y change according to the time when all processes are considered. The model accounts for the fact that stress applied on stressed faces changes over time. This is caused by the shrinking/growth of stressed area due to free-face dissolution/precipitation of the dissolved mineral on the unstressed surfaces. In a variable fracture the stress decreases considerably if either the stressed surface grows or more stressed crystals involve. The figure shows that crystal has fully dissolved after 15,862 years. However, the growth of crystal on the unstressed surfaces is negligible showing that the stress has remained nearly constant in this case.

7 CONCLUSION

In this study, a model is developed to study closure rate of a fracture in fractured rocks. It accounts for advection through the fracture, diffusion into the rock matrix as well as into the stagnant water existing in the fracture plane, pressure dissolution and free-face dissolution/precipitation of stressed and unstressed minerals in the fracture. Analytical expressions for the Laplace-transformed concentrations in the flowing channel and stagnant water zone have been derived. These analytical solutions can be used in a pseudo steady state approach, PSSA, to get closure/opening rate of fracture due to stress or the conditions that may cause growing of the fracture aperture. It should be noted that our analytical approach is for linear processes and nonlinear processes have to be studied in a different way which is not in the scope of the current paper. This rather simple model gives us some insights into which processes and mechanisms have the larger impact on the fracture aperture under different circumstances. The power of the analytical approach is that the rates of different competing processes can be summarized in a limited number of characteristic terms.

Some simulations have performed to illustrate contribution of different processes or mechanisms in closure rate of fracture. The results show that diffusion into the rock matrix that acts as a sink for dissolving minerals, strongly dominates the rate of concentration change in the fracture and as a result the fracture closure/opening rate. The pressure dissolution is the rate limiting process and is used to estimate the closing rate of fracture.

8 NOTATION

a_f/a_s Specific surface of dissolving crystals in rock matrix adjacent to the flowing fracture/the stagnant water zone (L^{-1})

b_f/b_s Half aperture of the flowing channel/stagnant water zone (L)

$C_{eq,1}$ Equilibrium solubility of mineral at reference stress 1 bar (ML^{-3})

$C_{eq,F}/C_{eq,P}$ Equilibrium solubility of mineral at stress σ_F/at effective stress σ'_P (ML^{-3})

C_f/C_s Concentration in the flowing channel/stagnant water zone (ML^{-3})

C_{pf}/C_{ps} Pore water concentration in the rock matrix adjacent to the flowing channel/adjacent to the stagnant water zone (ML^{-3})

D_{ef}/D_{es} Effective diffusivity in the rock matrix adjacent to the flowing channel/adjacent to the stagnant water zone (L^2T^{-1})

D_{pf}/D_{ps} Pore diffusivity in the rock matrix adjacent to the flowing channel/adjacent to the stagnant water (L^2T^{-1})

D_w Water Diffusivity (L^2T^{-1})

F_f Ratio of flow-wetted surface of the flowing channel to volumetric water flow rate (TL^{-1})

F_s Ratio of the stagnant-water-wetted surface to the diffusion conductance of the stagnant water zone (TL^{-1})

k_{diss} Dissolution rate constant at reference stress 1 bar (LT^{-1})

$M_{crystal}$ Molar mass of minerals, silica (MN^{-1})

N Ratio between the diffusion rate into the stagnant water zone and the mass flow rate through the channel (–)

s Laplace transform variable (T^{-1})

t Time (T)

u Groundwater velocity (LT^{-1})

V_m molar volume of silica (L^3N^{-1})

W_f Half width of the flowing channel (L)

W_s Half width of the stagnant water zone (L)

α Fraction of wetted surface in stagnant water demonstrating all lumped stressed surfaces

δ_f/δ_s Thickness of the rock matrix adjacent to flowing channel/adjacent to the stagnant water (L)

$\varepsilon_{pf}/\varepsilon_{ps}$ Porosity of the rock matrix adjacent to the flowing channel/adjacent to stagnant water (–)

$\rho_{crystal}$ Density of crystal, silica (ML^{-3})

σ_F Stress applied on crystal free faces ($ML^{-1}T^{-2}$)

σ' Effective stress ($ML^{-1}T^{-2}$)

σ'_P Effective stress applied on the crystal stressed faces ($ML^{-1}T^{-2}$)

$\tau_{Fdiss,f}$ Characteristic time of free-face dissolution in the flowing channel (T)

$\tau_{Fdiss,s}$ Characteristic time of free-face dissolution in stagnant water zone (T)

$\tau_{Pdiss,s}$ Characteristic time of pressure dissolution in stagnant water zone (T)

τ_f/τ_s Characteristic time of advection/diffusion through the stagnant water zone (T)

φ Ratio of the contact area between stressed crystals and fracture surface to the total cross-sectional area normal to the applied total stress (–)

REFERENCES

Crawford J. & Moreno L. 2006. Äspö hard rock laboratory, Äspö task force on modelling of groundwater flow and transport of solutes. Modelling of Task 6D, 6E and 6F, using CHAN3D, SKB Int. Progr. Rep. IPR-06-19. *Swedish Nuclear Fuel and Waste Management Company*, Stockholm.

de Hoog, F. R. Knight, J. H. Stokes, A. N. 1982. An improved method for numerical inversion of Laplace transforms, *SIAM J. Sci. Stat. Comput.*, 3(3): 357–366.

Elsworth, D. & Yasuhara, H. 2006. Short-timescale chemo-mechanical effects and their influence on the transport properties of fractured rock, *Pure Appl. Geophys*. 163: 2051–2070.

Elsworth, D. & Yasuhara, H. 2010. Mechanical and transport constitutive models for fractures subject to dissolution and precipitation, *Int. J. Numer. Anal. Meth. Geomech.* 34: 533–549.

Kahlweit, M. 1975. Ostwald ripening of precipitates, *Adv Colloid Interface Sci.* 5: 1–35.

Lin, W. N. & Daily, W. 1990. Hydrological properties of topopah spring tuff under a thermal-gradient-laboratory results, *Int. J. Rock Mech. Min. Sci. Geomech. Abstr.* 27: 373–386.

Lindgren, M. & Lindström F. 1999. Radio Nuclide Transport Calculations, SKB Tech. Rep. TR-99-23. *Swedish Nuclear Fuel and Waste Management Company*, Stockholm.

Liu, J. Sheng, J. Polak, A. Elsworth, D. Yasuhara, H. Grader, A. 2006. A fully-coupled hydrological–mechanical–chemical model for fracture sealing and preferential opening, *Int. J. Rock Mech. Min. Sci.*, 43: 23–36.

Mahmoudzadeh, B. Liu, L. Moreno, L. Neretnieks, I. 2013. Solute transport in fractured rocks with stagnant water zone and rock matrix composed of different geological layers—Model development and simulations, *Water Resources. Res.* 49: 1709–1727.

Monzurul Alam, M. Fabricius, I. L. Christensen, H. F. 2012. Static and dynamic effective stress coefficient of chalk, *Geophysics*, 77 (2): 1709–1727.

Moore, D. E. Lockner, D. A. Byerlee, J. D. 1994. Reduction of permeability in granite at elevated-temperature, *Science* 265: 1558–1561.

Neretnieks, I. 1985. Diffusivities of some constituents in compacted wet bentonite clay and the impact on radionuclide migration in the buffer, *Nucl. Technol.*, 71: 458–470.

Neretnieks, I. 2006. Channeling with diffusion into stagnant water and into a matrix in series, *Water Resour. Res.*, 42, W11418, doi:10.1029/2005WR004448.

Neuman, S. P. 2005. Trends, prospects and challenges in quantifying flow and transport through fractured rocks, *Hydrogeol. J.*, 13: 124–147.

Neretnieks, I. 2014. Stress-mediated closing of fractures—Impact of matrix diffusion, *J. Geophys. Res. Solid Earth*, 119, doi:10.1002/2013JB010645.

Park, J. B. Hwang, Y. Lee, K. J. 2001. Analytic solutions of radionuclide transport with the limited diffusion from the fracture into a porous rock matrix, *Ann. Nucl. Energy*, 28: 993–1011.

Polak, A. Elsworth, D. Yasuhara, H. Grader, A. S. Halleck, P. M. 2003. Permeability reduction of a natural fracture under net dissolution by hydrothermal fluids, *Geophys Res Lett.* 30: 2020.

Rimstidt, J. D. & Barnes, H. L. 1980. The kinetics of silica-water reactions, *Geochim, Cosmochim. Acta*, 44: 1683–1699.

Raj, R. 1982. Creep in polycrystalline aggregates by matter transport through a liquid phase, *J. Geophys. Res.*, 87(B6): 4731–4739, doi:10.1029/JB087iB06p04731.

Revil, A. 1999. Pervasive pressure-solution transfer: A poro-visco-plastic model, *Geophys. Res. Lett.* 26(2): 255–258, doi:10.1029/1998GL900268.

Revil, A. 2001. Pervasive pressure solution transfer in a quartz sand, *J. Geophys. Res.*, 106(B5): 8665–8686, doi:10.1029/2000JB900465.

Revil, A. Leroy, P. Ghorbani, A. Florsch, N. Niemeijer, A. R. 2006. Compaction of quartz sands by pressure solution using a Cole-Cole distribution of relaxation times, *J. Geophys. Res.*, 111, B09205, doi:10.1029/2005JB004151.

SR-Site, 2011. Long-term safety for the final repository for spent nuclear fuel at Forsmark. Main report of the SR-Site project, SKB technical report, TR-11-01, ISSN 1404-0344,

Stumm, W., & Morgan, J. J. 1996. Aquatic Chemistry, *John Wiley*. New York.

Tang, D. H., Friend, E. O. Sudicky, E. A. 1981. Contaminant transport in fractured porous media: Analytical solution for a single fracture, *Water Resour. Res.*, 17(3): 555–564.

Terzaghi, K. 1923. Die Beziehungen zwischen Elastizitat und Innendruck: Sitzungsberichte, Akademie der Wissenschaften, K I. IIa 132 (3–4): 105–121.

Watson, E. J., 1981. Laplace Transform and Applications. Van Nostrand Reinhold Co., London.

Yasuhara, H. Elsworth, D. Polak, A. 2004. Evolution of permeability in a natural fracture: significant role of pressure solution. *J Geophys Res.* 109: B03204.

Yasuhara, H. & Elsworth, D. 2008. Compaction of a rock fracture moderated by competing roles of stress corrosion and pressure solution, *Pure Appl. Geophys.* 165: 1289–1306.

Yasuhara, H. Polak, A. Mitani, Y. Grader, A. Halleck, P. Elsworth, D. 2006a. Evolution of fracture permeability through fluid-rock reaction under hydrothermal conditions, *Earth Planet Sci. Lett.* 244: 186–200.

Yasuhara, H. Elsworth, D. Polak, A. Liu, J. Grader, A. Halleck, P. 2006b. Spontaneous switching between permeability enhancement and degradation in fractures in carbonate: lumped parameter representation of mechanically- and chemically- medicated dissolution, *Transp Porous Media.* 65: 385–409.

Yasuhara, H. Elsworth, D. Polak, A. 2003. A mechanistic model for compaction of granular aggregates moderated by pressure solution, *J. Geophys. Res.* 108(B11), 2530, doi:10.1029/2003JB002536.

Yasuhara, H. & Elsworth, D. 2006. A numerical model simulating reactive transport and evolution of fracture permeability, *Int. J. Numer. Anal. Meth. Geomech.*, 30: 1039–1062.

Yasuhara, H. Kinoshita, N. Ohfuji, H. Lee, S. D. Nakashima, S. Kiyoshi, K. 2011. Temporal alteration of fracture permeability in granite under hydrothermal conditions and its interpretation by coupled chemo-mechanical model, *Appl. Geochem.*, 26: 2074–2088, doi:10.1007/s11431-011-4437-6.

Zhao, Z. Liu, L. Neretnieks, I. Jing, L. 2013. Solute transport in a single fracture: Impacted by chemically mediated changes, *Int. J. Rock Mech. Mining Sci.*, 66: 69–75.

Future Communication Technology and Engineering – Chan (Ed.)
© 2015 Taylor & Francis Group, London, ISBN: 978-1-138-02777-0

Formation of drainage waters in tailing dumps of Khrustal'noe deposit in Kavalerovsky District, and their effect on the hydrosphere (the Russian Far East)

V.P. Zvereva & A.M. Kostina
Far Eastern Federal University, Vladivostok, The Russian Federation
Far East Geological Institute, FEB of RAS, Vladivostok, The Russian Federation

A.D. Pyatakov
Far Eastern Federal University, Vladivostok, The Russian Federation

ABSTRACT: The paper presents the results of modelling of drainage waters at three tailing dumps of Khrustal'noe deposit (Fabrichnyi Settlement, Kavalerovsky District) within the temperature interval of 0 to +45°C. The Eh-pH parameters of the systems of the technogenic mineral formation in the tailings have been determined at different ratios of enclosing rock/sulphides (95:5, 90:10, 80:20, 60:40, and 20:80). The authors considered the process of the hypergene and technogene mineral crystallization from micropore solutions, determined their qualitative and quantitative ion composition, and showed their action on the surface and ground waters of the district. It was established that the solutions were enriched with elements of sulphide ores (Pb, Zn, As, S), and that rocks enclosing the mineralization (K, Na, Ca, Mg, Al, and Si) were highly mineralized. The tailing dumps significantly damage the hydrosphere and ecosphere as a whole.

1 INTRODUCTION

The objects of Kavalerovsky District (Khrustal'nensky mining-concentrating plant) are mostly situated in the Zerkal'naya River basin. There are fifteen deposits (Khrustal'noe, Silinskoe, Arsenyevskoe, Vysokogorskoe, Dubrovskoe, Ternistoe, Verkhnee, Iskrinskoe, and others) and four concentrating mills. Ores of the deposits belong to the cassiterite-silicate and cassiterite-sulphide veined and streaky-impregnated morphological types.

The ores were mined by both open (quarries) and closed (adits) ways. The ores of Khrustal'noe deposit were processed at the mining-concentrating mill of Fabrichnyi Settlement. They were kept in storage at three tailing dumps in the immediate vicinity from the Settlement. They occupy 15 ha of useful area. 34.8 million tons of wastes of the mining production (tailings) have been accumulated here. The tailings at all tailing dumps are represented by a fine-dispersed grey mass. They are composed of the following minerals: pyrite, pyrrhotite, galena, sphalerite, arsenopyrite, chalcopyrite, quartz, fluorite, tourmaline, chlorite, and others. Quantitative and semi-quantitative spectral analyses of seven samples taken at three tailing dumps of the first mill showed that the content of ore elements in the tailings changed within the following limits (%): Sn – 0.04-0.10, Cu – 0.006–0.26, Pb – 0.004–0.076, Zn – 0.08–1.00, As – 0.01–0.05, Ni – 0.001–0.003, Co – 0.0002–0.0009, Cr – 0.002–0.003, V – 0.004–0.010, Ag – 0.0003–0.003, Ga – 0.001–0.001, B – 0.01–0.05, Bi – 0.0001–0.0003, Sr – to 0.01 and Ca – to 0.1.

In Kavalerovsky District, the main and single mined element was always tin, although in recent years In and Ag were incidentally extracted, and sulphides, bearing Cu, Pb, and Zn in commercial amounts, went to the tailings and were kept in storage at the tailing dumps. In 2001 the tin-ore industry in Kavalerovsky District ceased to operate, though the ore reserves have not been mined. The reclamation measures in the district were not carried out.

The investigations by many workers showed that the tailing dumps actively influenced the atmosphere, lithosphere, biosphere, and hydrosphere of the district where they were situated (Arzhanov, 1976, Elpatyevsky et al., 1976, Kolotov et al., 1979, Krupskaya, 1998, Bortnikova, 2001, Zvereva, 2008a, and others).

With the increase of the contact surface of the fine-crushed sulphides, they are kept in storage at the tailing dumps. With the weathering agents, they result in activation of hypergene processes and pass into the technogenic stage. A wide set of hypergene and technogene minerals are crystallized from the highly concentrated solutions of both pore and mine waters. Among them are: posnjakite, rowvolfite, woodwardite, pitticite, fibroferrite, glockerite, hisingerite, allophane, and others. On the walls and roof of the mine workings they are represented by sintered stalactite-stalagmite formation of about 0.5 m. In some places on the tailing dump surface they form incrustations about 0.5 cm thick, whose areas are 5 square meters and more. The present-day technogene formations are more often badly crystallized and x-ray amorphous. They are white, blue, green, yellow, and brown in colour of various tints (Zvereva, 2008a, b).

The main aim of this work is the modelling of the composition of drainage waters formed under the action of hypergene and technogene processes proceeding in the wastes of the mining production (tailings) at the tailing dump of Fabrichnyi Settlement in a wide temperature interval of 0 to +45°C, and estimation of their effect on the hydrosphere. Therefore, we set the following tasks: 1. To trace the process of the sulphide oxidation at three tailing dumps within the temperature interval of 0 to +45°C. 2. To establish the technogene minerals, crystallized from the micropore solutions of drainage waters. 3. To determine their qualitative and quantitative ion composition. 4. To estimate their action on the hydrosphere of the district. 5. To carry out the verification of the data obtained.

According to the climatic characteristics (Nesterov & Nesterova, 1978), the temperature in the district varies from −35 to +35°C, but through the sulphide oxidation it increases, so the upper limit was taken for ten degrees higher.

2 MATERIALS AND METHODS

The method of the physicochemical modelling is used in our investigations. It allows the study of the process with the help of the physicochemical models, and testing their agreement with the object of investigation by verification. Modelling makes it possible to objectively estimate the geochemical processes proceeding in tailings, and to predict the object state in the future.

The physicochemical modelling of the technogenesis processes in the CCM tailings was performed with the help of the Selektor-Windows program complex based on the minimization of Gibbs free energy. Such tasks were successfully solved on the example of other systems and objects by many workers: I.K. Karpov (1981), E.M. Dutov (2005), V.A. Bychinsky and co-authors (2008), V.O. Khudolozhkin (2008), and others. The authors first carry out the modelling of the hypergene processes proceeding in the mine-industrial technogenic system of Kavalerovsky District.

Our previous articles (Zvereva et al., 2013a, b) describe the investigations of the system with 45 hypergene and technogene minerals. Later, they found or calculated (A.A. Tupitsin and V.A. Bychinsky, Siberian Branch of RAS) the thermodynamic parameters for allophane, glockerite, pitticite, serpierite, fibroferrite, hisingerite, and hydrogoethite, which were found in the district (Zvereva, 2008a). These thermodynamic data were included into the models under consideration. In addition, in this work the researchers studied the dependence of the element behaviour in the solution and the crystallization process of technogenic minerals on the increase of the sulphide component amount to 80%.

Table 1. Content of hypogene minerals in modelled systems of tailing oxidation in contact with enclosing rocks (%).

Mineral	1 td.	2 td.	3 td.
Sulphides			
Pyrite FeS_2	11.3	29.3	18
Pyrrhotite $Fe_{1-x}S_n$	1.3	8.7	4.7
Arsenopyrite FeAsS	14.7	6.7	5.3
Chalcopyrite $CuFeS_2$	9.3	6.7	5.3
Galena PbS	43.3	19.3	23.3
Sphalerite ZnS	20	29.3	43.3
Enclosing rock			
Chlorite $(Mg,Fe)_3(Si,Al)_4O_{10}(OH)_2{\cdot}(Mg,Fe)_3(OH)_6$		78.38	
Sericite $KAl_2(Si_3Al)O_{10}(OH)_2$		11.89	
Tourmaline $Na(Li,Al)_3Al_6[(OH)_4(BO_3)_3Si_6O_{18}]$		5.95	
Epidote $Ca_2Al_2Fe^{+3}(SiO_4)_3OH$		2.16	
Calcite $CaCO_3$		1.62	

Note: td. – tailing dump.

For the modelling of the system the authors chose the unified thermobarometric conditions: P = 1 atm. The rain water composition was as follows: N_3^-, N_2^- NH_4^+, $NH_4N_3^0$, HNO_2^0, $NH_4NO_3^0$, NH_4OH^0, $NH_4NO_2^0$, NH_3^0, $H_2CO_3^0$, HCO_3^-, CO_3^{2-}, $C_2O_4^{2-}$, CH_4^0, O_2^0, H_2^0, N_2^0, Ar^0, He^0, Kr^0, Ne^0, OH^-, H^+, H_2O^0, NO_3^-, HNO_3^0 (pH = 5,66). The models were open to the atmosphere. The chemical composition of the atmosphere was calculated after R. Horne (1972). The composition of 10 kg of the atmosphere includes (in moles): Ar – 3.209, C – 0.1036, N – 539.478, O – 144.8472. The calculations in the models of the hypergene processes in the tailings of the tailing dumps under consideration accounted for 19 independent components and 369 dependent ones including 283 dissolved particles, 18 gases, and 68 minerals and solid solutions.

3 RESULTS AND DISCUSSION

For each of the tailing dumps we modelled the systems with the following ratios of enclosing rock-sulphide: 95:5, 90:10, 80:20, 60:40, and 20:80, as in their different parts the wastes composition is different. The contents and ratios of the hypogene minerals of sulphide ores and rocks enclosing the mineralization used in models were determined from the literature sources (Radkevich, 1980, Finashin, 1986) and are given in Table 1.

The modelled systems have close values of the Eh-pH parameters of 0.42 to 0.86 V and 5.65–13.28 (tailing dump 1), 6.62–13.25 (2), and 6.98–13.24 (3).

Through the tailing oxidation in the modelled systems the following hypergene minerals are crystallized from the high-concentrated micropore solutions (Table 2): oxide and hydroxide classes – goethite α-FeOOH and tenorite CuO; sulphate class – woodwardite $Cu_4Al_2[SO_4](OH)_{12}{\cdot}4H_2O$, anglesite $Pb[SO_4]$, fibroferrite $Fe^{+3}[SO_4](OH){\cdot}5H_2O$,

Table 2. Hypergene minerals crystallized in the modelled systems of the tailing oxidation in contact with enclosing rock (g).

Mineral	1 td.	2 td.	3 td.
Goethite	21.88–24.16	22.81–26.89	21.98–24.90
Tenorite	0.20–0.82	0.0003–0.29	0.0004–0.23
Anglesite	3.56–32.75	7.71–11.53	0.0001–17.26
Woodwardite	0.13–0.82	0.10–0.60	0.06–0.47
Gypsum	0.002–2.07	0.0002–2.08	1.87–2.06
Alunogen	4.44–7.16	6.79–7.12	–
Magnesite	0.34–10.05	6.71–10.05	6.26–10.05
Duftite	0.38–17.0	0.22–12.34	0.20–9.74
Allophane	5.69–32.06	5.71–32.02	7.89–32.03
Montmorillonite	0.0009–26.88	0.0004–27.48	0.002–27.88
Fibroferrite	69.96	110.43	74.32

Note: td. – tailing dump – the mineral is absent from the modelled system.

alunogen $Al_2[SO_4]_3 \cdot 17H_2O$, and gypsum $Ca[SO_4]$ $2H_2O$; carbonate class – magnesite $Mg[CO_3]$; arsenate class – duftite $CuPb[AsO_4](OH)$; silicate class – montmorillonite $m\{Mg_3[Si_4O_{10}](OH)_2\} \cdot p\{Al,Fe^{+3})_2[Si_4O_{10}](OH)_2\} \cdot nH_2O$, and allophane $m\,Al_2O_3 \cdot nSiO_2 \cdot pH_2O$.

Woodwardite and tenorite are precipitated in the systems with the enclosing rock/sulphide ratios of 95:5 and 90:10 at all tailing dumps, alunogen – only at 80:20 ratio, anglesite and gypsum – at 40:60 and 80:20 ratios, fibroferrite – at 80:20 ratio, and in the systems of the 3d tailing dump it was also found at 40:60 ratio. Thus, the crystallization of anglesite, fibroferrite, alunogen, and gypsum requires higher contents of sulphides in the system that result in the formation of acid solutions whose pH is lowered to 5.61. In the modelled systems of the third tailing dump, where pH reaches 6.98, alunogen is absent, i.e. its crystallization requires more acid medium.

The elements of sulphide ores in the solutions are present as the following ions: $Cu(CO_3)_2^{2-}$, Cu^+, Cu^{2+}, $CuCO_3^0$, $CuHCO^{3+}$, CuO^0, CuO_2^{2-}, $CuOH^+$, $CuSO_4^0$, $HCuO_2^-$, As^{3+}, As^{5+}, AsO_4^{3-}, $H_2AsO_4^-$, $H_3AsO_4^0$, $HAsO_4^{2-}$, $NaAsO_4^{2-}$, Fe^{3+}, FeO^+, FeO_2^-, $FeOH^{2+}$, $HFeO_2^0$, $HZnO^{2-}$, $Zn(CO_3)_2^{2-}$, $Zn(SO_4)_2^{2-}$, Zn^{2+}, $ZnCO_3^0$, $ZnHCO_3^+$, ZnO^0, ZnO_2^{2-}, $ZnOH^+$, $ZnSO_4^0$, $HPbO_2^-$, $Pb(CO_3)_2^{2-}$, $Pb(OH)_2^0$, $Pb(SO_4)_2^{2-}$, Pb^{2+}, $PbCO_3$, $PbHCO_3^+$, $PbNO_3^+$, PbO, $PbOH^+$, $PbSO_4^0$, SO_4^{2-}, HSO_4^-. Destruction of the enclosing rocks and their interaction with sulphide components result in the formation of the following ions in the solutions: $Al(OH)^{2+}$, Al^{3+}, AlO^+, AlO_2^-, $HAlO_2^0$, $B(OH)_3^0$, BO_2^-, $Ca(HCO_3)^+$, Ca^{2+}, $CaCO_3^0$, $CaHSiO_3^+$, $CaOH^+$, $CaSO_4^0$, $H_2O_2^0$, HCO_3^-, $HSiO_3^-$, K^+, $KHSO_4^0$, KOH^0, KSO_4^-, $Mg(HCO_3)^+$, Mg^{2+}, $MgCO_3^0$, $MgHSiO_3^+$, Na^+, $NaHSiO_3$, $NaOH^0$, $NaSO_4^-$, SiO_2^0. It should be noted that, with the increase of temperature, the ion composition becomes more diverse.

In the solutions the content of the main elements of sulphide ores is high (g/l): Pb – 0.027 to 8.81, 0.003 to 1.98, and 0.003 to 12.7 at tailing dumps 1, 2, and 3, correspondingly; Zn – 6.87 to 11.3, 1.02 to 17.1, and 1.49 to 24.5; As – 0.143 to 0.658, 0.033 to 0.304, and 0.026 to 0.24, and S – 1.15 to 8.58, 1.54 to 10, and 1.45 to 12.7. Copper and iron were practically precipitated in full in the form of different hypergene and technogene minerals (Table 2). Mineralization of the interpore solutions is (g/l): 26.3 to 56 (tailing dump 1), 24.7 to 55.2 (2), and 24.9 to 72.7 (3).

The verification with ion and mineral compositions and with the help of hydrochemical analyses of the samples of technogene waters of Kavalerovsky District showed that the results obtained are close to the literature data (Elpatyevsky et al., 1976; Kolotov et al., 1979, & Zvereva 2008a).

4 CONCLUSION

The modelling we performed for the tailing dumps of the Khrustal'nenskaya Concentrating Mill in a wide interval of temperature and enclosing rock/sulphides ratios with the account for the new thermodynamic data showed that the Eh-pH parameters of the systems change from 0.42 to 0.86 V and 5.65 to 13. Through the oxidation of the tailing sulphide components from the highly mineralized interpore solutions, which form the drainage waters, the following hypergene and technogene minerals are precipitated: goethite, tenorite, woodwardite, anglesite, fibroferrite, alunogen, gypsum, magnesite, duftite, montmorillonite, and allophane whose amount reaches 110 g. The maximal summary amount of minerals precipitated from the solutions is 199 to 241 g. The solutions are enriched in the elements of both sulphide ores (Pb, Zn, As, S) and rocks enclosing the mineralization (K, Na, Ca, Mg, Al, Si), and their mineralization varies from 24.7 to 72.2 g/l. Each of the tailing dumps damages the hydrosphere and ecosphere as a whole. Taking into consideration the fact that all three tailing dumps are located one under another in the valley between Fabrichnyi Settlement and the Kavalerovka River, their negative action should be summarized.

Consequently, their negative action on the ecosphere increases significantly.

REFERENCES

Arzhanova, V.S. 1976. Indication of the dust atmospheric precipitations with method of washes of leaves of the wood vegetation. *Geochemistry of hypergenesis zone and human technical activity*: 87–97. Vladivostok: FESC of the USSR Acad. Sci.

Bortnikova, S.B. 2001. *Geochemistry of heavy metals in technogene systems (questions of formation, development, and interaction with the ecosphere components)*: Author's abstract of Doctor dissertation of geol.-miner. sci. Novosibirsk.

Bychinsky, V.A., Didenkov, Yu.N., Golovnykh, N.V., Shepelev, I.I., Tupitsin, A.A. & Chudnenko, K.V. 2008. Complex measures on lowering the pollution of natural waters in the district of slime storages of alumina integrated works. *Geoecology* (3): 22–231.

Dutova, E.M. 2005. *Hydrogeochemistry of hypergenesis zone of the Altai-Sayan folded area*: Author's abstract of Doctor dissertation of geol.-miner. Sci. Tomsk: TGASU Publishing House.

Elpatyevsky, P.V., Chudaeva, V.A., Zazykina, E.F. & Arzhanova, V.S. 1976. Role of technogenic factor in formation of ion discharge (on the example of one of the rivers in Primorye). *Geochemistry of hypergenesis zone and human technical activity*: 17–27. Vladivostok: Publishing House of FESC of the USSR Acad. Sci.

Finashin, V.K. 1986. *Tin deposits of Primorye.* Vladivostok: FESC of the USSR ACad. Sci.

Radkevich, E.A. 1980. *Geology, mineralogy, and geochemistry of Kavalerovo district.* Moscow: Nauka.

Karpov, I.K. 1981. *Physicochemical modeling with computer in geochemistry*: 372–383. Novosibirsk: Nauka.

Kolotov, B.A., Vagner, B.B., Minacheva, L.I., Spasskaya, T.S., & Sokolovskaya, L.A. 1979. Features of formation of water chemical composition of ore deposits of the North-East USSR. *Problems of formation of chemical composition of ground waters*: 20–37. Moscow: VSEGINGEO.

Krupskaya, L.T. 1998. Estimation of action of mining industry on soils of the Far East. *Action of the mining industry processes on objects of natural environment*: 80–86. Vladivostok: Dal'nauka.

Horne, R. 1972. *Marine chemistry*. Moscow: Mir.

Khudolozhkin, V.O. 2008. Physicochemical modeling of processes of formation of composition of surface waters of tailing dumps of Pb, Cu, and Zn deposits of Dal'negorsk ore district of the Far East. *Ecological Chemistry* 17(1): 35–46.

Nesterov, V. N. & Nesterova, E.S. 1978. Regime of temperature and humidity in the Volkovsky Creek basin. *Investigations of secondary biogeocoenoses of the Middle Sikhote-Alin*: 134–138. Vladivostok: FESC of the USSR Acad. Sci.

Zvereva, V.P. 2008a. *Ecological consequences of hypergene processes at tin-sulfide deposits of the Far East.* Vladivostok: Dal'nauka.

Zvereva, V.P. 2008b. Present-day mineral formation in hypergenesis zone of the exploited tin deposits of the Far East. *Transactions of Mineralogical Society* 6: 28–37.

Zvereva, V.P., Pyatakov, A.D. & Kostina, A.M. 2013a. Modeling of Sulfide Oxidation in Tailing Dumps of Kavalerovsky District and Their Impact on the Hydrosphere (Primorsky Krai, Russia). *Russian Journal of General Chemistry* 83(13): 2750–2753.

Zvereva, V.P., Pyatakov, A.D., Kostina, A.M. & Frolov, K.R., Lysenko, A.I. 2013b. Physicochemical Modeling of the Impact of Tailing Dumps on the Hydrosphere in the Kavalerovsky Tin-Ore District of the Russian Far East. *Russian Journal of General Chemistry* 83(13): 2663–2667.

Future Communication Technology and Engineering – Chan (Ed.)

The research on the influencing factors of carbon emission based on geographic detector model

R.N. Wu
School of Environment, Northeast Normal University, Institute of Natural Disaster Research, Northeast Normal University, Jilin Changchun, China
Inner Mongolia Key Laboratory of Remote Sensing and Geographic Information, Inner Mongolia Huhhot, China

J.Q. Zhang
School of Environment, Northeast Normal University, Institute of Natural Disaster Research, Northeast Normal University, Jilin Changchun, China

M.X. Huang & Y.H. Bao
Inner Mongolia Key Laboratory of Remote Sensing and Geographic Information, Inner Mongolia Huhhot, China

ABSTRACT: The research on the influencing factors of carbon emission has an important practical significance and provides an important theoretical basis for establishing regional carbon emission reduction policies, developing low-carbon economy and tackling climate change. Thus, based on geographic detector model, this article carries out quantitative analysis by studying on 6 social economic factors, including GDP, economic growth, urbanization rate, year-end total population, industrial structure and road density from energy consumption and carbon emission data of 101 counties in the Inner Mongolia. It turns out that Inner Mongolia is suffering from a great gap between the rich and the poor, carbon emission in this area is very uneven and regional difference is highly significant; apart from the strong positive autocorrelation of space of GDP and road density, other factors are relatively randomly distributed in space; GDP, economic growth rate, urbanization level and industrial structure have higher risks on carbon emission, while the risk of the total population at the end of the year and road density is relatively low; the influence of the interaction of GDP and economic growth rate on carbon emission is weaker than the interaction of other variables except their overall influence on each other, and their influence on carbon emission is enhanced non-linearly.

1 INTRODUCTION

Global climate change has been an indisputable fact, among which the global warming, especially the warming caused by carbon dioxide has drawn more and more attentions. How to effectively reduce carbon emission while maintaining fast economic development is the significant issue that all provinces and cities of China are trying to figure out. Moreover, it is not only an issue that national government pays attention to, but also an issue that needs to be cared about by the academic world. In 1970s, American ecologist Ehrlich and Comnoner brought up IPAT model for the evaluation of environment pressure, which indicated that population (P), average wealth (A) and technology (T) as well as the interaction among them would have tremendous influence on environment (Ehrlich et al., 1971; 1972). In 2009, Shoufu Lin analyzed the influence of Chinese environment based on STIRPAT model. According to the result, the factors that influenced Chinese environment was population, urbanization level, GDP per capita, industrial level and energy intensity, and population ranked the most influential one among all, followed by urbanization level, industrial level, GDP per capita and energy intensity (ShouLin et al., 2009). There are many specific categories and indicators of carbon emission factors, and different factors have different ways and levels of influence on carbon emission level, which is distributing in different levels of social economy, so it is inevitable to take multi-factors into consideration. However, domestic and international research is limited to 'single factor', which means the total amount of carbon dioxide release is only compared to one factor and the research lacks comprehensive analysis of combined action of multi-factors (Wang Q W et al., 2010).

2 DATA RESOURCES AND RESEARCH METHOD

2.1 *Data resources and choice of variable*

Considering the time lagging of variables, the samples used in this article is the average value of 2007, which can smooth the influence of emergent factors in a short period of time, so the statistics from it can more accurately reflect the carbon emission situation of all regions and the mechanism of action of influential factors. Moreover, the selected statistics are cross-section data from 101 counties in the Inner Mongolia, including carbon emission volume, gross regional production year-end total population, economic growth rate, urbanization rate, industrial structure and road density. Among all, carbon emission volume is calculated by the way provided in newly released *Guideline of Editing Provincial Greenhouse Gases List (trial)*, i.e. the answer of energy consumption volume multiples corresponding emission factor.

2.2 *Geographic detector model*

Assuming there is a factor in geographic space that can determine carbon emission, this factor must have certain similarity of carbon emission distribution in geographic space (Wang JF et al., 2010). Let's assume geographic space to be Ω and the geographic distribution of carbon emission to be H. Geographic space is divided into N_T units by regular grids and carbon emission in each unit is H_i $(1 \le i \le N_T)$. The space distribution of the factor that affects carbon emission is D, and the factor can be divided into n_D sub-domains in terms of geographic space. With the overlaying effects of carbon emission (H) and influential factor (D), it creates n_D sub-domains in the geographic space, and each sub-domain includes $n_{D,z}$ $(1 \le z \le n_D)$ grids, thus $N_T = \sum_{z=1}^{n_D} n_{D,z}$. The carbon emission of each gird unit in the sub-domain is:

$$H_{z,i}(1 \le z \le n_D, 1 \le i \le n_{D,z})$$

As a result, the average carbon emission of entire geographic space Ω is:

$$H_T = \frac{1}{N_T}\sum_{i=1}^{N_T} H_i \tag{1}$$

The sample variance of carbon emission of entire geographic space Ω is:

$$\sigma_T^2 = \frac{1}{N_T}\sum_{i=1}^{N_T}(H_i - \overline{H_T})^2 \tag{2}$$

(1) Risk Detection

If risk factor D divides geographic space into several sub-domains and mark two sub-domains among them as z1 and z2, the average carbon emission of these two sub-domains is:

$$\overline{H_{z1}} = \frac{1}{n_{D,z1}}\sum_{i=1}^{n_{D,z1}} H_{z1,i} \tag{3}$$

$$\overline{H_{z2}} = \frac{1}{n_{D,z2}}\sum_{i=1}^{n_{D,z2}} H_{z2,i} \tag{4}$$

The sample variance of carbon emission of these two sub-domains is:

$$\sigma_{z1}^2 = \frac{1}{n_{D,z1}}\sum_{i=1}^{n_{D,z1}}(H_{z1,i} - \overline{H_{z1}})^2 \tag{5}$$

$$\sigma_{z2}^2 = \frac{1}{n_{D,z2}}\sum_{i=1}^{n_{D,z2}}(H_{z2,i} - \overline{H_{z2}})^2 \tag{6}$$

We can calculate the average carbon emission of each sub-domain by using formula (3). As a result, we can also find out the carbon emission of each sub-domain by $\overline{H_{z1}}$ and $\overline{H_{z2}}$. Carbon emission of each sub-domain might be different if differences exist between $\overline{H_{z1}}$ and $\overline{H_{z2}}$. Next, we run some tests on the difference and find out if they are significant. We use t test to analyze their differences.

$$t_{\overline{H_{z1}}-\overline{H_{z2}}} = \frac{\overline{H_{z1}}-\overline{H_{z2}}}{[\sigma_{z1}^2 / n_{D,z1} + \sigma_{z2}^2 / n_{D,z2}]^{1/2}} \tag{7}$$

The DF is:

$$df = \frac{\sigma_{z1}^2 / n_{D,z1} + \sigma_{z2}^2 / n_{D,z2}}{\frac{1}{n_{D,z1}-1}(\sigma_{z1}^2 / n_{D,z1})^2 + \frac{1}{n_{D,z2}-1}(\sigma_{z2}^2 / n_{D,z2})^2} \tag{8}$$

In order to test null hypothesis $H_0 : \overline{H_{z1}} = \overline{H_{z2}}$, we need to provide confidence level α (which is normally 5%). If $|t_{\overline{H_{z1}}-\overline{H_{z2}}}| > t_{\alpha/2}$, then refuse H_0. This means there are significant differences of these two sub-domains on carbon emission. Otherwise, then we cannot confirm the existence of significant difference, and the difference between them may be caused by error.

(2) Factor Detection

Risk factor divides the entire geographic space into several sub-domains, so we can calculate layered population variance caused by the division of influence factor D.

$$\sigma_{TD}^2 = \frac{1}{N_{T,D}}\sum_{z=1}^{n_D}\sum_{i=1}^{n_{D,z}}(H_{z,i} - \overline{H_z})^2 \tag{9}$$

with $N_{T,D} = \sum_{z=1}^{n_D} n_{D,z}$. If the influence factor can completely affect carbon emission, then the sample variance of each sub-domain is close to 0 (Wang JF et al., 2010), so σ_{TD}^2 is close to 0. Here is the definition of power of determinants (PD):

$$PD = 1 - \frac{\sigma_{TD}^2}{\sigma_T^2} \tag{10}$$

We can tell from formula (10) that PD is between 0 and 1. When PD is closer to 1, σ_{TD}^2 is closer to 0, which means this factor has higher influence on carbon emission.

(3) Ecology Detection

How to analyze a factor's influence on carbon emission is higher than another factor and their difference is significant? Ecology detection can tackle the task. For example, as to GDP and urbanization level, how can we tell which one has higher influence on carbon emission? Let's assume these two factors to be D_1 and D_2. According to formula (9), the layered variance of population of calculating two different influence factors is σ^2_{TD1} and σ^2_{TD2}, and the number of gird units that divide these two risk factors are $n_{T,D1}$ and $n_{T,D2}$. In order to compare the difference of σ^2_{TD1} and σ^2_{TD2}, we use F for testing:

$$F = \frac{n_{T,D1}(n_{T,D1} - 1)\sigma^2_{TD1}}{n_{T,D2}(n_{T,D2} - 1)\sigma^2_{TD2}} \tag{11}$$

This calculation is nearly in compliance with the division $F(n_{T,D1} - 1,\ n_{T,D2} - 1)$, with $df = (n_{T,D1}, n_{T,D2})$. In order to test null hypothesis $H_0: \sigma^2_{TD1} = \sigma^2_{TD2}$, we need to provide confidence level α (which is normally 5%). If H_0 is refused under confidence level α, then it means there are significant differences of these two factors on carbon emission.

(4) Interactive Detection

We can apply interactive detection to analyze the influence of the interaction of two or more factors on carbon emission. For two risk factors D_1 and D_2, their influence on carbon emission might be independent or coefficient. If combined action exists, their influence on carbon emission after interaction will be stronger. We can use GIS software to overlay geographic image D_1 and D_2, and a new geographic image E will be created. By comparing the PD of D_1, D_2 and E, we can tell the influence of their interactive actions (Wang JF et al., 2010; Hu Y et al., 2011).

3 RESULT ANALYSIS

3.1 *Analysis of influencing factors of carbon emission based on geographic detector model*

3.1.1 *Risk detection*

Table 1 to 6 shows the result from risk detector. What we can know from Table 1 is that when GDP < 2 billion Yuan, carbon emission volume is 870,000 tons; with the increase of GDP, carbon emission volume is also increasing gradually. It indicates that areas with higher GDP have higher risk of carbon emission. Table 2 shows the influence of population on carbon emission. We can tell from this table that carbon emission volume is not increasing with the development of population. On the contrary, carbon emission volume reaches peak when the population is between 70,000 and 150,000, which is 2.686 million tons. When population increases above this range, carbon emission volume yet is decreasing slowly. This means population scale is not necessarily in direct proportion to carbon emission volume. Economic growth rate has

Table 1. Average carbon emissions of GDP in the hierarchical.

GDP (million RMB)	<20	20–26	26–42	42–57	57–94	>94
Carbon emission (million t)	8.7	10.6	67.4	46.9	262.3	841.7

Table 2. Average carbon emissions of population in the hierarchical.

Population (million people)	<7	7–15	15–23	23–31	31–38	>38
Carbon emission (million t)	6.6	268.6	68.9	72.7	59.1	54.1

Table 3. Average carbon emissions of economic growth rate in the hierarchical.

Economic growth (%)	<11	11–16	16–18	18–21	21–24	>24
Carbon emission (million t)	45.4	32.6	38.4	20.8	37.6	343.7

Table 4. Average carbon emissions of urbanization rate level in the hierarchical.

Urbanization rate (%)	<23	23–30	30–49	49–63	63–86	>86
Carbon emission (million t)	18.2	34.8	85.3	155.2	119.2	69.7

showed higher influence at two ends and lower influence in the middle (Tab 3). When economic growth rate is above 24% or below 11%, the average carbon emission volume of the region is relatively high, especially when economic growth rate is extremely high. However, the influence of urbanization rate on carbon emission volume is exactly the opposite – higher in the middle and lower at two ends (Tab 4). The influence of industrial structure on carbon emission volume is significant (Tab 5). When industrial structure takes smaller proportion of the whole industry, it doesn't have great influence on local carbon emission (<34%); however, when industrial structure continues to increase, carbon emission volume starts to grow crazily. As to road density, its influence on carbon emission is relatively random (Tab 6) and do not have distinct connection.

3.1.2 *Factor detection*

Factor detector is used to analyze the influence of 6 factors on carbon emission, which is sorted from high to low PD value as follows: GDP (0.56) > economic growth rate (0.26) > population (0.18), industrial structure (0.18) > road density (0.11) > urbanization

Table 5. Average carbon emissions of industry structure in the hierarchical.

Industrial structure (%)	<23	23–29	29–34	34–48	48–57	>57
Carbon emission (million t)	14.9	20.9	13.8	42.7	160.5	262.9

Table 6. Average carbon emissions of road destiny in the hierarchical.

Road density (km/km^2)	<0.002	0.002–0.007	0.007–0.012	0.012–0.017	0.017–0.047	>0.047
Carbon emission (million t)	15.1	159.4	61.3	227.3	69.6	368.4

Table 7. The result of the ecological detector.

	GDP	Population	Econolncre	IndustrySt	CityLevel
GDP					
Population	Y				
Econolncre	N	Y			
IndustrySt	Y	N	Y		
CityLevel	Y	N	Y	N	
RoadDensity	Y	N	Y	N	N

Table 8. The result of the interaction detector.

	GDP	Population	Econolncre	IndustrySt	CityLevel
Population	0.82↑				
Econolncre	0.67–	0.50↑			
IndustrySt	0.86↑	0.55↑	0.53↑		
CityLevel	0.70↑	0.34↑	0.44↑	0.26↑	
RoadDensity	0.71↑	0.37↑	0.56↑	0.45↑	0.47↑

level (0.03). It indicates that GDP has the highest influence on carbon emission volume, which is consistent with the result of risk detection.

3.1.3 *Ecological detection*

The application of ecology detector can find out the significant differences of the influence of 6 factors on carbon emission (Tab. 7). The result shows that GDP and economic growth rate do not have significant differences of the influence on carbon emission, and population, industrial structure, road density and urbanization level do not have significant differences on carbon emission either, and the variables in the first group and the variables in the second group have significant differences of the influence on carbon emission.

3.1.4 *Interaction detection*

At last, interactive detection is used to detect the influence of interaction of two factors on carbon emission. We can tell from Table 8 that the influence of interaction of GDP and economic growth rate on carbon emission is less than their total influence on carbon emission separately. Apart from this, the influence of other interactive factors on carbon emission is non-linearly enhanced.

↑ After interaction of two factors, the impact on carbon emission is more than the sum of their impact, namely the non-linear enhancement.

– After the two factors interaction, the influence of carbon emission is greater than the impact of individual. But it is less than the sum of their impact on carbon emission, namely, the linear increases.

4 DISCUSSIONS

With the development of GDP, carbon emission volume is also gradually increasing, but when population, economic growth rate, industrial structure and urbanization level reach to a certain number or a certain degree, there is no significant difference in carbon emission volume, and road density has no distinct influence on carbon emission. The influence of social economic factors on carbon emission is sorted as follows: GDP (0.56) > economic growth rate (0.26) > population, industrial structure > road density > urbanization level. Among all these social economic factors, GDP and economic growth rate have tremendous influence on carbon emission. The influence of interaction of GDP and economic growth rate on carbon emission is less than their total influence on carbon emission separately. Apart from this, the influence of other interactive factors on carbon emission is non-linearly enhanced.

ACKNOWLEDGEMENTS

This study was supported by "Twelfth five-year" science and technology support projects (2013BAK05B01).

REFERENCES

Ehrlich P R, Holden J P. 1971. Impact of population growth. Science, 171: 1212–1217.

Ehrlich P R, Holden J P. 1972. One dimensional economy. Bulleton of Atomic Scientists, 16: 18–27.

Shoufu Lin, Dingtao Zhao, Dora Marinova. 2009. Analysis of the environmental impact of China based on STIRPAT model. Environmental Impact Assessment Review, (29): 341–347.

Wang Q W, Zhou P, Zhou D Q. 2010. Research on dynamic carbon dioxide emissions performance, regional disparity and affecting factors in China. China Industrial Economics, 1(1): 46–54.

Wang JF, Li XH, Christakos G, Liao YL, et al. 2010. Geographical detectors-based health risk assessment and its application in the neural tube defects study of the Heshun Region, China. IJGIS, 24(1): 107–127.

Hu Y, Wang J, et al. 2011. Geographical Detector-Based Risk Assessment of the Under-Five Mortality in the 2008 Wenchuan Earthquake, China. PloS ONE, 6: e21427.

Future Communication Technology and Engineering – Chan (Ed.)
© 2015 Taylor & Francis Group, London, ISBN: 978-1-138-02777-0

Impact assessment of the processes occurring in tailing dumps of a Krasnorechenskaya concentrating mill on the hydrosphere of Dalnegorsky District, according to modelling evidence

Valentina P. Zvereva & Alyona M. Kostina
Far East Federal University, Sukhanova, Vladivostok, Russian Federation
Far East Geological Institute FEB RAS, Vladivostok, Russian Federation

Anastasia I. Lysenko
Far East Geological Institute FEB RAS, Vladivostok, Russian Federation

ABSTRACT: The authors conducted an impact assessment of the hypergene processes occurring in the tailing dumps of a Krasnorechenskaya concentrating mill on the hydrosphere of Dalnegorsky District, using a physicochemical modelling method. New thermodynamic data for some of the hypergene minerals we used in the calculations, was taken from the literature sources, and obtained by the program founders. The assessment shows that the solution mineralization used in all simulated systems is high, which leads to the crystallization of a wide range of hypergene and technogenic minerals. Highly concentrated micropore solutions form the drainage water whose content sulphide elements exceed the background characteristics in tens, hundreds, thousands, and even tens of thousands of times.

1 INTRODUCTION

Lead-zinc industry of Dalnegorsky District has been developing for more than 100 years. More than 20 skarn and polymetallic deposits are being processed here, by either opened or closed methods. There is Krasnorechenskaya concentrating mill (KCM) in the area, which dressed tin-polymetallic ores. Tailing dumps were formed and were active from 1956 to 1995. On the area of 27.4 hectares 14.3 million tons of tailings were accumulated. The main ore minerals are cassiterite, sphalerite, and galena, and minor – arsenopyrite, chalcopyrite, pyrite, and pyrrhotite.

Such a long and active development of the mining industry in the district has led to the presence of numerous ditches, clearings, quarries, adits, and tailing dumps, which contributed to the creation of mining technogenic system hypergene processes. They had been activated, and then switched to the technogenic stage. It happened due to the increase of sulphide contact surface with weathering agents (oxygen, water, carbon dioxide, etc.). Hypergene and technogenic processes facilitate geochemical stress on hydrosphere, as mine, slime, and drainage waters contain a wide range of toxic ore and rock elements: Zn, Pb, Cu, As, Fe, Mg, Al, Si, and others (Borisov & Yelpatievskiy, 1992, Zvereva & Krupskaya, 2012).

Studies conducted in the Dalnegorsky district of the Far East had shown an adverse effect of mining industry on the atmosphere, lithosphere, biosphere, and hydrosphere. The environmental situation in the district is unfavourable, and Rudnaya River is acknowledged as one of the most polluted rivers in the world (Zvereva, 2008, Borisova & Elpatievsky, 1992, Elpatievsky & Lutsenko, 2000, Elpatievsky, 2000, Zvereva, 2000, 2006, Zvereva & Zarubina, 2008).

Modelling of sulphide oxidation processes in contact with host rocks allows to fully review hypergene processes and identify possible hypergenic and technogenic mineral phases which crystallizes from highly concentrated solutions. It also provides opportunity to define composition of solutions, forms of dissolved particles, as well as the volume of elements, washed out to the hydrosphere from tailing dumps. In the recently calculated models, thermodynamic constants of hypergene minerals were entered (Tupitcin et al., 2009) in accordance with the full mineral composition of tailings (Bulavko, 2000, Tarasenko & Zinkov, 2001, Zvereva, 2001).

The modelling was performed using the program of chemical equilibria calculation Selektor, which is based on the minimization of thermodynamic potentials (developers are I.K. Karpov, K.V. Chudnenko and V.A. Bychinskiy).

The main purpose of this work is to assess the impact of technogenic and hypergene processes occurring in the KCM drained tailings of Dalnegorsky District on hydrosphere using physicochemical modelling. Based on the goal, the following tasks were formulated: 1) to simulate sulphides oxidation in contact with the host rock; 2) to identify possible technogenic minerals crystallized from drainage water micropore solutions; 3) to

Table 1. Mineral composition of oxidized tailings, %.

Minerals	Variants		
	1	2	3
Sphalerite ZnS	29.9	34.9	19.9
Galena PbS	20	10	30
Pyrite FeS_2	5	10	10
Pyrrhotite FeS	10	30	15
Chalcopyrite $CuFeS_2$	5	2.5	5
Arsenopyrite FeAsS	5	2.5	5
Argentite Ag_2S	0.05	0.05	0.05
Acanthite Ag_2S	0.05	0.05	0.05
Jamesonite $Pb_4FeSb_6S_{14}$	25	10	15
Axinite $Ca_2FeAl_2[BO_3][Si_4O_{12}](OH)$	20	20	20
Diopside $CaMg[Si_2O_6]$	17	17	17
Calcite $CaCO_3$	10	10	10
Albite $Na[AlSi_3O_8]$	20	20	20
Clinochlore $Mg_5Al[AlSi_3]O_{10}(OH)_8$	13	13	13
Feldspar $K[AlSi_3O_8]$	20	20	20

determine their qualitative and quantitative composition; 4) to assess the impact of tailing dumps on the district hydrosphere; 5) to verify the obtained data.

2 RESEARCH METHOD

With the Selektor program, the authors created a series of physicochemical models with temperature intervals from 0 to +45°C (increment – 5°C) and a pressure of 1 atm. The systems were opened to the atmosphere. 21 independent (Ag-Al-Ar-As-B-C-Ca-Cu-Fe-K-Mg-N-Na-Pb-S-Si-Sb-Zn-H-O-e) and 372 dependent components were taken into account in the calculations, 322 out of which – are dissolved particles, 18 are gases, and 32 are hypogene and hypergene minerals.

Chemical compositions and host rock – sulphides ratios were determined on the basis of literature (Bulavko 2000, Tarasenko & Zinkov 2001) and are presented in Table 1. The authors have taken the maximum, minimum, and average concentrations of sulphide minerals in the KCM tailings, therefore 3 variants of the mineral composition of the tailings were considered (Table 1) with 95: 5, 90:10, 80:20 and 60:40 host rock – sulphides ratios.

3 RESULTS AND DISCUSSION

Based on the models created, we assessed the impact of the processes occurring in KCM tailing dumps on the district hydrosphere.

Hypergene minerals crystallize in all three systems (Table 2), but their share varies from one system to another. In the first variant it reaches (%): 3.78 for valentinite, 5.06 – goethite, 14.36 – anglesite, 18.31 – gypsum, 40.24 – fibroferrite, in the second variant – 1.37, 6.42, 6.05, 16.85, and 40.24, and in the third one – 2.16, 5.37, 12.16, 16.07, and 31.03 respectively. The process of hypergene mineralization indicates a high sulphide solubility and, consequently, a potential danger to these waters.

At various host rock – sulphides ratios total (or sulphide) element concentration is the following (g/l): 8.23–29.60 (3.08–13.21) for the first variant, 7.66–30.42 (2.53–14.3) for the second one, and 7.81–27.50 (2.86–12.01) for the third one. Pore solutions of the KCM tailing dumps with mineral composition corresponding to 2 and 3 variants are acknowledged as the most mineralized ones; consequently these solutions in a form of drainage water will pollute surface waters more than others. The ratio of 60:40 provides the highest solution salinity with basic or sulphide elements. The basic sulphide ore elements contained in the waters here are (in the order of concentration increase, mg/l): Sb – 13.13–14.27, Ag – 4.399–36.57, Cu – 43.17–692, As – 57.12–915.9, S – 428.3–3495, Pb – 12.53–6611, Zn – 685.5–9925.

In addition, pore solutions and drainage water contain the following host rock elements (mg/l): K (0.08–0.14), Al (0.07–0.24), Si (2.59–4.83), Na (48.31–87.61), B (228.6–357.7), Ca (276.6–547.2) & Mg (0.05–554.5).

The simulation results demonstrate that the first variant solutions contain antimony at the ratio of 95:5, the second one – zinc, silver, sulphur at 60:40, lead – 80:20, and the third one – copper and arsenic at the ratio of 60:40. The iron concentration in all systems is low and approximately equal, i.e. it precipitates from the solution in the form of goethite. The content of host rock elements Si, Al, B, Mg, Na, and K in the modelled solutions is higher in case of a larger host rock share in the models. Axinite destruction contributes to the high boron content in the solution.

The larger sulphide component is, the larger amounts of potentially hazardous substances are oxidized, the higher solution salinity will be, and consequently the bigger damage will be dealt to hydrosphere, as we would expect.

It should be noted that in this case the greatest load on the surface water is in the second and third variants in comparison with the first one.

Sulphide component increase in the system simulated leads to a slight pH decrease from 9.29 to 6.85, and alkaline condition change from medium to neutral. Eh value varies from 0.66 to 0.81 V.

Consider next the effect of temperature factor (with the increase from 0 to +45°C) on the mineral formation processes and element content in the solution. Solution mineralization of the two system types with host rock-sulphides ratios of 95:5 and 60:40 depending on total (and sulphide) element content varies in the range (g/l): for the first variant from 7.52 to 8.01 (2.88–3.04), and for the second one – 27.28–27.49 (12.7–12.9). With the temperature increase, the total and sulphide element mineralization of the solutions also rises up.

Depending on the temperature and sulphide – host rock ratio, the element content in the solution is the following (mg/l) (in order of their reduction):

- atmospheric gases: C – from 0.11 to 237.5, N – 0.001–38.48, H – 0.19–104.9, and O – 0.74–3631;

Table 2. Modelling results of sulphide oxidation in the KCM tailing dumps.

	Host rock – sulphides ratio			
Solution parameters and minerals, %	95:5	90:10	80:20	60:40
Variant 1				
pH	9.2952	8.5993	8.3275	6.8561
Eh, V	0.6677	0.7088	0.7249	0.8118
Goethite FeOOH	3.79	4.61	5.06	–
Fibroferrite $Fe(SO_4)(OH)(H_2O)_5$	–	–	3.57	24.45
Gypsum $CaSO_4(H_2O)_2$	4.68	9.70	18.31	11.25
Anglesite $PbSO_4$	–	–	–	14.36
Valentinite Sb_2O_3	0.51	1.04	2.13	3.78
Variant 2				
pH	8.7885	8.2439	8.3279	7.2385
Eh, V	0.6976	0.7298	0.7248	0.7892
Goethite FeOOH	4.69	6.42	5.51	–
Fibroferrite $Fe(SO_4)(OH)(H_2O)_5$	–	–	11.53	40.24
Gypsum $CaSO_4(H_2O)_2$	5.59	11.98	16.85	10.47
Anglesite $PbSO_4$	–	–	–	6.05
Valentinite Sb_2O_3	0.19	0.40	0.79	1.37
Variant 3				
pH	9.0891	8.3029	8.2964	6.8540
Eh, V	0.6798	0.7263	0.7267	0.8119
Goethite FeOOH	4.16	5.37	4.81	–
Fibroferrite $Fe(SO_4)(OH)(H_2O)_5$	–	–	8.25	31.03
Gypsum $CaSO_4(H_2O)_2$	4.84	10.13	16.07	10.79
Anglesite $PbSO_4$	–	–	–	12.16
Valentinite Sb_2O_3	0.30	0.62	1.24	2.16

- host rocks: Al – <0.085, Mg – 0.019–0.299, K – $2.589 \cdot 10^{-7}$–0.50, Si – до 7.976, Na – <156.8, B – 0.02–357.7, and Ca – 0.01–576.1;
- sulphides: Ag – 4.43, Sb – 23.16, Cu – 85.16, As – 112.7, S – 485.8, Zn – 685.6, and Pb – 1643.

For solutions of the two system types with the host rock – sulphides ratios of 95:5 and 60:40 according to the content of all (and sulphide) elements – the mineralization varies in the range of (g/l): 7.52 to 8.01 (2.88–3.04) and 27.28–27.49 (12.7–12.9), respectively, and it increases slightly with a rise in temperature.

With the temperature increasing, the contents of Ag, As, Cu, and Zn remains constant, the amount of S and Sb – increases and Pb decreases slightly (within 0.01 %). On the content of sulphide elements within the specified criteria, the ratio between the host rock and sulphides is not affected.

In this case, the temperature acts only slightly as a catalyst for the destruction of primary minerals, their transformation into secondary minerals, and the dissolved form. The temperature increase from 0 to +45°C has almost no impact on the quantitative content of sulphide elements in the solution. Their content in the solution is constantly high. It indicates a significant removal of all sulphide ore and host rock elements with drainage solution to surface and ground water throughout the investigated temperature range. Minor fluctuations in the component concentrations in the solutions with temperature increase can be explained by a minor pH and Eh change in the solution. For example, with 95:5 host rock – sulphides ratio Eh potential reduces from 0.69 to 0.66, and pH – from 9.73 to 8.74. In the system with 60:40 ratio their parameters are reduced from 0.81 to 0.80 and from 7.54 to 6.41, respectively.

Modelling in the temperature range chosen showed that hypergene processes and mineral formations occur constantly and actively, and content of sulphide ore and host rock elements is consistently high in the solution all the time, which are washed out to the hydrosphere with drainage water.

Additionally, a series of physicochemical models characterizing similar systems with pressure changing in the range from 0.8 to 1.2 atm was created. The modelling shows that the pressure has insubstantial effect on the oxidation of hypogene minerals and crystallization of hypergene ones, as well as on the element content in the solution.

The assessment based on the results of hypergene processes modelling in all considered variants shows that a wide element range of ores and host rocks (Pb, Zn, Cu, Fe, As, S, Si, B, Ca, Mg, and others) is being washed out from KCM tailing dumps of Dalnegorsky District into the hydrosphere. Solution salinity in hypergene processes occurring in ore bodies and in the tailings is tens, or even hundreds of grams, per litre.

Modelling results are confirmed by the literature data on the mineral composition of hypergene and technogenic minerals (Bulavko, 2000, Zvereva, 2001, Tarasenko & Zinkov, 2001). The ionic composition of pore solutions is fully confirmed by experimental data on sulphide oxidation available in the literature

(Postnikoova & Yakhontova 1984, Zvereva 2008). The elements content in the solutions and their mineralization are verified by hydrochemical and literature data. The contents of these elements in natural mine, slime, and drainage waters are higher than the background values of the district even under the most significant dilutions (Zvereva, 2008, Borisova & Elpatievsky, 1992, Elpatievsky & Lutsenko, 2000, Elpatievsky, 2000, Zvereva, 2000, 2006, Zvereva & Zarubina, 2008).

Therefore, sulphide oxidation modelling in conjunction with the fullness of the thermodynamic data makes it possible to assess the ecological situation in the region at the moment, and also to show the impact of individual components of the mining technogenic system – the district tailing dumps on the hydrosphere.

The following results can be concluded from the abovementioned information. Temperature is the catalyst for primary mineral destruction, secondary mineral crystallization, and their accumulation in the dissolved form, but its role is insignificant. It has little effect on the quantitative content of sulphide elements in the solution, i.e. it is consistently high throughout the considered temperature range from 0 to +45°C. The crystallization process of hypergene and technogenic minerals in the district is active enough. However, before and after mineral crystallization such uncleaned solutions wash out a wide range of elements, such as sulphide ores and host rocks, contaminating surface and groundwater around the clock for decades.

Physicochemical modelling allows us to determine the volumes of ore and host rock elements washed out with mine, drainage, and slime solutions to ground and surface water. Consequently, it can be carried out not only for assessment, but also for establishment of the impact extent, hydrosphere pollution degree, the future prediction, and so on.

REFERENCES

Borisova, V.N. & Yelpatievskiy, P.V. 1992. The possibility to solve some of the environmental issues in prospecting, *J. Pacific geology 3*. 134–139.

Bulavko, N.V. 2000. *Mineralogy of Skarn Deposits of the Dal'negorsk Ore Field (Primorsky Krai)*. Vladivostok: Far East Publishing House.

Elpatievsky, P.V. & Lutsenko, T.N. 2000. Mining industry complex as a factor in the formation of water chemical composition. *Scientific and practical aspects of the production of non-ferrous and precious metals. Abstracts of the reports of Intern. conf. in Khabarovsk 2*. 407–415.

Elpatievsky, P.V. 2000. Metal-bearing of technogenic waters of ore deposits. *Geology and mining in the past, present and future. Abstracts of reports at the conference "Geol. Primorye"*. Vladivostok: Dal'nauka. 26–29.

Postnikova, V.P. & Yakhontova L.K. 1984. *Mineralogy of hypergene zone of tin deposits of Komsomolsky region*. Vladivostok: Far Eastern Scientific Center, Academy of Sciences of the USSR.

Tarasenko, I.A & Zinkov, A.V. 2001. *The environmental effects of mineralogical-geochemical transformation of the tailings of Sn-Ag-Pb-Zn ores*. Vladivostok: Dal'nauka.

Tupitcin, A.A., Mukhetdinova, A.V. & Bychinsky, V.A. 2009. *Preparation of the thermodynamic properties of individual substances to the physicochemical modeling of high-temperature processes*. Irkutsk: publishing house ISU.

Zvereva, V.P. & Krupskaya, L.T. 2012. Technogenic water of mining Komsomolsky, Kavalerovskiy and Dal'negorskiy district of the Far East and their impact on the hydrosphere, *J. Ecological chemistry 21*. 144–153.

Zvereva, V.P. 2000. Environmental effects of hypergenesis and technogenesis of tin deposits. *Proceedings of the conference "Geology and mining in Primorye in the past, present and future"*. Vladivostok: Dal'nauka. 29–31.

Zvereva, V.P. 2001. Morphology and mineralogy of hypergene zone of polymetallic deposits of Dalnegorsky ore district (Primorye). *Collection ore deposits of continental margins*. Vladivostok: Dal'nauka. 180–190.

Zvereva, V.P. 2006. Ecological situation in Dalnegorsky mining district (Primorye). *Mining Journal 4*. 78–80.

Zvereva, V.P. 2008. *Ecological effects of hypergene processes in tin deposits of the Far East*. Vladivostok: Dal'nauka.

Zvereva, V.P., Zarubina, N.V. 2008. Mining technogenic system of Dalnegorsky district of the Far East and its impact on the ecosphere. *Geoecology 6*. 500–505.

Future Communication Technology and Engineering – Chan (Ed.)
 ISBN: 978-1-138-02777-0

Space monitoring of the earth and Haar wavelet transform

M.L. Kazaryan
North-Ossetian State Pedagogical Institute, North-Ossetian State University, Russia
Financial University under the Government of the Russian Federation, Russia

M.A. Schahramanian
Institute of Aerospace Technologies and Monitoring, Russian State University of Oil and Gas n.a. I.M. Gubkin, Russia

A. Richter
Institute of Aerospace Technologies and Monitoring, Russia

ABSTRACT: We investigate the use of orthogonal transformations in Earth remote sensing techniques which allows a spacecraft to get information of medium and high resolution, and carry out hyperspectral measurements. While we perfect the photographs, the apparatus of discrete orthogonal transforms are used, especially the wavelet transform. Haar's wavelets are examined, and their research of correctness is conducted.

Keywords: Space technology, Remote sensing of earth, Space monitoring, Fourier transformation, Discrete wavelet transform, Haar series.

1 INTRODUCTION

Remote sensing of Earth technologies (RST) is an accessible, swift and effective method of environmental monitoring [1]. It is a method of recognition and change detection on aerospace imagery which is based on the spectral brightness of the image elements and dependent on external factors such as atmospheric conditions, seasonality, clouds, etc. [2].

To identify a more universal method of processing satellite images, a different approach is required, which is based on the idea that the structural properties of object images are more resistant to external influence. Therefore, algorithms based on these properties are more effective [3, 11].

Fourier and wavelet analysis methods are related to these ones [4, 5, 6, 13].

The idea of frequency and time localization is the basis of wavelet transformation, which gives the opportunity to analyze the results both in frequency and in a physical space. This is especially important when working with aerospace photos (inhomogeneous signals).

The study of the correctness of Haar's wavelet – transformation has already been conducted [13]. This article deals with the continuation of these investigations.

Preliminary information

It is known from [6] that the problem of summing of Fourier's series square-integrable on the $[a, b]$ interval of $f(t)$ function with approximate coefficients $\{c_k\}_{k=1}^{\infty}$ instead of the exact coefficients

$$a_k = \int_a^b f(t)\ \psi_k(t)\ dt, \qquad k = 1,2,\ldots$$

for some orthonormal system of functions $\{\Psi_k(t)\}$ is ill-posed. Namely, if

$$\sum_{k=1}^{\infty} (c_k - a_k)^2 \leq \delta^2, \qquad \delta > 0 \tag{1}$$

then the error, i.e. the deviation of the function and the sum of its Fourier series with $\{c_k\}_{k=1}^{\infty}$ coefficients instead of $\{a_k\}_{k=1}^{\infty}$, in the uniform metric can be arbitrarily large.

Definition 1. Define the Haar wavelet function as follows [6]:

$$\chi_1(t) \equiv 1$$

$$\chi_{mj}(t) = \begin{cases} 2^{\frac{m-1}{2}}, \text{при } t \in \left[\frac{j-1}{2^{m-1}}, \frac{2j-1}{2^m}\right) \\ -2^{\frac{m-1}{2}}, \text{ при } t \in \left[\frac{2j-1}{2^m}, \frac{j}{2^{m-1}}\right) \\ 0 \quad \text{in other cases} \end{cases}$$

where $m = 1,2,\ldots$; $j = 1,\ldots,2^{m-1}$, and at $j = 2^{m-1}$

the right of the segments is considered to be closed and right. On numbering functions of a single index k, $k = 2^{m-1} + j$ is assumed.

This definition differs from that of the Haar's [8] by values of Haar functions at points of discontinuity, however, the basic property of the Haar sys-tem is retained – the even desire of the Fourier–Haar continuous on [0, 1] the function $f(t)$ to $f(t)$.

In the case of the Haar system, the condition of uniform roundedness is not satisfied, and the method proposed in [6], does not provide a continuous function that approximates a continuous function. These studies have already been carried out [13].

Previously in [13] the consideration of the problem of regularization of wavelet – Haar's series with approximate coefficients was justified.

In solution of this task S_p, $1 \leq p \leq \infty$ classes play a significant role. They are introduced and described in detail by I.M. Sobol (see 9) for the study of multidimensional quadrature formulas and functions and contain functions with rapidly con-vergent wavelet – Haar's series.

Definition 2 [9]. Through S_p, $1 \leq p \leq \infty$ mark the class of functions $f(t)$, and satisfy the following conditions.

It can be represented by Haar's series:

1. $f(t) = \sum_{k=1}^{\infty} a_k \chi_k(t)$

2. $A_p(f) \equiv \sum_{m=1}^{\infty} 2^{\frac{m-1}{2}} \{\sum_{j=1}^{2^{m-1}} |a_{m\,j}|^p\}^{\frac{1}{p}} \leq A$,
   где $A = const,\ A \geq\ 0.$

Let

$$S_p = \bigcup_{A \geq 0} S_p(A).$$

Introduce some notation. Mark the space of functions continuous on segment [0, 1] through $C_{0,1}$, and the space of integral p-th power functions through $L^p_{0,1}$ [10].

We define a function $\varphi(t)$ that represents a generalized summation method, and it is analogous to the sum function [6], as follows.

Definition 3. A monotonic function $\varphi(t)$ continuous on the right at point 0 with

$$\varphi(0) = 1; \qquad \lim_{t \to \infty} \varphi(t) = 0\,;$$

$$\int_0^{\infty} \varphi(t)\, t^{-\frac{1}{2}}\, dt \ <\ \infty \tag{4}$$

We call a generalized summation function, and the method of summation of series by this function is called a generalized summation method. From the following relationship we can see that the classes S_p are sufficiently wide: the class of Lipschitz

$$H_{\propto'}(L),\ 0 < \propto' \leq 1,\ L > 0,\ L\ =\ const\,.$$

$$H_{\propto'}(L)\ = \left\{ \begin{matrix} f(t);\, |f(t_1) - f(t_2)| \leq L\, |t_1 - t_2|^{\propto} \\ \text{для любых точек } t_1, t_2 \text{ из } [0,1] \end{matrix} \right\}$$

belongs to S_p if $\propto' P > 1$ [9].

The following theorem is true.

Theorem 1 [13]. If Haar's series

$$\sum_{k=1}^{\infty} a_k\, \chi_k(t)$$

with arbitrary real coefficients converges uniformly on [0, 1], then it is the Fourier–Haar series for its sum.

Suppose instead of the exact coefficients $\{a_k\}_{k=1}^{\infty}$ of the function $f(t)$ their approximate values $\{c_k\}_{k=1}^{\infty}$ are known, so the correlation is satisfied (1). Then the following theorems are true.

Theorem 2 [13]. Let $f(t) \in S_p$, $1 \leq p \leq \infty$, $t \in [0, 1]$, $a = a(\delta)$ monotonically strives to zero and $\delta a^{-\frac{1}{2}}(\delta) \to 0$ at $\to 0$, then:

1. $f_\delta(a,t) = \sum_{k=1}^{\infty} \varphi(ak) c_k \xi_k(t)$ belongs to $S_{p''}$ $p' = \max(2,p)$;
2. $f_\delta(a,t)$ uniformly on [0, 1] approaches to $f(t)$.

Taking into account the condition (2), and the above theorem can be reformulated as follows:

$$f_\delta\,(\alpha, t)\ =\ \sum_{k=1}^{\infty} \varphi(\alpha k)\; c_k\, \chi_k(t).$$

Theorem 3 [13]. Let the sequence of real figures $\{a_k\}_{k=1}^{\infty}$ satisfy the condition (2) and be together with $\{c_k\}_{k=1}^{\infty}$ and it also satisfies the condition (1). Suppose further $\alpha = \alpha(\delta)$ that tends monotonically to zero and $\delta a^{-\frac{1}{2}}(\delta) \to 0$ for $\delta \to 0$. Then functions

1) $f(t)$ and $f_\delta(a,t) = \sum_{k=1}^{\infty} \varphi(ak) c_k \chi_k(t)$ belong to the classes S_p, $1 \leq p \leq \infty$ and $S_{p'}$, $p' = \max(2,p)$ accordingly;
2) $f_\delta(\alpha,t)$ uniformly on [0, 1] approaches to $f(t)$.

2 REGULARIZED SUMMATION OF THE CONTINUOUS FUNCTION

S_p, $1 \leq p \leq \infty$ classes contain a sufficiently large number of continuous functions. But not all continuous functions are included in the S_p classes, so the separate consideration of the problem for $C_{(0,1)}$ classes is expedient. Here we prove an analogue of Theorem 2 for functions from $C_{(0,1)}$. We see that, as in the previous theorem, approximating function belongs to the class $S_{p'}$, $2 \leq p' \leq \infty$.

Theorem 4. Suppose that $f(t)$ is continuous on [0, 1] function, $0 < \alpha \leq 1$, $\alpha = \alpha(\delta)$ monotonically strives to zero and $\delta\alpha^{-\frac{1}{2}}(\delta) \to 0$ for $\delta \to 0$. Then

$$f_\delta\,(\alpha, t)\ =\ \sum_{k=1}^{\infty} \varphi(\alpha k) c_k \chi_k(t).$$

1. Tend to $f(t)$ uniformly on the interval [0, 1];

2. At a fixed $\alpha, f_\delta(a,t) \in S_{p'}$, $2 \le p' \le \infty$,

Proof. Consider the deviation

$$|f(t) - f_\delta(\alpha,t)| \le$$

$$\le \left|\sum_{k=1}^{k(\propto)} [1-\varphi(\alpha k)]a_k \chi_k(t)\right| +$$

$$+\left|\sum_{k=k(\propto)+1}^{\infty} a_k \chi_k(t)\right| +$$

$$+\left|\sum_{k=k(\propto)+1}^{\infty} a_k \varphi(\alpha k)\chi_k(t)\right| +$$

$$+\left|\sum_{k=1}^{\infty} \varphi(\alpha k)\gamma_k \chi_k(t)\right| = S_1 + S_2 + S_3 + S_4$$

Separately estimate the amount of S_1, S_2, S_3, S_4. For S_1 we get

$$\left|\sum_{k=1}^{k(\propto)} [1-\varphi(\alpha k)]a_k \chi_k(t)\right| \le$$

$$\le [1-\varphi(\alpha k(\propto))] \sum_{k=1}^{k(\propto)} |a_k \chi_k(t)| \le$$

$$\le A\,[1-\varphi(\alpha k(\propto))] \sum_{m=1}^{m(\propto)-1} 2^{\frac{m-1}{2}},$$

where $A = const$, $k(\propto) \le 2^{m(\propto)-1} \le 2k(\propto)$

As

$$\sum_{m=1}^{m(\propto)-1} 2^{\frac{m-1}{2}} \le 2\sqrt{k(\propto)},$$

hence

$$S_1 \le A_1 [1-\varphi(\alpha k(\propto))] \sqrt{k(\propto)} =$$
$$= A_1 [1-\varphi(\alpha k(\propto))]\, k(\propto) \frac{1}{\sqrt{k(\propto)}},$$

where $A_1 = const$. By the continuity of the function $\varphi(t)$ at the right at the point $t = 0$, you can always choose $\propto$ and, accordingly, the number $k(\propto)$ so that the inequality is carried out

$$\varphi(\alpha k(\propto)) \ge 1 - \frac{C}{k(\propto)}$$

And, therefore, S_1 tends to zero at $\propto \to 0$.

The amount of S_2 represents the residue decomposition of the continuous function according to the system of Haar's functions and it is estimated with the help of the continuity of this function as follows [9]:

$$S_1 \le \frac{C}{k(\propto)}, \quad C = const.$$

Amount

$$S_2 \le C_0\, \omega\left(\frac{1}{k(\propto)}, f\right), \quad C_0 = const.$$

is a remnant of a uniformly convergent series. Indeed, since $f(t)$ is continuous, then the partial sums of its expansion

$$\sum_{k=k(\alpha)+1}^{\infty} a_k\, \varphi(\alpha k)\chi_k(t)$$

are limited and $\varphi(t)$ tends monotonically to zero at $t \to 0$. Therefore, according to the basis of Dirichlet series

$$\sum_{k=1}^{N} a_k\, \chi_k(t)$$

converges uniformly, which implies that S_3 tends to zero at $k \to \infty$.

We now estimate the amount S_4

$$\sum_{k=1}^{\infty} a_k\, \varphi(\alpha k)\chi_k(t)$$

$$\left|\sum_{k=1}^{\infty} \gamma_k\, \varphi(\alpha k)\chi_k(t)\right| =$$

$$= \left|\gamma_1\, \varphi(\alpha) \sum_{m=1}^{\infty}\sum_{j=1}^{2^{m-1}} \gamma_{mj}\, \varphi(\alpha(2^{m-1}))\chi_{mj}(t)\right|$$
$$\le |\gamma_1\, \varphi(\alpha)|$$

$$+ \sum_{m=1}^{\infty}\left|\sum_{j=1}^{2^{m-1}} \gamma_{mj}\, \varphi\big(\alpha(2^{m-1}+j)\big)\chi_{mj}(t)\right|.$$

Since

$$\left|\sum_{j=1}^{2^{m-1}} \gamma_{mj}\, \varphi\big(\alpha(2^{m-1}+j)\big)\chi_{mj}(t)\right| \le$$

$$\le \delta\left(\sum_{j=1}^{2^{m-1}} \varphi^2\big(\alpha(2^{m-1}+j)\big)\, \chi_{mj}^2(t)\right)^2 \le$$

then

$$\left|\sum_{k=1}^{\infty} \gamma_k\, \varphi(\alpha k)\, \chi_k(t)\right| \le$$

$$\le \delta + \delta\, \frac{\propto^{-\frac{1}{2}}}{ln2} \int_0^{\infty} \varphi(t)\; t^{-\frac{1}{2}}\, dt \le C_1\, \delta(\alpha)\, \alpha^{-\frac{1}{2}},$$

Hence, considering the conditions imposed on $\varphi(t)$ (see (3)), it is easy to verify that

$$\left|\sum_{k=1}^{\infty} \gamma_k\, \varphi(\alpha k)\, \chi_k(t)\right| \le$$

$$\le \delta + \delta\, \frac{\alpha^{-\frac{1}{2}}}{\ln 2} \int_0^{\infty} \varphi(t)\, t^{-\frac{1}{2}}\, dt \le C_1\, \delta(\alpha)\, \alpha^{-\frac{1}{2}},$$

where $C_1 = const$.

Thus, the first part of the theorem is provided, as the comparison of the appraisal of the amounts. S_1, S_2, S_3, S_4, show that uniform convergence $f_\delta(a,t)$ to $f(t)$ is ensured.

Let us prove the second part of the theorem. From the proof of the first part, it is obvious that the function $f_\delta(a,t)$ is the sum of a uniformly convergent series $\sum_{k=1}^{\infty} \varphi(ak)\ c_k \chi_k(t)$ and, according to Theorem 1, the function $f_\delta(a,t)$ is represented by its Fourier-Haar series. This means that the first condition of S_p class definition is held. It is not difficult to see that the coefficients of function $f_\delta(\propto, t)$ satisfy the condition (2). Namely, when $2 \le p'$ and α is fixed.

$$A_p\big(f_\delta(\propto, t)\big) =$$

$$= \sum_{m=1}^{\infty} 2^{\frac{m-1}{2}} \left\{ \sum_{j=1}^{2^{m-1}} \left|\varphi(\propto\ (2^{m-1}+j))C_{mj}\right|^p \right\}^{\frac{1}{p}} \le$$

$$\le \sum_{m=1}^{\infty} 2^{\frac{m-1}{2}} \varphi(\propto\ 2^{m-1}) \left\{ \sum_{j=1}^{2^{m-1}} \left|C_{mj}\right|^p \right\}^{\frac{1}{p}} \le$$

$$\le \sum_{m=1}^{\infty} 2^{\frac{m-1}{2}} \varphi(\propto\ 2^{m-1}) \left(\sum_{k=1}^{\infty} C_k^2 \right)^{\frac{1}{2}}.$$

Here we have used the following well-known inequality [10]: if

$0 < q < q^l$, then

$$\max_{1\le j\le M} |U_j| \le \left\{ \sum_{j=1}^{M} |U_j|^{q'} \right\}^{\frac{1}{q'}} \le \left\{ \sum_{j=1}^{M} |U_j|^{q} \right\}^{\frac{1}{q}}$$

Let's denote $\propto 2^x = t$, $\propto 2 \ln 2 \cdot 2^x\, dx = dt$ then for $A_p(f_\delta(\propto, t))$ we obtain the following estimate

Apfδ∝,t≤C2∝−120∞φtt−12dt=C3∝−12, C3=const.

Therefore, for each fixed $\propto$ function $f_\delta(\propto, t)$ belongs to S_2 the theorem is proved.

3 CONCLUSION

The article deals with the investigation of Haar wavelet series for correctness by using Tikhonov regularization method. The problem of regularized summation for classes $C_{(0,1)}$ is considered.

The theorem, based on the stability and uniform con-vergence of the regularized and generalized summation function wavelet Haar series of a class of continuous functions with approximate coefficients, is proved, and it also proves the statement that the approximating function belongs to the $S_{p'}$, $2 \le p' \le \infty$ class.

These studies can find its application while space monitoring of the Earth is conducted.

REFERENCES

Chui K. 2001. *Introduction to Wavelets*. Moscow: 412 p.

Dobechies I. 1992. *Ten Lectures on wavelets*. Philadelphia: SIAM. 120 p.

Fikhtengolts G.M. Course of differential and integral calculus. – Moscow: Nauka, 1964, Vol.2, 463 p.

Gordienko A.S. 2010. Using the wavelet analysis in the development of aerospace images *Proceedings of the universities. Surveying and aerial photography*. No 2. 68–71.

Haar A. 1910. Zur Theoria der orthogonalen Funktionsysteme. *Math. Fnn*. 69, hh. 331–371.

Kazaryan M.L. 2009. *The study of digital processsing of signals by discrete orthogonal transformations for stability*. Vladikavkaz. 81 p.

Kazaryan M.L. 2014. Haar wavelet transform in the telecommunications system and their research on sustainability. *Telecommunications*. No 9. p. 10–2.

Pratt W. 1982. *Digital image processing*. New York: Wiley.

Shahramanjan M.A. 2003. *New Information Technologies in the problems of national security of Russia (natural-technological aspects)*. Moscow: Moscow Institute of Civil Defense FC.

Shovengerdt R.A. 2010. *Remote sensing. Methods and image processing model*. Moscow: Techno-sphere.

Shtark G.G. 2007. *Application of wavelets for DSP*. Moscow: Technosphere. 192 p.

Sobol I.M. 1969. *Multidimensional quadrature formulas and Haar functions*. Moscow: Nauka, 288p.

Tikhonov A.N. 1964. About sustainable methods of summation of Fourier series. *Dokl. USSR Acade-my of Sciences*. Vol. 156. No 2. p. 268–271.

Future Communication Technology and Engineering – Chan (Ed.)
© 2015 Taylor & Francis Group, London, ISBN: 978-1-138-02777-0

Environmental problems of coal mining in the Southern Far East of Russia and soil remediation on mining dumps

L.T. Krupskaya
Pacific National University, Khabarovsk, Russian Federation
Far East Scientific-Research Institute of Forestry, Khabarovsk, Russian Federation

L.P. Gul
Far East Scientific-Research Institute of Forestry, Khabarovsk, Russian Federation

V.P. Zvereva
Far East Federal University, Vladivostok, Russian Federation
Far East Geological Institute FEB RAS, Vladivostok, Russian Federation

M.S. Onishchenko
Far East Scientific-Research Institute of Forestry, Khabarovsk, Russian Federation

ABSTRACT: The article presents the results of the research of coal deposits development environmental problems. It reveals that the mined lands and mining dumps are sources of negative impact on the environment. To reduce their negative impact a complex of activities is proposed. The necessity of the mining dumps soil cover restoration in the South of the Russian Far East is substantiated.

Keywords: Piles, Energy resources, Brown coal deposits, Reclamation, Industrial pollution.

1 INTRODUCTION

Mineral complex of the Far East Federal District is historically developing due to the mining of tin, lead, tungsten, zinc, silver and other chemical raw materials (fluorite, borosilicates) deposits and energy resources (coal and lignite). By the beginning of the 1990s in the region about 10 large ore mining and processing enterprises (GOK) and a number of mines were operating; more than 50 enterprises were developing gold alluvial deposits, and no more than 200 organizations engaged in mining of coal and lignite, construction materials and various raw materials. In recent years, the development of gold, platinum primary ore deposits and small-scale placer deposits have come to the fore in mining industry, as a result the number of mining sites have increased sharply, mainly due to the integration of technogenic gold placer deposit. During the last five years the licensed number of them is reaching several hundred. The development of coal deposits is also a priority.

The ecological peculiarity of the region under research is determined by its geomorphological structure. Orographic location of the territory creates the conditions for transfer of all kinds of contaminants from peripheral mining areas and their accumulation in the most populated flood plain and valley is part of the main waterway of the Far East, the Amur river, having biospheric significance. Since the Amur basin occupies more than 80 per cent of the territory of the Far East region, the majority of mining areas, including coal, are drained by its tributaries with transfer of toxic elements and their compounds in the lowland plains and the riverbed, contributing to technogenic pollution of the ecosphere. In connection with it, the aim of the study is to put forward the measures to decrease the negative impact of coal mining disturbed lands, including dumps, and the environment safety. Proceeding from the aim, the following tasks are identified: 1. Analyze, summarize and systematize data and of patent materials on this problem; 2. Identify the environmental problems associated with the coal deposits development; 3. Develop measures to reduce the negative impact on the disturbed lands ecosystem.

2 OBJECTS AND METHODS OF RESEARCH

The methodological basis of research was the teaching of academician V.I. Vernadsky (1978) about the biosphere and noosphere and the main theses given in the Program and Methods of Study of Technogenic Biogeocoenoses by B.P. Kolesnikov & L.V. Motorina (1978). The methods of scientific generalization and systematization, as well as modern instrumental physico-chemical and traditional, chemical and biological methods are used. Soil sampling is carried out in accordance with GOST (National Standard) 17.4.3.01-83 "Nature Protection. Soil. General requirements for sampling" and GOST 17.4.4.02-84 "Nature Protection. Soil. Methods for sampling and sample preparation for chemical, bacteriological, helminthological analysis". Geobotanic description is made, vegetation species diversity is studied, and the predominant plant species are selected for study of toxic chemicals accumulation in them according to GOST 27262-87.

Biological methods are used. The certified methods of component determination are used which are allowed for the purposes of state ecological control and included in the State Register of Environmental Control.

The obtained data are processed on a personal computer, by using software packages, which are based on generally accepted analysis of variance methods.

Coal development is carried out by applying underground and open-pit method, and the overburden is transported to outside dumps, resulting in formation of pit-dump and extractive technogenic landscapes with sharply broken relief.

The objects of research are dangerous technological processes of coal deposits development as well as disturbed lands, dumps formed by technological activities of a coal enterprise. The subject of research is a set of ecological factors with a negative impact on the environment.

3 RESULTS AND DISCUSSION

Analysis, generalization and systematization of literary sources and results of patent researches show that great factual material is accumulated, but it is still fragmentary. The researchers have found that in the process of coal mining an intensive transformation of all components of the biosphere occurs (Trubetskoy et al 2009, Trofimov et al 1970, Trofimov & Ragim-Zade 1970, Krupskaya 1997, Starozhilov et al 2011, Leonenko 2006). So the researchers believe that it is essential to ensure environmental safety in relation to the high ecosphere technogenic load connected with coal development. But the problem is not enough studied for the region under study (Trubetskoy et al 2009, Trofimov & Ragim-Zade 1970, Krupskaya 1997).

The following problems appear as follow as a result of coal deposits development: not only the transfer of large areas of forest lands, but also deforestation for overburden dumps; river diversion and violation of the hydrological regime of surface stream flows and water bodies and groundwater aquifers due to rock dumps location and quarry reclamation; pollution of the environment, namely, soil, surface and ground water with dumps and quarry dewatering; intensification of erosion processes on the dumps and development of modern geological processes (landslides) on ledges of the overburden dumps; formation of diverse technogenic relief, ugly violating aesthetic value of the Far East landscapes; dust and aerosol pollution of the atmosphere, often creating a greenhouse effect and detrimentally affecting the sanitary-hygienic conditions of air in mining towns and settlements; dust pollution of soil-vegetation layer with disintegrated coal and rock mass and its toxic substances.

Figure 1. "Salt heads" on the broken stems of grass vegetation formed due to mine waters rising.

Typically, these problems above lead to the complex impacts on environment. For example, the restoration of the ground water level in the disused mines of Partizansk town (Primorskiy Krai), combined with ground surface gravitational subsidence has contributed to underflooding of the town area and squeezing of methane mixtures – deoxygenated "dead" air in the living rooms and cellars, and it was destructive to the human body and caused people death.

Our research finds that technogenesis with mineralized mine waters and assist gases (mainly methane) converts and strengthens the transfer of soil and groundwater, leading to unusual phenomena – movement of liquid and vapour moisture with substances dissolved in it, deposition of mineral substances in plant vessels, fracture of plant stem in place of its narrowing, and formation of a "salt head" at breaking place (Deryagin 1950, Shein 2005). For example, in the zone of Nagornaya mine impact (Primorskiy Krai) a specific phenomenon is observed: white salt knots in the form of salt heads formed at the ends of the broken grass stems of cat's tail, reed, small reed, standing in water (fig. 1).

It should be noted that the major environmental problems associated with the development of coal deposits in the South of the Russian Far East are due to retarded technology, particularly rock mass processing and the commercial components extraction rather than the increasing of mining production and the growing number of energy companies. A radical change of the environment protection system is required, and at present it no longer meets requirements of the scale of coal production and modern environment to coal deposits development. In connection with the negative impact of damaged land, coal mining dumps on the environment in the South of the Russian Far East

(Krupskaya L.T. 1992, Starozhilov V.T. et al 2011), there is a need for research on land rehabilitation, reclamation, including those contaminated with heavy metals, their protection from water and wind erosion. However, there are no special studies on development of theoretical bases of disturbed lands remediation, and the creation of forest ecosystems and optimal conditions for formation of physico-chemical, hydrological and agrochemical properties of edaphic biogeocenosis environment has not yet been carried out. The possibilities for accelerating the creation of waste land to be well-functioning soil horizons are low, as the most important structural units of restored ecosystems have not been studied yet.

Development of the coal industry in the Russian Far East leads to a severe crisis in land use. Therefore, more than 70 per cent of soils covering the plain part are more or less transformed. Only in the south of the Far East region about 700 thousand hectares of productive lands (Krupskaya L.T.1992), including forest, are destroyed. Almost the fertile soil layer is completely destroyed. Large areas are occupied by waste storage. The greatest environmental problem of land use is dumps formation, which are large-scale stationary sources of air pollution, along with mining.

Contaminants are washed away and inevitably are flooded into rivers. According to experts, each year, about 350 tons of dust are carried away with the wind from industrial waste dumps and more than 7 tons of salts are washed away (Starozhilov V.T. et al 2011). Pollution of soil fertile layer, water, atmosphere, surface layers occurs under the influence of weathering, washing away high toxic waste particles with heavy metals content, and in turn it has a direct impact on forest ecosystems. Overgrowth of the most industrial waste dumps is impossible or difficult because of adverse agrophysical soil properties of dump material, toxicity of surface layers for plants, and lack of organic substance, etc. Otherwise, the majority of industrially damaged areas will always be ecoclines, which is territories sharply different from natural landscapes, common in the region.

Therefore, to restore a root layer with favorable soil-ecological properties and reduce the cost, the remediation activities and recreation of soil-ecological functions are necessary. For this purpose, a comprehensive scheme of the territory rehabilitation should be implemented, taking into account all environmental aspects. The main task of rehabilitation is, first of all, water shutoff in aquifers outcropped by residual borrow cuts, where lakes are formed due to hydraulic discharge. On recultivated surface it is necessary to create conditions for organic matter accumulation, humus layer formation and normal plants growth, otherwise annual losses of both surface and groundwater will be ensured.

Long ago academician V. I. Vernadsky (1978) said that a fertile soil layer was the best filter for surface water and groundwater purification. However, in most cases the pace of remediation in coal companies is considerably behind the land violation volumes, and reclamation technologies have a low level of research intensity and environmental efficiency, not allowing to reduce the negative impact of damaged land of the surrounding areas. It is clear that an innovative approach to the forest ecosystem remediation is required, particularly in designing. But the main thing is to involve the administrative authorities to solve the problem of disturbed lands reclamation.

As practice shows, in those regions, where the authorities are interested in disturbed land restoration and control this work, the reclamation is done purposefully and efficiently, such as in Kuzbass. The remediation problems are not limited only by a decision of organizational problems. The theoretical studies are of particular equal importance. They allow to develop the ways of assessing the efficiency of remediation activities and the environmental prediction of the theory of technogenic landscapes, as well as the practical implementation of innovative technologies of remediation in the Southern Far East of Russia (Patent of the Russian Federation, 10.07.2013), such as using mixed crust compost or micro-organisms, providing a high soil-ecological effect. As practice shows, their widespread application is greatly limited by the scarcity of material resources, indicating the need of great use in innovative technologies.

4 CONCLUSION

The identified environmental problems of the coal deposits development in the Southern Far East of Russia have necessitated the successful and effective solution of reclamation problem to create the forest ecosystems here. It must be carried out under the following conditions. The first thing is to develop an integrated patterns of the territories rehabilitation in which remediation and restoration of soil-ecological functions of the dumps should be subjected to the hydrology of the area restoration and participation of all types of waste, including wood, in the life cycle. The second thing is to develop technical regulations for the design work on the reclamation and recultivation as a whole. Comprehensive rehabilitation patterns and technical regulations must be developed and implemented at the regional level as the normative acts (laws). Certainly, it is necessary to develop the law for recultivation of disturbed lands in the course of coal deposits.

REFERENCES

Deryagin, B.V. & Melnikova, M.K. 1950. Study of water movement under the temperature gradient influence. A.F. Ioffe 70th anniversary collection, USSR Academy of Science 75–86.

Kolesnikov, B.P. & Motorina, L.V. 1978. Methods of study of biogeocoenoses in technogene landscapes, Program and methods of study of technogene biogeocoenoses, Moscow: Nauka, 5–21.

Krupskaya, L.T. 1992. Land protection and rational use at mining companies in Priamurye and Primorye. Priamurskoe Geographic Society. Khabarovsk. 175
Leonenko, A.V. 2006. Social-ecological study of mining enterprises development zones in the Southern Far East. Mining Journal 9: 78–80
Starozhilov, V.T. et al 2011. Assessment of technogenic landscapes state to ensure their environmental safety (coal development and mining in Primorsky krai). Ecology of industrial production, 4, 41–45
Shein, E.V. 2005. Course of soil physics. Textbook: Moscow: MGU 432
Trofimov. S.S. & Ovchinnikov, V.A. 1970. Anthropogenic Kuzbass relief: Recultivation in Siberia and Urals. Novosibirsk: Nauka, Siberian Branch. 5–25.
Trofimov, S.S. & Ragim-Zade, F.K. 1970. Hypergenesis and evolution of technogenic Kuzbass relief. Reclamation of technogenic Siberian landscapes. Novosibirsk: Nauka, Siberian Branch. 14–27.
Trubetskoy, K.N. et al. 2009. Main trends of mineral complex environment problems solution in the Far Eastern Region. Geoecology 483-489.
Vernadsky, V.I. 1978. Living matter, Moscow: Nauka, 357

Future Communication Technology and Engineering – Chan (Ed.)
© 2015 Taylor & Francis Group, London, ISBN: 978-1-138-02777-0

Nonylphenol, benzophenones and benzotriazoles in different environmental matrices and foods

A. Careghini, A. Mastorgio, S. Saponaro & E. Sezenna
Politecnico di Milano, DICA – Sezione Ambientale, Milan, Italy

ABSTRACT: Nonylphenol (NP), Benzophenones (BPs) and Benzotriazoles (BT) are critical Contaminants of Emerging Concern. They all are classified as endocrine disruptors, but they are not currently monitored in the environment. In this work, we review the information available in literature about their presence in the different environmental matrices (soil, sediments, surface water, and groundwater) and in foodstuff to address risks for human health and the environment. More investigation should be carried out to assess their presence in soils and, for BPs and BTs, in both surface water and groundwater. The packaging composition for foodstuff should be reformulated to avoid contamination from NP and BPs.

1 INTRODUCTION

Contaminants of emerging concern (CECs) are defined as chemicals not commonly monitored in the environment, though having the potential to enter soil and aquatic ecosystems and cause adverse effects in humans, wildlife and the environment. CECs include industrial, agricultural and household chemicals, such as flame retardants and plasticizers, pesticides, gasoline additives, fluorinated compounds and nanomaterials, as well as pharmaceuticals and personal care products, such as human and veterinary antibiotics and drugs, fragrances, sun-screen agents, anti-microbial cleaning agents and excipient (Focazio et al. 2008).

Nonylphenol (NP), Benzophenones (BPs) and Benzotriazoles (BT) are some of the most critical CECs, as they all are classified as endocrine disruptors and are widely spread in the environment and foodstuff. NP is used as a formulant in pesticides, as a lubricating oil additive, as a catalyst in epoxy resins curing, at industrial laundries and, in the past, to produce nonylphenol ethoxylates for consumer products (e.g., surfactants, detergents, wetting agents, dispersants, defoamers, de-inkers, antistatic agents) (CCME 2002, EC 2003, US EPA 2010). BPs are used as a flavor ingredient, a fragrance enhancer, a perfume fixative and an additive for plastics, coatings and adhesive formulations. They are also used in laundry and household cleaning products and in the manufacture of insecticides, agricultural chemicals, hypnotic drugs, antihistamines and other pharmaceuticals. BPs are used as an ultraviolet (UV)-curing agent in sunglasses and to prevent UV light from damaging scents and colors in products such as perfumes and soaps (Liu et al. 2012, Zhang et al. 2011). BTs are used as additives in fluids to inhibit corrosion of cooling systems and metal machines and in defrosting liquids (in particular for aircrafts). BTs are also used as UV stabilizers, pesticides and herbicides (Wolschke et al. 2011, Zhang et al. 2011).

In this work we review the information available in literature about the presence of these compounds in the different environmental matrices (soil, sediments, surface water, groundwater) and in foodstuff to address risks for human health and the environment.

2 NONYLPHENOL (NP)

Few studies are available in literature on NP occurrence in soil. Vikelsøe et al. (2002) studied the distribution of NP in dressed and fertilized agricultural soils in Denmark. NP concentration between 0.01 and 0.98 μg/kg d.w. were found in unamended soils, soils fertilized with manure or with limited amounts of sewage sludge and 34 μg/kg at run-off points. Higher concentrations of NP (1.45–2.43 mg/kg) were measured in soils exposed to high addition of sewage sludge. Moreover, significant concentrations of NP in soil persisted up to 8 years after amendment had ceased. CCME (2002) reports a concentration of 2.72 mg/kg d.w. of total NPs at an sludge-amended site in Canada. Gibson et al. (2010) measured values of NP between 59 and 299 μg/kg d.w. in Mexican agriculture fields irrigated with untreated wastewater for 10 to 90 years. There is little evidence of NP accumulation in soils and poor evidence of migration through the different horizons in soil.

Analytical data of sediment samples point out a variability which is up to four orders of magnitude: the reported concentrations of NP range from 3.6 μg/kg d.w. to 72 mg/kg d.w. Bennett et al. (1997) investigated NP concentrations in sediment samples from the Great Lakes (USA and Canada). The highest values were detected in sediments from the most industrialized and urbanized areas; the most contaminated site was located along the Detroit River near the outflow of the west Windsor sewage treatment plants (37 mg/kg d.w.). Sampling locations far from outflows generally exhibited concentrations below 0.20 mg/kg d.w. The same trend in the Great Lakes' sediments was observed by Bennie et al. (1997), where nine locations were investigated in heavy industrial areas. At seven sites NP concentrations were below 2.2 mg/kg d.w. But higher values (up to 72 mg/kg d.w.) were detected at other sites close to discharge points of sewage treatment plants. Kannan et al. (2003) did not detected NP in six sediment samples upstream of the discharge of a Waste Water Treatment Plant (WWTP) in the Kalamazoo River (Michigan, USA). But measurable concentrations (5.8–15.3 μg/kg d.w.) were found in samples collected from the downstream of the effluent pipe. Kawahata et al. (2004) investigated the relationship between pollution and human activities on Okinawa and Ishigaki Islands. On Okinawa Island (1238 thousand people) NP was detected in six out of nine samples, while on Ishigaki Island (44 thousand people) NP was quantified in two out of six samples. The detected concentrations ranged between 12 and 46 μg/kg d.w, with the lowest values at rural areas. NP pollution was ascribed to domestic and/or industrial waste. Yamashita et al. (2000) studied the vertical distribution of NP in a sediment core (0–80 cm) collected from the Tokyo Bay. NP ranged between 5.54 and 3.23 mg/kg d.w. in the surface sediments (0–14 cm) and between 0.34 and 1.96 mg/kg d.w. in sediments at depth of 14–25 cm. No detectable value (<10 μg/kg d.w.) was found in deeper sediments. High levels of NP have been measured in China as the result of constant increase of NP production. In sediments from the Pearl River system (China), Gong et al. (2011) measured NP concentrations between 31 μg/kg d.w. and 21.9 mg/kg d.w. Wu et al. (2007) observed concentrations of NP between 3.54 and 32.43 mg/kg d.w. in sediments collected from two urban lakes in Wuham (China). Fu et al. (2007) analyzed sediment samples from the Jiaozhou Bay (China) and from five surrounding rivers; concentrations between 3.6 and 299 μg/kg d.w. were measured in the bay samples and between 23.8 and 39700 μg/kg d.w. in the inflowing rivers. Micic and Hofmann (2009) detected concentrations of NP between <0.02 and 2.83 mg/kg d.w. in sediment samples collected along the Danube river; important point sources were industrial sites, especially oil refineries, and drains of untreated wastewater. Bester et al. (2001) found NP at concentrations between 10 and 153 μg/kg d.w. in 16 open-sea sediment samples out of 19 from the German Bight of the North Sea. In sediments from the Elbe river and from some of its tributaries (Germany), the measured concentrations of NP varied between 367 and 1378 μg/kg d.w. (Heemken et al. 2001). In Italy, concentrations of NP were measured in the Rieti district (44–567 μg/kg d.w.) along the Tiber river (50–970 μg/kg d.w.) and in the Venice Lagoon (47–192 μg/kg d.w.) (Vitali et al. 2004, Patrolecco et al. 2006, Pojana et al. 2007). The Rieti district included sampling points with different soil uses (urban, industrial, agricultural, open country) and several rivers and lakes in a limited zone (2750 km2). The Venice lagoon has a highly urbanized coastal water ecosystem that receives both industrial and municipal wastewater effluents.

NP was measured in groundwater only by Lacorte et al. (2002) and Latorre et al. (2003) in two agricultural areas of Spain, with values between below 0.01 (detection limit) and 0.9 μg/l.

The measured concentrations of NP in surface waters span between $3 \cdot 10^{-4}$ and 37.3 μg/l. Bennie et al. (1997) observed measurable levels of 4-NP in 9 out of 38 water samples from the Great Lakes (USA and Canada) at concentrations between 0.013 and 0.92 μg/l. The highest recorded value was found at the combined outflow from two WWTPs. Kannan et al. (2003) detected NP at 1.1 μg/l only in one water sample collected from the downstream of a WWTP in the Kalamazoo River (Michigan, USA). Basheer et al. (2004) analyzed surface seawater samples collected from 28 locations along the Singapore coastline (both inshore and offshore sampling points) and measured concentrations of NP between 0.02 and 2.76 μg/l. The trend of NP was similar to BPA, with concentrations from the offshore locations lower than in the inshore samples. Kawahata et al. (2004) found NP concentrations above the detection limited in 4 out of 14 water samples in Okinawa and Ishigaki Islands (Japan), with values between 0.12 and 0.17 μg/l. In Wu et al. (2007), high concentrations (1.94 to 32.85 μg/l) were detected in the two urban lakes of Wuham city (China), while low values were found in areas where submerged macrophytes had been restored. Fu et al. (2007) found NP between 0.02 and 0.269 μg/l in water samples taken from the Jiaozhou Bay (China) and between 0.91 and 28.6 μg/l in samples from five rivers inflowing the bay. A seasonal trend was observed, with the highest values in summer and the lowest in winter. In Azevedo et al. (2001), NP was detected in 79% of the river and coastal water samples analyzed in Portugal, with values between 0.3 and 30 μg/l. But in two river samples, the concentrations were above 10 μg/l. Micic and Hofmann (2009) detected NP in 3 out of 19 water samples collected from the Danube river; the measured values ranged between 0.10 and 0.13 μg/l. Fries and Puttmann (2003) analyzed water samples collected from six German rivers (Rhein, Elbe, Main, Oder, Nidda and Schwarzbach), resulting in concentrations between 0.028 and 1.22 μg/l. The concentrations decreased from approximately 10 μg/l in 1984 to less than 2 μg/l in 2000. Heemken et al. (2001) found NP values between 0.001 and 0.052 μg/l in water samples from the Elbe river and from some tributaries, between 0.078 and 0.221 μg/l in samples collected from the Weisse Elster and between 0.0003 and 0.84 μg/l in samples collected from the North Sea near the estuary of the Elbe river. Similar values were also measured by Bester et al. (2001) in the German Bight of the North Sea, where NP concentrations ranged from 0.033 μg/l in the estuary to 0.0007 μg/l in the central North sea; the concentrations decreased significantly with the distance from the coast. Cespedes et al. (2005) studied the Llobregat basin (Spain), a fluvial system with major agricultural, urban and industrial impacts. NP was detected at values between 0.22 and 37.3 μg/l in surface waters, with a general trend of increasing concentrations from the upper to the lower course of the river. The highest level of NP was found in a small stream with many industrial and domestic wastewater discharges. Vitali et al. (2004) analyzed water samples collected in the upper part of the Tiber river basin (Rieti district, Italy); NP concentrations were generally below 0.1 μg/l, although concentrations up to 1.6 μg/l were occasionally measured in samples collected from the downstream of the wastewater inputs. Patrolecco et al. (2006) measured NP concentrations between 0.13 and 0.58 μg/l in water samples collected from the Tiber river (Italy). Concentrations above 0.30 μg/l were found in the rural areas heavily exploited for agriculture. NP concentrations remained almost constant in the investigated seasons (0.15–0.34 μg/l in summer, and 0.13–0.58 μg/l in winter), suggesting that waters from the WWTP discharges and urban drainage were significant sources of NP in the urban environment. Pojana et al. (2007) measured NP concentrations between 0.004 and 0.211 μg/l in water samples collected from the Venice Lagoon (Italy).

Significant NP concentrations were found in different foods, with values between 0.1 and 100 μg/kg f.w. In Guenther et al. (2002), NP concentrations in packed foodstuff purchased from supermarkets in Germany spanned between 0.1 and 19.4 μg/kg f.w. The concentration was not related to the food fat content and NP migrating into food occurred at different stages of the food production. Ferrara et al. (2005) investigated the occurrence of different alkylphenols, including NP, in eight edible marine species from the Adriatic Sea sampled at 11 Italian sites. High levels were found in all the species and sites (2.7–1431 μg/kg f.w.), with an average NP concentration of 384 μg/kg f.w. In the following paper, Ferrara et al. (2008) investigated NP concentrations in aquatic species from the Tyrrhenian Sea at three Italian sites, resulting in values in the range 5–1220 μg/kg f.w. and an average concentration of 147 μg/kg f.w. Basheer et al. (2004) measured concentrations of NP between 46.6 and 197.0 μg/kg f.w. in seafood samples purchased from a local supermarket in Singapore; the highest values was measured in prawns. Shao et al. (2007) found NP in 21 samples of different types of meat purchased from supermarkets in Beijing (China), at levels from 0.49 to 55.98 μg/kg f.w. Cacho et al. (2012) measured the concentration of nine endocrine disrupter chemicals, including NP, in different vegetables bought at a local supermarket in Spain; values above the detection limit were found only in arugula samples (48 μg/kg f.w.). Gyllenhammar et al. (2012) measured NP values in Swedish foods (20 μg/kg f.w.), fruits (25–65 μg/kg f.w.), cereal products (28–71 μg/kg f.w.), vegetables (20 μg/kg f.w.) and potatoes (24 μg/kg f.w.). Lu et al. (2013) measured concentrations of 4-n-NP plus 4-NP in vegetables and fruits which are commercially available in Florida. Significant values were found in carrots and pumpkins (10.4 and 11.3 μg/kg f.w., respectively), in apples and citruses (17.1 and 29.5 μg/kg f.w., respectively). Loyo-Rosales et al. (2004) investigated the presence of NP in commercial water bottled in different materials (HDPE, PET, PVC). NP was found in water contained in HDPE and PVC bottles at values of 29–180 ng/l and 15–300 ng/l respectively. Based on the concentrations measured in food and the expected consumption rates, the average daily intake of NP varies between 0.067 and 0.370 μg/kg bw/day for adults (Guenther et al. 2002, Ferrara et al. 2005, Ferrara et al. 2008, Shao et al. 2007, Gyllenhammar et al. 2012, Lu et al. 2013). Loyo-Rosales et al. (2004) calculated an average daily intake of 360 and 600 ng/day from water in HDPE and PVC bottles, respectively. Diet seems to be the major way for humans' NP taken study.

3 BENZOPHENONES (BPs)

Concentrations of BP-types in soils range between 0.73 and 16.55 μg/kg d.w. In Jeon et al. (2006), BP-3 was detected in 5 soil samples out of 33, at concentrations between 0.73 and 3.88 μg/kg d.w. Sánchez-Brunete et al. (2011) investigated the concentration of different BPs (4HB, BP-1, BP-3, BP-8 and 2,2-dihydroxy-4.4-dimethoxybenzophenone, BP-6) in soil samples from two agricultural fields and from one industrial site in Spain. Only BP-1 was found at 5.7 μg/kg in the industrial soil and BP-6 at 0.6 μg/kg in the agricultural soil amended with sewage sludge.

Concentrations of BPs in sediments vary between 0.5 and 105 μg/kg d.w., being BP-3 between 0.27 and 4.7 μg/kg d.w. Jeon et al. (2006) measured concentrations of different UV filters in sediment samples collected from lakes and rivers in Korea affected by municipal and recreational wastewaters. The parent compound BP detected most frequently had the highest concentrations (1.52 to 9.73 μg/kg d.w) among the investigated BPs; BP-3 was not detected in any sediment sample. Sánchez-Brunete et al. (2011) analyzed sediments collected in rivers and along the Mediterranean coast in Spain at bathing or recreational areas. BP-6 was at values between 1.2 and 6.1 μg/kg d.w. Zhang et al. (2011) monitored the concentration of five UV filters (BP-1; 2,2′,4,4′-tetrahydroxybenzophenone, BP-2; BP-3; BP-8; 4HB) in sediments from Songhua River (China), Saginaw River and Detroit River (USA). BP-3 was detected at concentrations between 0.272 and 0.545 μg/kg d.w. in samples from Songhua River and between 0.728 and 4.66 μg/kg d.w in the sediments from the Saginaw River and the Detroit River. Kameda et al. (2011) measured BP concentration in sediments collected from rivers (polluted by industrial and domestic wastewaters) and lakes (used as the background sites) in Saitama Prefecture, Japan. Values between 2.7 and

58 μg/kg d.w. were found at the polluted sites and between 3.8 and 105 μg/kg d.w. at the background sites. Dry and wet deposition from atmosphere and the recreational activities were thought to be the source of BP-3 at the background sites.

Stuart et al. (2011) monitored BPs concentration in groundwater in England, resulting in values between 0.03 and 2.8 μg/l.

Concentrations of BP-types in surface waters are reported between 1 and 390 ng/l, BP-3 between 2 and 550 ng/l. Jeon et al. 2006 detected significant concentrations of BPs in one sample from a river (BP-1, 47 ng/l) and one sample from a lake (4HB, 85 ng/l) in South Korea. Balmer et al. (2005) detected concentrations of BP-3 above the detection limit (2 ng/l) in water samples from four Swiss lakes out of five (10–35 ng/l); these lakes were selected because of the near municipal wastewater plants or the local recreational activities. Grabicova et al. (2013) investigated the occurrence of BPs in water samples collected from ponds and rivers near recreational area in the South Bohemia by comparing them with not polluted background areas. The concentrations of both BP-3 and BP-4 in the recreational areas (12–550 ng/l and 4.0–390 ng/l, respectively) were higher than those in the background areas (14–20 ng/l and 3.4–37 ng/l, respectively). Rodil et al. (2012) measured the concentration of many emerging pollutants (including BP-3 and BP-4) in water samples collected in the metropolitan area of La Coruna (Spain). BP-4 was detected in 75% of the surface water samples at concentrations between 2.5 and 70 ng/l; BP-4 was also detected in several drinking water samples, up to a maximum level of 62 ng/l. BP-3 was not found in samples of surface or drinking water. Yoon et al. (2010) investigated the occurrence of wastewater-derived contaminants including BP in water samples collected along the Han River and near the northwest Seoul metropolitan area. BP was detected in six sampling sites out of ten, at concentration between 50 and 130 ng/l. Kameda et al. (2011) measured the concentrations of BP and BP-3 in water samples collected from the polluted sites and background sites in the Saitama Prefecture (Japan). Values in the range 2–68 ng/l and 1–57 ng/l were found for BP and 4–12 ng/l and 2–10 ng/l for BP-3.

Balmer et al. (2005) detected BP-3 in fish collected from three Swiss lakes out of five. In particular, roach fish had values between 0.66 and 3.31 μg/kg f.w.

4 BENZOTRIAZOLES (BTs)

Few data are available on BT soil and sediment contamination; the ranges of concentration are 4.1–660 μg/kg and 0.4–13000 μg/kg respectively. McNeill and Cancilla (2009) investigated soils collected from three different US airports. But BT was detected just in one sample (4.1 μg/kg d.w.). In the old Oslo Airport (Fornebu), BT was detected in 19 out of 20 topsoil samples beside the runway, with an average concentration of 0.33 mg/kg d.w. A soil sample taken at 1.2 m depth had a concentration of 0.51 mg/kg d.w. (Breedveld et al. 2002).

Breedveld et al. (2002) observed high concentrations of BT (13000 μg/kg d.w.) in a sediment sample of a small drainage ditch draining the snow disposal site of the Fornebu Oslo airport. BT was detected at the average concentration of 420 μg/kg d.w. in three surface sediment samples and not detected (<100 μg/kg d.w.) in three other samples collected at a depth of 0.1 m in the upper organic layer of a wetland area receiving the drainage water from the snow disposal site. In sediment samples taken from the Songhua River (China), Zhang et al. (2011) detected BT just in one sampling location (out of six) at a value of 0.39 μg/kg d.w.; on the contrary, BT was detected in all sediment samples from the Saginaw River and the Detroit River (USA) at values between 0.424 and 33.4 μg/kg d.w.

BT concentrations in groundwater range between <0.001 and 1100 μg/l, but the values measured by Breedveld et al. (2002) at the Oslo airports were three order of magnitude higher than the maximum value measured by other authors. Excluding this paper, BT in groundwater ranges between <0.001 and 3.242 μg/l. Breedveld et al. (2002) detected concentrations of BT between 1.2 and 1100 μg/l and between 0.11 and 20 μg/l in groundwater samples from the old (Fornebu) and new (Gardermoen) Oslo airports respectively. Groundwater samples were taken from the snow disposal site, the drainage ditch, the de-icing pad and the regeneration plant of aircraft de-icing/anti-icing fluids. The highest values were measured in groundwater samples taken from the sand and gravel deposits which were used to backfill the de-icing pad during construction. The areas with the original peat and clay deposits had the lowest BT values. Loos et al. (2010) performed a survey at European scale; BT was detected in 53% groundwater samples collected at contaminated sites. The maximum concentration was of 1.032 μg/l and the average of 0.024 μg/l. Kahle et al. (2009) detected BT at values between 0.016 and 0.077 μg/l in four out of six groundwater pumping stations located in a densely populated area of the Canton of Zurich (Switzerland); the detection performed in the aquifers affected by significant water infiltration from rivers receiving considerable discharge from WWTPs. Reh et al. (2013) measured significant BT concentrations (0.0049 to 3.242 μg/l) in 67 out of 163 groundwater samples collected from a karstified aquifer in Germany under an urban area of about 65 km2; this area was characterized by intensive industrial activities and several waste disposal sites. Liu et al. (2011) measured BT in groundwater collected near a WWTP in Australia at a concentration of 0.280 ± 0.0175 μg/l.

The concentrations of BT in surface water are reported between 0.011 and 33 μg/l. Breedveld et al. 2002 measured concentrations of BT between 1.5 and 33 μg/l in samples collected in the area of the Fornebu Oslo airport. Giger et al. (2006) and Vousta et al.

(2006) investigated the Glatt river and its valley, a densely populated region of 260 km^2 with 240000 inhabitants in the northeastern part of Switzerland. The Glatt river catchment included the northern part of the city of Zurich; ten municipal wastewater treatment plants discharged their final effluents into the river. Furthermore, the international airport of Zurich was located on the east of the Glatt River. Similar values were found in the two studies (0.06–6.3 μg/l in Giger et al. (2006) and 0.636–3.69 μg/l in Vousta et al. (2006)). The mass balance clearly indicated the input from the Zurich airport (28% of the total BT load), where BT was used as an anticorrosive component. Giger et al. (2006) also analyzed waters collected from three prealpine lakes in the Canton of Zurich (Greifensee, Zurich and Geneva). Greifensee lake is located northeast of Zurich and outflows into the Glatt River; a population of about 100000 inhabitants live in the catchment discharging to the Greifensee. Eight municipal WWTPs discharged their treated effluents into Lake Zurich, while Lake Geneva received wastewaters from Lausanne city. BT concentrations measured in the three lakes were 0.9–1.1 μg/l, 0.12–0.4 μg/l and 0.18–0.2 μg/l respectively. Kahle et al. (2009) measured BT concentration in water samples taken from 8 lakes of the Swiss Midland region at values between 0.11 and 0.917 μg/l. In water samples collected from lake Tegel (Germany), BT was found at 0.9 μg/l (Weiss and Reemtsma, 2005). Nodler et al. (2011) measured an increase of BT of one order of magnitude between water samples collected from upstream (0.034–0.0176. μg/l) and downstream (0.248–0.845 μg/l) of the WWTP discharge in the Leine river (Germany). Kiss and Fries (2009) assessed the occurrence of BT in three German rivers (Main, Hengstbach, and Hegbach). Main and Hengstbach rivers suffered effluents from domestic WWTPs and were influenced by the Frankfurt International Airport; the concentrations ranged between 0.025 and 1.474 μg/l. Due to the absence of wastewater effluents in the Hegbach river, a significant concentration (0.038 μg/l) was observed just in one sample. Heeb et al. (2012) investigated the 175 km stretch of the Wenyu river and the North Canal from Beijing to Tianjin (China), resulting in BT concentrations between 0.5 and 4.5 μg/l. Wastewaters from industrial areas and from the Beijing international airport were the most probable sources of pollution.

5 CONCLUSIONS

Further research activities should be carried out to overcome some relavent issues. Treatment technologies for waters must be implied to improve the removal efficiency of NP in order to reduce their discharge into water bodies from wastewater treatment plants. More investigation should be carried out to assess the presence and distribution of all these CECs in soils and of BPs and BTs in both surface water and groundwater. Specific studies should be carried out to study about the possible pollution of soils due to the use of biosolids and/or treated wastewater in agriculture practices, particularly to understand the transport of CECs in the edible parts of plants and the risk for humans due to their consumption. The packaging composition for foodstuff should be reformulated to avoid contamination from NP and BPs. Moreover, the use of BPs as a flavor additive should be avoided.

REFERENCES

Azevedo, D.A., Lacorte, S., Viana, P. & Barcelo, D. 2001. Occurrence of nonylphenol and bisphenol A in surface waters from Portugal. *Journal of the Brazilian Chemical Society* 12: 532–537.

Balmer, M.E., Buser, H.R., Muller, M.D. & Poiger, T. 2005. Occurrence of Some Organic UV Filters in Wastewater, in Surface Waters, and in Fish from Swiss Lakes. *Environmental Science & Technology* 39: 953–962.

Basheer, C., Lee, H.K. & Tan, K.S. 2004. Endocrine disrupting alkylphenols and bisphenol-A in coastal waters and supermarket seafood from Singapore. *Marine Pollution Bulletin* 48: 1145–1167.

Bennett, E.R. & Metcalfe C.D. 1997. Distribution of Alkylphenol Compounds in Great Lakes Sediments, United States and Canada. *Environmental Toxicology and Chemistry* 17(7): 1230–1235.

Bennie, D.T., Sullivan, C.A., Lee, H.B., Peart, T.E. & Maguire, R.J. 1997. Occurrence of alkylphenols and alkylphenol mono- and diethoxylates in natural waters of the Laurentian Great Lakes basin and the upper St. Lawrence River. *Science of the Total Environment* 193: 263–275.

Bester, K., Theobald, N. & Schroder, H.F. 2001. Nonylphenols, nonylphenol-ethoxylates, linear alkylbenzenesulfonates (LAS) and bis(4-chlorophenyl)-sulfone in the German Bight of the North Sea. *Chemosphere* 45: 817–826.

Breedveld, G.D., Roseth, R. & Hem, L. 2002. *Triazoles in the terrestrial environment. 20001103-1. Final Report.* Norwegian Geotechnical Institute, Oslo: Norway.

Cacho, J.I., Campillo, N., Vinas, P. & Hernandez-Cordoba, M. 2012. Determination of alkylphenols and phthalate esters in vegetables and migration studies from their packages by means of stir bar sorptive extraction coupled to gas chromatography–mass spectrometry. *Journal of Chromatography A* 1241: 21–27.

CCME 2002. Canadian soil quality guidelines for the protection of environmental and human health: Nonylphenol and its ethoxylates. In: *Canadian environmental quality guidelines*. Canadian Council of Ministers of the Environment, Winnipeg: Canada.

Cespedes, R., Lacorte, S., Raldua, D., Ginebreda, A., Barcelo, D. & Pina, B. 2005. Distribution of endocrine disruptors in the Llobregat River basin (Catalonia, NE Spain). *Chemosphere* 61: 1710–1719.

EC Directive 2003/53/EC of European Parliament and Council of 18 June 2003 amending for the 26th time Council Directive 76/769/EEC relating to restrictions on the marketing and use of certain dangerous substances and preparations (nonylphenol, nonylphenol ethoxylate and cement). *Official Journal of the European Union* 17.7.2003. L 178/24.

Ferrara, F., Ademollo, N., Delise, M., Fabietti, F. & Funari, E. 2008. Alkylphenols and their ethoxylates in seafood from the Tyrrhenian Sea. *Chemosphere* 72: 1279–1285.

Ferrara, F., Fabietti, F., Delise, M. & Funari, E. 2005. Alkylphenols and alkylphenol ethoxylates contamination of crustaceans and fishes from the Adriatic Sea (Italy). *Chemosphere* 59: 1145–1150.

Focazio, M.J., Kolpin, D.W., Barnes, K.K., Furlong, E.T., Meyer, M.T., Zaugg, S.D., Barber, L.B. & Thurman, E.M. 2008. A national reconnaissance of pharmaceuticals and other organic wastewater contaminants in the United States—II. Untreated drinking water sources. *Science of the Total Environment* 402(2–3): 201–216.

Fries, E. & Puttmann, W. 2003. Occurrence and behaviour of 4-nonylphenol in river water of Germany. *Journal of Environmental Monitoring* 5: 598–603.

Fu, M., Li, Z. & Gao, H. 2007. Distribution characteristics of nonylphenol in Jiaozhou Bay of Qingdao and its adjacent rivers. *Chemosphere* 69: 1009–1016.

Gibson, R., Durán-Álvarez, J., León Estrada, K., Chávez, A. & Jiménez Cisneros, B. 2010. Accumulation and leaching potential of some pharmaceuticals and potential endocrine disruptors in soils irrigated with wastewater in the Tula Valley, Mexico. *Chemosphere* 81: 1437–1445.

Giger, W., Schaffner, C., Petere, H. & Kohler, H.-P. 2006. Benzotriazole and Tolyltriazole as Aquatic Contaminants. Input and Occurrence in Rivers and Lakes. *Environmental Science & Technology* 40: 7186–7192.

Gong, J., Yang, Y., Chen, D.Y. & Ran, Y. 2011. Sequential ASE extraction of alkylphenols from sediments: Occurrence and environmental implications. *Journal of Hazardous Materials* 192: 643–650.

Grabicova, K., Fedorova, G., Burkina, V., Steinbach, C., Schmidt-Posthaus, H., Zlabek, V., Kroupova, H.K., Grabic, R. & Randak, T. 2013. Presence of UV filters in surface water and the effects of phenylbenzimidazole sulfonic acid on rainbow trout (Oncorhynchus mykiss) following a chronic toxicity test. *Ecotoxicology and Environmental Safety* 96: 41–47.

Guenther, K., Heinke, V., Thiele, B., Kleist, E., Prast, H. & Raecker, T. 2002. Endocrine Disrupting Nonylphenols Are Ubiquitous in Food. *Environmental Science & Technology* 36: 1676–1680.

Gyllenhammar, I., Glynn, A., Darnerud, P.O., Lignell, S., van Delft, R. & Aune, M. 2012. 4-Nonylphenol and Bisphenol A in Swedish food and exposure in Swedish nursing women. *Environment International* 43: 21–28.

Heeb, F., Singer, H., Pernet-Coudrier, B., Qi, W., Liu, H., Longree, P., Muller, B. & Berg, M. 2012. Organic Micropollutants in Rivers Downstream of the Megacity Beijing: Sources and Mass Fluxes in a Large-Scale Wastewater Irrigation System. *Environmental Science & Technology* 46: 8680–8688.

Heemken, O.P., Reincke, H., Stachel, B. & Theobald, N. 2001. The occurrence of xenoestrogens in the Elbe river and the North Sea. *Chemosphere* 45(3): 245–259.

Jeon, H., Chung, Y. & Ryu, J. 2006. Simultaneous determination of benzophenone-type UV filters in water and soil by gas chromatography–mass spectrometry. *Journal of Chromatography A* 1131: 192–202.

Kahle, M., Buerge, I.J., Muller, M.D. & Poiger, T. 2009. Hydrophilic anthropogenic markers for quantification of wastewater contamination in ground-and surface waters. *Environmental Toxicology and Chemistry* 28(12): 2528–2536.

Kameda, Y., Kimura, K. & Miyazaki, M. 2011. Occurrence and profiles of organic sun-blocking agents in surface waters and sediments in Japanese rivers and lakes. *Environmental Pollution* 159: 1570–1576.

Kannan, K., Keith, T.L., Naylor, C.G., Staples, C.A., Snyder, S.A. & Giesy, J.P. 2003. Nonylphenol and nonylphenol ethoxylates in fish, sediment, and water from the Kalamazoo River, Michigan. *Archives of Environmental Contamination and Toxicology* 44: 77–82.

Kawahata, H., Ohta, H., Inoue, M. & Suzuki, A. 2004. Endocrine disruptor nonylphenol and bisphenol A contamination in Okinawa and Ishigaki Islands, Japan e within coral reefs and adjacent river mouths. *Chemosphere* 55: 1519–1527.

Kiss, A. & Fries, E.F. 2009. Occurrence of benzotriazoles in the rivers main, Hengstbach, and Hegbach (Germany). *Environmental Science and Pollution Research* 16: 702–710.

Lacorte, S., Latorre, A., Guillamon, M. & Barcelò, D. 2002. Nonylphenol, Octyphenol, and Bisphenol A in Groundwaters as a Result of Agronomic Practices. *The Scientific World Journal* 2: 1095–1100.

Latorre, A., Lacorte, S. & Barcelo, D. 2003. Presence of nonylphenol, octyphenol and bisphenol-A in two aquifers close to agricultural, industrial and urban areas. *Chemosphere* 57: 111–116.

Liu, Y.S., Ying, G.G., Shareef, A. & Kookana, R.S. 2011. Simultaneous determination of benzotriazoles and ultraviolet filters in ground water, effluent and biosolid samples using gas chromatography–tandem mass spectrometry. *Journal of Chromatography A* 1218: 5328–5335.

Liu, Y.S., Ying, G.G., Shareef, A. & Kookana, R.S. 2012. Biodegradation of the ultraviolet filter Benzophenone-3 under different redox conditions. *Environmental Toxicology and Chemistry* 31(2): 289–295.

Loos, R., Locoro, G., Comero, S., Contini, S., Schwesig, D., Werres, F., Balsaa, P., Gans, O., Weiss, S., Blaha, L., Bolchi, M. & Gawlik, B. 2010. Pan-European survey on the occurrence of selected polar organic persistent pollutants in ground water. *Water Research* 44: 4115–4126.

Loyo-Rosales, J.E., Rosales-Rivera, G.C., Lynch, A.M., Rice, C.P. & Torrents, A. 2004. Migration of Nonylphenol from Plastic Containers to Water and a Milk Surrogate. *Journal of Agricultural and Food Chemistry* 52: 2016–2020.

Lu, J., Wu, J., Stoffella, P.J. & Wilson, P.C. 2013. Analysis of Bisphenol A, Nonylphenol, and Natural Estrogens in Vegetables and Fruits Using Gas Chromatography-Tandem Mass Spectrometry. *Journal of Agricultural and Food Chemistry* 61: 84–89.

McNeill, K.S. & Cancilla, D.A. 2009. Detection of Triazole Deicing Additives in Soil Samples from Airports with Low, Mid, and Large Volume Aircraft Deicing Activities. *Bulletin of Environmental Contamination and Toxicology* 82(3): 265–269.

Micic, V. & Hofmann, T. 2009. Occurrence and behaviour of selected hydrophobic alkylphenolic compounds in the Danube River. *Environmental Pollution* 157: 2759–2768.

Nodler, K., Licha, T., Fischer, S., Wagner, B. & Sauter, M. 2011. A case study on the correlation of micro-contaminants and potassium in the Leine River (Germany). *Applied Geochemistry* 26: 2172–2180.

Patrolecco, L., Capri, S., De Angelis, S., Pagnotta, R., Polesello, S. & Valsecchi, S. 2006. Partition of nonylphenol and related compounds among different aquatic compartments in Timber River (Central Italy). *Water, Air, & Soil Pollution* 172: 151–166.

Pojana, G., Jonkers, N. & Marcomini, A. 2007. Natural and synthetic endocrine disrupting compounds (EDCs) in water, se-diment and biota of a coastal lagoon. Environment Internatio-nal 33: 929–936.

Reh, R., Licha, T., Geyer, T., Nodler, K. & Sauter, M. 2013. Occurrence and spatial distribution of organic micropollutants in a complex hydrogeological karst system during low flow and high flow periods, results of a two-year study. Science of The Total Environment 443: 438–445.

RIKZ 2001. Chemical study on Bisphenol A. RIKZ/2001.027. Technical Report. Rijksinstituut voor Kust en Zee, Ministerie van Verkeer en Waterstaat, The Netherlands.

Rodil, R., Quintana, J.B., Concha-Grana, E., Lopez-Mahia, P., Muniategui-Lorenzo, S. & Prada-Rodriguez, D. 2012. Emerging pollutants in sewage, surface and drinking water in Galicia (NW Spain). Chemosphere 86: 1040–1049.

Sánchez-Brunete, C., Miguel, E., Albero, B. & Tadeo, J. 2011. Analysis of salicylate and benzophenone-type UV filters in soils and sediments by simultaneous extraction cleanup and gas chromatography-mass spectrometry. Journal of Chromatography A 1218: 4291–4298.

Shao, B., Han, H.,Li, D.,Ma, Y.,Tu, X. & Wu, Y. 2007. Analysis of alkylphenol and bisphenol A in meat by accelerat-ed solvent extraction and liquid chromatography with tandem mass spectrometry. Food Chemistry 105:1236–1241.

Stuart, M.E., Manamsa, K., Talbot, J.C. & Crane, E.J. 2011. Emerging contaminants in groundwater. OR/11/013. British Geological Survey Open Report. 123 pp. Keyworth, Notting-ham: UK.

US EPA 2010. Nonylphenol (NP) and Nonylphenol Ethoxylates (NPEs) Action Plan. RIN 200-ZA09. Technical report. US Environ. Protection Agency, Washington, DC: USA.

Vikelsøe, J., Thomsen, M. & Carlsen, L. 2002. Phthalates and nonylphenols in profiles of differently dressed soils. Sci-ence of The Total Environment 296: 105–116.

Vitali, M., Ensabella, F., Stella, D. & Guidotti, M. 2004. Nonylphenols in freshwaters of the hydrologic system of an Italian district: Association with human activities and evalua-tion of human exposure. Chemosphere 57: 1637–1647.

Vousta, D., Hartmann, P., Schaffner, C. & Giger, W. 2006. Benzotriazoles, Alkylphenols and Bisphenol A in Municipal Wastewaters and in the Glatt River, Switzerland. Environmen-tal Science and Pollution Research 13(5): 333–341.

Weiss, S. & Reemtsma, T. 2005. Determination of benzotriazole corrosion inhibitors from aqueous environmental sam-ples by liquid chromatography-electrospray ionization-tandem mass spectrometry. Analytical Chemistry 77(22): 7415–7420.

Wolschke, H., Xie, Z., Moller, A., Sturm, R. & Ebinghaus, R. 2011. Occurrence, distribution and fluxes of benzotriazoles along the German large river basins into the North Sea. Water Research 45: 6259–6266.

Wu, Z.B., Zhang, Z., Chen, S.P., He, F., Fu, G.P. & Liang, W. 2007. Nonylphenol and octylphenol in urban eutrophic lakes of the subtropical China. Fresenius Environmental Bulletin 16: 227–234.

Yamashita, N., Kannan, K., Imagawa, T., Villeneuve, D.L., Hashimato, S., Miyazaki, A. & Giesy, J.P. 2000. Vertical profile of polychlorinated dibenzo-p-dioxins, dibenzofurans, naph-thalenes, biphenyls, polycyclic aromatic hydrocarbons, and al-kylphenols in a sediment core from Tokyo Bay, Japan. Environ. Science & Technology 34: 3560–3567.

Yoon, Y., Ryu, J., Oh, J., Choi, B.G. & Snyder, S.A. 2010. Occurrence of endocrine disrupting compounds, pharmaceuti-cals, and personal care products in the Han River (Seoul, S. Korea). Science of The Total Environment 408:636–643.

Zhang, Z., Ren, N., Li, Y.F., Kunisue, T., Gao, D. & Kannan, K. 2011. Determination of Benzotriazole and Benzophenone UV Filters in Sediment and Sewage Sludge. Environmental Science & Technology 45: 3909–3916.

Future Communication Technology and Engineering – Chan (Ed.)
 ISBN: 978-1-138-02777-0

Lombardy region (Italy) priorities list of contaminated sites: A new relative risk assessment procedure

A. Careghini, A. Mastorgio, S. Saponaro & E. Sezenna
Dipartimento di Ingegneria Civile e Ambientale, Politecnico di Milano, Milano, Italy

M. Bellotti
Regione Lombardia – Direzione Generale Ambiente, Energia e Reti, Unitá Operativa Tutela Ambientale, Milano, Italy

S. Anelli & A. Trozzi
Ente Regionale per i Servizi all'Agricoltura e alle Foreste Regione Lombardia, Milano, Italy

ABSTRACT: This paper presents a methodology (SER-APHIM) developed to prioritize efforts into investigation and/or remedial actions at contaminated sites at regional scale (Lombardy Region – Italy). The approach is a two-tiered procedure based on the "source-pathway–receptor" paradigm. Tier 1 (SER – Short Environmental Radar) is a screening tool that runs with very few site-specific data and leads to a list of critical sites to sort by using Tier 2 (APHIM – Assessment of Priorities for Human Health and Environmental Impacts). Tier 2 needs more detailed information about sites and their surrounding areas, including contaminant concentrations in primary sources and secondary sources, geology, hydrogeology, data from a Geographic Information System (GIS) developed for the purpose (which relates the georeferenced sites and their surrounding areas to regional cartography and census data providing climate conditions, land use, natural resources, hydrographic networks and human receptors). The Tier 1 application resulted in a list of 50 sites affected by primary sources, polluted soil or groundwater; sufficient data were available only for 43 sites, which were sorted by the Tier 2 application.

1 INTRODUCTION

Chemical pollution can hamper soil functions, affect groundwater and surface water quality and water supplies, and threaten human health and ecosystems. Soil contamination at local scale is a worldwide problem due to past or current industrial activities, spill events, or improper waste disposal. In the European context, poor environmental management practices have left an estimated legacy of about 340,000 contaminated sites that need remediation (EEA 2014). As far as Italy is concerned, in 2012 the national environmental protection agency reported approximately 15,000 potentially contaminated sites (ISPRA 2012). In Lombardy, the regional list includes 4,292 sites, among which 1,235 have been already remediated, 786 are classified as contaminated sites and 1,873 as potentially contaminated sites (Regione Lombardia 2012).

In Europe, the estimated effort to manage contaminated sites is considerable, about 6,500 million Euros per year; the vast majority (about 81%) is spent for remediation measures, whilst about 15% is spent on site investigations (Panagos et al., 2013). Limited economic resources, especially from public agencies, do not allow to deal with all of them in a short time frame; thus attention has to be paid primarily to the sites that pose a major risk to human health and the environment, or where available information is not sufficient to exclude relevant impacts.

The implementation of a priority list of sites to investigate and/or to remediate can be accomplished with a sound approach capable of assessing potential impacts and risks according to a Relative Risk Assessment Procedure (RRAP). RRAPs are scoring systems whose results do not have meaning in terms of absolute health/ecological risk, but are allowed to rank sites according to the scores they have gained. The risk-scoring systems generally follow a "nested" architecture, where parameters are used to get factor scores, factor scores are combined into indicator scores and indicator scores are combined, usually according to a weighted linear combination or a multiplicative algorithm, to give the overall site score (EEA 2004). The various RRAPs differ in the data needed to get scores. The choice of the most suitable approach for a specific purpose has to consider the information available for the overall list of sites to rank, as they should be as homogeneous as possible. In the RRAP, spatial analysis is an important component, as pollution attenuates whilst migrating from the source through the surrounding environment and receptors are variously distributed in space according to demographic preferences and land use (Pizzol et al., 2011).

This paper presents the risk-based methodology (SER-APHIM, Short Environmental Radar – Assessment of Priorities for Human Health and Environmental Impacts) developed to prioritize efforts into investigations and/or remedial actions at polluted sites at regional scale in the Lombardy Region (Italy). The approach is a two-tiered procedure based on the "source-pathway–receptor" paradigm. Tier 1 (SER) is a screening tool that runs with very few site-specific data and leads to a list of critical sites to sort by using Tier 2 (APHIM).

2 MATERIALS AND METHODS

2.1 *Tier 1*

Tier 1 is used to order "orphan" sites (i.e. areas where responsible for the contamination is unknown, or unwilling or unable to clean it up and thus the public administration must take charge of the sites). whose remediation has not started yet, or sites where no action has been undertaken for the last three years.

It takes into consideration the following issues (Di Nuzzo 2004):

- if the preservation state of storage facilities at the site is not satisfactory, a score is assigned based on: a) presence/potential presence of primary sources (waste, non-aqueous phase liquid, leaking tanks or pipelines), with values of 4/2; b) release/potential release of volatile hazardous chemicals, with values of 4/2; c) release/potential release of volatile hazardous chemicals, with values of 2/1;
- if soil is polluted, scores of 4 or 2 are assigned when the polluted area is wider than 10,000 m^2, or in the range of 1,000 m^2–10,000 m^2, respectively; a multiplying factor up to 1.1 can be used, based on the number of hazardous pollutants in soil;
- polluted groundwater in shallow aquifers or in deep aquifers results in a score of 4 for each case; an additional score of 6 or 3 is due when drinking water has been affected or drinking water wells are close enough to be potentially reached by pollution; if the plume extends beyond the site boundary or this event cannot be excluded, scores of 2 or 1 are added; a final multiplying factor up to 1.1 can be used, based on the number of hazardous pollutants in groundwater;
- a score of 2 is due if vapour or dust emission has occurred; the score is 1 if the event cannot be excluded;
- if water in rivers or canals is chemically affected or organoleptically altered, a score of 4 or 2 is due in case of irrigation use or different use, respectively; a score of 2 is added when the overflow frequency is more than once every ten years;

Table 1. Data sheet to fill in for Tier 2.

Site name
Report code
Gauss Boaga coordinates (centroid of the site)
Site surface (m^2)
Reference table values (green/residential use, commercial/industrial use)
Recreational use (yes/no)
Agricultural use (yes/no)
Site accessibility (yes/no)
Ground surface completely paved (yes/no)
Groundwater containment as emergency action (yes/no)
Primary sources:
– tanks (volume in m^3)
– non-aqueous phase liquid in groundwater (yes/no)
– waste (volume in m^3, area in m^2, cover: complete and permeable/complete and impermeable/ uncompleted or absent, pollutant concentrations)
Secondary sources:
– regulation in force at the investigation time;
– unsaturated surface soil/subsoil (investigated – yes/no, polluted area in m^2, depth from ground surface of the top and the bottom of the soil source, pollutant concentrations);
– groundwater (investigated – yes/no, plume length, pollutant concentrations)
Soil:
– lithology from ground surface to the bottom of the soil source (clay/silt/sand/gravel);
– lithology from the bottom of the soil source to the capillary fringe (clay/silt/sand/gravel);
– capillary fringe/saturated zone lithology (clay/silt/sand/ gravel/other impermeable material/other permeable material);
– soil organic carbon from ground surface to the bottom of the soil source;
– groundwater depth from ground surface;
– hydraulic gradient in the saturated zone.

- when human beings at the polluted site or within 1 km exhibit signs of disease, a score of 4 is due for each of these two situations.

For each site, the total score is calculated as the summation of the scores related to the different issues above.

2.2 *Tier 2*

The first n sites (n = 50) in the list from Tier 1 are sorted with Tier 2; a data sheet has to be filled in for each of them according to Table 1. Conservative default values are used when numerical input values are not known.

Tier 2 takes into consideration the issues reported in Table 2, whose score can span in the range specified in the squared brackets on the right part of the table. The total score is calculated as the summation of the scores related to the different issues and can span in the range [0, 100]. Whenever waste, soil or groundwater have not been investigated from a chemical point of view, so that their pollutant concentrations are not known, the total

Table 2. Issues and scores of Tier 2.

Issue	Score
Primary sources	Score: **[0, 20]**
Tanks	Score: **[0, 0.5]**
Non-aqueous phase liquids	Score: **0 or 2.5**
Waste	Score: **[0, 17]**, of which ◆ [0, 0.5] for waste volume; ◆ [0, 16.5] due to pollutants quantified in waste, of which: • [0, 10] for human health risk ([0, 5] for carcinogenic effects, • [0, 5] for chronic effects); [0, 5] for volatilization to ambient air; • [0, 1.5] for leaching to groundwater.
Secondary sources	Score: **[0, 50]**
Soil and/or groundwater	◆ [0, 30] for human health risk ([0, 15] for carcinogenic effects, [0, 15] for chronic effects); ◆ [0, 10] for unsaturated soil leaching to groundwater; ◆ [0, 10] for transport of dissolved pollutants in groundwater.
Surface water A (primary and secondary rivers, torrents, canals)	Score: **[0, 7.5]**
Surface water B (lakes, quarry lakes, ponds)	Score: **[0, 7.5]**
Agricultural areas	Score: **[0, 5]**
Pasture	Score: **[0, 5]**
Natural resources	Score: **[0, 5]**

score shown in graphs has an "error bar", whose length is as long as scores 16.5, 25 and 25, respectively.

2.2.1 *Tanks, non-aqueous phase liquid, waste volume*

If tanks are present at the site j, whose total volume is V_j^T (m^3), the score is calculated as $0.5 \times V_j^T/\mathrm{Max}[V_j^T]_{j=1,\dots,n}$. Therefore, the maximum score for this issue is 0.5 and it is assigned to the site with the biggest tank volume.

A similar approach is used for waste volume.

If non-aqueous phase liquid has been detected in groundwater, a score of 2.5 is added, otherwise the score is 0.

2.2.2 *Human health risk from waste and secondary sources*

In human health risk calculation, receptors can be residents or recreational users (both divided into up to 9-years-old children and adults) or workers. The land around the site is divided into five zones, according to the distance from the site boundary (Zone 1: site area; Zone 2: from Zone 1 external boundary to 30 m far from site boundary; Zone 3: from Zone 2 external boundary to 100 m far from site boundary; Zone 4: from Zone 3 external boundary to 1000 m far from site boundary; Zone 5: from Zone 4 external boundary to 3000 m far from site boundary). Each zone is subdivided into census districts according to the National Institute of Statistics Census of 2001 (ISTAT, 2001), which also provides the number of residents (children and adults) and workers therein; the number of child and adult recreational users is calculated at provincial level as 5% of child residents and 1% of adult residents, respectively. In each district (or province for the recreational use), the area of the different soil uses is obtained from the Lombardy Region tool Use of Agricultural Land and forestry (DUSAF 2.1) (Regione Lombardia 2007). From these data, a receptor density is calculated at district level for residents and workers and at provincial level for recreational users, which is used to estimate the number of different receptors in each zone taking into consideration the area of the districts inside the zone.

Different exposure pathways are potentially active in the different zones, according to Table 3. Direct exposure pathways (dermal contact, and soil ingestion) are excluded whenever the site is not accessible. A complete pavement interrupts direct exposure pathways and particulate matter inhalation. The exposure parameters used in human health risk calculations are reported in APAT (2008).

Though Table 3 groups waste and secondary sources, a different approach is used to calculate the exposure to vapours from the two types of source. In fact, whilst for vapour migration from groundwater/soil through soil, simplified analytical models are commonly used to calculate air concentration for absolute site-specific risk assessment APAT (2008), and they were actually used in this work, those tools can hardly be used to describe vapour transport in heterogeneous waste. Therefore, in Table 2 vapour impact on human health from waste is apart from human health risk and it is called "volatilization to ambient air"; for each site j, the score is calculated as:

$$5 \times \frac{W_j^a}{Max[W_j^a]_{j=1,\dots,n}} \tag{1}$$

with

$$W_j^a = \sum_z \left(\sum_k \left(N_{j,z,k} \sum_i \left(S_j^w \times C_{j,i}^w \times C_i^{a,sat} \times \frac{1}{C_{k,i}^a} \times \eta_z \right) \right) \right) \tag{2}$$

where z is the zone number, k is the kind of receptor (k = lifelong resident, adult resident, worker, lifelong recreational user, adult recreational user), $N_{j,z,k}$ is the number of receptors k in the zone z around the site, i is the index for pollutants quantified in waste, S_j^w is the area covered by the waste at the site, $C_{j,i}^w$ is the pollutant concentration in the waste at the site, $C_i^{a,sat}$ is the saturation concentration of pollutant i in air, $C_{k,i}^a$ is the screening concentration of pollutant i in air (that is the CSC^{air} for moderate-intensity physical activity from ISPRA-INAIL 2010), and η_z is the attenuation

Table 3. Exposure pathways potentially active for human receptors in the different Zones.

Zone	Source	Exposure pathways
1	Waste and unsaturated surface soil	Dermal contact, soil ingestion, particulate matter inhalation (the worst scenario between indoor and outdoor exposure)
	Waste, unsaturated soil and groundwater (for pollutants affecting both soil and groundwater, only the worst scenario resulting from the two is scored)	Vapour inhalation (the worst scenario between indoor and outdoor exposure)
2	Waste and unsaturated surface soil	Particulate matter inhalation (the worst scenario between indoor and outdoor exposure)
	Waste and unsaturated soil	Vapour inhalation (the worst scenario between indoor and outdoor exposure)
3	Waste and unsaturated surface soil	Outdoor particulate matter inhalation
	Waste and unsaturated soil	Outdoor vapour inhalation
4	Waste and unsaturated soil	Outdoor vapour inhalation (air concentration attenuated according to the Gaussian dispersion model prediction at 550 m from site boundary)
5	Waste and unsaturated soil	Outdoor vapour inhalation (air concentration attenuated according to the Gaussian dispersion model prediction at 2000 m from site boundary)

factor from the Gaussian dispersion model in the zone z ($z = 4, 5$).

For the other exposure pathways from waste at the site j, the chronic risk and the carcinogenic risk are calculated according to the formulas for surface soil from APAT (2008). Then, two different scores are calculated as:

$$5 \times \frac{W_j^{ch}}{Max[W_j^{ch}]_{j=1,\dots,n}} \tag{3}$$

with

$$W_j^{ch} = \sum_z \left(\sum_{k'} \left(N_{j,z,k'} \times \begin{cases} \sum_i HI^w{}_{j,k',i} & \text{if } \sum_i HI_{j,k',i} > 1 \\ 0 & \text{if } \sum_i HI_{j,k',i} \le 1 \end{cases} \right) \right) \tag{4}$$

$$5 \times \frac{W_j^{carc}}{Max[W_j^{carc}]_{j=1,\dots,n}} \tag{5}$$

with

$$W_j^{carc} = \sum_z \left(\sum_k \left(N_{j,z,k} \times \sum_i R^w{}_{jk,i} \right) \right) \tag{6}$$

where z is the zone number, k' is the kind of receptor for chronic effects (k' = child resident, adult resident, worker, child recreational user, and adult recreational user), k is the kind of receptor for carcinogenic effects (k = lifelong resident, adult resident, worker, lifelong recreational user, and adult recreational user), $N_{j,z,k'}$ and $N_{j,z,k}$ are the number of receptors k' and k in the Zone z of site j, $HI^w_{j,k',i}$ is the hazard index for pollutant i (sum of contributes from all the active exposure pathways but vapour inhalation) and receptor k' from waste, $R^w_{j,k,i}$ is the carcinogenic risk for pollutant i (sum of contributes from all the active exposure pathways but vapour inhalation) and receptor k from waste.

For health risk from secondary sources at the site j, the chronic risk and the carcinogenic risk are calculated according to the formulas from APAT (2008). Then, two different scores are calculated as:

$$15 \times \frac{SS_j^{ch}}{Max[SS_j^{ch}]_{j=1,\dots,n}} \tag{7}$$

with

$$SS_j^{ch} = \sum_z \left(\sum_{k'} \left(N_{j,z,k'} \times \begin{cases} \sum_i HI^{SS}{}_{j,k',i} & \text{if } \sum_i HI_{j,k',i} > 1 \\ 0 & \text{if } \sum_i HI_{j,k',i} \le 1 \end{cases} \right) \right) \tag{8}$$

$$15 \times \frac{SS_j^{carc}}{Max[SS_j^{carc}]_{j=1,\dots,n}} \tag{9}$$

with

$$SS_j^{carc} = \sum_z \left(\sum_k \left(N_{j,z,k} \times \sum_i R^{SS}{}_{j,k,i} \right) \right) \tag{10}$$

where z is the zone number, k' is the kind of receptor for chronic effects (k' = child resident, adult resident, worker, child recreational user, and adult recreational user), k is the kind of receptor for carcinogenic effects (k = lifelong resident, adult resident, worker, lifelong recreational user, and adult recreational user), $N_{j,z,k'}$ and $N_{j,z,k}$ are the numbers of receptors k' and k in the zone z of site j, $HI^{SS}_{j,k',i}$ is the hazard index for pollutant i (sum of contributes from all the active exposure pathways) and receptor k' from secondary sources, $R^{SS}_{j,k,i}$ is the carcinogenic risk for pollutant i (sum of contributes from all the active exposure pathways) and receptor k from secondary sources.

2.2.3 *Groundwater risk from waste and polluted unsaturated soil*

Groundwater risk is related to the infiltration rate of water in waste and in the polluted unsaturated soil, which is based on the annual rainfall at the site, obtained from Ceriani & Carelli (2006). The infiltration rate in waste is 30% of the annual rainfall if waste is not completely covered, 20% of the annual rainfall in the case of permeable coverage, and 3% in the case of impermeable coverage. For the unsaturated soil, the infiltration rate is calculated according to APAT

(2008), and takes into consideration soil lithology and a 90% reduction in the case of paved soil.

The score for groundwater risk from waste at site *j* is calculated as:

$$1.5 \times \frac{W_j^{gw}}{Max[W_j^{gw}]_{j=1,\ldots,n}} \quad (11)$$

with

$$W_j^{gw} = \sum_i \left(V_j^w \times C_{j,i}^w \times Sol_i \times I_j^w \times \frac{1}{LDF_j^w} \times \frac{1}{CSC_i^{gw}} \right) \quad (12)$$

where *i* is index for pollutants in waste, V_j^w is the waste volume, $C_{j,i}^w$ is the pollutant concentration in waste (mg/kg), Sol_i is the pollutant solubility in water, I_j^w is infiltration rate of water in waste, LDF_j^w is the groundwater dilution factor for waste from APAT (2008), and CSC_i^{gw} is the regulatory concentration limit for the pollutant *i* in groundwater.

The score for groundwater risk from unsaturated soil at site *j* is calculated as:

$$10 \times \frac{S_j^{gw}}{Max[S_j^{gw}]_{j=1,\ldots,n}} \quad (13)$$

with

$$S_j^{gw} = Max\left[\frac{C_{j,i}^{gw-calc}}{CSC_i^{gw}} \times \frac{1}{T_{j,i}^s} \right]_i \quad (14)$$

where *i* is the index for pollutants in soil, $C_{j,i}^{gw\text{-}calc}$ is the pollutant concentration expected in groundwater according to APAT (2008) (Soil Attenuation Model included), CSC_i^{gw} is the regulatory concentration limit for the pollutant *i* in groundwater, and $T_{j,i}^s$ is the pollutant average migration time in the unsaturated soil (based on water seepage velocity and pollutant retardation factor in the unsaturated soil).

The score for groundwater risk from groundwater at site *j* is calculated as:

$$10 \times \frac{GW_j^{gw}}{Max[GW_j^{gw}]_{j=1,\ldots,n}} \quad (15)$$

with

$$GW_j^{gw} = Max\left[\frac{C_{j,i}^{gw}}{CSC_i^{gw}} \times \frac{1}{R_{j,i}^{gw}} \right]_i \quad (16)$$

where *i* is the index for pollutants in groundwater, $C_{j,i}^{gw}$ is the pollutant concentration in groundwater, CSC_i^{gw} is the regulatory concentration limit for the pollutant *i* in groundwater, and $R_{j,i}^{gw}$ is the pollutant retardation factor in the saturated soil ($R_{j,i}^{gw} = 10^{10}$ in case of groundwater containment).

2.2.4 *Surface water "A" and "B"*

Primary rivers, secondary rivers, torrents and canals (surface water A) or lakes, quarry lakes and ponds (surface water B) close to the site *j* result in two scores calculated as:

$$7.5 \times \frac{A_j}{Max[A_j]_{j=1,\ldots,n}} \quad (17)$$

with

$$A_j = S_j \cap A_j \quad (18)$$

$$7.5 \times \frac{B_j}{Max[B_j]_{j=1,\ldots,n}} \quad (19)$$

with

$$B_j = S_j \cap B_j \quad (20)$$

where A_j and B_j are the overlapping areas of the site surface S_j and the surface water area extended of its protection zone (150 m from each bank for primary rivers in the Regional Technical Cartography, 50 m for the other kinds of surface water A, 300 m from banks of primary lakes in the Regional Technical Cartography, and 100 m for the other kinds of surface water B).

2.2.5 *Agricultural areas, pasture, natural resources*

Agricultural areas, pasture, or natural resources around the site *j* result in three scores calculated as:

$$5 \times \frac{C_j}{Max[C_j]_{j=1,\ldots,n}} \quad (21)$$

with

$$C_j = C_{j,1} + C_{j,2} + 0.8 \times C_{j,3} + 0.5 \times C_{j,4} + 0.1 \times C_{j,5} \quad (22)$$

$$5 \times \frac{D_j}{Max[D_j]_{j=1,\ldots,n}} \quad (23)$$

with

$$D_j = D_{j,1} + D_{j,2} + 0.8 \times D_{j,3} + 0.5 \times D_{j,4} + 0.1 \times D_{j,5} \quad (24)$$

$$5 \times \frac{E_j}{Max[E_j]_{j=1,\ldots,n}} \quad (25)$$

with

$$E_j = E_{j,1} + E_{j,2} + 0.8 \times E_{j,3} + 0.5 \times E_{j,4} + 0.1 \times E_{j,5} \quad (26)$$

where z is the zone number, and $C_{j,z}$, $D_{j,z}$ and $E_{j,z}$ are the areas of agricultural soil, pasture and natural resources inside each zone z of site j.

2.2.6 *Geographic information system*

In order to relate the georeferenced sites and their surrounding areas to regional cartography and census data, climate conditions, land use, natural resources, hydrographic networks and human receptors, a Geographic Information System (GIS) has been developed.

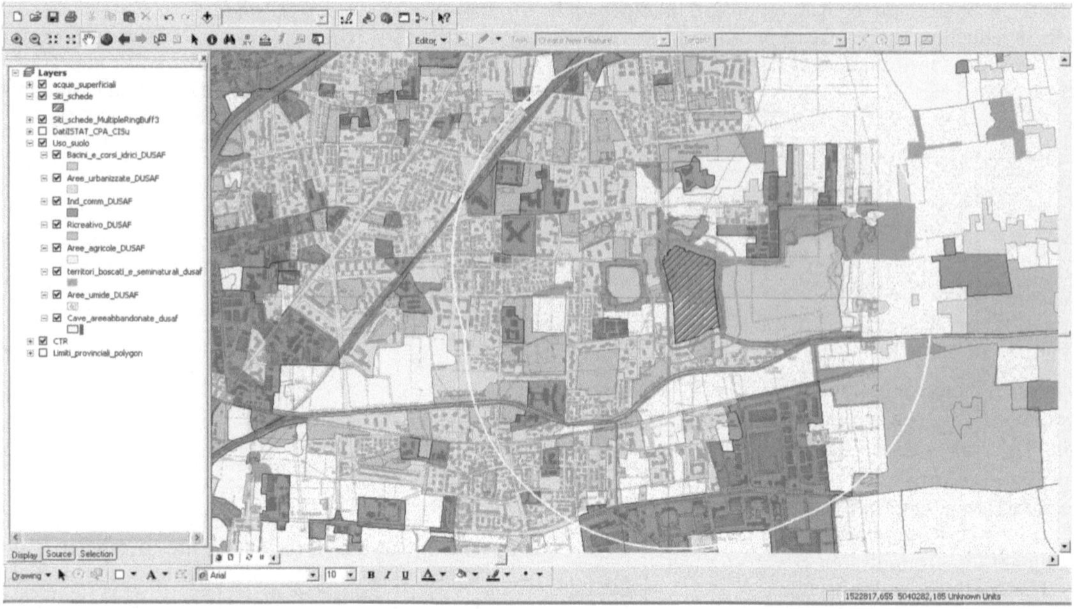

Figure 1. Soil use around the site and surface waters by the Geographical Information System.

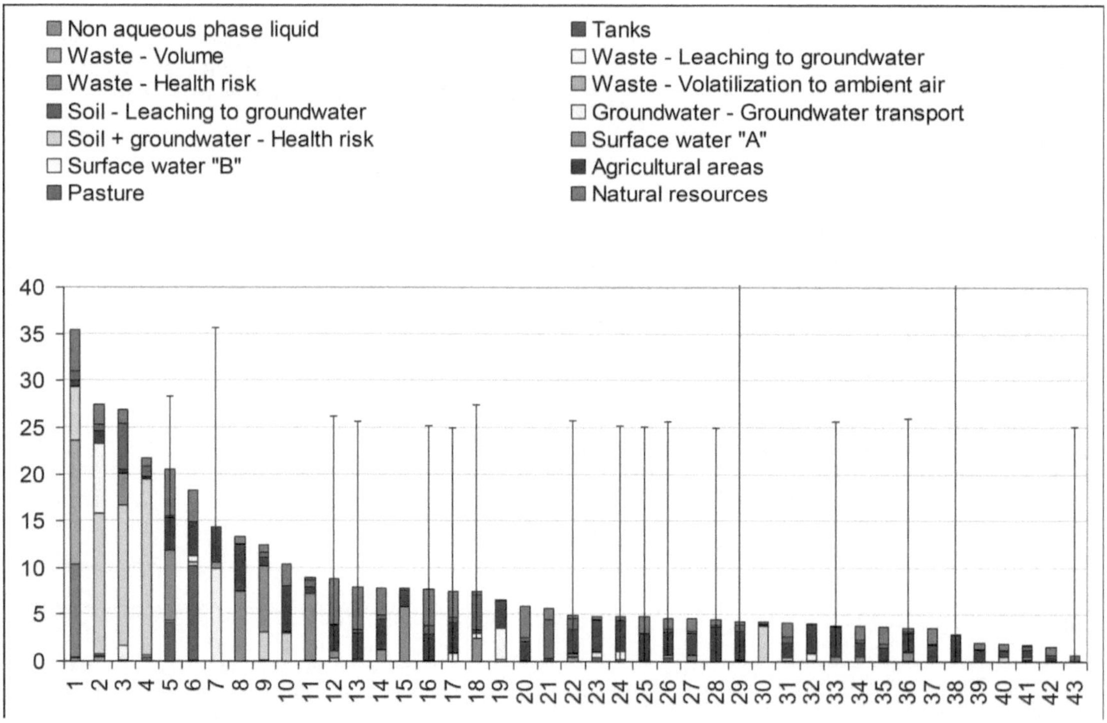

Figure 2. Total score of the sites from Tier 2. The "error bar" suggests that waste or secondary sources have not been characterized from the chemical point of view.

3 RESULTS AND DISCUSSION

For 43 sites among the 50 resulting from Tier 1, enough data were available to fill in the data sheet necessary for Tier 2; the remaining seven sites will be dealt with in the future.

Figure 1 shows, as an example, how the Geographical Information System enables the identification of the different soil use around the site and surface waters.

The results of Tier 2 are shown in Figure 2, ordered according to the total score. Twenty-two sites have total score below 5.0, eleven sites between 5.0 and 10.0 and ten sites over 10.0 (up to 35.5). However, from the "error bar", it is clear that most of the sites with medium to low score have been poorly characterized from the chemical point of view. None of the sites are affected by non-aqueous phase liquid and only one site has tanks. As far as waste is concerned, the major contributions are due to its volume and potential leaching to groundwater, whilst health risk and volatilization to ambient air are significant only for site n. 1. Health risk is negligible also from secondary sources, with the exception of sites n. 10, n. 5, and from n. 3 to n. 1. Soil leaching to groundwater is significant at sites n. 4, 3 and 2. Soil leaching to groundwater is significant at sites n. 6 and 5, whilst transport of pollutants dissolved in groundwater is significant only at site n. 7. Positive scores have frequently come out for agricultural areas, pasture, and natural resources.

REFERENCES

APAT 2008. Criteri metodologici per l'applicazione dell'analisi assoluta di rischio ai siti contaminati – Rev. 2 (marzo 2008), Agenzia per la Protezione dell'Ambiente e per i Servizi Tecnici, http://www.isprambiente.gov.it/it/temi/siti-contaminati/analisi-di-rischio.

Di Nuzzo, N. 2004. I nuovi orientamenti normativi della regione Lombardia in materia di bonifica di siti contaminati. Proceedings of "Nuovi indirizzi nella bonifica dei siti contaminati. La prassi, la normativa, le nuove tecnologie", Milano, 3 dicembre 2004

EEA 2004. Towards an EEA Europe-wide assessment of areas under risk for soil contamination. Volume II – Review and analysis of existing methodologies for preliminary risk assessment, European Environmental Agency, December 2004.

EEA 2014. A Progress in management of contaminated sites (LSI 003) – Assessment published May 2014. European Environmental Agency, http://www.eea.europa.eu/data-and-maps/indicators/progress-in-management-of-contaminated-sites-3/assessment.

ISPRA 2012. Annuario dei dati ambientali – Edizione 2012. Istituto Superiore per le Protezione e la Ricerca Ambientale, http://annuario.isprambiente.it/content/schedaindicatore/?v=10&id_ind=4314&id_area=A10&id_tema=T50.

ISPRA-INAIL 2010. Protocollo ISPRA-INAIL (ex-ISPESL) per la valutazione del rischio associato all'inalazione di vapori e polveri, in ambienti aperti e confinati nei siti di bonifica (Rev. 0 – Ottobre 2010).

ISTAT 2001. 14° censimento della popolazione e delle abitazioni, http://www.istat.it/it/censimento-popolazione-e-abitazi/popolazione-2001.

Panagos, P., Van Liedekerke, M., Yigini, Y. & Montanarella, L. 2013. Contaminated sites in Europe: review of the current situation based on data collected through a European Network, Journal of Environmental and Public Health. Article ID 158764, 11 pages, http://dx.doi.org/10.1155/2013/158764.

Pizzol, L., Critto, A., Agostini, P. & Marcomini, A. 2011. Regional risk assessment for contaminated sites Part 2: Ranking of potentially contaminated sites. Environment International 2011; 37: 1307–1320.

Ceriani, M. & Carelli, M. 2006. Carta delle piogge medie annue, Programma di tutela e uso delle acque, Regione Lombardia, Marzo 2006.

Regione Lombardia 2007. Destinazione d'Uso dei Suoli Agricoli e Forestali – DUSAF 2.1, http://www.territorio.regione.lombardia.it/cs/Satellite?c=Page&childpagename=DG_Territorio%2FDGLayout&cid=1213441282211&p=1213441282211&pagename=DG_TERRWrapper.

Regione Lombardia 2012. Programma Regionale di Gestione dei Rifiuti – Aggiornamento del Piano Regionale di Bonifica delle Aree Inquinate (Art. 199, comma 5 del D.Lgs. 152/2006) – 1° Conferenza di Valutazione – Valutazione Ambientale Strategica (VAS), 14 marzo 2012, http://www.arca.regione.lombardia.it/shared/ccurl/490/702/Slide%20struttura%20Bonifiche.pdf.

Future Communication Technology and Engineering – Chan (Ed.)
© 2015 Taylor & Francis Group, London, ISBN: 978-1-138-02777-0

Research on an evaluation approach of environmental flow for rivers

J.J. Tu & X.S. Ai
State Key Laboratory of Water Resources and Hydropower Engineering Science, Wuhan University, Wuhan, Hubei, China

Z.Y. Gao
Hubei Urban Construction Vocational and Technological College, Wuhan, Hubei, China

J. Zhu
Wangcheng Water Authority, Changsha, Hunan, China

ABSTRACT: The river regime downstream is under a greatly effection of the construction of dams, while the natural flow is recognized as the most appropriate environment flow(E-flow). In this paper, two aspects have been chosen for analysis, based on downstream discharge data after a hydropower station dam was built. Firstly, we established a river E-flow characteristics processing by analysing a long series of the natural process of daily traffic, and then drew the characteristic environmental flow diagram as the reference of river E-flow. Secondly, we proposed the evaluation system and method of indexes including 7-days-Flow Deviation rate, 7-days-E-flow Guaranteed Rate, monthly ecological water dispersion coefficient, and comprehensive coefficient, to determine the satisfaction between environmental flow and natural flow. The case study shows that the characteristic environmental flow diagram can be used as a representation of the river's natural characters and the indexes we used can get a satisfaction result in the assessment of river ecosystem health.

1 INTRODUCTION

The ecological destruction problem that river dams can cause, river is becoming the focus of the whole society (Postel S., 2005). The dams are used to store water, intercept and slow down floods, generate electricity, irrigate, plan, and manage the water supply (Poff N.L., 1997). Whilst they also reshape the nature flow regime by changing the flow total magnitude, the time and space distribution, the flow peak, and frequency, the operation of the reservoir is an important factor, especially large cascade reservoirs that mainly generate electricity, where the downstream flow process is hugely different from the original natural flow (Tharme R.E., 2003). If the purpose of E-flow process downstream is given the only one, and the reservoirs regulate rationally, it is possible to meet the natural flow and protect the river ecosystem health (Helsel D.R., 2002). So, here are the key problems: how to obtain the E-flow process, and how to prove its reasonability.

The integrity of the river ecosystem largely depends on the continuity dynamic characteristics of the river's natural flow, namely river's hydrological regime, which is the driving force behind the river ecosystem (Poff N.L., 2013). In order to determine the river environmental flow process, the first task is to analyse the river's hydrological regime. However, the regime is random, and we need to analyse the daily flow regime over several years to find out certain regularities and obtain the river environmental flow regime, which includes magnitude, duration, timing, rate of change, and other characteristics.

- The river flow grade and frequency. In order to characterize different conditions, the daily flows can be categorized into extremely low flows, low flows, high flows, small floods and large floods according to the flow magnitude. Different levels create different habitats; each plays an irreplaceable role on the river ecosystem (Poff N.L., 2013).
- Duration. Because different species have different adaptive capacity, a particular flow duration often has a significant impact on the survival of species.
- The occurrence time distribution. The life cycle of many species is closely related to the flow change, which decides the occurrence time distribution and it is critical for the ecological environment.
- The rate of change. The rate of flow change is the rate at which a flow changes from one to another, including rising and falling rates, which can be more intuitive to reflect the changes in the river hydrological regime.

The ecological flow should not only be defined as a way of ensuring that enough water is left, but also to reshape the natural flow regime to some extent.

We can maintain the ecological flow by extracting the eigenvalues that reflect the situation of the natural flow regime through the historical information of the natural river flow, and reproduce the process by reservoir operation.

2 CHARACTERISTICS OF ENVIRONMENTAL FLOW

2.1 *Environmental flow characteristics*

Analysis of the flow magnitude, time, duration, and frequency of occurrence of the five characteristics flow includes dry flow, base flow, high flow, small floods, and large floods in each month, based on the long sequences of daily flow data and the Indicators of Hydrologic Alteration (IHA) (Richter, 1996) proposed indicator system.

1) The Extreme Low Flow

The Extreme Low Flow is a low flow that occurs during the dry season, which often restricts the size of the city, the irrigated area, and the navigable capacity and time, and is also an important factor to ensure hydropower. Choose the medium flow of the smaller flows rather than the low flow as the extreme low flow, and the medium duration corresponds to the extreme low flows selected as the duration of extreme low flow.

2) Low flow

Low flow, namely basic ecological flow, which is the fundamental flow, should be saved in the river to maintain and protect the basic ecological functions, and select medium flow from the flows that are between low flow and high flow in each month, as the low flow.

3) High flow

High flow occurs almost every month with different frequency and duration. The high flow process could be simplified as a high current pulse. Analyse the monthly high flows, take the statistic medium high flow peak and the corresponding frequency and duration as the high flow plus peak.

4) Small flood

Set the 2 year return period flood in the historical daily flow data series as a small flood (adjust the return period when it is needed), select the medium flow pulse from the flows which are larger than the small flood flow threshold and smaller than the large flood threshold value, as a small flood peak, and the medium duration of all the small floods as the small flood duration.

5) Large flood

Large floods usually bring disaster to coastal people, but sometimes may have a positive effect on the river ecosystem. Set the 20 year return period flood in the historical daily flow data series as a large flood, and the selection of the large flood peak, duration is similar to the small flood. Due to the large flood's lower frequencies, the ecological characteristics flow in general year cannot be considered only if the frequency is greater than 20 years, in which condition discharge would be enlarged in accordance with the inflow appropriately.

2.2 *Environmental flow diagram characteristics*

Based on all the methods of the environmental characteristics flow above, we can get the monthly environmental characteristics flow. Draw the environmental characteristics flow in a vertical axis, and the time in a horizontal axis, namely the ecological characteristics flow diagram which could be used as an important basis of power reservoir discharged for the ecological flow process. This is because the flag can perform the flow and other characteristics, like timing and duration intuitively, and the area surrounded by the ecological flow and the time line is the year's total water demand.

3 EVALUATING INDEX OF ENVIRONMENTAL FLOW

As the ecological characteristics flow diagram could be the basic reference for reservoir ecological discharge, an evaluation approach should also be built to judge whether the discharges of release intentionally can meet the basic ecological flow requirements or not. Therefore, in this paper we establish the evaluation system which includes three indexes: the 7-days-Flow Deviation rate, the 7-days-E-flow Guaranteed Rate, and the monthly ecological water dispersion coefficient, and switch the three indexes to a comprehensive coefficient, to determine the deviation between the assumption flow and environmental flow.

3.1 *7-days-flow deviation rate*

Calculate the medium flow of seven consecutive daily natural flows (use the next month's flow when there are less than seven consecutive days in this month), and take the medium flow of the statistic 7-days-flow as the standard. The 7-days-Flow Deviation rate is the deviation rate between the standard 7-days environmental flow processes and actual ones, which can reflect the bias degree of the ecological flow and the historical natural flow in the same period:

$$C_i = \frac{Q_{ei}(t)}{Q_{7i}(t)} \times 100\%$$

$Q_{ei}(t)$ is the medium flow of the actual 7-days environmental flow processes in the ith month, $Q_{7i}(t)$ is the standard environmental flow processes in the ith month. The value of C_i is more close to 1, the better.

3.2 *7-days-E-flow Guaranteed Rate*

Take the ratio of the number of days that the actual 7-days environmental flow process is larger than the standard in the ith month to the total days in the ith month as the 7-days-E-flow Guaranteed Rate, to reflect the satisfaction of ecological flow:

$$P_{ti} = \frac{N(Q_{7i} > Q_{ei})}{N_i} \times 100\%$$

$N(Q_{7i} > Q_{ei})$ is the number of days that the ecological flow process is larger than the standard in the ith month. N_i is the total days of this month. P_{ti} is the 7-days-E-flow Guaranteed Rate, $P_{ti} \in [0,1]$, the closer P_{ti} is to 1, the better the guaranteed rate is.

3.3 *Monthly ecological water dispersion coefficient*

The monthly ecological water dispersion is the sum of the dispersion values between the actual mid-value of the ecological flow and the natural flow, and the dispersion values of the characteristics extreme in the month, which reflect the degree of dispersion of the environmental flow and the natural flow:

$$F_i' = \left(\frac{Q_{mi} - \overline{Q_{mi}}}{Q_{mi}}\right)^2 + \left(\frac{Q_{Exi} - \overline{Q_{Exi}}}{Q_{Exi}}\right)^2$$

Q_{mi} is the E-flow in the ith month, $\overline{Q_{mi}}$ is the natural flow in the ith month, Q_{Exi} is the characteristics extreme flow in the ith month, and $\overline{Q_{Exi}}$ is the characteristics extreme of natural flow in the ith month.

The monthly ecological water dispersion coefficient is:

$$F_i = 1 - F_i' / 10$$

The indicator reflects the desirability of the monthly E-flow with the natural flow. When F_i' equals 1, it indicates that the E-flow is completely desirable, while F_i' equals 0, the desirability is none. When F_i' is larger than 10, let F_i' equal 10, shows the E-flow is fully discrete to the natural flow.

3.4 *Comprehensive coefficient*

The comprehensive coefficient is the geometric average of the above performance criteria that summarizes the results, facilitating comparison between the E-flow with the natural inflow:

$$W = \sqrt[3]{C \cdot P \cdot F}$$

C, P, F are the deviation rate, the guaranteed rate, and the dispersion coefficient respectively; $W \in [0,1]$, the closer to 1 W is, the healthier the river is.

According to W the river ecosystem health can be classified into five grades (as show in Table 1).

To sum up, the comprehensive coefficient can reflect and detect the reasonability of the E-flow.

4 STUDY SITE

4.1 *The Dongjiang reservoir*

Dongjiang reservoir is located in Zixing City, Hunan Province in China (see Figure 1). It is a multi-year regulating storage reservoir with the tasks of hydropower electricity, shipping, water available, etc. The environmental flow has not been taken as an integral part of the water management in its Routine Operation Drawing. Two of the main objectives of this study are: 1) Characterize and compare large flood events, small floods, high flows, low flows and extreme low flows in Dongjiang, and get the ecological characteristics flow diagram; 2) Use the indexes to evaluate the effect that the Dongjiang reservoir has on the ecosystem health.

Table 1. The classification of river ecosystem health grade.

Grade	better	good	well	bad	worse
W	1−0.8	0.8−0.6	0.6−0.4	0.4−0.2	0.2−0

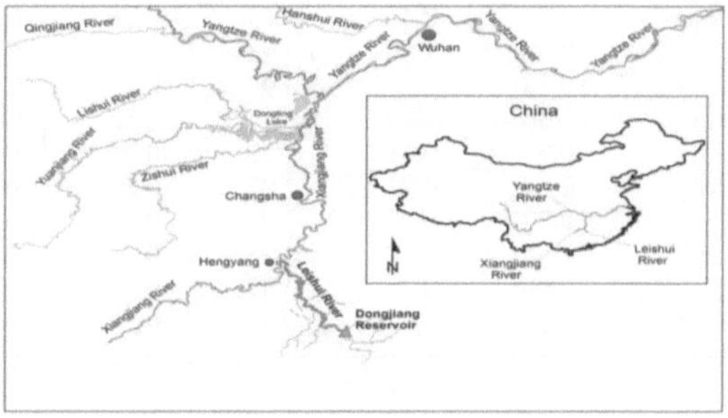

Figure 1. The position of the Leishui River System.

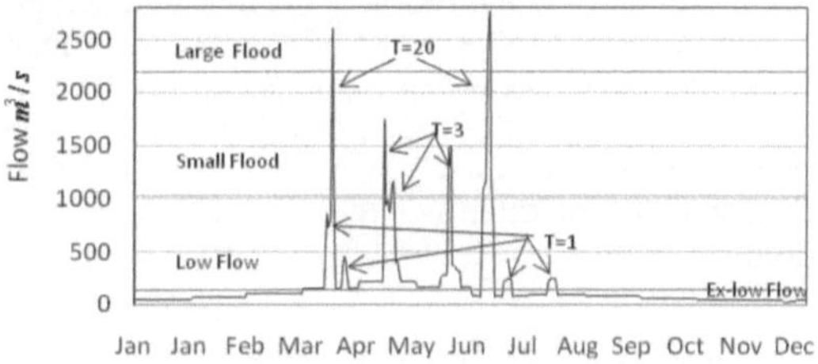

Figure 2. The environmental flow characteristics diagram of the Dongjiang Reservoir.

4.2 *Result*

With the data from 1952–2010 of the natural inflow and the 1993–2010 data of the discharge flow of the Dongjiang reservoir, and the IHA software, we analysed the environmental characteristics flow set based on the foregoingrules: the thresholds of high flow, small flood, large flood are 133 m^3/s, 1020 m^3/s, and 2192 m^3/s, respectively. We then drew the ecological characteristics flow in the environmental characteristic flow diagram (Figure 2).

using the statistical characteristics of natural flow as the evaluation criteria, calculated the indexes mentioned above, and evaluated the reservoir discharge process in the 1993–2010 data over the ecological characteristics flow diagram, and the river ecosystem health grade are shown in Table 2.

The results show that after the building of the Dongjiang dam, the actual discharge changed the river's natural flow regime greatly. For example C and

Table 2. The environmental flow evaluation in different discharge situations.

	Actual reservoir discharge					Ecological characteristics flow diagram				
Month	C	P	F	W	Grade	C	P	F	W	Grade
Jan	3.30	0	0.44	0	Worse	0.82	1	1.00	0.93	Better
Feb	1.94	0	0.92	0	Worse	0.66	1	0.97	0.86	Better
Mar	0.73	0.48	0.99	0.28	Bad	0.77	1	0.99	0.92	Better
Apr	0.57	1	0.98	0.47	General	0.72	1	0.98	0.89	Better
May	0.48	1	0.97	0.51	General	0.94	1	0.99	0.97	Better
Jun	0.36	1	0.96	0.52	General	0.58	1	0.95	0.82	Better
Jul	0.91	0.61	1.00	0.21	Bad	0.50	1	0.90	0.77	Good
Aug	0.87	0.97	1.00	0.24	Bad	0.51	1	0.91	0.77	Good
Sep	0.77	0.53	0.99	0.35	Bad	0.70	1	0.98	0.88	Better
Oct	0.58	0	0.98	0	Worse	0.73	1	0.99	0.90	Better
Nov	0.00	0	0.87	0	Worse	0.67	1	0.98	0.87	Better
Dec	0.00	0	0.66	0	Worse	0.68	1	0.97	0.87	Better

P are both 0 in October and December, and the values of C in January and February are much higher; the highest is 3.3. The value of F in July and August equals 1, and the comprehensive coefficient shows there is no "better" or "good", 3 times the "General", 5 times "bad" and 5 times "worse". While the ecological flow characteristics diagram is better to meet the natural flow.

5 CONCLUSION

Because of the huge storage capacity of the Dongjiang reservoir, the process of reservoir discharge can be largely controlled by managers. This paper proposes the ecological characteristics flow diagram to determine the ecological water requirements process, and establishes a comprehensive coefficient synthesized by the evaluation system, and which includes three indexes: the 7-days-Flow Deviation rate, the 7-days-E-flow Guaranteed Rate, and the monthly ecological water dispersion coefficient, to determine the deviation between the environmental flow and the natural flow. By applying these methods to the Dongjiang reservoir, the result shows the e-flow process by ecological flow characteristic diagram is more suitable for the natural flow process than the real operation, and it can be seen as an important reference method to the multi-year regulating storage reservoir ecological regulation. Meanwhile, the approach proposed in this paper can be used on any river sites to evaluate environmental flow.

REFERENCES

Arthington AH, Bunn SE, Poff NL, Naiman RJ. The challenge of providing environmental flow rules to sustain river ecosystems. Ecol Appl [J], 2006, 16: 1311–1318.

Brian D. Rochter, Richard E. etc. The natural flow regime: A Paradigm for river conservation and restoration, BioScience [J], 1997: 769–784.

Helsel DR, Hirsch RM. Techniques of Water-Resources Investigations of the United States Geological Survey, Hydrologic Analysis and Interpretation [J], 2002: 171–194.

Poff NL, Allan JD, Bain MB, Karr JR, Pressboard KL. The natural flow regime- a paradigm for river conservation and restoration. Bioscience [J], 1997, 47: 769–784.

Poff NL, Matthews JH. Environmental flows in the Anthropocene: past progress and future prospects. Current Opinion in Environmental Sustainability [J], 2013: 667–675.

Postel S, Richter B. Rivers for Life: Managing Water for People and Nature [M], Washington, D.C, Island Press, 2005: 35–100.

Tharme RE. A global perspective on environmental flow assessment emerging trends in the development and application of environmental flow methodologies for rivers. River Research and applications [J]. 2003, 19: 397–441.

Future Communication Technology and Engineering – Chan (Ed.)
© 2015 Taylor & Francis Group, London, ISBN: 978-1-138-02777-0

Productivity of a five-spot mixed well pattern in a dual porosity reservoir

J.J. Li, H.Q. Jiang, Y.X. Wang & H.X. Liu
China University of Petroleum (Beijing), Beijing, China

ABSTRACT: Since the late 1980s, the number of horizontal well has developed rapidly around the world, and it has become a very promising technology to develop oil and gas fields and enhance oil recovery. Carbonate reservoirs are typical dual porosity reservoirs. Moreover, their most fractures are high-angle or even vertical. Compared with conventional vertical wells, they have some obvious advantages that use horizontal well technology to develop such reservoirs. For example, fracture can improve flow capacity near a wellbore reservoir zone of the horizontal wells, and therefore it ensures a higher liquid production capability at a lower production pressure. As for the dual porosity reservoir developed in the vertical-horizontal mixed well pattern, it is also necessary to consider the impact of the matrix and fracture on horizontal well productivity when carrying productivity evaluation. In this paper, taking a five-spot well pattern as an example, based on the method of images and the principle of superposition, a horizontal well steady productivity semi-analytical model is established under the condition of any mix well pattern and validated by a case study. This considered the effect of the well interference within the mixing well pattern, and the pressure drop caused by wellbore friction and fluid acceleration on the productivity of the horizontal well. With this case study, the model is proved to be accurate and easy to use.

1 INTRODUCTION

With the rapid development of drilling and production technology, a vertical-horizontal mixed well pattern is more and more often used to coordinate with different development stages. An appropriate well pattern increases production. Therefore, productivity evaluation of different well patterns is always under the extreme attention of researchers. Song Wenling set the productivity prediction model of a horizontal-vertical and multilateral horizontal-vertical mixed well pattern based on the equivalent filtrational resistance method. Ye Shuangjiang established a calculation model of infill horizontal wells in a mixed well pattern. Hou Jiagen raised the deliverability formulas for a vertical-horizontal well pattern in a low permeability oil reservoir. Zhao Haiyang established the production formula of vertical fracture wells in a low permeability dual porosity reservoir. Meng Yujuan raised a theoretical model on the study of production performance of low speed non-Darcy flow in double porosity reservoirs. Liu Yongliang set up a well test model of vertical fractured wells in a double porosity reservoir with low permeability with consideration of the startup pressure gradient.

However, the reviewed literature is all limited to some extent. Some of them only calculated productivity of a vertical-horizontal mixed well pattern in porous media, whilst others only calculated single well productivity in dual porosity reservoirs.

Therefore, to take both factors into consideration, we established a new productivity evaluation model in this article, we raised the productivity evaluation model of a vertical-horizontal mixed well pattern in a matrix-fracture dual porosity reservoir, and validated it by a case study.

2 ESTABLISH A MODEL

2.1 *Assumptions*

In dual porosity reservoirs, the flow of fluid in areas far from the wellbore is considered the same as flow in a hypothetical continuously homogeneous system consisting of fractures and matrix. This is quite reasonable as the flow in these areas is only connected with the horizontal wellbore by wellbore fractures and matrix which are nearby. However, in near wellbore area, fluid flows into the horizontal wellbore both from fractures and matrix, making flow behaviour complex. In addition, the pressure difference between the neighbouring wellbore fractures and matrix leads to flow from the matrix to the fractures and the coupling effect of flow amongst the fractures, the matrix, and the horizontal wellbore is significant.

As is shown in Figure 1, we consider a five-spot mixed well pattern consisting of four vertical injection wells and a horizontal production well. The horizontal well is located in the middle of the quadrate and the four injection wells occupy the four vertexes.

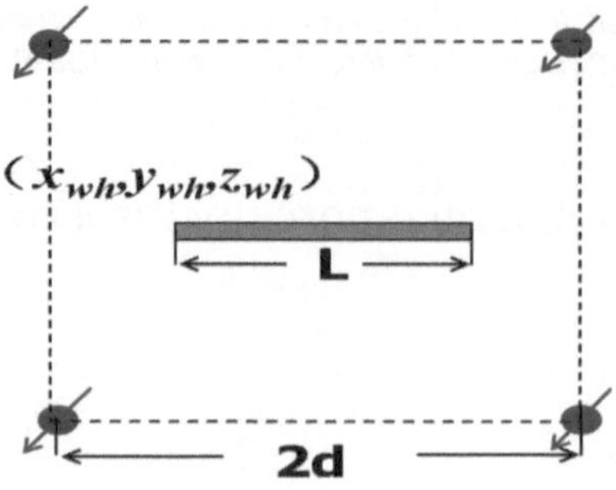

Figure 1. Diagram of a five-spot horizontal-vertical mixed well pattern.

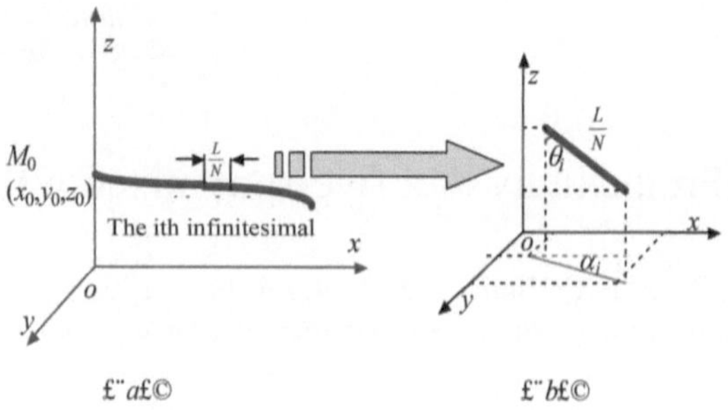

Figure 2. Coordinates of a horizontal well.

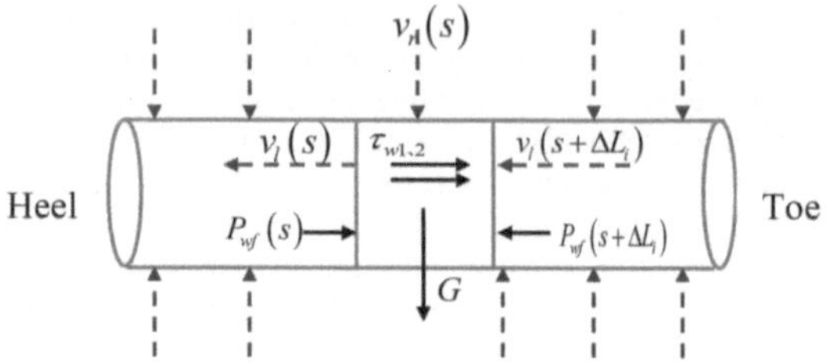

Figure 3. Pressure drop model in a horizontal well.

2.2 *Horizontal well system*

2.2.1 *Coordinates of horizontal well*

The horizontal production well is divided into many infinitesimals, and their coordinates are described first, as is shown in Figure 2 and Formula 1.

$$
\begin{aligned}
x = {} & x_{wh} + \sum_{j=0}^{Nv}\left[\frac{Lv}{Nv}\sin\left(j\frac{\theta_{kop}}{Nv}\right)\cos\left(j\frac{\alpha_{kop}}{Nv}\right)\right] + \\
& \sum_{j=0}^{N_h}\left[\frac{L_h - L_{kop}}{N_b}\sin\left(\theta_{kop} + j\frac{(\theta_h - \theta_{kop})}{N_b}\right)\cos\left(\alpha_{kop} + j\frac{(\alpha_h - \alpha_{kop})}{N_b}\right)\right] + \\
& \sum_{k=0}^{i}\left\{\sum_{s=0}^{j_2}\left[\frac{L_m(k+1) - L_m(k)}{N_p}\sin\left(\theta_m(k) + s\frac{(\theta_m(k+1) - \theta_m(k))}{N_p}\right)\cos\left(\alpha_m(k) + s\frac{(\alpha_m(k+1) - \alpha_m(k))}{N_p}\right)\right]\right\} \\
y = {} & y_{wh} + \sum_{j=0}^{Nv}\left[\frac{Lv}{Nv}\sin\left(j\frac{\theta_{kop}}{Nv}\right)\sin\left(j\frac{\alpha_{kop}}{Nv}\right)\right] + \\
& \sum_{j=0}^{N_h}\left[\frac{L_h - L_{kop}}{N_b}\sin\left(\theta_{kop} + j\frac{(\theta_h - \theta_{kop})}{N_b}\right)\sin\left(\alpha_{kop} + j\frac{(\alpha_h - \alpha_{kop})}{N_b}\right)\right] + \\
& \sum_{k=0}^{i}\left\{\sum_{s=0}^{j_2}\left[\frac{L_m(k+1) - L_m(k)}{N_p}\sin\left(\theta_m(k) + s\frac{(\theta_m(k+1) - \theta_m(k))}{N_p}\right)\sin\left(\alpha_m(k) + s\frac{(\alpha_m(k+1) - \alpha_m(k))}{N_p}\right)\right]\right\} \\
z = {} & z_{wh} + \sum_{j=0}^{Nv}\left[\frac{Lv}{Nv}\cos\left(j\frac{\theta_{kop}}{Nv}\right)\right] + \sum_{j=0}^{N_h}\left[\frac{L_h - L_{kop}}{N_b}\cos\left(\theta_{kop} + j\frac{(\theta_h - \theta_{kop})}{N_b}\right)\right] + \\
& \sum_{k=0}^{i}\left\{\sum_{s=0}^{j_2}\left[\frac{L_m(k+1) - L_m(k)}{N_p}\cos\left(\theta_m(k) + s\frac{(\theta_m(k+1) - \theta_m(k))}{N_p}\right)\right]\right\}
\end{aligned}
\tag{1}
$$

where,

(x, y, z)—coordinate of any infinitesimal of the horizontal well

(x_{wh}, y_{wh}, z_{wh})—coordinate of horizontal wellhead

L_v—length of vertical wellbore

L_{kop}—depth of kick off point

θ_{kop}—inclination angle of kick off point

L_h—depth of horizontal section

θ_h—inclination angle

α_h—azimuth

θ_m—measured inclination angle

L_m—measured depth

α_m—measured azimuth

2.2.2 *Pressure drop inside the horizontal wellbore*

When fluids flow through the horizontal wellbore, pressure drop occurs along the production section, which falls into two categories: a surface friction pressure drop and an acceleration pressure drop. The pressure drop along the horizontal section is shown in Figure 3.

According to the mass conservation equation and momentum theorem, the pressure drop along the wellbore could be represented as follows (Formula 1-2):

$$
\begin{aligned}
\Delta P_{wf}(i) = {} & \frac{L}{N}\left\{\frac{2\rho}{\pi^2 D^5}\left[f_2\phi + f_1(1-\phi)\right]\left[2q_l(s) - q_r(s)\frac{L}{N}\right]^2 + \rho g + \right. \\
& \left. \frac{16\rho}{\pi^2 D^4} q_r(s)\left[2q_l(s) - q_r(s)\frac{L}{N}\right]\right\}(1 \le i \le N)
\end{aligned}
\tag{2}
$$

2.2.3 *The potential of the horizontal well*

According to the method of images and the principle of superposition, the potential of the whole horizontal wellbore in a reservoir with a closed boundary is represented as Formula 3:

$$
\Phi_h(x,y,z) = \sum_{i=1}^{N}\Phi_i(x,y,z) = -\frac{N}{4\pi L}\sum_{i=1}^{N}(q_l(i)\Phi_{hi}) + C \tag{3}
$$

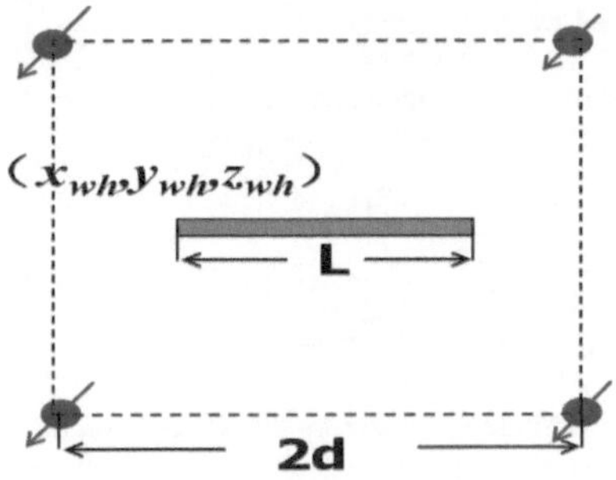

Figure 4. Coordinates of vertical wells.

2.3 *Vertical well system*

2.3.1 *Coordinates of vertical wells*

Similarly, the coordinates of the vertical wells are described by Formulae 4-1.

$$\begin{cases} x_{v1} = x_{wh} + \frac{L}{2} - d \\ y_{v1} = y_{wh} + d \\ z_{v1} = z_H + (1+t)H \end{cases} \begin{cases} x_{v2} = x_{wh} + \frac{L}{2} - d \\ y_{v2} = y_{wh} - d \\ z_{v2} = z_H + (1+t)H \end{cases}$$

$$\begin{cases} x_{v3} = x_{wh} + \frac{L}{2} + d \\ y_{v3} = y_{wh} + d \\ z_{v3} = z_H + (1+t)H \end{cases} \begin{cases} x_{v4} = x_{wh} + \frac{L}{2} + d \\ y_{v4} = y_{wh} - d \\ z_{v4} = z_H + (1+t)H \end{cases} \quad (4\text{-}1)$$

Assuming that the middle of the horizontal section is exactly the centre of the well pattern and the four vertical injection wells are in the four vertexes of the quadrate well pattern, the coordinates of the wellhead of the horizontal well are (x_{wh},y_{wh},z_{wh}), and the formation thickness and the top depth are H and H_Z. If the vertical wells are all strictly perpendicular to the water level, their coordinates would be:

$$\begin{cases} x_{v1} = x_{wh} + \frac{L}{2} - d \\ y_{v1} = y_{wh} + d \\ z_{v1} = z_H + (1+t)H \end{cases} \qquad \begin{cases} x_{v2} = x_{wh} + \frac{L}{2} - d \\ y_{v2} = y_{wh} - d \\ z_{v2} = z_H + (1+t)H \end{cases}$$

$$\begin{cases} x_{v3} = x_{wh} + \frac{L}{2} + d \\ y_{v3} = y_{wh} + d \\ z_{v3} = z_H + (1+t)H \end{cases} \qquad \begin{cases} x_{v4} = x_{wh} + \frac{L}{2} + d \\ y_{v4} = y_{wh} - d \\ z_{v4} = z_H + (1+t)H \end{cases} \quad (4\text{-}2)$$

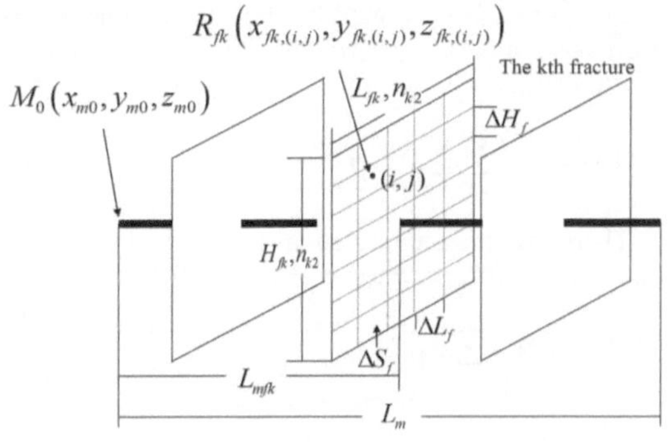

Figure 5. Coordinates of fractures.

2.3.2 *The potential of vertical wells*

According to the method of images and the principle of superposition, the potential of four vertical injection wells in an arbitrary point of the reservoir is represented as Formula 5:

$$\Phi_v(x,y,z) = \sum_{j=1}^{4} \Phi_{vj}(x,y,z) + C = \frac{1}{4\pi H} \sum_{j=1}^{4} q_v(j) \Phi_{vj}(x,y,z) + C \quad (5)$$

2.4 *Fracture system*

2.4.1 *Coordinates of fractures*

The coordinates description of fractures in the dual porosity reservoir is shown in Figure 5.

Assuming that there are N fractures intersecting with the horizontal production section, each fracture is regarded as a plane, and the kth fracture ($k = 1, 2, 3 \ldots, N$) is L_{fk} in length and H_{fk} in height. Divide this fracture into N_{k1} segments along the length direction and N_{k2} segments along the height direction. Set the initial point of the horizontal section $M_0(x_{m0}$, y_{m0}, $z_{m0})$ as a reference point, and the (i,j) infinitesimal's coordinates on the kth fracture is Rfk($x_{fk,(i,j)}$, $y_{fk,(i,j)}$, $z_{fk,(i,j)}$) (Formula 6):

$$x_{fk,(i,j)} = x_{m0} + \Delta L_m \sum_{s=1}^{v} (\sin\theta_{ms} \cdot \cos\alpha_{ms}) + \delta_{xk} + i \cdot \Delta L_f \cdot \sin\alpha_{fk} - j \cdot \Delta H_f \cdot \cos\theta_{fk} \cdot \cos\alpha_{fk}$$

$$y_{fk,(i,j)} = y_{m0} + \Delta L_m \sum_{s=1}^{v} (\sin\theta_{ms} \cdot \sin\alpha_{ms}) + \delta_{yk} + i \cdot \Delta L_f \cdot \cos\alpha_{fk} + j \cdot \Delta H_f \cdot \cos\theta_{fk} \cdot \sin\alpha_{fk}$$

$$z_{fk,(i,j)} = z_{m0} + \Delta L_m \sum_{s=1}^{v} \cos\theta_{ms} + \delta_{zk} + j \cdot \Delta H_f \cdot \sin\theta_{fk} \quad (6)$$

2.4.2 *Flow in fracture system*

In the dual porosity reservoir, fluid exchange occurs between the fractures and matrix, as well as between the fractures and wells. According to the method of images, the potential of arbitrary point (x,y,z) generated by the (i,j) infinitesimal of the kth fracture in formation with the top and bottom closed boundaries is as follows (Formula 7):

$$\phi_{fk,(i,j)} = \frac{-q_{fk,(i,j)}}{4\pi \Delta L_{fi} \Delta H_{fj}} \sum_{n=-\infty}^{+\infty} \begin{bmatrix} \xi(2nh + z_{fk,(i,j)}, x_0, y_0, z_0) + \\ \xi((2n-1)h - z_{fk,(i,j)}, x_0, y_0, z_0) \end{bmatrix} + c \quad (7)$$

The potential generated by the kth fracture in an arbitrary point (x_0, y_0, z_0) in infinite formation is (Formula 8):

$$\phi_{fk}(x,y,z)=\sum_{j=1}^{N_{k2}}\sum_{i=1}^{N_{k1}}\phi_{fk,(i,j)}+c \tag{8}$$

Therefore, the total potential generated by N fractures in infinite formation is (Formula 9):

$$\Phi_F(x,y,z)=\sum_{k=1}^{N}\phi_{fk}+c \tag{9}$$

2.5 *Coupled model*

Based on the fracture density, the horizontal wellbore is divided into several segments. The segment length is equal to the fracture interval, and there is one and only one fracture in each infinitesimal.

Assuming that these fractures are all with infinite conductivity, the pressure in the fractures is equal to the pressure at the borehole wall, $p_{wfk}=p_{wfk,(i,j)}$. In addition, the fluid in the fractures is supplemented by a matrix system. The flow rate in the kth fracture is shown as follows (Formula 10):

$$q_{fk}=\sum_{j=1}^{N_{k2}}\sum_{i=1}^{N_{k1}}q_{fk,(i,j)} \tag{10}$$

The fluid which flows into the wellbore could be divided into two parts: fluid from fractures and fluid from oil zones. As a result, the flow rate of the ith infinitesimal could be defined as (Formula 11):

$$q_{li}=\sum_{k=1}^{i-1}q_{fk}+\sum_{s=1}^{i-1}q_{rs} \tag{11}$$

According to the principle of superposition, the potential of an arbitrary point in the formation generated by the whole well pattern and fracture system is (Formula 12):

$$\Phi(x,y,z)=\Phi_F(x,y,z)+\Phi_h(x,y,z)+\Phi_v(x,y,z)+c \tag{12}$$

If the potential on the borehole wall of an injection well is Φ_{wv}, then (Formula 13):

$$\Phi_{wv}=\Phi_F(x_v,y_v,z_v)+\Phi_h(x_v,y_v,z_v)+\Phi_v(x_v,y_v,z_v)+c \tag{13}$$

According to the above two formulas, the potential of an arbitrary point in the formation generated by the whole system is (Formula 14):

$$\begin{aligned}\Phi(x,y,z)=\Phi_{wv}&+\sum_{k=1}^{N}\sum_{j=1}^{N_{k2}}\sum_{i=1}^{N_{k1}}\frac{-q_{fk,(i,j)}}{4\pi\Delta L_{fk}\Delta H_{fk}}\varphi_{fk,(i,j)}(x,y,z)\\&+\sum_{k=1}^{N}\frac{-q_r(k)}{4\pi\Delta L}\varphi_{hk}(x,y,z)+\sum_{j=1}^{4}\frac{q_v(j)}{4\pi H}\varphi_{vj}(x,y,z)\\&-\sum_{k=1}^{N}\sum_{j=1}^{N_{k2}}\sum_{i=1}^{N_{k1}}\frac{-q_{fk,(i,j)}}{4\pi\Delta L_{fk}\Delta H_{fk}}\varphi_{fk,(i,j)}(x_v,y_v,z_v)\\&+\sum_{k=1}^{N}\frac{-q_r(k)}{4\pi\Delta L}\varphi_{hk}(x_v,y_v,z_v)+\sum_{j=1}^{4}\frac{q_v(j)}{4\pi H}\varphi_{vj}(x_v,y_v,z_v)\end{aligned} \tag{14}$$

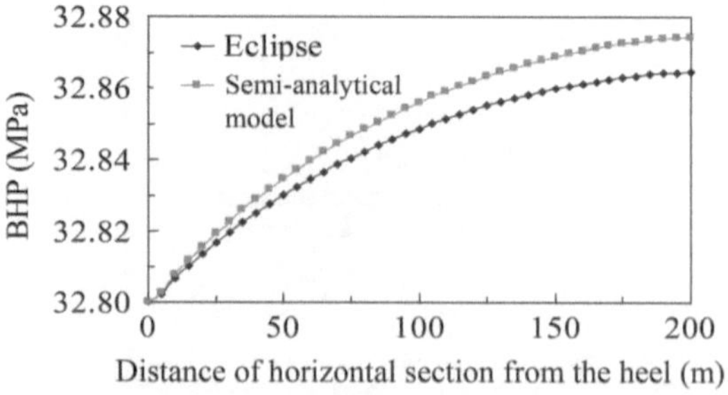

Figure 6. Distribution of wellbore pressure in a horizontal section.

3 MODEL SOLUTION

According to the above formula, we pick special points on each infinitesimal of the fractures and of the horizontal wellbore to establish an equation set about the fracture infinitesimal flow rate and horizontal wellbore infinitesimal radial flow rate. By solving the equation set, the producing rate of the horizontal well is obtained, which is shown as follows (Formula 2-1):

$$Q=q_{l1}=\sum_{k=1}^{N}\sum_{i=1}^{N_{k1}}\sum_{j=1}^{N_{k2}}q_{fk,(i,j)}+\sum_{k=1}^{N}q_{rk} \tag{15}$$

4 RESULT ANALYSIS

There is an actual reservoir which is of dual porosity and developed by a five-spot vertical-horizontal mixed well pattern, with a burial depth of 3680 m, a net pay of 5.3 m, a horizontal wellbore length of 280 m, a wellbore radial of 0.1 m, a reservoir permeability of 0.208 um^2, a porosity of 17.5%, a reservoir pressure of 43 MPa, and a BHP (Bottom Hole Pressure) of 40 MPa. Putting these parameters into our semi-analytical model, we obtained the bottomhole pressure distribution and flow rate distribution of the horizontal well, as well as the radial flow distribution, as shown in Figure 6 and Figure 7. In the meantime, we built up a numerical model by Eclipse and calculated the well productivity. The results are shown in Table 1.

As is shown in Figure 6, the BHP at the heeling is the lowest, and the BHP becomes greater as the distance from the heeling increases. In addition, the BHP calculation result of our semi-analytical model is lower than that of Eclipse.

According to the linear flow rate and radial flow rate distribution, we could work out that the two ends of the horizontal wellbore are both have a relatively higher radial flow rate, and the middle part has a lower radial flow rate. Again, the radial flow rate calculated by our semi-analytical model is lower than that of Eclipse. As the distance from the heeling increases, the linear flow rate will decrease. The linear flow rate calculated by the semi-analytical model is also lower than the result of Eclipse.

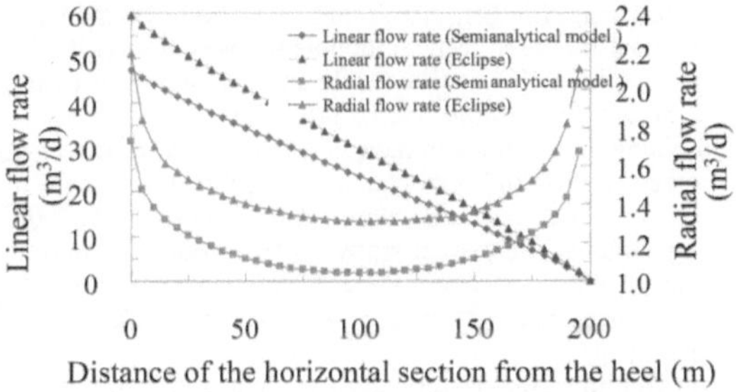

Figure 7. Distribution of linear flow rate and radial flow rate.

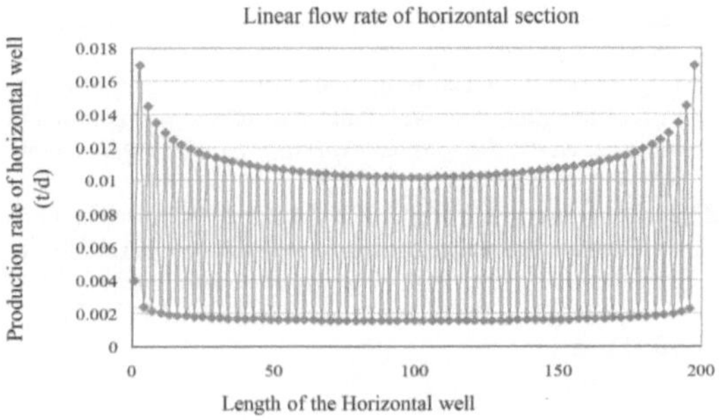

Figure 8. Linear flow rate distribution under the condition that for every three intervals there is only one which contains fractures.

Under the condition that in every three intervals there is only one which contains fractures, the linear flow rate distribution is shown in Figure 8. From the results, we could find that the overall trend is also high at the two ends and low in the middle. However, there is one difference compared to the condition that each interval contains fractures: the linear flow rate fluctuates within limits, and the intervals containing fractures have a higher flow rate; those without fractures have a lower flow rate.

Based on the principle of superposition, the potential function and the flow function in the well spot could be superimposed and the potential distribution and flow distribution of the whole well pattern could be obtained.

For the area near the ends of the horizontal well, the streamlines are concentrated, which indicates strong productivity.

The nearer to the middle part of the well, the sparser the streamlines, until no streamline exists at all, which indicates that it is easy to accumulate remaining oil in this area.

For the five-spot vertical well pattern, the remaining oil area presents a "cross" shape, and the area is centred by the production well. Therefore, the vertical-horizontal mixed well pattern turns out to be more effective.

The streamline distributions for horizontal wells of 200 m and 400 m are drawn separately, as is shown in

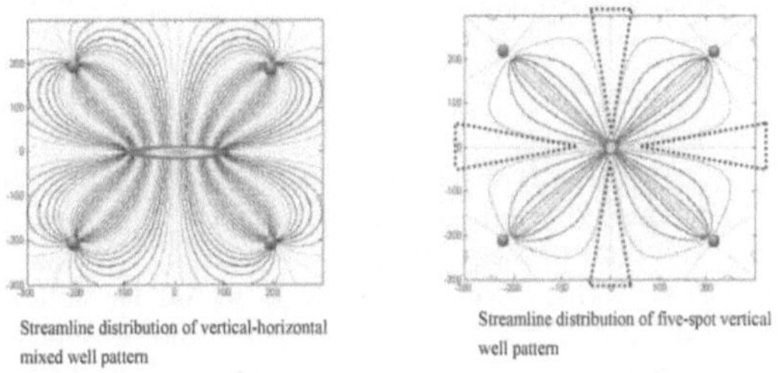

Figure 9. Streamline distribution of a vertical-horizontal mixed well pattern.

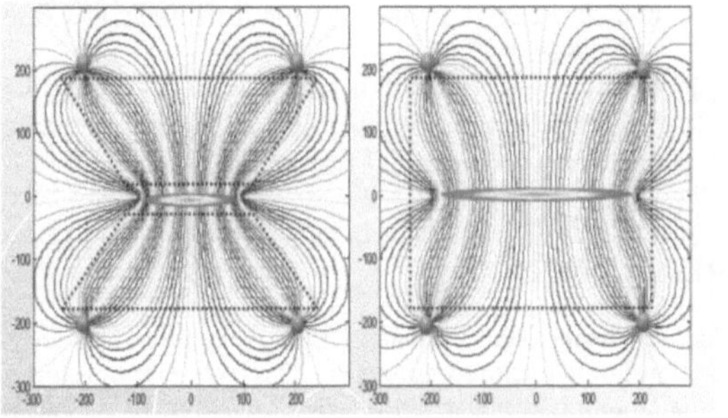

Figure 10. Streamline distribution of different horizontal lengths.

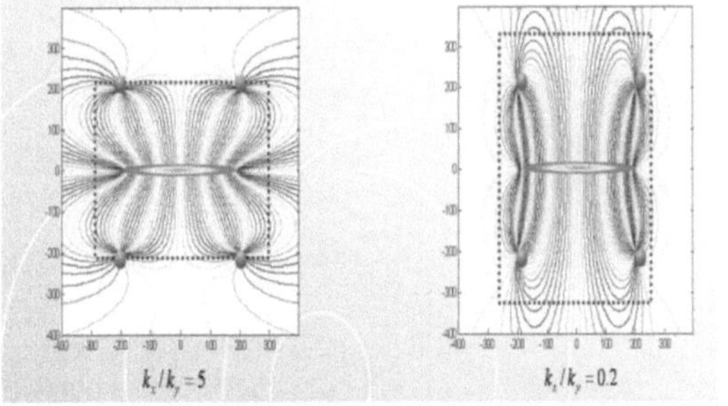

Figure 11. Streamline distribution of different dominant permeability directions.

Figure 10. The results indicate that the oil drainage area is larger if the horizontal well is longer.

The streamline distribution is relevant to the reservoir permeability. The streamline always stretches along a highly permeable area, showing strong directionality. Therefore, when the horizontal well is perpendicular to the primary permeability direction, the streamline will converge into the production well uniformly and form linear displacement, leading to a good displacement effect. However, if the horizontal well is parallel with the primary permeability, some streamlines will concentrate at the two ends of the well and form nonlinear displacement, leading to a relatively poor displacement effect.

5 CONCLUSIONS

(1) A mathematical model of the five-spot vertical-horizontal well pattern in the dual porosity reservoir was established and a horizontal well productivity calculation method for a five-spot well pattern was obtained.
(2) The productivity formula was tested by actual field data from an inland reservoir and presented high accuracy. The calculation result obtained by the formula also presented good consistency with that of Eclipse.
(3) As the distance from the heeling of the horizontal well increases, the corresponding bottomhole pressure also increases, whilst the linear flow rate decreases.
(4) The two ends of the horizontal well are both have a relatively high radial flow rate, and the middle part has a lower radial flow rate.

REFERENCES

Ban Fansheng, Gao Shusheng, Xiong Wei, et al. Fluid cross flow law in fractured-porous dual media reservoir. *Journal of Liaoning Technical University (Natural Science)*. 27(5): 677–679, 2008.

Du Dianfa, Hou Jiagen, Li Dongdong, et al. Deliverability formulas for vertical-horizontal well pattern in low permeability oil reservoir. *Petroleum geology & recovery efficiency*. 19(2): 64–66, 2012.

Liu Yongliang, Liu Bin, Xu Yanmei, et al. Well test model of vertical fractured wells in double porosity reservoir with low permeability with consideration of startup pressure gradient. *Well testing*. 19(5): 5–8, 2010.

Liu Xiangping, Jiang Zhixiang, Jiang Ruyi, et al. Effect of pressure drop in horizontal well on fluid flow in/near it. *Oil drilling & production technology*. 22(1): 42–45, 2000.

Meng Yujuan. Theoretical study on the production performance of low speed non-Darcy flow in the double porosity reservoirs. *Inner Mongolia petrochemical industry*. 11: 143–145, 2010.

Song Wenling, Feng Fengping, Zhao Chunsen, et al. Productivity of mixed well patterns of the horizontal wells or multilateral horizontal wells with vertical wells. *Journal of daqing petroleum institute*. 28(2): 107–109, 2004.

Sun Zhixue, Yao Jun, Tang Yongliang, et al. Study of horizontal well pattern development in low permeability reservoirs. *Petroleum geology and recovery efficiency*. 18(5): 74–77, 2011.

Zhao Chunsen, Xiao Danfeng, Song Wenling, et al. Optimization of the staggered well pattern of mixed horizontal and vertical wells. *Petroleum exploration and development*. 32(1): 119–122, 2005.

Zhao Hai yang, Jia Yonglu, Cai Mingjin, etc. Deliverability analysis of vertical fracture wells in low permeability dual porosity reservoir. *Journal of southwest petroleum university (science & technology edition)*. 31(2): 71–73, 2009.

Future Communication Technology and Engineering – Chan (Ed.)
© 2015 Taylor & Francis Group, London, ISBN: 978-1-138-02777-0

Removal of volatile organic compounds from water using a novel adsorbent

P. Zhang, A. Smith & R. Ding
Department of Earth and Atmospheric Sciences, City College of New York, New York, NY, USA

T.J. Bandosz
Department of Chemistry, City College of New York, New York, NY, USA

ABSTRACT: In this work an activated carbon obtained from polystyrene sulfonic acid-based organic salt was examined to simultaneously adsorb eleven Volatile Organic Compounds (VOCs) from aqueous solution. Adsorption data were modeled with different isotherms to determine the equilibrium adsorptive capacities for these compounds. The performance of the polymer-based activated carbon was then compared to that of a commercial wood-based activated carbon. The results indicate that the polymer-based activated carbon has almost twice the adsorption capacity of the wood-based activated carbon for VOCs.

1 INTRODUCTION

One of the most important groups of volatile organic compounds (VOCs) to be addressed in drinking water treatment include 8 carcinogens (benzene, carbon tetrachloride, 1,2-dichloroethane, 1,2-dichloropropane, dichloromethane, tetrachloroethylene (PCE), trichloroethylene (TCE), and vinyl chloride) and 8 suspected carcinogens (aniline, benzyl chloride, 1,3-butadiene, 1,1-dichloroethane, nitrobenzene, oxirane methyl, 1,2,3-trichloropropane and urethane) (USEPA, 2011). This group of VOCs has a great impact on human heath when inhaled or ingested. Acute health effects can be manifested as nervous system damage, pulmonary edema, and others, while chronic health effects include cancer, birth defects, immune systems deficiencies, and others (Metzger, 2005).

A novel type of activated carbon was produced by the carbonization of a commercially available polystyrene sulfonic acid-co-maleic acid salt containing iron as a metal cation. This activated carbon has very high surface area, porosity, and adsorption capacities for toxic gases (Hines et al, 2004).

The objectives of this study were to: 1) determine the adsorptive capacity of the polymer based activated carbon (PBAC) for eleven different VOCs (carbon tetrachloride, 1,2-dichloroethane, dichloromethane, 1,1-dichloroethane, 1,2-dichloropropane, 1,2,3-trichloropropane, PCE, TCE, 1,3-butadiene, benzene, and benzyl chloride) by performing batch adsorption experiments, and 2) compare the adsorptive capacity to that of a commercial wood-based activated carbon (WBAC).

2 MATERIALS AND METHODS

The PBAC was made by the carbonization of a polystyrene sulfonic acid-based organic salt using a horizontal furnace with nitrogen as an inert gas. The furnace was heated at a rate of 50°C/min and held at 800°C for 40 min (Hines et al, 2004). The resultant material has a well-developed porous structure (micro- and mesopores), and the porous structure was further enhanced by washing the material with hydrochloric acid and cleaning with distilled water. The final material has the appearance of foam and possesses a low density, a high degree of porosity and a large surface area. The WBAC was obtained from Westvaco (WVA 900).

Batch adsorption experiments were performed to determine the adsorptive capacity of the PBAC for VOCs. Five samples were made by weighting 8 mg of the activated carbon into five different 10.5 mL amber glass vials and mixing with 10.5 mL solution containing different concentrations of the 11 VOCs (250, 500, 1000, 2500, or 5000 ppb). Samples were shaken on an orbital shaker at room temperature for 15 hours, and analyzed using a gas chromatograph/mass spectrometer (GC/MS) with a headspace autosampler. Due to limited amount of PBAC available, only duplicate experiments were run for each solution concentration.

Similar batch adsorption experiments were also performed with the WBAC (at six different VOC concentrations: 100, 250, 500, 1000, 2500, and 5000 ppb). Triplicate experiments were run for each concentration.

The equilibrium data were fitted to Langmuir and Freundlich isotherms.

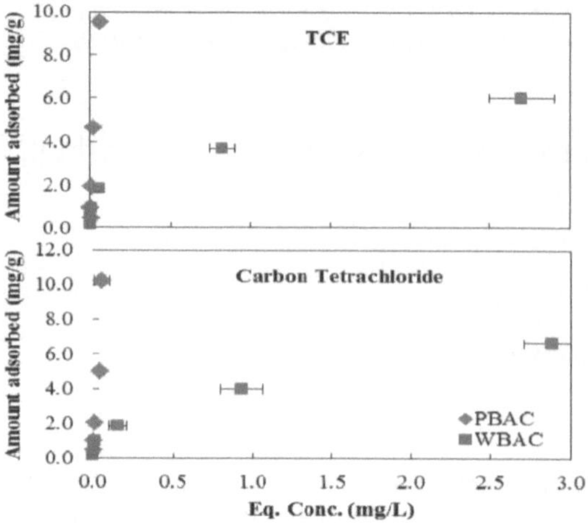

Figure 1. Comparison between PBAC and WBAC for the adsorption of TCE and carbon tetrachloride.

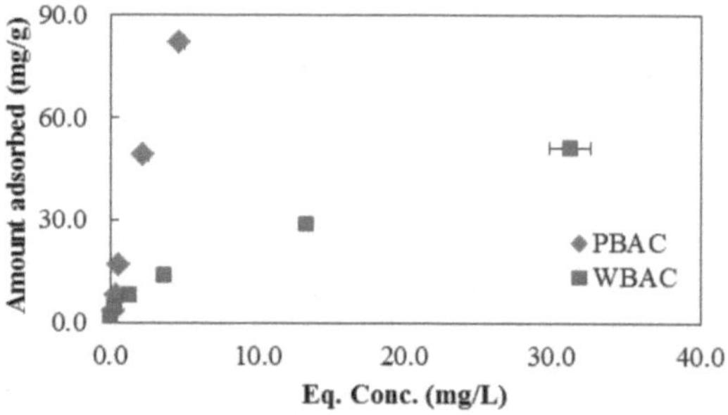

Figure 2. Plot of total amount of 11 VOCs adsorbed against total equilibrium solution concentration for both PBAC and WBAC.

3 RESULTS AND DISCUSSION

An example of adsorption isotherms for TCE and carbon tetrachloride with the two activated carbons is presented in Figure 1. It is evident that the PBAC has much higher adsorption capacities for these two VOCs than the WBAC. For instance, at the highest contaminant loading, the amount of TCE adsorbed by PBAC is about 9.5 mg/g, 1.6 times the amount adsorbed by WBAC (~6.0 mg/g). Likewise, the maximum amount of carbon tetrachloride adsorbed by the PBAC (~10.2 mg/g) is also about 1.6 times the amount adsorbed by WBAC (~6.6 mg/g). This is indeed the case for all 11 VOCs examined here.

A plot of the total amount of the 11 VOCs adsorbed against the total equilibrium solution concentration, for both activated carbons, is shown in Figure 2. The total maximum amount of VOCs adsorbed at the highest contaminant loading by the PBAC is about 82 mg/g, 1.6 times of that adsorbed by the WBAC (51 mg/g).

Both the Langmuir and the Freundlich isotherms fit the experimental data very well with the latter having slightly better R^2 values than the former. For the majority of the compounds (e.g., 1,2,3-trichloropropane (26.20 mg/g for PBAC and 8.87 mg/g for WBAC), TCE (36.85 mg/g for PBAC and 5.45 mg/g for WBAC), carbon tetrachloride (23.06 mg/g for PBAC and 8.04 mg/g for WBAC), and benzene (14 mg/g for PBAC and 3.44 mg/g for WBAC)) the adsorption capacity obtained from the Langmuir fit is four to five times higher for PBAC with respect to WBAC. The total maximum adsorption (sum of 11 VOCs) from the Langmuir fit is 208.7 mg/g and 87.2 mg/g for the PBAC and WBAC, respectively.

Table 1. Structural parameters for both carbons.

	S_{BET} (m^2/g)	V_t (cm^3/g)	V_{meso} (cm^3/g)	$V_{<0.7\,nm}$ (cm^3/g)	$V_{<1\,nm}$ (cm^3/g)
PBAC	950	0.655	0.327	0.164	0.216
WBAC	1519	1.265	0.923	0.095	0.096

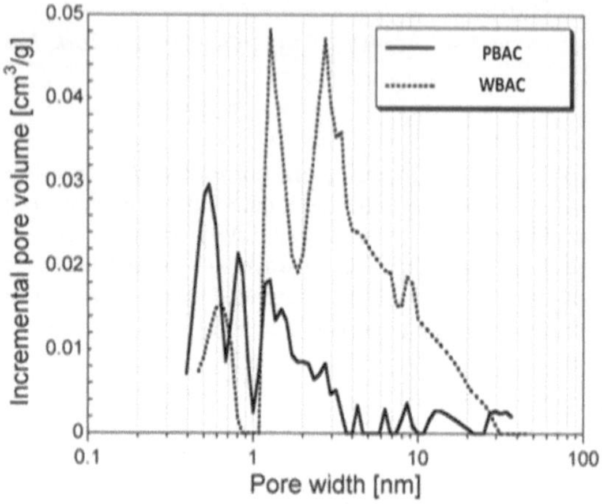

Figure 3. Pore size distribution for both carbons.

The structural parameters were collected from the nitrogen adsorption isotherm, and are presented in Table 1 and Figure 3. It can be seen that PBAC has a higher pore volume of micropores than WBAC, especially in the range between 0.4 nm and 1 nm (about twice of WBAC). The mesopore distribution (20 Å to 500 Å) is much better for WBAC, having almost three times more mesopore volume than PBAC. The surface area is larger for WBAC as well as the total pore volume (Hines et al., 2004).

The molecular diameters for all compounds were calculated using the software ChemBioOffice version 12 from Cambridgesoft. It was found that all the VOCs examined here have an average diameter of less than 4 Å. Therefore these molecules could be preferentially adsorbed to the micropores of < 10 Å, explaining the high adsorption capacities of the PBAC for these compounds with respect to the WBAC.

4 CONCLUSIONS

Batch adsorption experiments were carried out to examine the adsorption capacities of PBAC and WBAC for VOCs. PBAC has almost twice the capacity of the WBAC. This is attributed to the fact that PBAC has well-developed micropores suitable for the adsorption of compounds with small diameters, such as the VOCs examined here. The WBAC has a high mesopore volume but low micro-pore volume, making it less effective in capturing these VOCs.

REFERENCES

USEPA. 2011. Basic Questions and Answers for the Drinking Water Strategy Contaminant Groups Effort; EPA 815-F-11-002; Office of Water, US Environmental Protection Agency: Washington, D.C.

Meltzger M. 2005. *Volatile Organic Compounds Health Effects*. Cleveland: National Testing Laboratories, Ltd.

Hines, D., Bagreev, A., & Bandosz, T. J. 2004. Surface properties of porous carbon obtained from polystyrene sulfonic acid-based organic salts. *Langmuir*, 20(8), 3388–3397.

Future Communication Technology and Engineering – Chan (Ed.)
© 2015 Taylor & Francis Group, London, ISBN: 978-1-138-02777-0

Bioecology of Hyla japonica (Gunther, 1859) in Priamurye, Russia

V.T. Tagirova
Far Eastern State University of Humanities, The Russian Federation

ABSTRACT: The article presents the results of ecological research on Hyla japonica (Gunther, 1859) in Priamurye (the Amur River basin) in Russia, testifying to the fact that the number of species has been decreasing during the last decade; its biological peculiarities have been researched poorly.

Keywords: Far Eastern Hyla, seasonal activity, population figures, fertility, ontogenesis, metamorphosis.

1 INTRODUCTION

The first data about Hyla japonica became famous from the time when R. Maak travelled along the Amur River and the valley of the Ussuri (1859, 1961). It was named on the analogy with the ordinary tree frog (Hyla arborea). Later, the Far Eastern Hyla became known as East-Siberian or Eastern Hyla stepheni from F.M. Nikolsky's works (1905, 1907, 1918), who was the first to describe the main morphological characters of Hyla inhabiting the south of the Far East. Since 1940, the Far-Eastern Hyla became known as Hyla japonica Gunther 1859 (Terentyev, Chernov, 1940). However, its bio-ecological peculiarities have been researched poorly. That is why the purpose of the research was to study the capabilities of the breeding habitat, as well as growth, development, fertility, numbers, and safety of the type in order to find out its biocenosis role in the regional ecosystems. Proceeding from the aim, the following tasks were identified:

1. To analyse, summarize, and systematize data of literary sources on the designated issue.
2. To study arcal possibilities of the Hyla japonica, and its bioecological peculiarities.
3. To develop the suggestions on preservation of the species on the territory under study.

2 OBJECT AND METHODS OF RESEARCH

The object of the research was Hyla japonica and the structure of ecological relationships with the environment. The method of the research was based on academician V.I. Vernadski's theory on bio-sphere and noosphere, and also the main principles, described in A.M. Nikolski's works (1918), as well as S.L. Kusmin, I.V. Maslova,(2005) and others.

We have been collecting the data on the Far Eastern Hyla since 1973. The research has covered the main natural complexes. 1,368 larvae (tadpoles) of different stages of development, 1,187 underyearlings (after metamorphosis), and 157 adult specimens were examined. Ontogenesis was also studied during the period of spawn and larva genesis of Hyla in water including their ontogenesis under the influence of heavy metals, contained in water (Tagirova, Statsenko, 2005). During a period of several years, in the second part of June–July we studied their diet according to S.L. Kusmin's methods. The number of grown-up specimens was counted along linear routes and areas (Kashkarov, 1928). The research was conducted under field conditions together with postgraduates and undergraduates.

3 RESULTS AND DISCUSSION

The morphological research showed that it is an easily distinguishable species, identified by the expanded adhesive suckers on its toes. The abdomen is white or yellow with dark spots along the body sides and under the eye. It changes its colour depending on the structure of the substrate's surface, from brownish-clay to green. In cool weather, Hyla become dark and greyish-brown; in warm weather they are in a "light" dress – from greenish to milk-fawn, light green, and other colours. The weight of an adult Hyla does not exceed 15 grams with a body length of 48–51 mm. Males are shorter than females. The male vocal sacs are in the throat in the form of transparent bags. In a calm atmosphere they give group concerts not only at night, but in the day time as well, especially before the rain in slight humidity, or in cloudy weather. They did not lose swimming skills, in spite of the fact that they live an air – land way of life; their main support when moving in water is the swimming membranes on their back limbs (Kusmin, Maslova, 2005). Hyla, similar to the majority of Priamurye amphibian does not possess selective trophic ability (polyphage) and

Table 1. Periods of reproducing and metamorphosis of the Far Eastern Hyla, in ten day periods.

May			June			July			August			September			October		
1	2	3	1	2	3	1	2	3	1	2	3	1	2	3	1	2	3
	...	xx	xx	xx	xx	xx	xx	xx	--	++	++	++	++	++	++	++	

Explanation: ...– the beginning of reproducing; xx – mass reproducing; - - – the end of reproducing; ++ – metamorphosis.

eats invertebrates amongst which flying and crawling species prevail. The composition of diet basis changes with the season (Tagirova, Vaseneva, 1979). Beetles, small butterflies and moths, ants, spiders, bloodsucking diptera make up to 85% of the general number of the invertebrates in their diet. From dissected stomachs of 49 grown-up specimens during the period July 1973–2007, the representatives of coleopterous insects made up 37%, scale-winged insects (epidopterous insects) – 20%, ants – 17%, spiders – 7%, orthopterous insects – 5%, mosquitoes and other diptera – 4%, and mites – 2%. Other invertebrates made up only 8%.

Studies of seasonal population structure of the Far Eastern Hyla show that it leaves for winter from the end of September – until the middle of October, burying itself in falling leaves, in shallow roots of the trees, under snags, in hollows, under paper, under boards, and in manure. The last autumn calls of some males are heard near Khabarovsk until the third ten-day period of September. This anabiotic state lasts 190–210 days, from the middle of October until April. Hyla is weak at the end of April and October. They begin to awaken in the end of the first – the beginning of the second 10 days in May. Their active life lasts from May to the end of September – the beginning of October (130–150 days). Females appear in reservoirs 5–10 days later than males (Table 1).

In summer during the day time they are often on the bark or on leaves of plants. In twilight, Hyla go deep into the water of the reservoirs that are nearby. At sunrise they "bathe" in early dew, moving from one plant to another. Animals are not very visible on the green background of foliage.

The first laying of the Far Eastern Hyla starts from 20 May. Mass reproducing (laying of spawn in water) starts at the end of May until the end of July, seldom at the beginning of August (Table 1). Spawn is located on the water surface in one layer in 10–40 eggs (Figure 1). May layings, as a rule, in most puddles are doomed to die because of a small amount of precipitation. Sharp changes in temperature influence the future progeny negatively. Hyla prefer stagnant well-warmed puddles and less often, slowly running reservoirs, covered with riverside gramineous plants. Adult females at the age of 4–5 years are capable of laying about 1,500 eggs (Table 2).

Fertility depends on the body length and, consequently, on the age.

Table 2. Fertility, sizes, and diameter of eggs of the Far Eastern Hyla.

Body length, mm	Body mass, g	Number of eggs	Egg diameter, mm
37–38	4.900–5.350	360–380	1.3–1.5
39–40	5.470–6.090	456–987	1.4–1.3
40–41	6.780–7.865	341–800	1.3–1.6
42–43	8.990–10.720	1043–1465	1.2–1.4
45.7–49	7.250–11.030	600–1409	1.1–1.5

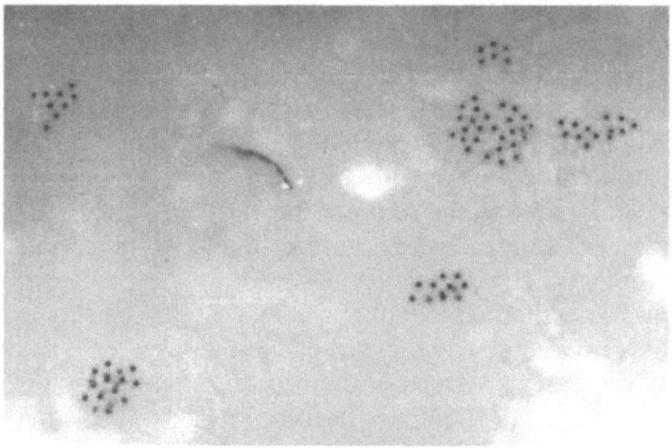

Figure 1. Swimming spawn of the Far Eastern Hyla on the water surface.

The first stages of embryonic development happen at the temperature 16–20°C for 3–4 days. They leave protein shell when they reach the length of no less than 6–7 mm. In this stage head, body and tail are clearly seen, the embryo can be considered as the primary larva stage (the 21st). Oral sucker is kept during three-four day. After being released from the protein shell larvae fall down, and their activity is poor. (Figure 2).

In the 23rd stage oral fissures appear. Corneous "teeth" (pseudodonts) are seen on the lower and upper lips. In the 25th–26th stage the tail and hind limbs are getting larger, and the number of horny teeth increases up to 234 on the upper lip, and up to 250 on the lower lip, if the length of a tadpole is 35 mm. The gill opening is on the left and is directed back and up. The dorsal crista of its caudal fin hardly reaches the vertical line of the gill opening. Hyla tadpoles are different from larvae of other amphibians. In the 27th stage, at a length of 37 mm, the hind limbs are differentiated. The

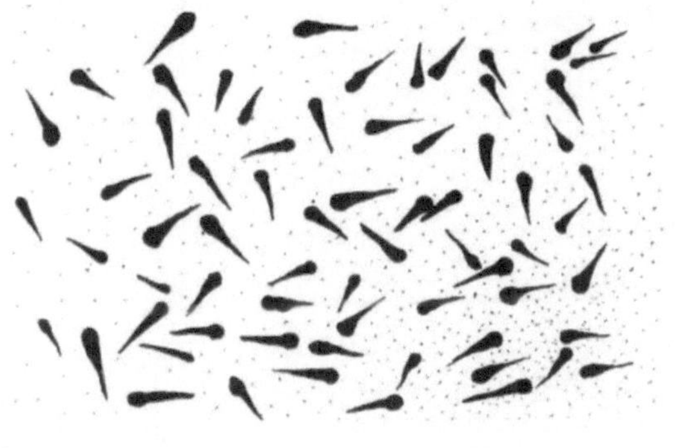

Figure 2. Larvae of the Far Eastern Hyla on the bottom of reservoir.

oral sucker with a complete series of corneous "teeth" is well-expressed. One can recognize and differentiate big Hyla tadpoles from other amphibians according to the adhesive suckers on their toes even during the period when only the hind limbs are free and the front limbs are still under the skin. These larvae have a greenish body and a high, sharp tail fin. The longest tadpole is 45 mm (28th–29th stage of development). After that, the tail begins to dissolve, and the front limbs under the skin are noticeable. When a tadpole turns into an underyearling, its tail is hardly seen and looks like a tail "bud." The majority of the larvae metamorphose in August. In the centre of Vyazemsk on 15.08.2006, 32.8% of tadpoles (n = 122) were 40 mm in length with front and hind limbs. Out of thirteen larvae we found on 31 August 2007, 30 km to the southwest of Khabarovsk, nine species (62.3%) were with two pairs of limbs including one species with a half-shortened tail, whose weight was 305 mg. The whole process of larva development is drawn out and can often take up to ninety days to complete. The length of an underyearling is 10–13 mm, and by autumn it is 16–21 mm. A one-year-old specimen reaches 21–23 mm in length, a two-year-old, 29 mm, a three-year-old, 37–39 mm, a four-year-old, 42–46 mm, and at five years or more, they reach up to 51 mm. They become mature at the age of three.

During the 1970s–1980s, the Far Eastern Hyla was quite common in the territory from Khabarovsk and Bolshoi Khekhtsir State Nature Reserve in the south, up to the Komsomolsk State Nature Reserve in the north-east: its density was two or more per hectare. Calculations of adult Hyla species between 1968–1978 in the Khabarovsk Territory showed that in some places to the south of Khabarovsk, Hyla turned out to be numerous (about eight per hectare); in the north districts (Verkhnebureinskiy, Solnechniy, Komsomolskiy, and Ulchskiy) – it was five times less than that (Tagirova, Vaseneva, 1979). During subsequent years, the number of Hyla was much lower in Priamurje.

The Far Eastern Hyla is not protected in the territory of the district under research. To be held in captivity, Hyla requires an aquarium-terrarium of not less than 70 × 60 cm. The Far Eastern Hyla is a part of diet for many wild animals such as raccoon dogs, badgers, foxes, mink, otters, ermine, Siberian weasels, and weasels. Birds, such as omnivorous crows, herons, storks, day and night beasts, and others also feed on Hyla.

4 CONCLUSION

Analysis and summarizing of the data show that bioecological peculiarities of the Far Eastern Hyla in Priamurye have been under-researched. The author's long-term experiments gave the possibility to expand the knowledge about the area, bioecology, and also of its population. A decrease in the number of Hyla directly relates to the economic activities of people. The measures to protect this species are being worked out.

REFERENCES

Vernadsky V.I. Biosphere. – Moscow: Mysl', 1967. 367 p.

Kashkarov D.N. Method of quantitative study of the vertebrata fauna and analysis of the data obtained // Proceedings of the Central Asia State University. Series 8a, issue 1. Tashkent, 1928. 127 p.

Kuz'min S.L., Maslova I.V. The Amphibia of the Russian Far East. Moscow: Tovarishchestvo nauchnykh izdaniy KMK, 2005. P. 189–225.

Maak, R. Travel on the Amur River made by order of the Siberian department of the Emperor's Russian Geographical Society in 1855. SPb, 1859. 286 p/

Maak, R. Travel in the valley of the Ussury River. / R. Maak. – SPb, 1861. P. 189.

Nikolskiy A.M. The Vermigrade and Amphibia of the Russian Empire. (Herpetologia Rossica) // Transaction of the Imperial Academy of Science on the Physical-Mathematical Department, 1905. 8-series. Vol. 17, No 1. Peretburg. 518 p.

Nikolskiy A.M. Identification Guide of the Vermigrade and Amphibia of the Russian Empire. Kharkov, 1907. 217 p.

Nikolskiy A.M. Fauna of Russia and neighboring countries. Amphibia. – Petrograd, 1918. 311 p.

Tagirova V.T., Vaseneva A.Y. Number of Amphibia and their part in the diet of the mustelids of the flood plain complex // Protection and efficient use of the flora and fauna of the Nizhnee Priamurie (Lower Amur River Basin) and Sakhalin. Khabarovsk: Khabarovsk State Pedagogical Institute, 1979. P. 105–121.

Tagirova, V.T., Statsenko L.A. To the question of the hydrochemical characteristics of the water bodies of reproduction of the Far Eastern tree frog (Hyla) in the suburbs of Khabarovsk // International scientific readings "Primorskie zori-2005" [Primorie Dawns-2005] devoted to the 10th anniversary of the Pacific Academy of Sciences for Ecology and Safety. Vladivostok. 2005. P. 181–183.

Terent'ev V.P., Chernov S.A. Brief identification guide of the Vermigrade and Amphibia of the USSR. L.: Uchpedgiz, 1940. 154 p.

Shchepina N.A., Borisova N.G., Baldanova D. R., Rudneva L.V. Amphibia of Buryatia. Ulan-Ude: Publishers of the Buryat Research Center at the Siberian Branch of the Russian Academy of Sciences, 2009. 148 p.

Future Communication Technology and Engineering – Chan (Ed.)
© 2015 Taylor & Francis Group, London, ISBN: 978-1-138-02777-0

Prediction of organ donation after irreversible brain damage

L. Torres
CCTC, Universidade do Minho, Braga, Portugal

H. Vicente
Departamento de Química, Centro de Química de Évora, Escola de Ciências e Tecnologia, Universidade de Évora, Évora, Portugal

P. Novais, A. Abelha & J. Neves
CCTC, Universidade do Minho, Braga, Portugal

ABSTRACT: An assessment of a patient, who may die within one hour following the removal of life-sustaining care, is the key to a successful donation of organs after cardiac death. Indeed, the accurate prediction of potential organ donation is paramount, due to the narrow time window in which different tasks must be done in parallel, and in a quick and effective way. In this work, it is stated that through a novel approach to knowledge representation and reasoning, based on an extension to the language of Logic Programming, and complemented with a view to computing based on Artificial Neural Networks, it is possible to determine if a patient who suffers from irreversible brain damage may stand as a future candidate for organ donation.

Keywords: prediction of organ donation, brain damage, knowledge representation and reasoning, logic programming, artificial neural networks.

1 INTRODUCTION

Over the last years, advances in immunosuppressive therapeutics, better patient selection, and improved technical expertise, have decisively contributed to the attainment of organ transplantation, which has proven to be a successful treatment for patients with end-stage organ failure. Despite its good results, this treatment has a problem, such as lack of resources, i.e. organs, so that the demand far exceeds the number of available donors. The major source of organs comes from brain dead patients but, unfortunately, (for potential organ recipients) this is not a common form of death. Furthermore, this is an undesirable outcome, since one of the goals of neurocritical care is preventing brain death from occurring (de Groot et al., 2010, Kompanje et al., 2013).

One possible way to expand the donors group is by the use of organs from Donation after Cardiac Death (DCD), also known as non-heart-beating donors. In general, these donors are patients whose deaths occur in the context of Withdrawing Life Sustaining Treatment (WLST). Therapeutic failure, meaningless outcome due to poor prognosis, and patient's autonomy/suffering are some of the most common reasons for WLST. The most frequent DCD candidates are patients who suffer from irreversible brain injury, but do not meet criteria for brain dead diagnosis (Coleman et al., 2008).

Unfortunately, not all potential DCD became actual donations. The success DCD resides in the period between WLST and death, which is often associated with hypotension and poor organ perfusion that generally result in warm ischemic injury to the organs.

A few tools have been developed (Coleman et al., 2008, Rabinstein et al., 2012) to predict this timing, but none are yet established as 'standard'. By analysing this variety of models and indicators, it is clear that the key to predict successfully death after WLST lies with the identification of the correct clinic variables. Yet, the identification of the set of factors that best can characterize this problem seems to be something that still needs further analysis.

In this article, based on some of the most important variables described so far, we make a start on the development of a system that can predict the potential organ donation after irreversible brain damage. We will be focusing on a logic programming-based approach to knowledge representation and reasoning, complemented with a computational framework based on Artificial Neural Networks.

2 KNOWLEDGE REPRESENTATION AND REASONING

Many approaches for knowledge representation and reasoning have been proposed using the Logic

Programming (LP) archetype, namely in the area of Model Theory (Kakas et al., 1998, Pereira & Anh, 2009), and Proof Theory (Neves, 1984, Neves et al., 2007). We follow the proof theoretical approach and an extension to the LP language, to knowledge representations and reasoning. An Extended Logic Program is a finite set of clauses in the form:

$$p \leftarrow p_1, \cdots, p_n, not\ q_1, \cdots, not\ q_m \tag{1}$$

$$?\,(p_1, \cdots, p_n, not\ q_1, \cdots, not\ q_m)\ \ (n, m \geq 0) \tag{2}$$

where "?" is a domain atom denoting falsity, the p_i, q_j, and p are classical ground literals, i.e. either positive atoms or atoms preceded by the classical negation sign $\neg$ (Neves, 1984). Under this representation formalism, every program is associated with a set of abducibles (Kakas et al., 1998, Pereira & Anh, 2009) given here in the form of exceptions to the extensions of the predicates that make the program. Once again, LP emerged as an attractive formalism for knowledge representation and reasoning tasks, introducing an efficient search mechanism for problem solving.

Due to the growing need to offer user support in decision-making processes, some studies have been presented related to the qualitative models and qualitative reasoning in Database Theory and in Artificial Intelligence research (Kovalerchuck & Resconi, 2010). With respect to the problem of knowledge representation and reasoning in LP, a measure of the *Quality-of-Information* (*QoI*) of such programs has been object of some work with promising results (Machado et al., 2010). The *QoI* with respect to the extension of a predicate *i* will be given by a truth-value in the interval [0,1].

It is now possible to engender the universe of discourse, according to the information given in the logic programs that endorse the information about the problem under consideration, according to productions of the type:

$$predicate_i(x_1, \cdots, x_n) :: QoI \tag{3}$$

and evaluate the *Degree of Confidence* given by $DoC = V_i\ (x_1, \ldots, x_n)/n$, which denotes one's confidence in a particular term of the extension of $predicate_i$. To be more general, let us suppose that the Universe of Discourse is described by the extension of the predicates:

$$f_1(\cdots), f_2(\cdots), \cdots, f_n(\cdots)\ \ where\ (n \geq 0) \tag{4}$$

Assuming that a clause denotes a happening, a clause has as argument all the attributes that make the event. The argument values may be of the type unknown, or members of a set, or may be in the scope of a given interval, or may qualify a particular observation. Let us consider the following clause where the second argument value may fit into the interval [10,20] with a domain of [5,30], the value of the first argument is unknown, which is represented by the symbol $\perp$, with a domain that ranges in the interval [5,10], and the third argument stands for itself, with a domain that ranges in the interval [10,20]. Let us consider that the case data is given by the extension of predicate f_1, given in the form:

$$f_1 : x_1, x_2, x_3 \rightarrow \{True, False\} \tag{5}$$

where "{" and "}" is one's notation for sets. One may have:

$$\{$$
$$\neg f_1(x_1, x_2, x_3) \leftarrow not\ f_1(x_1, x_2, x_3)$$
$$f_1 \underbrace{\underbrace{(\ \perp,\ [10,20],\ 15)}_{attribute`s\ values} :: 0.95}$$
$$\underbrace{[5,10][5,30][10,20]}_{attribute`s\ domains}$$
$$\cdots$$
$$\}$$

Once the clauses or terms of the extension of the predicate are established, the next step is to transform all the arguments, of each clause, into continuous intervals. In this phase, it is essential to consider the domain of the arguments. As the first argument is unknown, its interval will cover all the possibilities of the domain. The third argument speaks for itself. 0.95 is a reference value taken from the interval [0,1], denoting the *QoI* of the term $f_1(\perp, [10,20], 15)$. Therefore, one may have:

$$\{$$
$$\neg f_1(x_1, x_2, x_3) \leftarrow not\ f_1(x_1, x_2, x_3)$$
$$f_1 \underbrace{([5,10], [10,20], [15,15])}_{attribute`s\ values\ ranges} :: 0.95$$
$$\underbrace{[5,10]\ \ [5,30]\ \ [10,20]}_{attribute`s\ domains}$$
$$\cdots$$
$$\}$$

Now, one is in position to calculate the *Degree of Confidence* for each attribute that makes the term's arguments (e.g. for attribute two it denotes one's confidence that the attribute under consideration fits into the interval [10,20]). Next, we set the boundaries of the arguments intervals to be fitted in the interval [0,1] according to the normalization procedure given in the procedural form by $(Y - Y_{min})/(Y_{max} - Y_{min})$, where the Y_s stand for themselves:

$$\{$$
$$\neg f_1(x_1, x_2, x_3) \leftarrow not\ f_1(x_1, x_2, x_3)$$
$$x_1 = \left[\frac{5-5}{10-5}, \frac{10-5}{10-5}\right], x_2 = \left[\frac{10-5}{30-5}, \frac{20-5}{30-5}\right], x_3 = \left[\frac{15-10}{20-10}, \frac{15-10}{20-10}\right]$$
$$f_1 \underbrace{([0,1],\ \ [0.2,0.6],\ \ [0.5,0.5])}_{attribute`s\ values\ ranges\ once\ normalized} :: 0.95$$
$$\underbrace{[0,1]\ \ [0,1]\ \ [0,1]}_{attribute`s\ domains\ once\ normalized}$$
$$\cdots$$
$$\}$$

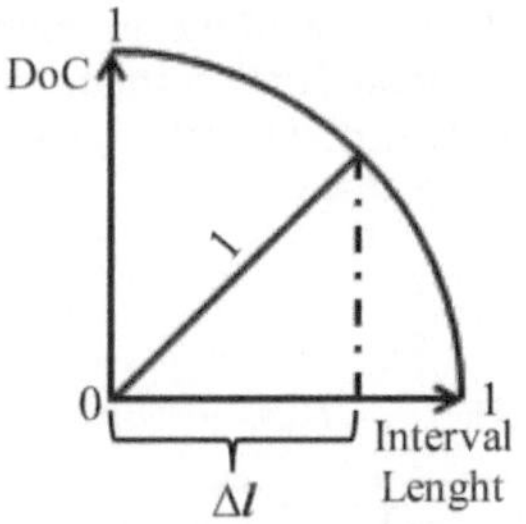

Figure 1. Evaluation of the degree of confidence.

The *Degree of Confidence* (*DoC*) is evaluated using the equation $DoC = \sqrt{1 - \Delta l^2}$, as illustrated in Figure 1, where Δl stands for the length of the argument's intervals, once normalized:

$$\{$$
$$\neg f_1(x_1, x_2, x_3) \leftarrow not\ f_1(x_1, x_2, x_3)$$
$$f_{1_{DoC}} \underbrace{\underbrace{\underbrace{(0, \quad 0.92, \quad 1)}_{attribute\text{'}s\ confidence\ values}}_{\substack{[0,1] \quad [0.2, 0.6] \quad [0.5, 0.5] \\ attribute\text{'}s\ values\ ranges\ once\ normalized}}}_{\substack{[0,1] \quad [0,1] \quad [0,1] \\ attribute\text{'}s\ domains\ once\ normalized}} :: 0.95$$
$$\dots$$
$$\}$$

3 A CASE STUDY

In order to exemplify the applicability of our model, we will look at the relational database model, since it provides a basic framework that fits into our expectations, and is understood as the genesis of the *LP* approach to Knowledge Representation and Reasoning (Neves, 1984).

As a case study, consider the scenario where a relational database is given in terms of the extensions of the relations depicted in Figure 2, which stands for a situation where one has to manage information about patients in a neurocritical state.

Under this scenario, some incomplete data is available. For instance, in *General Information* database the *Use of Inotropes* regarding the last patient is unknown, whilst in *Diagnosis* database the pH value ranges in the interval [7.30, 7.40].

In *General Information* database the domain of *Ventilation Mode* column is in the range [0,2], wherein 0 denotes Pressure Support; 1 stands for Synchronised Intermittent Mandatory Ventilation; and 2 denotes Pressure Control/Volume Control/Pressure regulated Volume Control. Regarding to *Use of Inotropes* column 0 and 1 denote, respectively, the non-use and the using of these agents.

The *FOUR Score* column in *Potential Donor* database is obtained by the sum in the row, which corresponds to each of its testable components. The *GCS* column works in the same way, but do not test brain steam reflexes. Despite this, it is still included on this model because it is the most commonly used. The *DCD-N* database comes from the already mentioned Rabinstein et al., (2012) model, once it provides a simple way to weight some of the most important factors, i.e. for the absence of cough reflex it is given 2 points, and for the absence of each of the other 1 point. The *DCD-N* column in *Potential Donor* database is obtained by the sum in the row.

Finally, in the *Diagnosis* database the domain of *Causes of Neurological Injury* column is in the range [0, 3], wherein 3 (three) stands for intracranial trauma; 2 (two) denotes intracranial haemorrhage; 1 (one) stands for hypoxia; and 0 (zero) denotes other. Then, one may have:

$$pot\_donor: Age,\ U_{se\ of}Ino_{tropes},\ FOUR_{Score},\ G_{lasgow}C_{oma}\ S_{core},\ DCD - N_{Score},$$
$$pH,\ FiO_{2Ratio},\ C_{ause\ of}\ N_{eurological}I_{njury},\ S_{pecialist}O_{pinion} \rightarrow \{0,1\}$$

where *pot_donor* stands for the predicate *potential donor*, 0 (zero) and 1 (one) denote, respectively, the truth values *false* and *true*. It is now possible to give the extension of the predicate *pot_donor*, in the form:

$$\{$$
$$\neg\, pot\_donor(Age, UIno, FOUR, GCS, DCD - N, pH, FiO_2, CNI, SO) \leftarrow$$
$$not\ pot\_donor(Age, UIno, FOUR, GCS, DCD - N, pH, FiO_2, CNI, SO)$$
$$pot\_donor\left(\underbrace{\underbrace{40, \ 0, \ 4, \ 4, \ [0,1], \ \bot, \ 0.25, \ 0, \ 0}_{attribute\text{'}s\ values}}_{\substack{[3,75][0,1][0,16][3,15][0,5][7.25,7.40][0.15,0.5][0,3][0,2] \\ attribute\text{'}s\ domains}}\right) :: 1$$
$$\dots$$
$$\}$$

In this program, the first clause denotes the closure of predicate *pot_donor*. The next clause corresponds to patient 1, taken from the extension of the *Potential Donor* relation presented in Figure 2. Moving on, the next step is to transform all the argument values into continuous intervals and then normalize the predicate's arguments in order to obtain the *Degree of Confidence* of the *pot_donor* predicate:

$$
\begin{array}{l}
\{ \\
\neg\, pot\_donor(Age, UIno, FOUR, GCS, DCD-N, pH, FiO_2, CNI, SO) \leftarrow \\
\qquad not\ pot\_donor(Age, UIno, FOUR, GCS, DCD-N, pH, FiO_2, CNI, SO) \\
pot\_donor_{DoC}\left(\underbrace{\underbrace{\underbrace{1, \quad 1, \quad 1, \quad 1, \quad 0.98, \quad 0, \quad 1, \quad 1, \quad 1}_{attribute's\ confidence\ values}}_{[0.51,0.51][0,0][0.25,0.25][0.08,0.08][0,0.2][0,1][0.28,0.28][0,0][0,0]}}_{attribute's\ values\ ranges\ once\ normalized}\right) :: 1 \\
\qquad \underbrace{[0,1]\ [0,1]\ [0,1]\ [0,1]\ [0,1]\ [0,1]\ [0,1]\ [0,1][0,1]}_{attribute's\ domains\ once\ normalized} \\
\ldots \\
\}
\end{array}
$$

where its terms make the training and test sets of the Artificial Neural Network given in Figure 3.

4 ARTIFICIAL NEURAL NETWORKS

Several studies have shown how Artificial Neural Networks (ANNs) could be successfully used to model data and capture complex relationships between inputs and outputs (Caldeira et al., 2011, Vicente et al., 2013). ANNs simulate the structure of the human brain being populated by multiple layers of neurons. As an example, let us consider the last case presented in Figure 2, where one may have a situation in which is needed an assessment of a potential organ donation after irreversible brain damage, given in the form:

$$
pot\_donor\left(\underbrace{\underbrace{33, \quad \bot, \quad 1, \quad 3, \quad 3, \quad [7.30,7.40], \quad 0.3, \quad \bot, \quad 1}_{attribute's\ values}}_{[3,75][0,1][0,16][3,15][0,5][7.25,7.40][0.15,0.5]\ [0,3][0,2]}\right) :: 1
$$

$$
\underbrace{\qquad\qquad\qquad\qquad}_{attribute's\ domains}
$$

According to the formalism presented above, after the transition to continuous intervals, it is possible to have the arguments of the predicates extensions normalized to the interval [0, 1] to evaluate the pot_donor_{DoC}:

$$
pot\_donor_{DoC}\left(\underbrace{\underbrace{\underbrace{1, \quad 0, \quad 1, \quad 1, \quad 1, \quad 0.75, \quad 1, \quad 0, \quad 1}_{attribute's\ confidence\ values}}_{[0.42,0.42][0,1][0.06,0.06][0,0][0.6,0.6][0.33,1][0.43,0.43][0,1][0.5,0.5]}}_{attribute's\ values\ ranges\ once\ normalized}\right) :: 1
$$

$$
\underbrace{[0,1]\ \ [0,1]\ \ [0,1]\ \ [0,1]\ [0,1]\ \ [0,1]\ \ [0,1]\ \ [0,1]\ \ [0,1]}_{attribute's\ domains\ once\ normalized}
$$

Figure 3 shows how the normalized values of the interval boundaries and their *DoC* and *QoI* values work as inputs to the ANN. The output translates the chance of the patient death within one hour of Withdrawing Life Sustaining Treatment (WLST), and *DoC* the confidence that one has on such a happening. In this study were considered 200 samples, of which 70% were used to train the ANN, and 30% were considered for testing (the data came from a main health care centre in the north of Portugal). The back propagation algorithm was used in the learning process of the ANN. The activation function used in the pre-processing layer was the identity one. In the other layers was used the sigmoid activation function.

5 CONCLUSIONS AND FUTURE WORK

On the one hand, classifying the patients who will die in a period of sixty minutes after withdrawing life-sustaining care, as potential organ donors, is a hard and complex task. Indeed, it is necessary to consider countless different factors with complex relationships amongst them. These entire characteristics highlight the benefits that Artificial Intelligencebased techniques can bring to this field in order to achieve better prognostics.

On the other hand, in this work, departing from the conclusions of some good existing models, the founding of a computational framework that uses

General Information			
#	Age	Ventilation Mode	Use of Inotropes
1	40	0	0
...	---	---	---
n	33	2	⊥

Full Outline of UnResponsive (*FOUR*)				
#	Eye Response	Motor Response	Brainstem Reflexes	Respiration
1	0	1	2	1
---	---	---	---	
n	0	0	1	0

Glasgow Coma Score (*GCS*)			
#	Eye Opening Response	Verbal Response	Motor Response
1	1	1	2
---	---	---	---
n	1	1	1

Potential Donor									
#	Age	Use of Inotropes	FOUR Score	GCS	DCD-N Score	pH	FiO_2 Ratio	Cause of Neurological Injury	Specialist Opinion
1	40	0	4	4	[0,1]	⊥	0.25	0	0
...	...	...	...	...	...	...	...	...	...
n	33	⊥	1	3	3	[7.30,7.40]	0.3	⊥	1

DCD-N				
#	Corneal Reflex	Cough Reflex	Motor Response to Pain	Respiration
1	0	0	0	⊥
...	---	---	---	---
n	1	0	1	1

Diagnosis			
#	pH	FiO_2 Ratio	Cause of Neurological Injury
1	⊥	0.25	0
---	...	...	...
n	[7.30,7.40]	0.3	⊥

Figure 2. Extension of the relational database model.

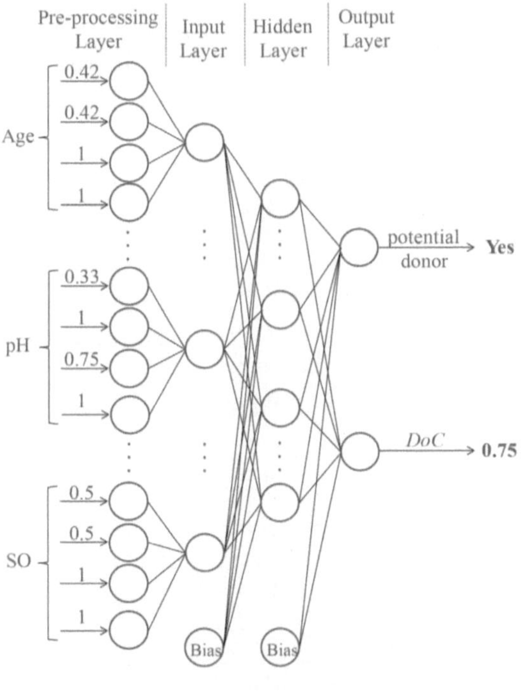

Figure 3. The Artificial Neural Network topology.

powerful knowledge representation and reasoning techniques to set the structure of the information and the associate inference mechanisms based in ANNs is presented. This representation is, above everything else, very simple and versatile, and capable of covering every possible instance by considering incomplete, contradictory, and even unknown data, a marker that is not present in existing systems.

As future work it will be considered in the application of others computational frameworks, e.g. Case-Based Reasoning (Carneiro et al., 2013) or Particle Swarm (Mendes et al., 2004) to handle the same problem.

ACKNOWLEDGEMENTS

This work is funded by National Funds through the FCT – Fundação para a Ciência e a Tecnologia (Portuguese Foundation for Science and Technology) within project PEst-OE/EEI/UI0752/2014.

REFERENCES

Caldeira, A.T., Arteiro, J., Roseiro, J. Neves, J. & Vicente, H. 2011. An Artificial Intelligence Approach to Bacillus amyloliquefaciens CCMI 1051 Cultures: Application to the Production of Antifungal Compounds. *Bioresource Technology* 102: 1496–1502.

Carneiro, D., Novais, P., Andrade, F., Zeleznikow, J. & Neves, J. 2013. Using Case-Based Reasoning and Principled Negotiation to provide decision support for dispute resolution. *Knowledge and Information Systems* 36: 789–826.

Coleman, N., Crowfoot, E. & Brieva, J. 2008. Prediction of death after withdrawal of life-sustaining treatments. *Critical Care and Resuscitation* 10: 278–284.

de Groot, Y., Jansen, N., Bakker, J., Kuiper, M., Aerdts, S., Maas, A., Wijdicks, E., van Leiden, H., Hoitsma, A., Kremer, B. & Kompanje, E. 2010. Imminent brain death: point of departure for potential heart-beating organ donor recognition. *Intensive Care Medicine* 36: 1488–1494.

Kakas, A., Kowalski, R. & Toni, F. 1998. The role of abduction in logic programming. In D.M. Gabbay, C.J. Hogger & J.A. Robinson (eds.), *Handbook of Logic in Artificial Intelligence and Logic Programming*, Volume 5: *Logic Programming*: 235–324. Oxford: Oxford University Press.

Kompanje, E., Jansen, N. & de Groot, Y. 2013. 'In plain language': uniform criteria for organ donor recognition. *Intensive Care Medicine* 39: 1492–1494.

Kovalerchuck, B. & Resconi, G. 2010. Agent-based uncertainty logic network. In *Proceedings of the IEEE International Conference on Fuzzy Systems – FUZZ-IEEE*: 596–603.

Machado, J., Abelha, A., Novais, P. & Neves, J. 2010. Quality of service in healthcare units. *International Journal of Computer Aided Engineering and Technology* 2: 436–449.

Mendes, R., Kennedy, J. & Neves, J. 2004. The Fully Informed Particle Swarm: Simpler, Maybe Better. *IEEE Transactions on Evolutionary Computation* 8: 204–210.

Neves, J. 1984. A logic interpreter to handle time and negation in logic data bases. In R.L. Muller & J.J. Pottmyer (eds.), *Proceedings of the 1984 Annual Conference of the ACM on the Fifth Generation Challenge*: 50–54.

Neves, J., Machado, J., Analide, C., Abelha, A. & Brito, L. 2007. The halt condition in genetic programming. In J. Neves, M.F. Santos & J. Machado (eds.), *Progress in Artificial Intelligence – Lecture Notes in Computer Science, Volume 4874*: 160–169. Springer: Heidelberg.

Pereira, L. & Anh, H., 2009. Evolution prospection. In. Nakamatsu, G. Phillips-Wren, L. Jain & R. Howlett (eds.), *New Advances in Intelligent Decision Technologies – First KES International Symposium IDT 2009, Studies in Computational Intelligence, Volume 199*: 51–64. Berlin: Springer.

Rabinstein, A., Yee, A., Mandrekar, J., Fugate, J., de Groot, Y., Kompanje, E., Shutter, L., Freeman, W.D., Rubin, M. & Wijdicks, E. 2012. Prediction of potential for organ donation after cardiac death in patients in neurocritical state: a prospective observational study. *The Lancet Neurology* 11: 414–419.

Vicente, H., Dias, S., Fernandes, A., Abelha, A., Machado, J. & Neves, J. 2012. Prediction of the Quality of Public Water Supply using Artificial Neural Networks. *Journal of Water Supply: Research and Technology – AQUA* 61: 446–459.

Future Communication Technology and Engineering – Chan (Ed.)
© 2015 Taylor & Francis Group, London, ISBN: 978-1-138-02777-0

Interactivity for facilitating and promoting traditional handicraft design

D. Shi
Nanjing University of Arts, Gulou District, Jiangsu Province, China

ABSTRACT: Based on the analysis of the problems which have been faced by handicrafts, this paper comes up with a project to transform the problems into opportunities. We will get close to this transformation if we can make full use of the new design resources to meet the new needs which come from a post-industrial society. This paper proposed one possible research frame and described its' potential result. Hopefully, after this attempt, there will be more research focusing on the innovation and development of handicrafts in apost-industrial society.

1 RESEARCH QUESTION

1.1 *Background*

The phrase 'post-industrial society' was first proposed in a book, *The Coming of Post-industrial Society*, written by Daniel Bell, and was used to describe the new changes in society structure and life conditions after the 1950s. (Trosow, 2000) These changes happened with the development of technology. Numerous new artefacts were created and the needs of people were changed with the development of a post-industrial society. Information technology got rid of the industrial bondage from design and turned design from material to non-material (D. Mark, 1998). This brings new resources to design theory and practice. People pay more attention to the dynamic part of a product, such as interaction, emotional communication with the product, and the experience of using the product. The need for personality and emotional components of products is growing with this change (C. Escallon, 2010).

Meanwhile, traditional handicraft can meet these new needs in a post-industrial society very well due to its characteristics of personality, culture, small batch production, and so on. Therefore, we can see that traditional handicraft has many new possibilities in a post-industrial society (N. M. Germann, 2011).

Firstly, according to the new changes in society, many new theories and methods in design were created to cope with the new problems of modern life, such as those in interaction design (Hansen, 2008) and service design (Morelli, 2006). Secondly, there are many new technologies which have created many opportunities for handicraft products and service systems to be redesigned by using technologies such as social computing (Wang et al., 2007). Thirdly, the continuous development of new materials makes the redesigning of handicrafts easier than ever before.

1.2 *Existing problems*

Even though there is a need for products with more characteristics of personality or culture, the shape, function or consumption modes of traditional handicraft have not changed according to the new needs, so these products still cannot meet the needs of modern life very well. Taking the Chinese fan produced by Wang Xingji as an example, almost 70% of these are sold by physical stores, and any other consumption modes have almost never been used.

The study of new design theory and methods pays more attention to modern products and seems to neglect the handicrafts, even though some new products are designed with a new shape or made of new materials, such as bamboo products designed by Sandeep Sangaru (Mir and Ain, 2010) [9]. Up to now, some researchers have noticed the potential of handicraft in a post-industrial society (Risatti, 2009), but few traditional products have been designed using the methods of interaction or service design (Romeiro Filho, 2013, Wiberg, 2014).

1.3 *The proposition and meaning of the research topic*

Based on the analysis of the background and existing problems, we found many problems in traditional handicraft design. In fact, a problem sometimes can be transformed into a solution under appropriate conditions. If we can meet the new needs in a post-industrial society with the help of new design theory and methods, we can build a bridge between modern design and traditional handicraft, and turn the problems into opportunities. Therefore, this research is focused on the study of new opportunities for traditional handicrafts (Figure 1).

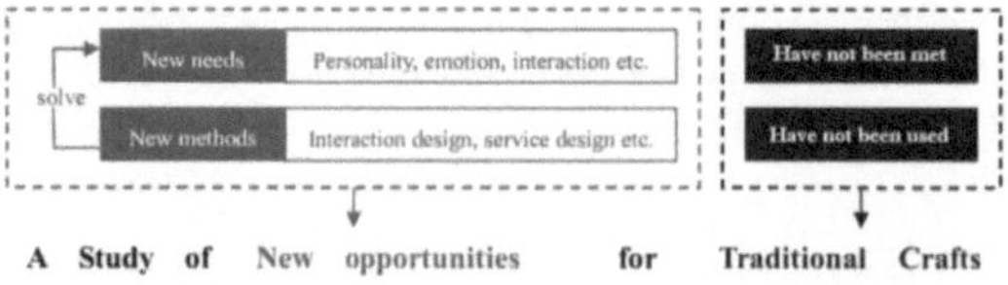

Figure 1. Research topic.

2 RESEARCH CONTENT AND METHODS

The research content can be summarized in four parts: identifying the interaction pattern; design conceptualization; prototyping; evaluation and reflection.

The main task of the first part is to investigate the interaction patterns (Hassenzahl et al., 2013). The producers and consumers of one type of handicraft will be investigated, and the design opportunity and the interaction pattern will be identified (Hassenzahl et al., 2013) through field research. This includes the investigation of design opportunities in the intersection between consumers' life style and handicrafts through interview, observation, and empathic approaches. At the same time, an investigation on the handicrafts and related service will be conducted to identify the potential stakeholders, recombine resources, and finally extract the interaction patterns.

The main task of the second part is to determine the final design concept. The final design concept will be an intelligent product or an App according to the final pattern, which can be used to promote the communication, interaction, and inspiration between handicraft producers and consumers (Figure 2 shows one possible design concept).

The main task of the third part is to make the prototype by using the interaction pattern determined in the second stage of study. This includes completing the prototype in a laboratory environment by inviting the stakeholders to participate in the design process.

The main method in this study is research through design. (Zimmerman et al., 2007). Research through design is used as a form of research to contribute to a design activity.

3 RESEARCH RESULT

3.1 *Expected results from practice*

The expected results will be an intelligent product or an App, which can be used to promote communication, interaction, and inspiration between handicrafts producers and consumers, according to the final design concept from the second stage of the study.

One possible solution can be an intelligent product (Product A in Figure 2), which can promote communication and inspiration between the consumers and the producers.

The consumers can get all-around information about the handicrafts and the producers through product A. They can judge the types of crafts used in the handicraft product by scanning the information on the surface of the product and matching the information with the database. After that, the all-around information will be shown on the end user application for the consumer (such as a computer or a mobile advice) (Figure 2).

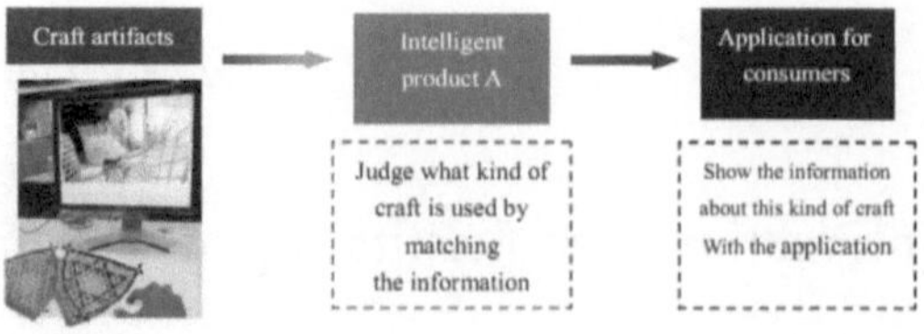

Figure 2. The information from producer to user.

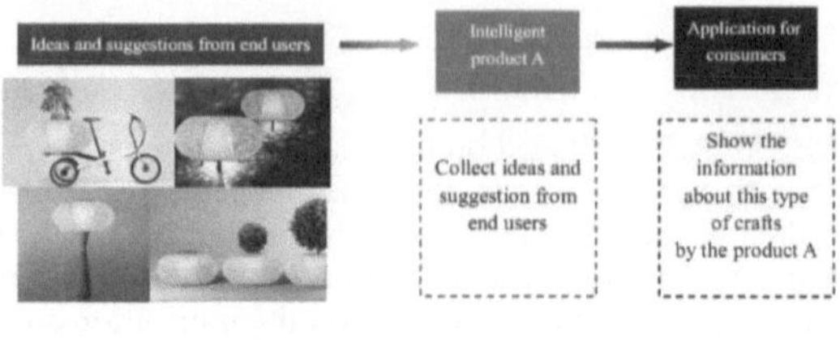

Figure 3. The information from user to producer.

The craftsman can obtain information about users' needs and the creative use of this particular handicraft product by using product A. It can collect ideas and suggestions about redesign from the users, and send this information to the producers who can make new products based on the ideas and suggestions from the end users (Figure 3).

3.2 *Expected results about theory*

Most interaction design focuses on the interaction between users and products, but there is a need to explore interaction design that focuses more on the interaction between users. In addition, through this study, it will be possible to build one framework on how to design products and/or service systems to strengthen the connection between traditional crafts and potential needs in a post-industrial society.

4 CONCLUSIONS

Based on the analysis of the problems which have been faced by handicrafts, this paper comes up with a project to transform the problems into opportunities. We will get close to this transformation if we can make full use of the new design resources to meet the new needs that come from a post-industrial society. This paper proposed one possible research frame and result. We hope that there will be more research following this study, which will focus on the innovation and development of handicrafts in a post-industrial society.

ACKNOWLEDGEMENT

This research is supported by Plans for Graduate Research and Innovation for Colleges and

Universities in Jiangsu Province. The project number is KYLX_1011.

REFERENCES

Hansen, H. 2008. Interaction design: Beyond human-computer interaction. *Technical Communication*, 55, 289–290.

D. Mark. 1998 Non-material Society–Design Culture and Technology of Post-industrial World. *Sichuan People's Publishing Press, Sichuan.*

C. Escallon. 2010. Droog Design: Sense and Experience. University of Notre Dame.

N. M. Germann. 2011. *An Argument for Modern Craftsmanship: A Philosophy of Design, Materials, and Process in a Post-Industrial Environment*: University of Cincinnati, 2011.

Hassenzahl, M., Eckoldt, K., Diefenbach, S., Laschke, M., Lenz, E. & Kim, J. 2013. Designing Moments of Meaning and Pleasure. Experience Design and Happiness. *International Journal of Design*, 7, 21–31.

Mir, F. A. & Ain, F. 2010. Legal Protection of Geographical Indications in Jammu and Kashmir – A Case Study of Kashmiri Handicrafts. *Journal of Intellectual Property Rights*, 15, 220–227.

Morelli, N. 2006. Developing new product service systems (PSS): methodologies and operational tools. *Journal of Cleaner Production*, 14, 1495–1501.

Risatti, H. 2009. NeoCraft: Modernity and the Crafts. *Winterthur Portfolio-a Journal of American Material Culture*, 43, 143–144.

Romeiro Filho, E. 2013. Design and Craftsmanship: The Brazilian Experience. *Design Issues*, 29, 64–74.

Trosow, S. E. 2000. The coming of post-industrial society: A venture in social forecasting. *Library Quarterly*, 70, 397–399.

Wang, F.-Y., Zeng, D., Carley, K. M. & Mao, W. 2007. Social computing: From social informatics to social intelligence. *Ieee Intelligent Systems*, 22, 79–83.

Wiberg, M. 2014. Methodology for materiality: interaction design research through a material lens. *Personal and Ubiquitous Computing*, 18, 625–636.

Zimmerman, J., Forlizzi, J., Evenson, S. & ACM 2007. *Research Through Design as a Method for Interaction Design Research in HCI.*

Future Communication Technology and Engineering – Chan (Ed.)
© 2015 Taylor & Francis Group, London, ISBN: 978-1-138-02777-0

Author index